ELECTRICAL

Grounding and Bonding

Electrical Grounding and Bonding, 3rd ed.
by Phil Simmons

Vice President, Career and Professional Editorial: Dave Garza

Director of Learning Solutions: Sandy Clark

Acquisitions Editor: Stacy Masucci

Managing Editor: Larry Main

Senior Product Manager: John Fisher

Editorial Assistant: Andrea Timpano

Vice President, Career and Professional Marketing: Jennifer Baker

Marketing Director: Deborah Yarnell

Marketing Manager: Kathryn Hall

Marketing Coordinator: Mark Pierro

Production Director: Wendy Troeger

Production Manager: Mark Bernard

Content Project Manager: Barbara LeFleur

Art Director: David Arsenault

Technology Project Manager: Joe Pliss

Cover Credits:

Electronic schematic: © Shane White/ iStockphotoLightbulb illustration: © Joseph Villanova electrical-relays-breakers-and-ballasts: © xtrekx/ Shutterstock

Compositor: MPS Limited, a Macmillan Company

Library of Congress Control Number: 2010939188

ISBN-13: 978-1-4354-9832-7

ISBN-10: 1-4354-9832-1

Delmar
5 Maxwell Drive
Clifton Park, NY 12065-2919
USA

Cengage Learning is a leading provider of customized learning solutions with office locations around the globe, including Singapore, the United Kingdom, Australia, Mexico, Brazil, and Japan. Locate your local office at: **international.cengage.com/region**

Cengage Learning products are represented in Canada by Nelson Education, Ltd.

To learn more about Delmar, visit **www.cengage.com/delmar**

Purchase any of our products at your local college store or at our preferred online store **www.CengageBrain.com**

Printed in the United States of America
1 2 3 4 5 6 7 13 12 11 10

BASED ON THE 2011
NATIONAL ELECTRICAL CODE®

ELECTRICAL Grounding and Bonding

3RD EDITION

PHIL SIMMONS

Australia Canada Mexico Singapore Spain United Kingdom United States

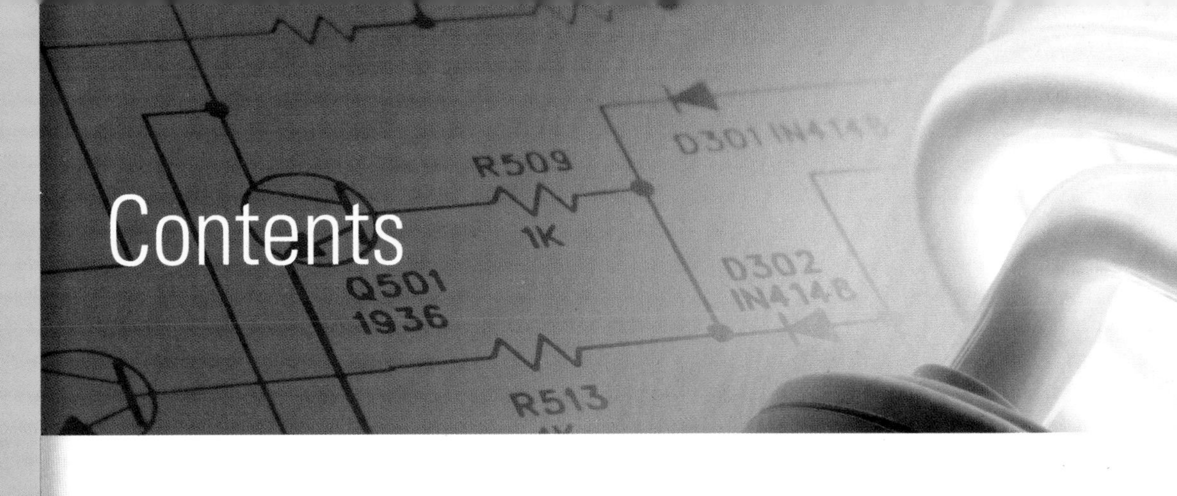

Contents

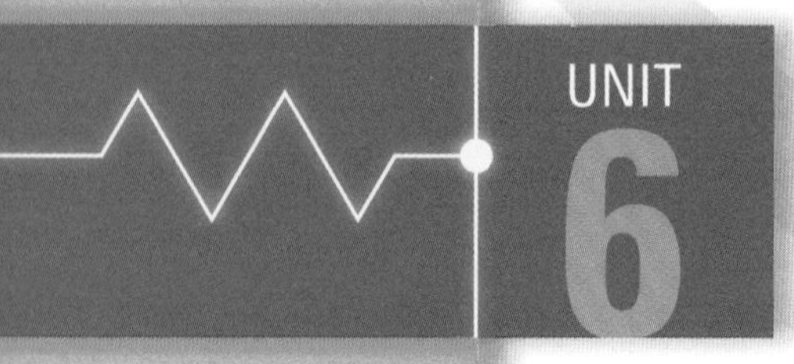
UNIT
6

UNIT 8

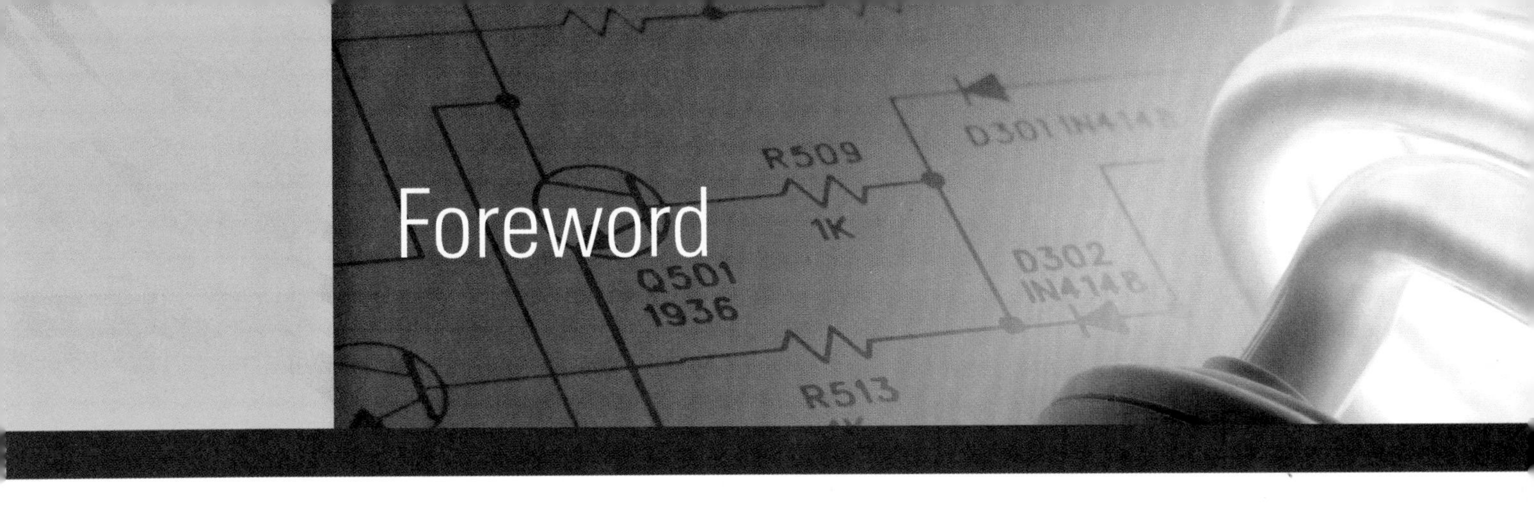

Foreword

Just who is Phil Simmons, the author of this book, *Electrical Grounding and Bonding*? I have known Phil for a long time. He has been involved in the electrical industry for many years. Phil is one of the most knowledgeable individuals I know when it comes to the electrical industry and the *National Electrical Code*® (*NEC*®).*

Phil's credentials are endless. He has held many high-level positions. He was the Executive Director of the International Association of Electrical Inspectors (IAEI), where he orchestrated the writing, producing, editing, revising, and reviewing of many of the technical manuals available from the IAEI. As editor-in-chief of the *IAEI News*, he saw to it that this bimonthly magazine became one of the finest technical publications available today. He was Chief Electrical Inspector for the State of Washington. He has worked with many state and national code officials.

Phil served as Acting Chairman of Code Making Panel 5 a few *Code* cycles back and was responsible for the major reorganization of *Article 250* in the *NEC*. *Article 250* is all about grounding and bonding.

Phil served on the Underwriters Laboratories Electrical Council and their Board of Directors.

Phil has conducted numerous seminars for many large organizations, such as General Motors, Ford Motor Company, Daimler/Chrysler, the University of Wisconsin, the University of Missouri, the U.S. Navy, and the U.S. Marines. He has conducted continuing education seminars for electrical inspectors, electrical contractors, electricians, and electrical apprentices.

Phil also is a master electrician and was an electrical contractor. When electrical installations and the *NEC* are discussed, he can say: "Been there . . . done that." This tremendous experience in the real world is the basis for his outstanding knowledge of the *NEC*.

In addition, Phil has made major contributions to efforts at standardization within the electrical industry. Many of you may not be aware that Phil brought the IAEI Soares grounding book up to date and contributed to the IAEI *Analysis of the NEC*, the *Neon Sign Manual*, *Ferm's Fast Finder*, and videos on the *NEC*. Over the years, he has served on the *Code* panels at IAEI Section and Chapter meetings. He has conducted innumerable seminars under the auspices of the National Fire Protection Association and the IAEI.

As most of you know, the *NEC* is not the most "user-friendly" document. Some individuals know the *Code* but find it difficult to teach others. Phil has the innate ability to explain the *NEC* in words and diagrams that can be understood by everyone.

**National Electrical Code*® and *NEC*® are registered trademarks of the National Fire Protection Association, Inc., Quincy, MA 02169.

Phil has written textbooks and examinations and has developed diagrams, PowerPoint presentations, slides, and transparencies. All of this knowledge is apparent in this excellent textbook on the difficult subject of electrical grounding and bonding.

Like the three legs of a stool that provide its stability, Phil's expertise has three essential components, each of which ensures the effectiveness of the others. "Leg one" is the experience he has had in the electrical field. "Leg two" is his incredible knowledge of electrical codes and standards. "Leg three" is his ability to share this wealth of experience and knowledge with the reader of this book.

I congratulate Phil for a job well done in writing this book. It will be an important addition to your collection of electrical books.

Ray C. Mullin
Author/Owner, Ray C. Mullin Books
Electrical Wiring—Residential
Electrical Wiring—Commercial
Smart House
Illustrated Electrical Calculations

Preface

Electricity follows the basic laws of physics, regardless of whether it is current flow over ungrounded ("hot") conductors, over grounded conductors (sometimes neutral conductors), or in the grounding system. So, if we can understand basic circuit flow, we can understand the requirements and performance rules for grounding and bonding of electrical systems and equipment. You will find several of the illustrations in this book to be fairly basic and uncomplicated. This complements the overall effort to make the rules for grounding and bonding as easy to understand as possible and to take the concepts of grounding and bonding back to the basics.

I want to mention here and applaud the efforts of Ronald P. O'Riley, who wrote a book titled *Electrical Grounding: Bringing Grounding Back to Earth,* through the sixth edition. Mr. O'Riley is now deceased. Although this book is not based on or intended to be a continuation of Mr. O'Riley's efforts, our goals in presenting a book on grounding and bonding of electrical systems are very similar. Quoting from the preface to Mr. O'Riley's sixth edition:

> *The author's wish is for this book to be a learning experience for members, and those in training for a career in the electrical industry. It is the author's hope that simplifying, illustrating, reasoning through, and coordinating the grounding requirements, as contained in* Article 250 *of the* National Electrical Code®, *will promote better understanding and use of the* Code. *This can result in safer, cleaner electrical installations and maintenance. The first rule is to make it safe: the second is to make it work. Both can be done. With this thought in mind, this book is directed at vocational instructors of electricity, electrical engineers, design engineers, construction electricians making installations in the field, maintenance electricians at factories or buildings, electrical inspectors, and many other members of the electrical industry. It is also the author's hope that the apprentice or person preparing for a career in the electrical industry and studying the* National Electrical Code® *will find the detailed explanations and accompanying diagrams in this book to be an interesting learning experience."*

Electrical Grounding and Bonding is based on my many years of experience in teaching subjects related to the *NEC,* field experience in the electrical construction industry, and association with the International Association of Electrical Inspectors (IAEI).

Other than the Introduction, which includes an explanation of many definitions applicable to electrical grounding and bonding along with a brief review of electrical theory, this book is organized by section number of the *NEC*. For example, if you're interested in learning about requirements for a grounding electrode system, you can follow the rule from the *NEC Section 250.50* to an identical code reference in this book.

Other features of the organization of this book are as follows:

1. The requirement from the *NEC* is included. Note that in most cases, the requirement is paraphrased rather than being a direct quote.
2. The requirement is discussed and explained.
3. An illustration of the requirement is provided.
4. Where appropriate, there is an explanation of how to comply with the rules, such as determining the appropriate size system bonding conductor.

ABOUT THE AUTHOR

Phil Simmons is owner of Simmons Electrical Services. The firm specializes in training, writing and illustrating of technical publications related to the electrical industry, inspecting complex electrical installations, and consulting on electrical systems and safety.

Phil has served the International Association of Electrical Inspectors (IAEI) in the following capacities:

- Secretary-treasurer of the Puget Sound Chapter
- Secretary-treasurer of the Northwestern Section
- International President in 1987
- Executive Director (1990–1995)
- Codes and Standards Coordinator (1995–1999)
- Editor of *IAEI News*

Phil is a licensed master electrician and former electrical contractor in the State of Washington. He is also a licensed journeyman electrician in the states of Alaska and Montana. He was chief electrical inspector for the State of Washington from 1984 to 1990, after serving as an electrical plans examiner and field electrical inspector.

Phil has had extensive experience for several years in preparing and presenting training at many locations on the following and additional subjects:

- Update to the *National Electrical Code*
- Electrical Systems in One- and Two-Family Dwellings
- Grounding and Bonding of Electrical Systems
- Wiring for Hazardous Locations
- Motors and Transformers
- Electrical Safety in Employee Workplaces

Phil is author of several technical articles that have been published in the *IAEI News*. He is also the author and illustrator of several technical books on electrical codes and safety, including these:

- IAEI *Analysis of the National Electrical Code*
- *Ferm's Fast Finder Index*
- *Electrical Systems in One- and Two-Family Dwellings*
- *IAEI Soares Book on Grounding of Electrical Systems*
- *NJATC Significant Changes in the National Electrical Code*

For the editions related to the 2011 *NEC*, Phil is coauthor, with Ray Mullin, of *Electrical Wiring Commercial* and *Electrical Wiring Residential*. Both are published by Delmar Cengage Learning.

Phil has served the National Fire Protection Association (NFPA) in several capacities, including the following:

- Standards Council (six years)
- *National Electrical Code* Technical Correlating Committee
- Code Making Panel (CMP)-17, Chairman of CMP-19, CMP-1, Acting Chair of CMP-5, and CMP-5 member
- Chair of the NFPA Electrical Section
- Electrical Codes Coalition Committee
- Instructor at NFPA's *NEC*, Electrical Safety in the Workplace (NFPA 70E), and related seminars

Phil is a past member of the Underwriters Laboratories Electrical Council and is a past UL Trustee. He is a member of the Standards Technical Panel for Grounding and Bonding Equipment. He is also a retired member of the International Brotherhood of Electrical Workers.

Phil has passed the electrical inspector certification examinations for One- and Two-Family, General, and Plan Review. He served for several years on the Educational Testing Service Electrical Advisory Committee for national electrical inspector certification examinations. He also served on the joint IAEI/NFPA Electrical Inspector Certification Examination development committee.

SUPPLEMENTS

Instructor's Resources to include Instructor's Manual, Testbank, Unit Presentations in PowerPoint, Image Gallery (order # 1435498313).

ACKNOWLEDGMENTS

I first want to thank the Lord Jesus Christ for giving me the abilities I have as well as the tremendous opportunities I have enjoyed over my career. I feel blessed in so many ways. "In Him we live and move and have our being."

I have been blessed with several friends who deserve special mention for their assistance, encouragement, contributions, wisdom, and support over the years: Ray Mullin, Mike Holt, Richard Loyd, and Mike Johnston.

Finally, I want to thank my wife Della for her support while I had to spend so much time devoted to this project, as well as keep the consulting business going. I promise I will really get after the "Honey-Do" list as soon as this project is completed!

Phil Simmons

CHANGES TO *NEC ARTICLES 100* AND *250* FOR THE 2011 EDITION

Once again, many changes were made to the portions of the *NEC* Code for which Panel 5 is responsible. Several proposals and comments were processed relative to the incorrect use of the term *grounding conductor*, particularly in *Chapter 8* of the *NEC*. *Chapter 8* includes the *Articles on Communications Systems*. The term *grounding conductor* was being used in place of the correct terms, *bonding conductor or jumper* or *grounding electrode conductor*. To emphasize this point, Code Panel 5 deleted the definition and use of the term *grounding conductor* in the articles under the jurisdiction of CMP-5.

Another significant change included creating a new term that will be used for both services and separately derived systems. The term is *supply-side bonding jumper*. This component is used for bonding conduits and other raceways on the line or supply side of the service equipment and at separately derived systems.

Most of the other work of CMP-5 involved reorganizing a few sections of *Article 250*.

Significant Changes in the *NEC* 2011 Edition *Articles 100* and *250*

Section	Change
Article 100: Grounding Conductor	Deleted the definition, as the term is no longer used in the *NEC.*
Article 100: Intersystem Bonding Termination	Simplified to indicate the purpose of the *Intersystem Bonding Termination* is to connect communications systems to the grounding electrode system.
Article 100: Separately Derived System	Significant changes made to address the issue of circuit connections that are present through equipment grounding conductors and the earth.
Article 100: System Bonding Jumper	Revised to include the new term *supply-side bonding jumper.*
250.2, Supply-Side Bonding Jumper	A new definition has been added to cover this conductor that is used for bonding equipment and raceways on the supply side of service equipment and between metallic enclosures of separately derived systems.
250.21(B)	Specifies the location of ground-fault detectors for ungrounded systems.
250.21(C)	Ungrounded systems are required to be legibly marked at the source or the first disconnecting means of the system in a method that will withstand the environment involved.
250.24(C)	The section has been reorganized to improve application of rules for sizing grounded conductors that are brought to the service. Included are clear rules for sizing a neutral in one raceway and in parallel raceway installations. The requirements for sizing the grounded conductor for 3-phase, 3-wire, delta-connected systems have been relocated to new *250.24(C)(3).*
250.30(A)	The requirements for grounding a grounded separately derived system have been reorganized. Requirements are incorporated for the new term "supply-side bonding jumper."
250.30(B)	For ungrounded separately derived systems, a new requirement was added to assure a low-impedance ground-fault return path.
250.30(C)	A new requirement has been added to cover grounding of separately derived systems where the source, often a generator or transformer, is located outdoors.
250.32(B)(1), Exception	The exception has been revised to permit the neutral to be used for grounding at separate buildings located on the premises only if the installation complied with the *NEC* requirements at the time the installation was made. Compliance with the conditions is required to continue using the neutral as the ground-fault return path.
250.32(B)(2)(a) and *(b)*	New requirements cover buildings and structures supplied from a separately derived system. Those buildings or structures supplied by a system that has overcurrent protection shall comply with *250.32(B)(1).* Those systems that are supplied by separately derived system where overcurrent protection is not provided where the conductors originate are required to comply with the general rules in *250.30(A).*
250.35(B)	Revised requirements for supply-side bonding jumper if a generator is not a separately derived system and overcurrent protection is not integral to the generator assembly.
250.52(A)(1) and *Exception*	The previous provisions in this section that allowed interior metal water piping for a length not greater than 5 feet to be used for connecting grounding electrode bonding jumpers has been relocated to a new *250.68(C).* Section *250.52(A)(1)* now includes only the description of a water pipe grounding electrode.
250.52(A)(2)	The description of the metal frame of a building or structure being used as a grounding electrode has been revised and simplified. Two methods are now provided to ensure the metal frame of a building or structure qualifies as making an earth connection.
250.52(A)(3)	The description of concrete-encased grounding electrodes has been significantly revised. This includes ensuring that at least one 20-ft length of metallic component is included in the concrete foundation or footing or within vertical structural components that are in direct contact with the earth.
250.53(A)	A single rod and pipe or plate electrode is required to be supplemented by an additional grounding electrode specified in *250.52(A)(2)* through *(A)(8).* An exception permits a single rod pipe or plate grounding electrode to be used, provided it has a resistance to earth of 25 ohms or less. This requirement was relocated and revised from *250.56* in the 2008 *NEC.*

(Continued)

Section	Change
250.60	Revised terminology from "air terminal" to "strike termination devices" for correlation with NFPA 780.
250.64(B)	This section covers securing and protecting grounding electrode conductors against physical damage. A new sentence was added to clearly permit grounding electrode conductors to be installed on or through framing members. Additional changes covers protecting grounding electrode conductors by some metal or nonmetallic raceway.
250.64(C)	This section covers continuity of the grounding electrode conductor. It has been expanded to cover bolted, riveted, or welded connections of structural metal frames of buildings or structures, as well as threaded, welded, brazed, soldered, or bolted-flange connections of metal water piping.
250.64(D)(1)	The permission has been added to this section to use a copper or aluminum busbar not less than ¼ in. × 2 in. (6 mm × 50 mm) for making connections of grounding electrode conductor and bonding conductors if the service equipment consists of more than one enclosure.
250.68(C)	This new section has been added and contains material formally located in *250.52(A)(1)* regarding the use of interior metal water piping for bonding grounding electrode conductors along with new provisions for using structural metal frames of buildings has grounding electrode and bonding conductors.
250.92(B)	This section has been reorganized and emphasizes bonding around impaired connections at service equipment. Examples of impaired connections include reducing washers, oversized, concentric, or eccentric knockouts.
250.94	The requirements for intersystem bonding has been reorganized into list format and redundancies removed.
250.102(C)	Extensive revisions have been made to this section including changing the title and providing requirements for supply-side bonding jumpers. This section was formerly improperly titled equipment bonding jumpers. Requirements are included for sizing a single supply-side bonding jumper and for installations made with two or more raceways where conductors are connected in parallel.
250.104(B)	A new informational note has been added to refer to *NFPA 54*, the *National Fuel Gas Code*, where additional requirements for bonding gas piping systems are found.
250.118(5) and *(6)*	These sections were revised to require the installation of an equipment grounding conductor through flexible metal conduit or liquidtight flexible metal conduit if used to connect equipment where flexibility is necessary to minimize the transmission of vibration from equipment or to provide flexibility for equipment that requires movement after installation.
250.118(10)	The conditions were revised under which Type MC cable is recognized as an equipment grounding conductor.
250.121	A new section requires that equipment grounding conductors are not permitted to be used as grounding electrode conductors.
Table 250.122	Rows for 30- and 40-ampere overcurrent devices have been deleted as being unnecessary. For the 4000-ampere overcurrent device the aluminum equipment grounding conductor has been reduced from 800 kcmil to 750 kcmil.
250.190	Requirements for grounding of equipment and systems rated over 1 kV have been revised both editorially and substantively.

Introduction to Grounding and Bonding

OBJECTIVES

After studying this introduction, the reader will understand

- the importance of using accepted definitions of terms applicable to grounding and bonding.
- the importance of providing a low-impedance path of proper capacity to ensure the operation of overcurrent protective devices.
- the various components of the grounding and bonding system.
- Ohm's law and basic electrical theory.
- electric shock hazards and the effect of electricity on the human body.

THE MYSTERY

Over the years, the subject of grounding of electrical systems, circuits, and equipment has been among the most misunderstood of topics related to electricity. Great debates were held around the turn of the twentieth century over whether electrical systems were safer if they were grounded or if they were left ungrounded. (This was also the era of debate between Thomas Edison and George Westinghouse over whether direct current [dc] was more economical and safer than alternating current [ac]. We all recognize ac distribution systems were adopted as the de facto standard primarily because of the ease of increasing or reducing voltages by means of a transformer.) As we shall see later, the *National Electrical Code (NEC)* makes clear which systems are required to be grounded and which systems are permitted to be operated ungrounded, as well as which circuits are *not* permitted to be grounded. With regard to these grounding issues, the *NEC* has ended the debate.

Speaking The Same Language

There are several reasons for the confusion about grounding and bonding of electrical systems and equipment. Inconsistent use of terms related to grounding and bonding no doubt plays a major role. Heated debates are often held about grounding when the parties may not clearly understand the definition of the term being used to describe a component of the grounding system. Use of terms that are closely related may add to the confusion. What does *grounded* mean? Does *bonded* mean the same thing as *grounded*? If a piece of equipment is *grounded*, is it also *bonded*? Can something be *grounded* and not be *bonded*? Conversely, can something be *bonded* and not be grounded? Consider a grounding-type receptacle located on the twenty-seventh floor of an office building: Is the receptacle still grounded even though the grounding electrode connection to the service equipment from which it is supplied is located many stories below? The *NEC* definition of *grounded* is "connected (connecting) to ground or to a conductive body that extends the ground connection."* Is the grounding terminal of the receptacle in fact connected to earth? If so, how? What "extends the connection to ground or earth"?

Answers to these and other questions about grounding and bonding are explored and explained throughout this book.

*Reprinted with permission from NFPA 70-2011.

DEFINITIONS

Let's look at some definitions next. An excellent grasp of the terms that are used in the *NEC*, as well as in this book, is essential for understanding the subject of grounding and bonding of electrical systems for safety. Unless noted otherwise, all definitions are from the *NEC* and are reprinted with permission from NFPA 70-2011.

As we look at these definitions, we find several have been changed in recent editions of the *NEC*. The *NEC* Technical Correlating Committee created a Task Group that was directed to review the use of terms related to grounding and bonding. The Task Group met several times between the 2005 and 2008 *NEC* revision cycles and reviewed the use of terms throughout the *NEC*. Many improper uses of existing terms were found. In addition, it was determined that several terms related to grounding and bonding were not clear or required revision for clarity. This review resulted in proposals and comments from the Task Group for many sections in the *NEC*, including for *Article 250*. In these proposals and comments, the Task Group made an effort to be more prescriptive (state what is required) in a *Code* rule and to rely less on defined terms. These changes make the requirements on grounding and bonding easier to understand and the *NEC* more user-friendly.

For the 2011 edition of the *NEC*, this work continued with a focus on getting the requirements related to grounding and bonding in *Chapter 8* of the *NEC* harmonized with the definitions in *Article 100* as well as with *Article 250*.

Ground *(NEC Article 100)*. "The earth."*

Discussion: The definition of this term was changed to its simplest form during the processing of the 2008 *NEC*. Coordinating changes were made throughout the *NEC*. Although electrical codes in other countries may use the terms *earthed* or *connected to earth*, as

FIGURE I-1 Ground (the earth), *Article 100.* (*© Cengage Learning 2012*)

equivalent to *ground* or *grounded*, the Code Making Panel responsible for *NEC* rules on grounding and bonding has not accepted proposals to include those terms.

It is recognized the earth consists of many different types of soil, from sandy loam, to rock. The photo in Figure I-1 shows several types of soil, from what appears to be soil that is clear of rocks, to solid rock. Obviously, the ability to make an electrical connection to organic matter like "dirt" can vary from being fairly easily accomplished to almost impossible.

Soil resistivity is determined largely by its content of electrolytes, which consist of moisture, minerals, and dissolved salts. Soil resistivity also is determined by its ability to retain moisture. Soil with high organic content, such as black dirt, is usually a good conductor because it retains a higher moisture level and thus electrolyte level. Sandy soils, which drain faster, tend to have a much lower moisture content and electrolyte level. As a result, they tend to have a higher resistance. Solid rock and volcanic ash contain or retain virtually no moisture and have high resistivity.

The resistance of soil varies from an average of 2370 ohm-centimeters for ashes, cinders, brine, or waste, to 94,000 ohm-centimeters for gravel, sand, and stones with little clay or loam. As a result, the earth's ability to carry current varies widely. See an additional explanation of the resistance of several types of soil in Appendix C in this book as well as for determining soil electrical resistance.

As discussed in several locations in this text, electrical connections to earth are not made for the purpose of carrying neutral or fault current. Primarily, earth connections are made for dissipating overvoltages from lightning or mishaps caused by contact between electric utility system higher-voltage lines and lower-voltage lines.

As shown in Figure I-2, to ground the electrical system, it is connected to earth by one or more grounding electrodes. Very often, electrical equipment is also connected to ground or is grounded in the same manner. Often a metallic object like a ground rod is used to attempt to make an electrical connection to organic matter—to dirt!

What is the electrical system? Although not defined in the *NEC*, this is the source voltage and current usually present at the electrical service as supplied by the electric utility. Perhaps more scientifically, the electrical system is often the source of electromotive force, the strength or intensity of which is measured in volts. Even though the actual production of electrical energy by the electric utility usually takes place some distance from the premises served, we consider the connection to the electric utility transformer(s) as connecting to or the transformer providing the electrical system. (See the definitions of *service* and *premises wiring system* in *Article 100* of the *NEC*.)

Another example of an electrical system is the supply voltage and current from a separately derived system. Such systems are power production sources or equipment that usually are located on the premises. These systems include but are not limited to a battery (or bank of batteries), transformers, generators, solar photovoltaic systems, and fuel cells. Granted, electrical energy is not "produced" by the transformer as it is by a generator, solar photovoltaic system, or fuel cell. The transformer simply increases or decreases voltage, depending on the ratio of the primary to the secondary windings. However, until the recent change in the definition of the term *separately derived system,* the *NEC* defined a transformer (depending on how it is connected) as a separately derived system. For discussion of grounding and bonding, we also consider a separately derived system to be an "electrical system."

Additional discussion on the subject and purposes of connecting the electrical system to ground

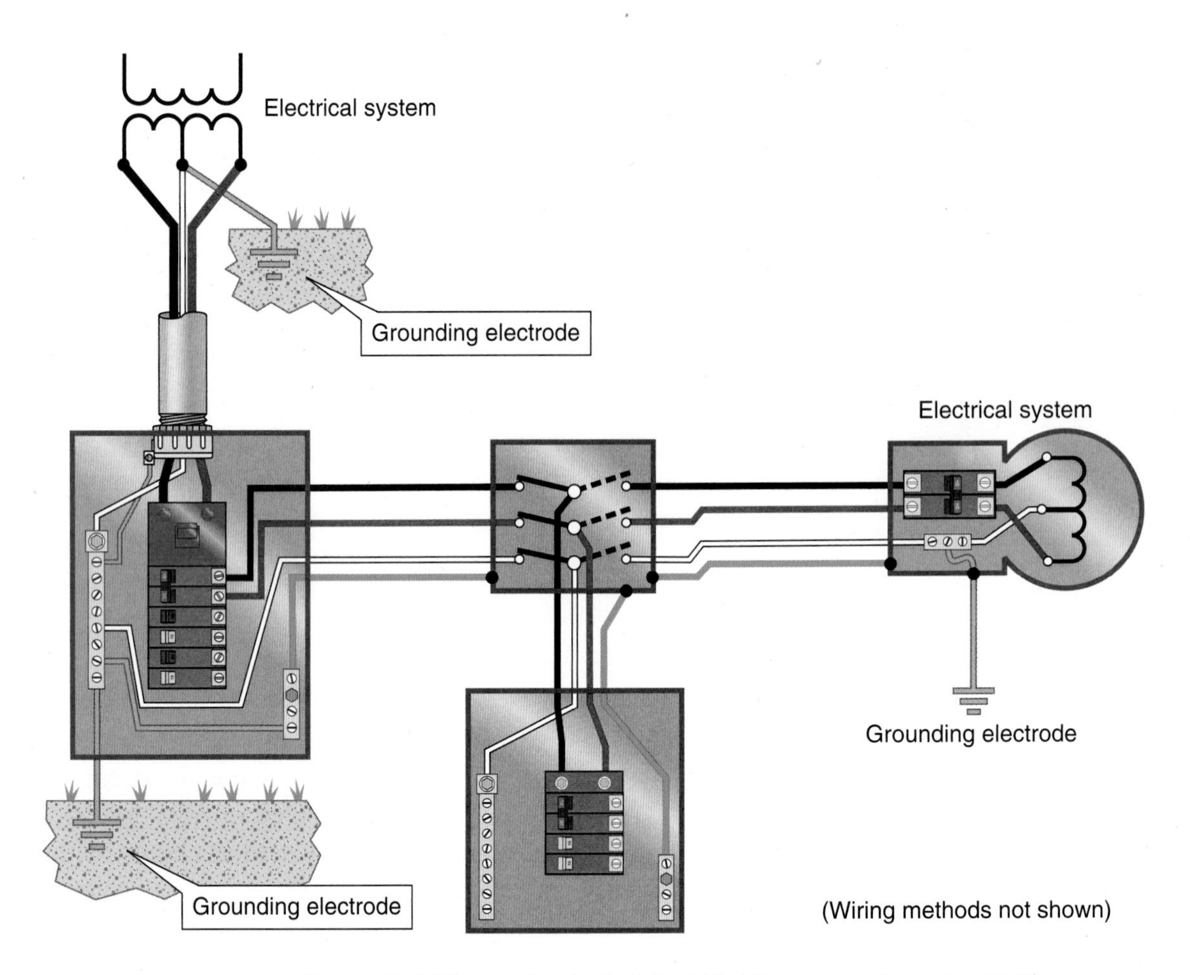

FIGURE I-2 Grounded (Grounding), *Article 100.* (*© Cengage Learning 2012*)

can be found in this book at *250.4(A).* A discussion on the purposes of grounding electrical equipment can be found in this book at *250.4(A)(2)* for grounded systems and at *250.4(B)(1)* for ungrounded systems.

Grounded (Grounding) ***(NEC Article 100).*** "Connected (connecting) to ground or to a conductive body that extends the ground connection."*

Discussion: Figure I-2 illustrates the concept of grounding electrical systems and equipment. The phrase "connected to ground" is not defined. It is generally accepted that equipment or systems are connected to *ground* (earth) through one or more grounding electrodes that make a satisfactory earth connection. The requirements for connecting a service and separately derived system to a grounding electrode (system) are found in *Part II* of *Article 250.* The requirements for installing or creating a grounding electrode system are found in *Part III* of *Article 250.* The reasons for grounding electrical systems are found in *250.4(A)(1),* where the performance goals of grounding electrical systems are stated. The purpose of connecting equipment to earth is found in *250.4(A)(2)* for grounded systems and in *250.4(B)(1)* for ungrounded systems.

Likewise, the phrase "a conductive body that extends the earth connection" is not defined. Although this phrase can perhaps have more than one meaning, for our purposes it means the following: A metallic conductive element, such as a raceway, cable, wire, or enclosure, that extends from the point where an earth connection is made at one or more grounding electrodes to another point on the electrical system where equipment such as a switchboard, panelboard, junction or pull box, or grounding-type receptacle is properly connected to it. As shown in Figure I-3, this equipment, even if located on an upper floor of a multistory building, is considered

*Reprinted with permission from NFPA 70-2011.

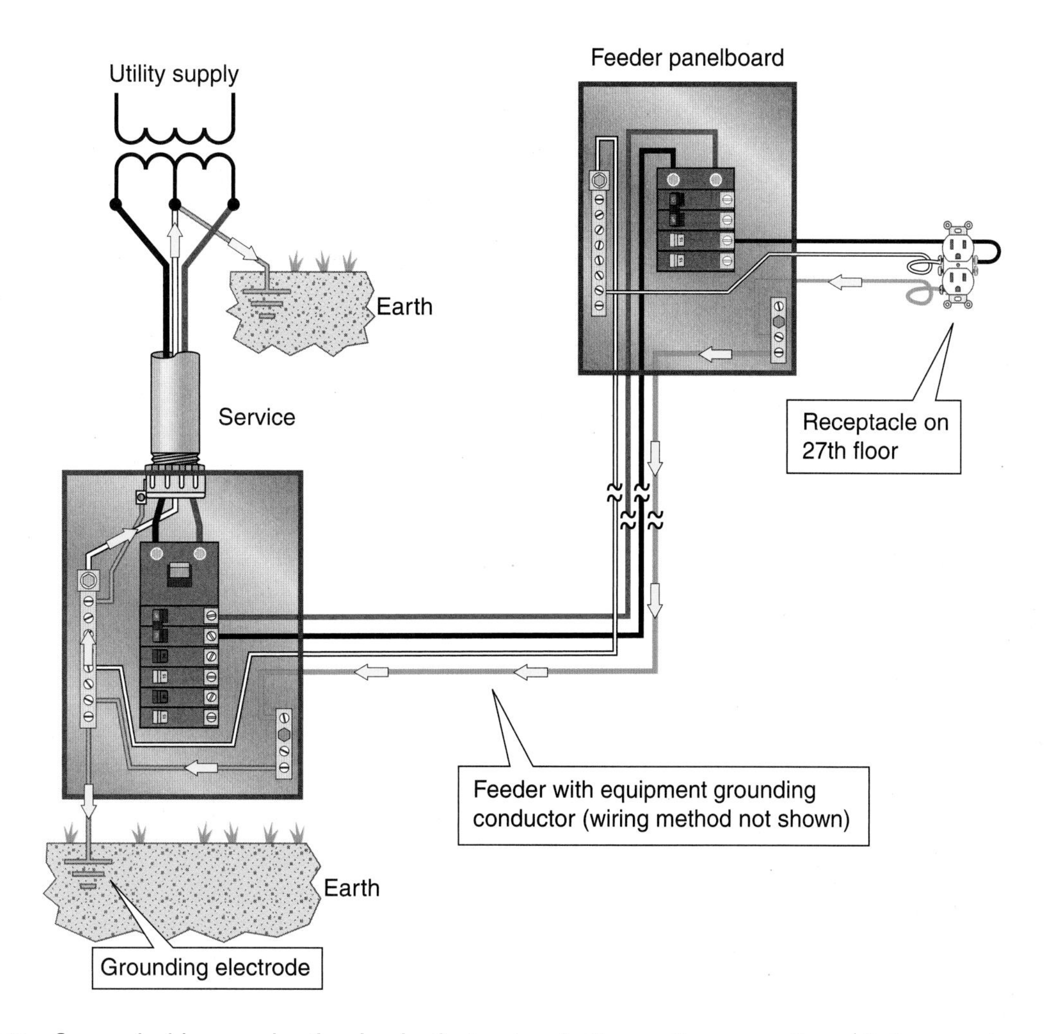

FIGURE I-3 Grounded by conductive body that extends the earth connection. (*© Cengage Learning 2012*)

to be *grounded* or *connected to earth.* It should be fairly easy to follow this grounding path through one or many points or connections to the grounding electrode (system). Requirements for providing an "effective ground-fault current path" are contained in *250.4* as well.

What does the word *equipment* mean here? This term as defined in *Article 100* of the *NEC* includes essentially anything and everything that is used in an electrical installation: fittings, devices, appliances, luminaires, apparatus (switchboards, panelboards, and motor control centers), and so forth.

As defined in *Article 100*, grounding electrodes are used to make the electrical connection to earth. Consider the metal frame of a building. If a metal structural column is driven into the earth for 10 ft (3.0 m) or more, it qualifies as a grounding electrode as described in *250.52(A)(2)*. The portion of the metal column that extends above the earth acts as a grounding electrode conductor. A structural metal grounding electrode often extends many stories above the point where it is either bonded to a concrete-encased grounding electrode or itself makes an earth connection. Separately derived systems can then use the structural metal as a grounding electrode conductor even though located many stories above grade level. Metal water pipe grounding electrodes are also recognized as "extending the earth connection" for certain occupancies where continuity of the water pipe can be verified. Extensive discussion of the rules for grounding separately derived systems is found in *250.30* in this book.

A system or equipment can be grounded (connected to ground or earth) and not be in compliance with *Code* rules. Figure I-4 shows a branch circuit that is extended to a metal pole that supports

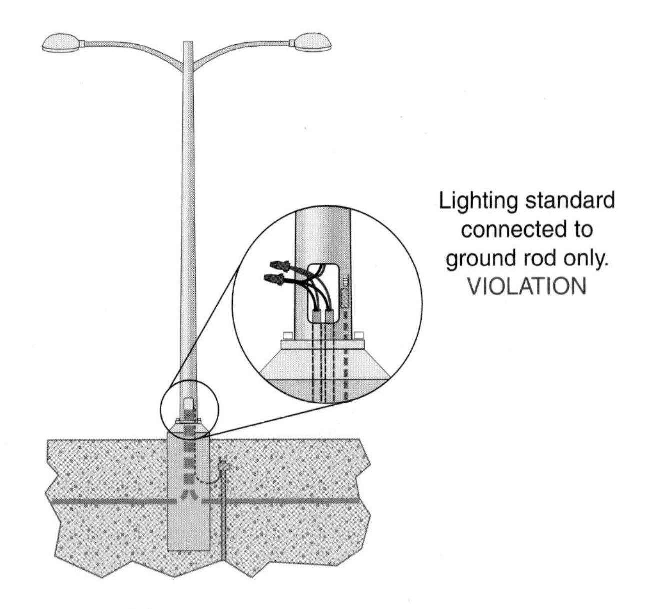

FIGURE I-4 Grounded improperly. (*© Cengage Learning 2012*)

a parking lot luminaire. An equipment grounding conductor is not run with the circuit. A ground rod is installed at the pole to ground the metal pole and the luminaire. According to the definition of *grounded,* the metal pole and luminaire are in fact grounded because the installation is literally "connected to earth" as specified in the definition. However, the pole and the luminaire are not grounded properly and cannot be considered to be in compliance with *Code* rules, because an effective ground-fault current path as required by *250.4(A)(5)* and *250.4(B)(4)* has not been provided. As seen in *250.4(A)(5),* the earth is not permitted to be used as the sole (only) equipment grounding conductor and as a result is not considered to be an effective ground-fault current path.

Therefore, when we use the word *grounded*, we imply that a connection is being made to the earth or to a conductive body that extends the earth connection in compliance with the applicable rules.

Effectively Grounded. This term was deleted during the processing of the 2008 *NEC*. The Code Panel concluded the term was vague and unenforceable. The term was thought to be an effort to add emphasis on the quality of grounding connections, not unlike being "very grounded." *Section 250.4* contains performance goals for grounding and bonding. The sections of *Article 250* following *250.4* contain "prescriptive" rules that provide the how-to requirements for installing grounding and bonding conductors, systems, and equipment.

Bonding (Bonded) ***(NEC Article 100).*** "Connected to establish electrical continuity and conductivity."*

Discussion: The term *bonded* is used as the past tense of *bonding*. This is another term related to grounding and bonding that was revised during processing of the 2008 *NEC*. *Bonding* implies a present action and *bonded* represents equipment that has been connected together. The phrase about having adequate capacity was removed because several sections of *Article 250* provide the minimum size of conductor required to provide adequate capacity to carry fault current.

Although there is more than one type of bonding, *bonding* in its simplest form means connecting metallic parts together. The conductor and connections provide the path or complete the path for fault current to flow (see Figure I-5).

Imagine a conduit connection to a panelboard or metal wireway for which a low-impedance connection cannot be ensured because of a painted surface or concentric or eccentric knockouts. A bonding jumper can be placed around the connection to ensure the continuity of the grounding path from the conduit to the enclosure.

*Reprinted with permission from NFPA 70-2011.

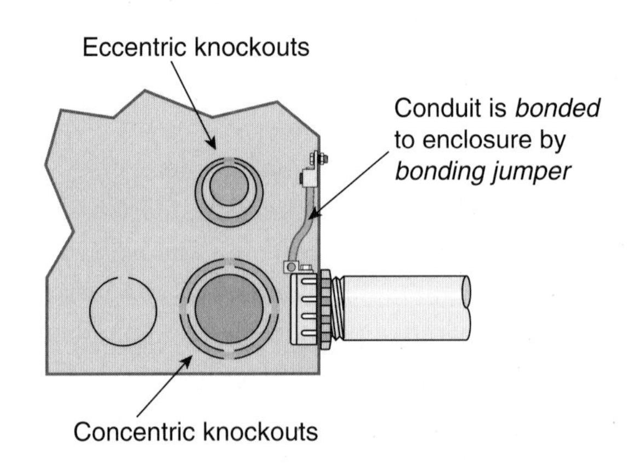

FIGURE I-5 Bonded. (*© Cengage Learning 2012*)

Another essential element of bonding is the size of the bonding jumper. Though the concept of sizing has been removed from the definition, the bonding conductor must be large enough to "conduct safely any current likely to be imposed." Obviously, a bonding jumper is of little value if it is so small that it melts while carrying fault current. Specific requirements are found in the *NEC* for bonding of service equipment in hazardous locations as well as in other applications. The proper sizing and connections of the bonding jumper are discussed in this book at the applicable *NEC* section.

As can be observed in Figure I-6, a conduit or equipment grounding conductor in a metal-clad (MC) cable bonds metallic enclosures together. It can also be observed that once bonded to an enclosure that is connected to earth (grounded), the bonding conductors extend the earth connection to the connected enclosures. Because the bonding conductor serves dual purposes, we can conclude that the functions of grounding and bonding become inseparable. Note as well that the definition of *equipment grounding conductor* has been revised to recognize the dual action of grounding and bonding.

Another type of bonding is equipotential bonding. The term *equipotential bonding* is not defined in the *NEC*, but the concept is that a bonding conductor is installed and connected between equipment so there is no shock hazard if a person or animal contacts both equipments at the same time. This equipotential bonding attempts to keep the equipment at the same voltage level even if a line-to-ground fault has occurred in one of the pieces of equipment. A voltage will be impressed on the equipment until the overcurrent device opens due to voltage drop of the circuit.

Equipotential bonding requirements appear in at least four locations in the *NEC*. Equipotential bonding is used in patient care areas of health care facilities, though the term is not used in *Article 517,* in *Article 680* for swimming pools and related equipment, in *Article 682* for certain metal equipment near natural and artificially made bodies of water, and in *Article 547* for agricultural buildings. Equipotential bonding is done to keep the potential (voltage) equalized so far as practicable so there is no shock hazard to people or animals that have contact with more than one conducting surface or are immersed in water. In this book, the words *bonded* and *bonding* are not used to mean equipotential bonding unless stated otherwise.

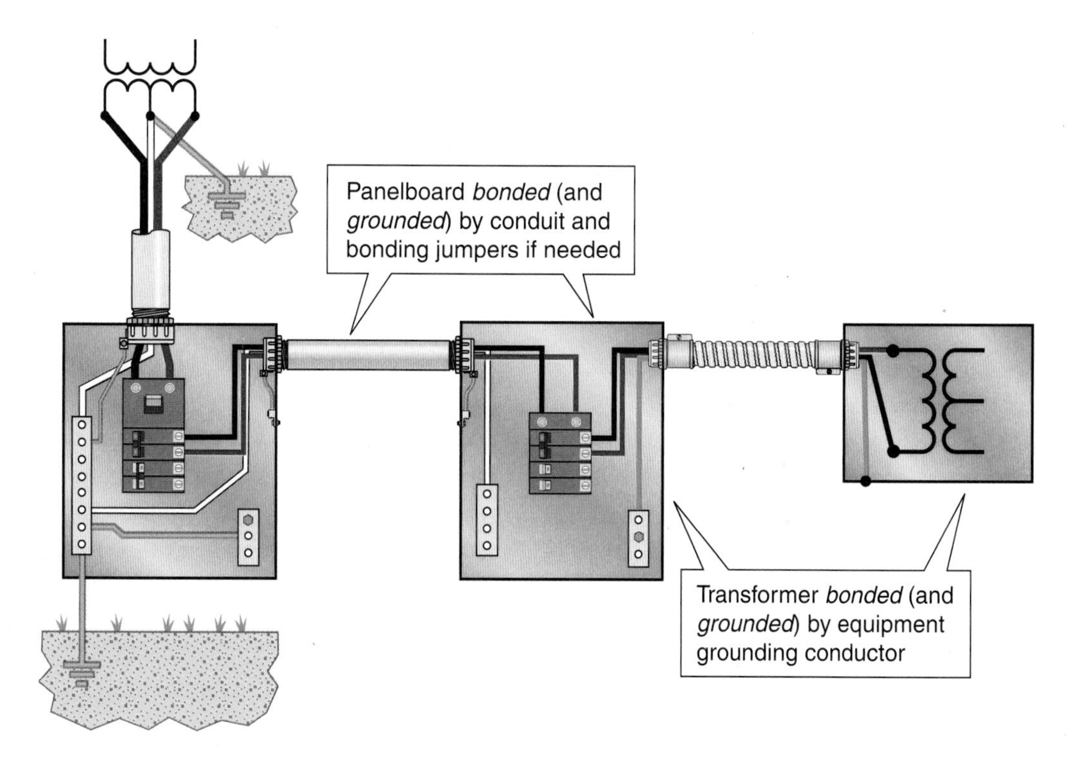

FIGURE I-6 Bonding with wiring methods and equipment grounding conductor. (*© Cengage Learning 2012*)

Bonding Conductor or Jumper ***(NEC Article 100).*** "A conductor to ensure the required electrical conductivity between metal parts required to be electrically connected."*

Discussion: A bonding conductor or jumper is a conductor, made of copper or aluminum, used to ensure electrical conductivity between metal parts that must be electrically connected (see Figure I-7). Obviously, a bonding jumper is used to comply with the requirement for bonding equipment or completing the path for current to flow. For some installations, the *NEC* requires the bonding jumper to be copper; for others, an aluminum conductor is permitted to be used. The *NEC* specifies the minimum size of copper or aluminum conductor permitted for the particular application. In every case, the conductor must be large enough to safely carry the fault current imposed on it without becoming so hot the connectors or conductors are damaged.

It is also obvious that a piece of conduit or other metal raceway functions as a bonding conductor. When properly connected to each metallic enclosure, the metallic raceway connects or bonds the metal enclosures together. In reality, the connection of the conduit to the first enclosure bonds the conduit to the enclosure. The connection of the metal conduit to the second enclosure makes a bonding connection. Thus the metal enclosures are connected together, or bonded, by the proper connection of the metal raceway between them.

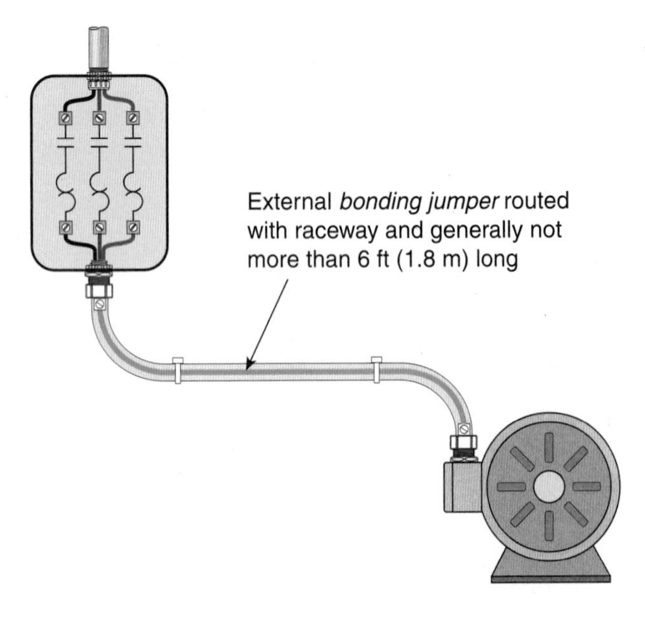

FIGURE I-7 Bonding conductor or jumper, *Article 100.* (*© Cengage Learning 2012*)

Equipment Bonding Jumper (Bonding Jumper, Equipment) ***(NEC Article 100).*** "The connection between two or more portions of the equipment grounding conductor."*

Discussion: Equipment bonding jumpers are used to complete the path provided by the equipment grounding conductors should it be impaired in any way such as a suspect connection of conduit or EMT to an enclosure (see Figures I-5 and I-7). Equipment grounding conductors are permitted to consist of conduit or metallic tubing, in addition to the wire-type conductors. A list of acceptable equipment grounding conductors is found in *NEC 250.118.* An equipment bonding jumper sometimes is needed to provide a reliable and low-impedance connection between conduit fittings, or between conduit and enclosure. An equipment bonding jumper should be installed if the integrity of connection of the equipment grounding conductor, be it conduit or otherwise, is not assured or is suspect in any way. Remember, we need to provide the effective ground-fault return path.

Equipment grounding conductors and bonding jumpers are permitted to be installed outside a raceway or cable assembly in certain situations. These equipment grounding or bonding jumpers generally are limited to not more than 6 ft (1.8 m) in length and must be routed with the raceway; an example is shown in Figure I-7. See *250.102(E), 250.130(C),* and *250.134(B).* Bonding jumpers are permitted to be longer than 6 ft (1.8 m) for outside installations on poles.

For conduit and metallic tubing installations, a single loose locknut, connector, or coupling connection will interrupt the ground-fault return path, creating a shock hazard if a ground fault occurs beyond the loose connection. (This cautionary note is not intended as a criticism of properly made conduit or metallic tubing installations.) As can be seen in Figure I-8, a ground fault on the downstream side of a loose conduit connection will create a shock hazard, as equipment on the load side of the fault will rise to the voltage of the system. This voltage will appear across the point of the loose connection. This is identical to measuring

*Reprinted with permission from NFPA 70-2011.

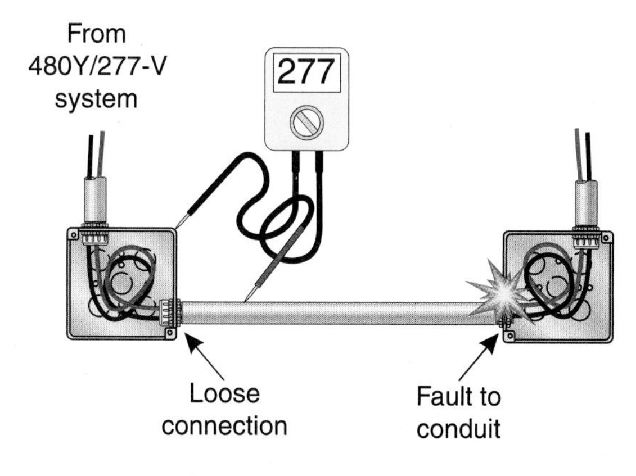

FIGURE I-8 Incomplete path.
(© Cengage Learning 2012)

the voltage across a switch. You will find system voltage across a switch that is opened.

Main Bonding Jumper (Bonding Jumper, Main) ***(NEC Article 100).*** "The connection between the grounded-circuit conductor and the equipment grounding conductor at the service."*

Discussion: *Article 250* has specific requirements for the size of the main bonding jumper. As with other bonding jumpers, the main bonding jumper must be large enough not to melt while carrying current. The main bonding jumper is located within the service equipment or the service disconnecting means and is permitted to be a wire, a bus, or a screw (see Figure I-9). It is identical in function to the system bonding jumper that is located in separately derived systems. The main bonding jumper and the system bonding jumper are critical to the safety of the electrical system because they provide the path for ground-fault current to return to the source. As noted previously, a complete circuit is required for current to flow. The main bonding jumper and the system bonding jumper are essential because each is a key component in completing the ground-fault current return path.

Generally, the ground-fault current returns to the service equipment over the equipment grounding conductor. The main bonding jumper transfers the fault current to the neutral, which then serves as the conductor to return the fault current to the source.

*Reprinted with permission from NFPA 70-2011.

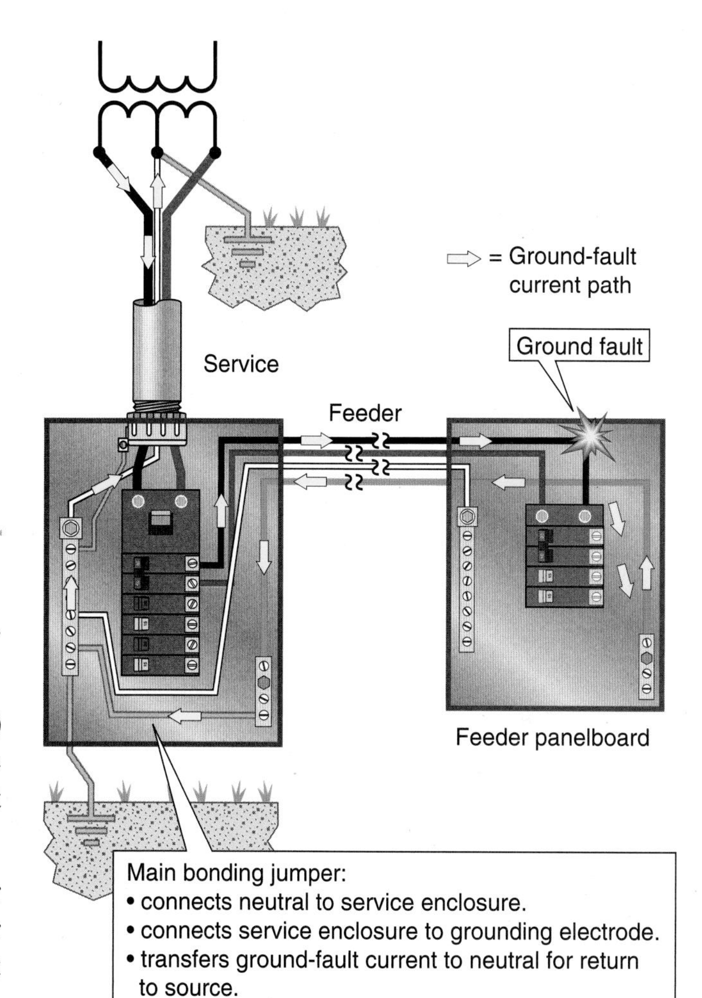

FIGURE I-9 Main bonding jumper, *Article 100.*
(© Cengage Learning 2012)

Supply-Side Bonding Jumper (Bonding Jumper, Supply-Side) ***(250.2).*** "A conductor installed on the supply side of a service or within a service equipment enclosure(s), or for a separately derived system, that ensures the required electrical conductivity between metal parts required to be electrically connected."*

Discussion: This definition was added to the 2011 *NEC*. As can be seen in Figure I-10, it is intended that this conductor be used for both services and separately derived systems. As is indicated in the title, this conductor is used for bonding and raceways on the line or supply side of the service and between enclosures that are required to be bonded for a separately derived system.

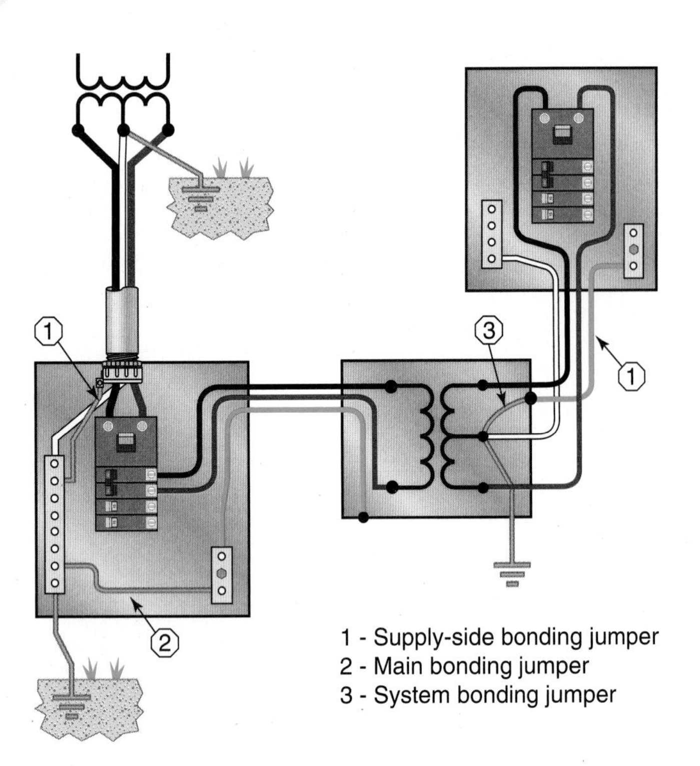

FIGURE I-10 Supply-Side Bonding Jumper, *250.2.* (*© Cengage Learning 2012*)

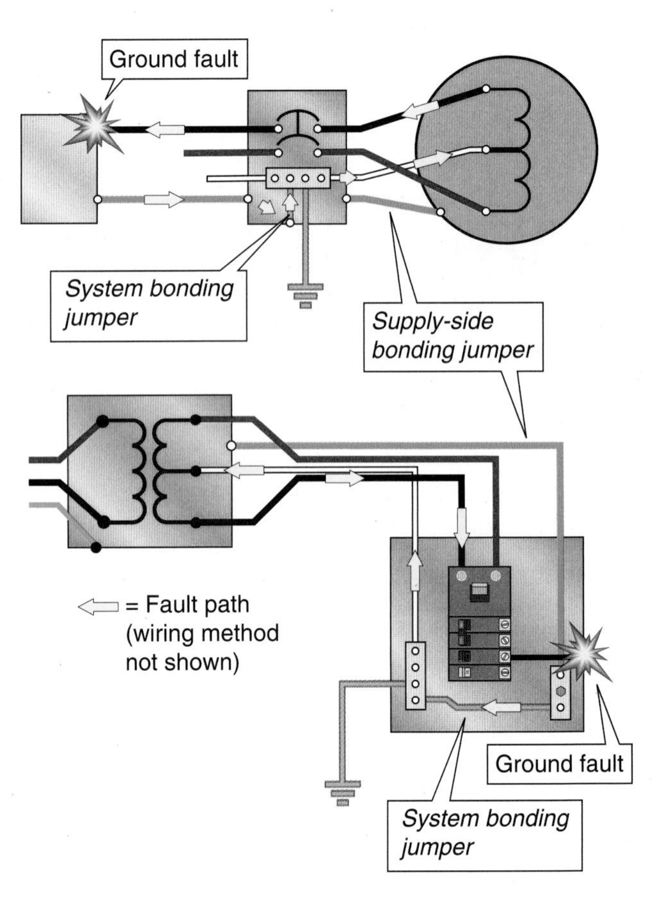

FIGURE I-11 System bonding jumper, *250.2.* (*© Cengage Learning 2012*)

Because this conductor is located on the supply side of a service or separately derived system, there is no overcurrent device ahead of it. The sizing of the supply-side bonding jumper is determined from *Table 250.66,* based on the size of the supply conductors at the point where the bonding is to occur. You determine the size of the ungrounded, or "hot," conductor of the system at the point where bonding is required, and then go to *Table 250.66* to find the size of the bonding conductor. Don't be confused by the title of *Table 250.66*. The title of the table says it is for determining the minimum size of a grounding electrode conductor. That's true, but the table is also used for many other purposes.

System Bonding Jumper (Bonding Jumper, System) ***(NEC Article 100).*** "The connection between the grounded-circuit conductor and the supply-side bonding jumper, or the equipment grounding conductor, or both, at a separately derived system."*

Discussion: Note that the definition of *system bonding jumper* is similar to that of the main bonding jumper, because they are similar in function. Like a main bonding jumper, the system bonding jumper is required to be of copper or other corrosion-resistant material and is permitted to be a wire, a bus, or a screw (see Figure I-11). Also, like the main bonding jumper, the system bonding jumper provides the critical path for ground-fault current to return to the source.

Generally, a system bonding jumper is permitted to be installed at only one location in a separately derived system. This approaches the concept of single point grounding. These steps are taken to eliminate creating parallel paths for neutral current.

The definition incorporates a reference to the new term, *supply-side bonding jumper*. As can be seen in Figure I-11, the supply-side bonding jumper connects directly or indirectly through a metal enclosure to the system bonding jumper. Additional information on how the two conductors function together to provide electrical safety is covered in this text in *250.30.*

*Reprinted with permission from NFPA 70-2011.

Solidly Grounded (Grounded, Solidly) ***(NEC Article 100).***

"Connected to ground without inserting any resistor or impedance device."*

Discussion: The vast majority of electrical systems are solidly grounded. This definition has previously been located in *230.95,* where it was used as a condition for determining when ground-fault protection of equipment is required. It is recognized that all conductors have resistance or impedance. The application of this term *solidly grounded* suggests that the resistance or impedance of the grounding electrode conductor is ignored. The system is considered solidly grounded when it is connected to the grounding electrode with a copper or aluminum grounding electrode conductor, so long as an impedance device such as a resistor or inductor is not located in the path from the service equipment or separately derived system to the grounding electrode. Another common type of system grounding is impedance or high-impedance grounding. A resistor or an inductor is inserted in the grounded conductor path from the grounding electrode to the service equipment or feeder equipment. This limits the fault current on the first ground fault. See *250.36* for additional information on this subject. Figure I-12 shows various types of grounding.

Grounded Conductor ***(NEC Article 100).*** "A system or circuit conductor that is intentionally grounded."*

*Reprinted with permission from NFPA 70-2011.

Discussion: This is a broad term that includes both neutral conductors and grounded conductors that are not neutral conductors. (New definitions of *Neutral Conductor* and *Neutral Point* were added to the 2008 *NEC*.) An example of a grounded conductor that is not a neutral is from a corner-grounded delta-connected system. See the discussion in *250.20* for systems that are required to be grounded, *250.21* for systems that are not required to be grounded but are permitted to be grounded, and *250.22* for circuits that are not permitted to be grounded.

It should be noted that the grounded conductor is defined as one that is intentionally grounded or connected to earth. This occurs at the electrical service and at or near the source of a separately derived system as required or provided for in *250.20, 250.21, 250.24,* or *250.30.* Therefore, it can be properly concluded that a conductor that is unintentionally connected to an enclosure by a fault is not a *grounded conductor* in the sense of this definition. This can occur when an ungrounded conductor from a high-impedance grounded neutral system or from an ungrounded system faults to a metallic enclosure.

Figure I-13 shows a grounded conductor from electrical systems that are grounded in various ways.

Grounding Conductor ***(formerly in NEC Article 100).***

Discussion: This term was deleted during the processing of the 2011 *NEC*. It was determined that the term was too close in meaning to other grounding and bonding conductors such as equipment

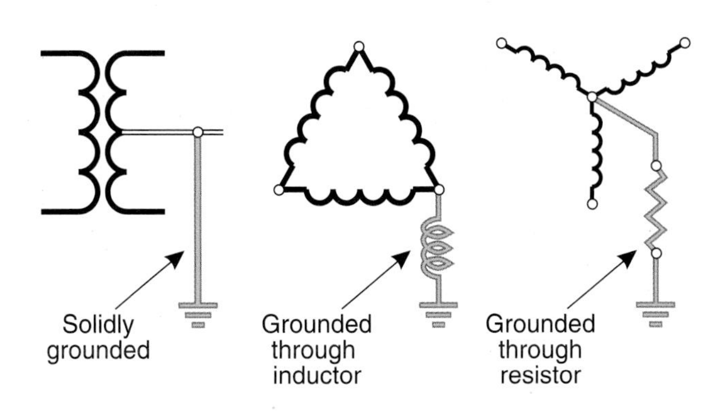

FIGURE I-12 Types of grounding. (*© Cengage Learning 2012*)

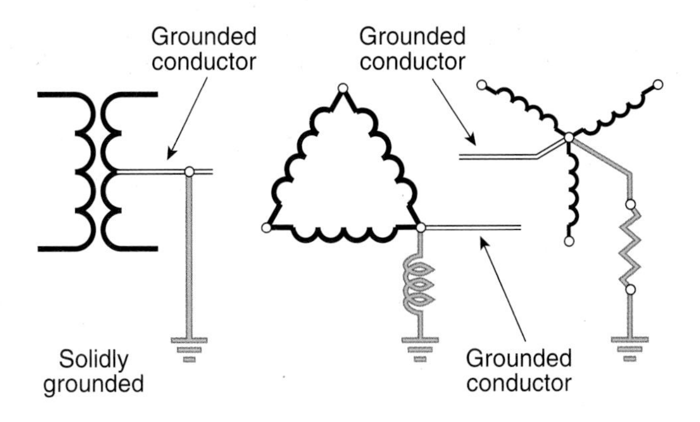

FIGURE I-13 Grounded conductor, *Article 100.* (*© Cengage Learning 2012*)

grounding conductors and equipment grounding jumpers. In addition, the term was being used incorrectly in *Chapter 8* for communications systems.

Equipment Grounding Conductor ***(EGC) (NEC Article 100).*** "The conductive path(s) installed to connect normally non-current-carrying metal parts of equipment together and to the system grounded conductor or to the grounding electrode conductor or both.

> "**Informational Note No. 1:** It is recognized that the equipment grounding conductor also performs bonding.
>
> "**Informational Note No. 2:** See 250.118 for a list of acceptable equipment grounding conductors."*

Discussion: Very important changes were made to the definition of this term for the 2008 *NEC*. The definition recognizes a conductive path or paths are provided by the equipment grounding conductor. The previous definition used the word *conductor*. When the word *conductor* is used, the connotation is that of a wire or busbar. Conduit and other metallic raceways as well as cable armor or metallic sheaths can provide the conductive path required to perform as a fault current return path.

The word *normally* before "non-current-carrying metal parts of equipment" clarifies that fault current does not "normally" flow over these conductive paths. The equipment grounding conductor is intended to carry current only during a line-to-ground fault. When that occurs, the equipment grounding conductor must carry adequate current long enough for the overcurrent protective device on the line side of the fault to open. (Current "normally" flows on ungrounded and grounded [often neutral] conductors.) If current is regularly flowing over equipment grounding conductors, it may be considered "objectionable current" as covered in *250.6*. It is intended that fault current will flow over equipment grounding conductors only while a ground fault exists.

The word *together* was added following the phrase "connect normally non-current-carrying metal parts of equipment" to recognize the function of bonding performed by the equipment grounding conductor. As discussed earlier in this book, the concepts and functions of grounding and bonding are merged in most installations. The equipment grounding conductor not only extends the earth connection but also bonds conductive parts together. (In fact, the Grounding and Bonding Task Group for the 2008 *NEC* attempted to create a phrase that embodies both the concepts and functions of grounding and bonding, but it was not successful.) A new *Informational Note* was added to remind the *Code* user that the equipment grounding conductor also performs bonding.

A list of acceptable equipment grounding conductors is included in *250.118.* Included are copper, aluminum, or copper-clad aluminum conductors and several types of conduit and metallic tubing, as well as the armor of various cables (see Figure I-14). Also see the discussion in *250.122* for the minimum required size of equipment grounding conductors.

Grounding Electrode ***(NEC Article 100).*** "A conducting object through which a direct connection to earth is established."*

Discussion: This definition was new in the 2005 *NEC*, although the term was used in several articles of the *NEC* for many editions. It was revised for the 2008 *NEC* to indicate the conducting object is used to make a direct connection to the earth. A description of the conducting objects recognized as grounding electrodes is found in *250.52(A). NEC 250.50* contains a requirement for connecting grounding electrodes to form a grounding electrode system. Figure I-15 shows the symbol (sometimes referred to as an upside-down Christmas tree) used for grounding electrodes in this book.

Note that the definition of the term does not include a concept of using a grounding electrode to clear ground faults. A grounding electrode is used for making an earth connection of electrical equipment and the electrical system for the purposes indicated in *250.4.* A path through the earth is created by two or more grounding electrodes that are in parallel with the grounded conductor. Though some ground-fault current will flow back to the source through the earth, the earth circuit almost always

*Reprinted with permission from NFPA 70-2011.

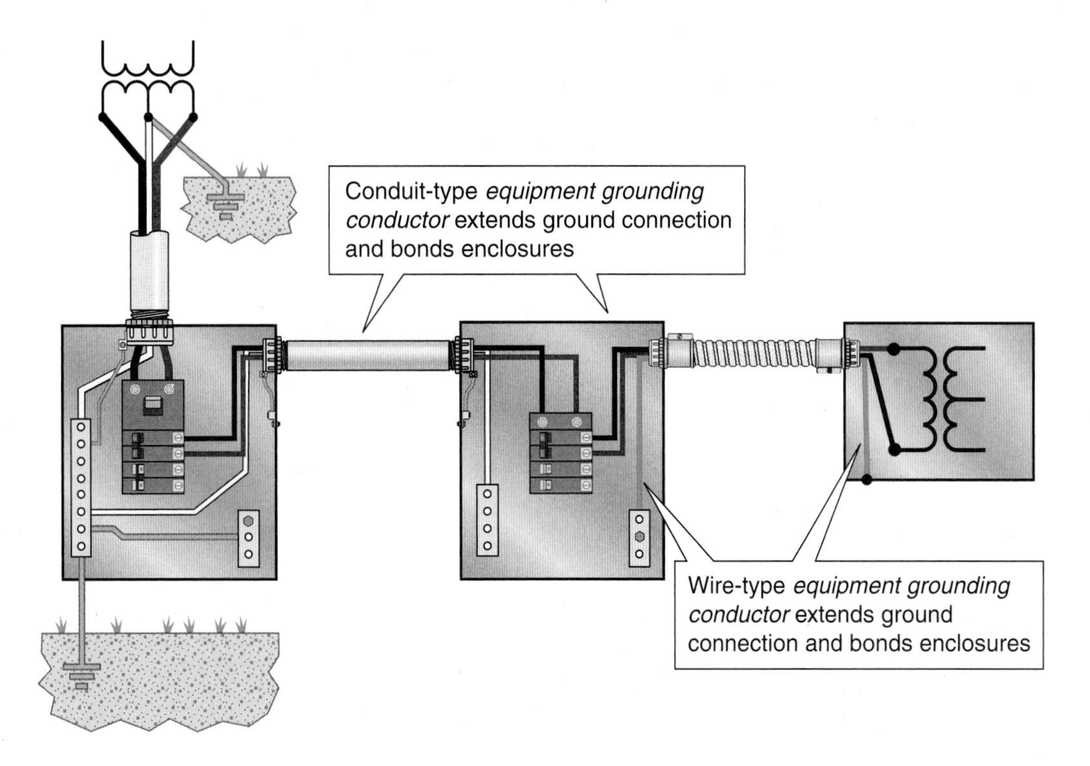

FIGURE I-14 Equipment grounding conductor, *Article 100.* (*© Cengage Learning 2012*)

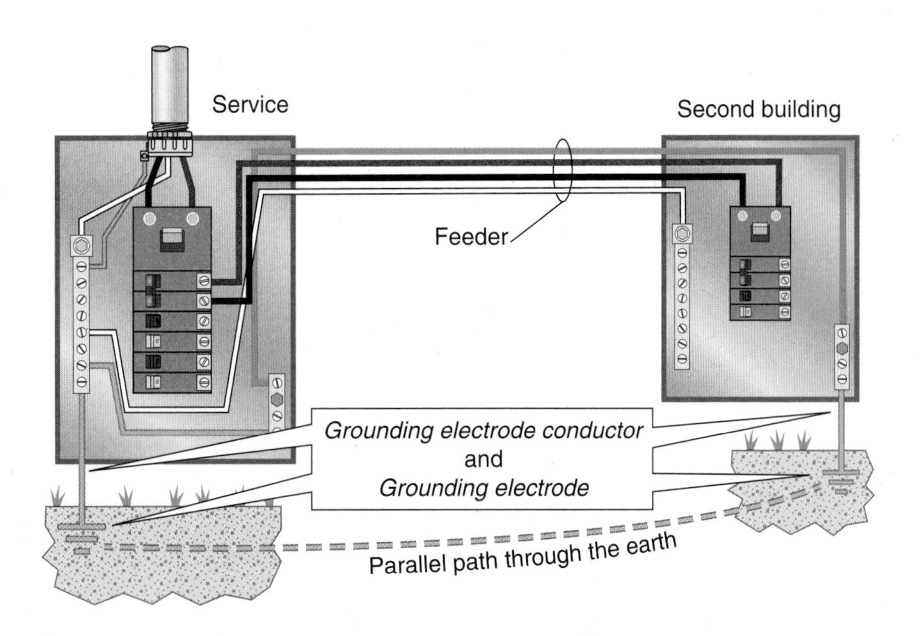

FIGURE I-15 Grounding electrode and grounding electrode conductor, *Article 100.* (*© Cengage Learning 2012*)

represents a high-impedance path and will carry a small percentage of the total current returning to the source.

See the information on testing of grounding electrodes, particularly those of the rod and pipe type, in Appendix C of this book.

Grounding Electrode Conductor ***(NEC Article 100).*** "A conductor used to connect the system grounded conductor or the equipment to a grounding electrode or to a point on the grounding electrode system."*

*Reprinted with permission from NFPA 70-2011.

Discussion: The definition of this term was simplified and broadened for the 2008 *NEC*. The examples of locations where the grounding electrode conductor is used to connect systems or equipment to a grounding electrode were deleted. These examples were unnecessary because specific rules are provided in the *NEC* for where a connection to a grounding electrode is required.

In addition, the phrase "or to a point on the grounding electrode system" is in recognition that a connection of the grounding electrode conductor is permitted to be made to a busbar, as provided for services in *250.24(A)(4)* and for separately derived systems in *250.30(A)(5), Exception No. 1,* and *250.64(F),* as well as directly to one or more grounding electrodes.

The grounding electrode conductor is fairly easy to identify because it connects the grounding electrode to the grounded system conductor, to the enclosures for the service equipment and the equipment grounding conductor for grounded systems, and to the enclosures for the service equipment and the equipment grounding conductor for ungrounded systems. Figure I-15 shows grounding electrode conductors.

Figure I-15 also illustrates an unavoidable parallel path for current that is established through the earth. This can be observed by the dashed line from grounding electrode to grounding electrode. The equipment grounding conductor also connects the grounding electrodes together. Most all current will flow through the equipment grounding conductor, as the path through the earth is normally of a higher resistance and impedance.

Specific rules are provided in *Article 250* for the sizing and installation of grounding electrode conductors, as well as where they are required to be connected to the electrical system or equipment. In some cases, specific requirements must be met; in others, considerable flexibility on installation methods is permitted.

Intersystem Bonding Termination *(NEC Article 100).* "A device that provides a means for connecting bonding conductors for communications systems to the grounding electrode system."*

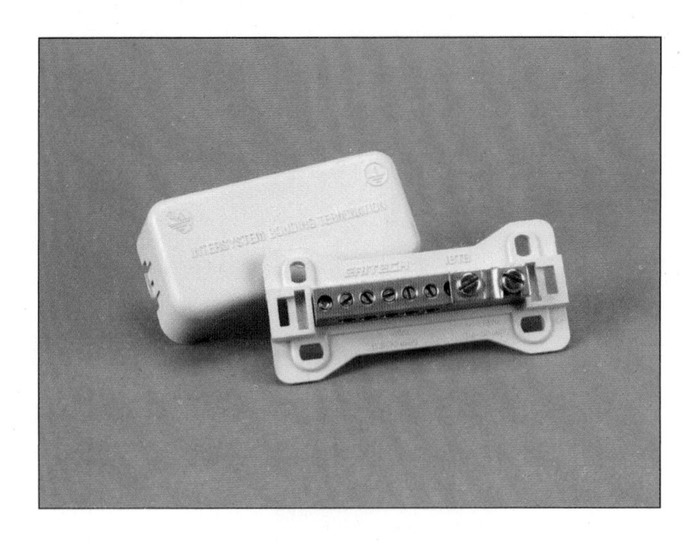

FIGURE I-16 Device for connecting conductors for intersystem bonding termination. (*Photo courtesy of Erico*)

Discussion: The definition has been revised and simplified from the 2008 *NEC*. Installation requirements are found in *250.94*. In that section, you will find specific rules for installing the intersystem bonding termination device at the service equipment as well as near the disconnecting means if the premises has a building or structure that is supplied by a feeder or branch circuit. Several types of devices to accomplish intersystem bonding are available. Figure I-16 shows one such device.

The intersystem bonding termination device provides the ability to connect all of the electrical systems that supply a building or structure together. This is aimed primarily at communications systems such as radio and TV antennas, including dish antennas, to ensure that all of these systems are grounded and bonded at the same point. This is critical to prevent flashover from overvoltage events such as nearby lightning strikes from damaging electrical equipment.

The owner or user of electrical equipment should carefully consider providing additional levels of protection by installing surge protective devices in accordance with *NEC Article 285*.

Neutral Point. "The common point on a wye connection in a polyphase system or midpoint on a single-phase, 3-wire system, or midpoint of a single-phase portion of a 3-phase delta system, or midpoint of a 3-wire, direct current system."*

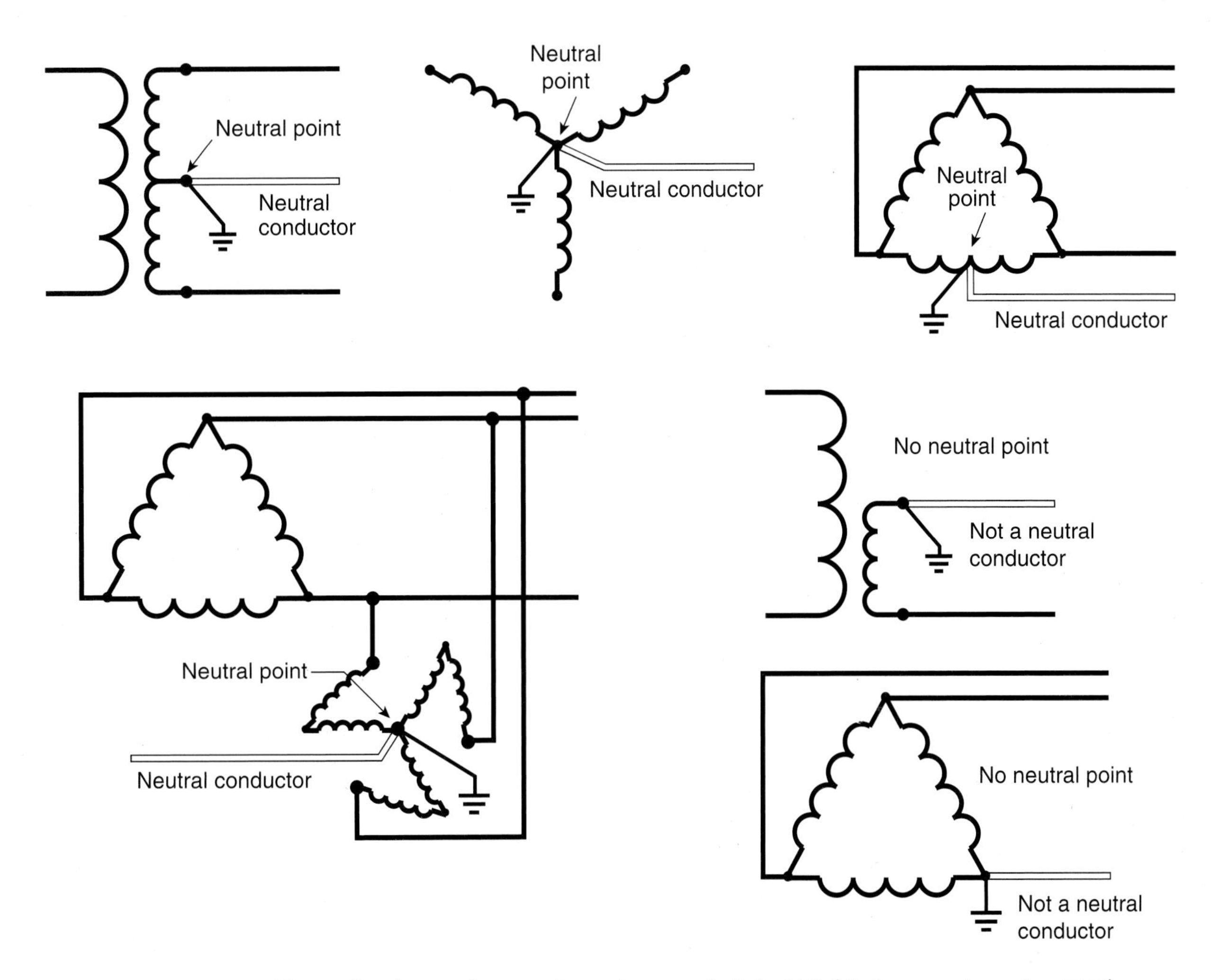

FIGURE I-17 Neutral point and neutral conductor, *Article 100.* (*© Cengage Learning 2012*)

Informational Note: "At the neutral point of the system, the vectoral sum of the nominal voltages from all other phases within the system that utilize the neutral, with respect to the neutral point, is zero potential."*

Discussion: This was a new definition for the 2008 *NEC*. The term *neutral* has been used in the *NEC* for many editions but was never defined. The electrical industry previously relied on definitions in other standards such as the International Electrical and Electronic Engineers' (IEEE) *Dictionary of Standard Electrical and Electronic Terms,* IEEE-100, or on common understanding of the term. The term *neutral point* is not used in the IEEE *Dictionary*. The definition of *neutral conductor* in the *NEC* is different from that in the IEEE *Dictionary*. As used in the *NEC*, the term *neutral point* is used to define the point of origin of the neutral conductor.

As shown in Figure I-17, an electrical system has a neutral point only if it has more than one phase ("hot") or ungrounded conductor. This includes the following systems:

120/240-volt, single-phase, 3-wire
120/240-volt, 3-phase, 4-wire (delta-connected)
208Y/120-volt, 3-phase, 4-wire (wye-connected)
480Y/277-volt, 3-phase, 4-wire (wye-connected)

Under the definitions, a 120-volt, single-phase, 2-wire system will not have a neutral point or neutral conductor. Although this system is required to be grounded by *250.21(B)(1),* it will have one ungrounded ("hot") and one grounded conductor, but no *neutral.*

*Reprinted with permission from NFPA 70-2011.

Neutral Conductor *(NEC Article 100).* "The conductor connected to the neutral point of a system that is intended to carry current under normal conditions."*

Discussion: This also was a new definition for the 2008 *NEC* and is illustrated in Figure I-17. The definition is the result of a Task Group that met between the 2005 and 2008 *NEC* cycles. Rather than including all the concepts necessary to properly define the term *neutral conductor* in this definition, a companion definition for *neutral point* was added. Commonly used systems that have a neutral conductor include these:

120/240-volt, single-phase, 3-wire
120/240-volt, 3-phase, 4-wire (delta-connected)
208Y/120-volt, 3-phase, 4-wire (wye-connected)
480Y/277-volt, 3-phase, 4-wire (wye-connected)

However, the grounded conductor of a 2-wire circuit consisting of one ungrounded conductor and one grounded conductor from a system having a neutral point is now defined as a neutral conductor. (See *250.26* for the system conductor that is required to be grounded.) Although many electricians have long referred to a 2-wire circuit with a grounded conductor as a neutral, most electrical "theorists" have stated that a neutral conductor must carry the unbalanced current from ungrounded conductors to be considered a neutral. Based on the new definitions of *neutral point* and *neutral conductor,* the grounded conductor of a 2-wire circuit is indeed a neutral if it originates from an electrical system that has a neutral point. However, a grounded conductor from a 120-volt, single-phase, 2-wire system is not a neutral as defined.

An important distinction is made in the definition between neutral conductors and equipment grounding conductors. The neutral conductor is expected to carry current under normal conditions, whereas the equipment grounding conductor is intended to carry current only under abnormal conditions. Excessive unintended current on the equipment grounding conductor should be considered as objectionable current as covered in *250.6.*

Most neutral conductors are also grounded conductors as defined in *Article 100.*

*Reprinted with permission from NFPA 70-2011.

Separately Derived System *(NEC Article 100).* "A premises wiring system whose power is derived from a source of electric energy or equipment other than a service. Such systems have no direct electrical connection from circuit conductors of one system to circuit conductors of another system, other than connections through the earth, metal enclosures, metallic raceways, or equipment grounding conductors."

Discussion: Note that the term *premises wiring system* used in this definition is also defined in *Article 100* and basically means everything on the load side of the *service point.* The *service point* is the point of demarcation between the electric utility system and the premises wiring. Separately derived systems, therefore, are not supplied directly from the electrical utility such as the service is, but are installed as a part of the premises electrical equipment.

The phrase "whose power is derived from a source of electric energy or equipment other than the service" can be confusing insofar as a transformer-type separately derived system is concerned. A transformer obviously does not produce energy from another source, as does a generator or solar photovoltaic system. A transformer simply increases or decreases voltage based on the ratio of the primary and the secondary windings. However, the source of electrical energy for a transformer is from the service if the transformer is not on the load side of an on-premises generator or other local source of electrical energy.

For the purposes of the *NEC* and this book, transformers are considered to be separately derived if there is no direct electrical connection, including a solidly connected ungrounded ("hot") or grounded circuit conductor, between the primary and the secondary of the transformer. A *direct electrical connection* is one that would be made at the terminals for the transformer windings, such as from the primary neutral to the secondary neutral (usually the XO terminal), as in the case of a wye–wye-connected transformer. Three-phase transformers are often connected delta on the primary and wye on the secondary. As a result, there are no direct electrical connections from primary to secondary. Connection of equipment grounding and bonding connections in the transformer does not constitute

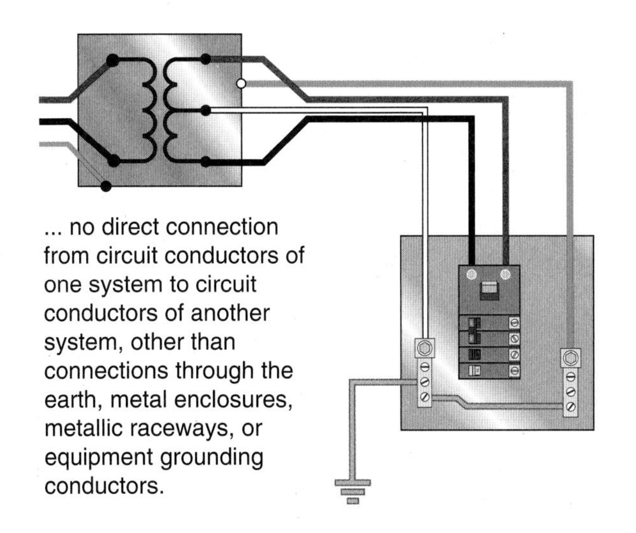

FIGURE I-18 Transformer-type separately derived system. (*© Cengage Learning 2012*)

a direct electrical connection and are permitted in the definition. Figure I-18 shows a transformer-type separately derived system.

Note that the revised definition recognizes that an unintended connection can exist between two or more systems by interconnection of equipment grounding conductors. As shown in Figure I-19, the neutral conductor from the utility transformer connects the system to the service disconnecting means. The equipment grounding conductor in the feeder from the service to the transformer connects the service equipment to the transformer. Finally, the system bonding jumper in the transformer completes the interconnection of the derived system to the utility system. Recognizing that this interconnection is commonplace in electrical installations, the previous definition was revised during processing of the 2011 *NEC* to be accommodating.

For generator-supplied separately derived systems, the key for determining whether the system is separately derived or not is to observe how the grounded conductor is treated in the transfer switch(es). If the grounded conductor (which may be a neutral) *is* switched with the ungrounded or phase conductors, the generator-supplied system is in fact separately derived. If the grounded conductor (which may be a neutral) *is not* switched with the ungrounded or phase conductors, the generator-supplied system is not separately

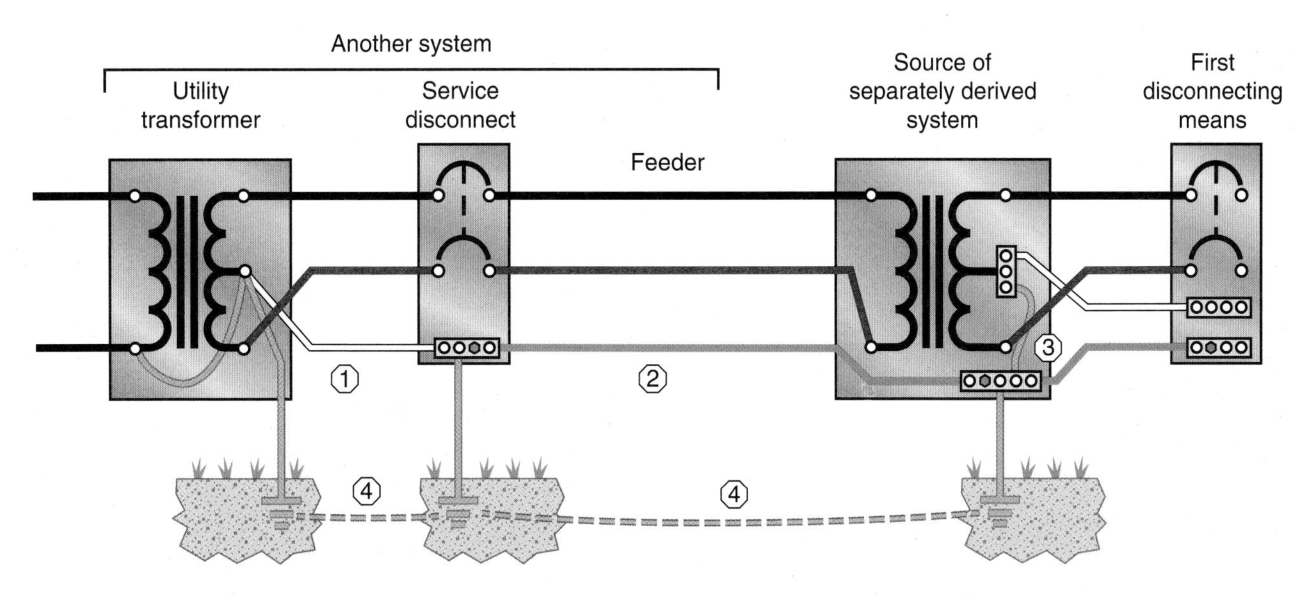

FIGURE I-19 Revised definition of separately derived system recognizes that interconnection of systems occurs through equipment grounding conductors and through the earth. (*© Cengage Learning 2012*)

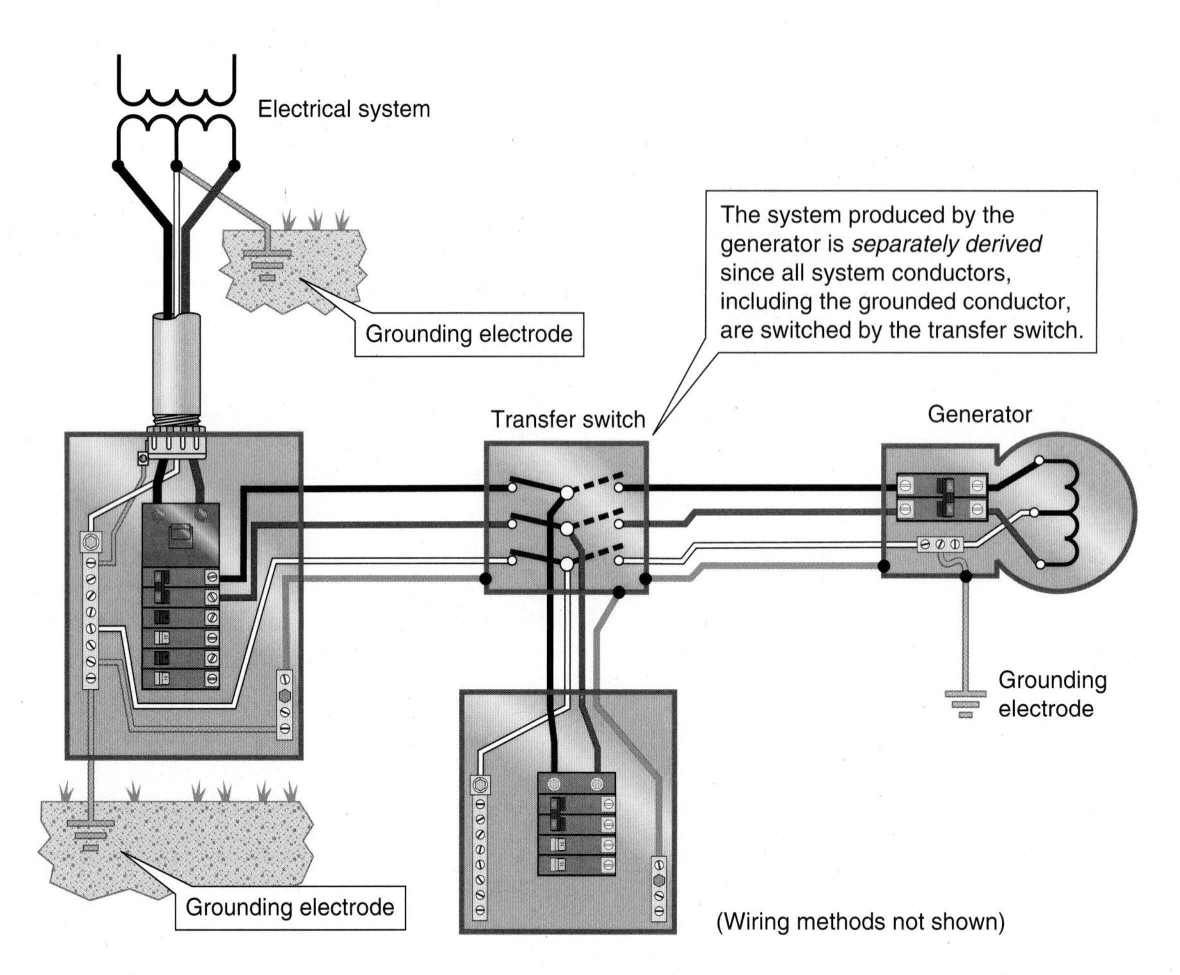

FIGURE I-20 Generator-type separately derived system. (*© Cengage Learning 2012*)

derived. Figure I-20 shows a generator-type separately derived system.

This is an important issue as the grounding and bonding rules change significantly depending on whether or not the system is separately derived. If the system is separately derived because the neutral is switched, the neutral and equipment grounding conductor must be bonded to the frame of the generator. Generally, connection to a grounding electrode system at the generator location is required. If the system is not separately derived because the neutral is connected solidly in the transfer equipment, the neutral to frame connection must be removed at the generator. This prevents creating a parallel path for neutral current to flow in.

Fuel cells and solar photovoltaic systems produce dc. An inverter and other conversion equipment, such as capacitors, are installed in the control equipment to produce ac when this is needed or desired for premises wiring. The grounded conductor of the premises wiring system commonly connects directly to the output grounded conductor from the inverter for the fuel cell or solar photovoltaic system. Systems connected in this way are not considered to be separately derived.

The Path

NEC 250.2 contains definitions that are specifically applicable to *Article 250.* Very important concepts are contained in these definitions. Careful adherence to these concepts is critical in installing the grounding system to ensure the safety of the installations and to prevent a shock hazard to people or animals. By following these concepts, a low-impedance path for ground-fault current to flow in will be created.

This low-impedance path facilitates operation of an overcurrent protective device. The intent of this requirement is to provide a path through which a lot of current flows in a short period of time, so the overcurrent device operates (opens) quickly to remove the fault from the system. Quickly removing the fault reduces the thermal and magnetic stresses placed on the system by the excessive current and removes a shock hazard.

To summarize, the electrician must install a ground-fault return path that allows the overcurrent device to clear the fault in the least amount of time.

Effective Ground-Fault Current Path *(250.2).* "An intentionally constructed, low-impedance, electrically conductive path designed and intended to carry current under ground-fault conditions from the point of a ground fault on a wiring system to the electrical supply source and that facilitates the operation of the overcurrent protective device or ground-fault detectors on high-impedance grounded systems."*

Discussion: This definition incorporates many of the essential elements of a "safety ground," or the equipment grounding conductor path (see Figure I-21). The fault-current path that is installed must comply with each element of the definition. The path must be as follows:

1. The path must be *intentionally constructed,* or built or made according to plan. Creation of an effective ground-fault current path is not necessarily something that "just happens" because another trade installed structural elements of a building, for example. To comply with this rule, we install conduit, cables, and equipment grounding conductors properly, using locknuts, couplings, and connectors made up tightly, with the equipment grounding path in mind as the installation is made.
2. The path must be *low-impedance.* (The term *impedance* is used to describe the total opposition to current flow for ac circuits, just as the term *resistance* is used to describe the opposition to current flow in dc circuits. More on these terms is presented later in this chapter.) To comply with this rule, all conductors of the circuit—including the ungrounded ("hot"), grounded conductor (which may be a neutral), and equipment grounding conductors—must be installed in close proximity with each other (see Figure I-22). Following this requirement

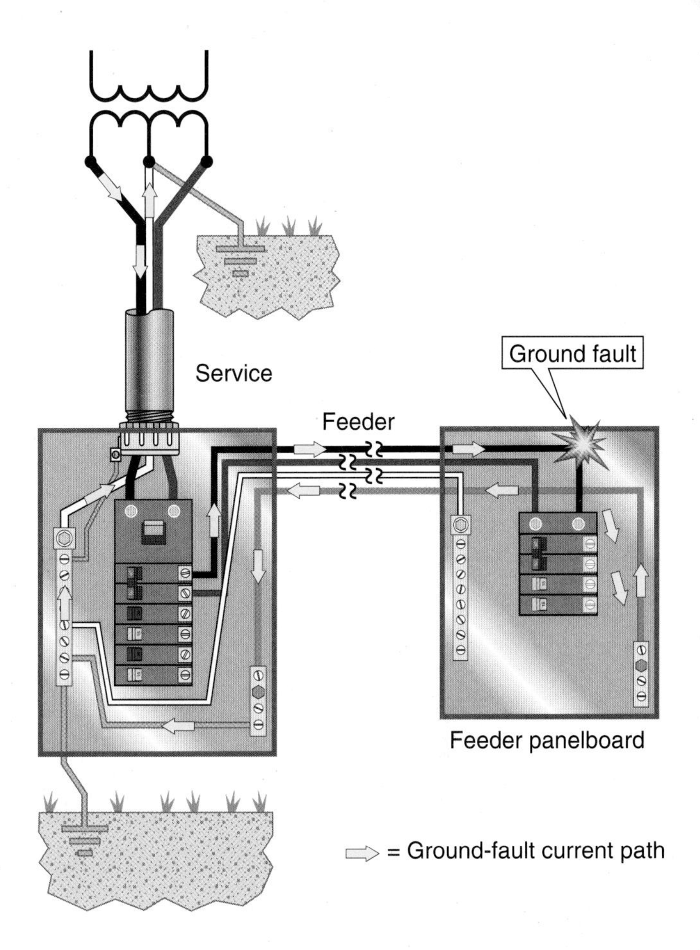

FIGURE I-21 Effective ground-fault current path, *250.2.* (*© Cengage Learning 2012*)

keeps the inductive reactance of the circuit as low as possible. (See *250.134* and *300.3(B)* for specific requirements.)

Figure I-23 shows the expanding and collapsing magnetic lines of force around ac circuits. In the United States, the default frequency of ac systems is 60 cycles per second, also referred to as 60 Hz. The strength of the

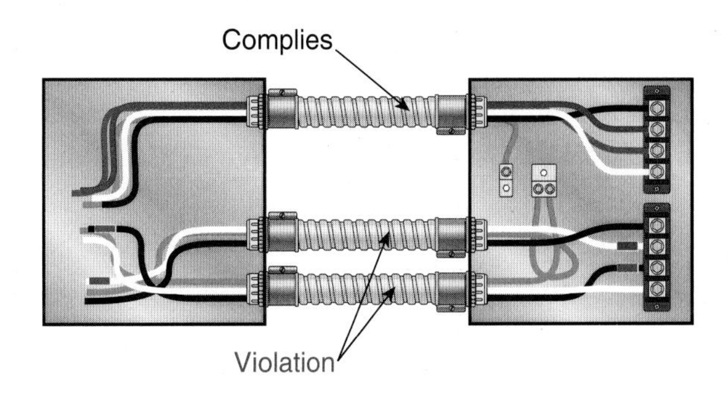

FIGURE I-22 All conductors of same circuit together. (*© Cengage Learning 2012*)

*Reprinted with permission from NFPA 70-2011.

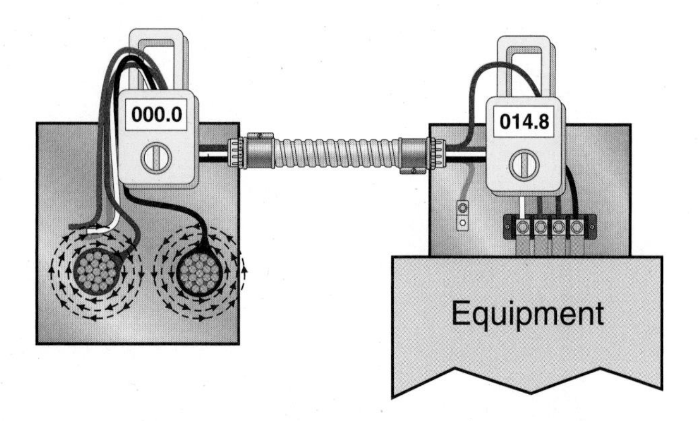

FIGURE I-23 Expanding and collapsing magnetic lines of force around ac circuits. (*© Cengage Learning 2012*)

magnetic field around the conductors varies in proportion to the current. This can be illustrated by a snap-on ammeter, which measures the current flowing through conductors by measuring the magnetic field around the conductor. This measurement can accurately be taken of any single conductor of the circuit. However, if both conductors of a 2-wire circuit are inserted in the meter at the same time, the magnetic fields cancel each other and the meter will read zero current. If these ac circuit conductors are separated, the magnetic fields cannot cancel each other and a high-impedance path is established.

Keeping all circuit conductors and equipment grounding conductor paths in the same raceways, cables, wireways, auxiliary gutters, cable tray, and flexible cord is essential to creating an effective path. If individual conductors are installed, such as in trenches, the conductors must be installed in close proximity with each other—again, to keep the inductive reactance and impedance as low as possible. See Appendix A for an exhaustive explanation of the importance of keeping all conductors of an ac circuit in the same cable or enclosure.

3. The path must *facilitate the operation of the overcurrent device.* Most circuit breakers and fuses operate according to the principle of "inverse time." This means the more current that flows through the breaker or fuse, the faster the circuit breaker or fuse opens. For example, a 50-ampere, 2-pole circuit breaker is rated and tested by the manufacturer to carry 50 amperes indefinitely. This capability is also a requirement of the product safety standard. (*NEC* rules require that the breaker carry only 80% of its rated current in normal operation unless it is designed and rated for continuous duty at 100% of its rating.) By examining the time–current curve for the circuit breaker (see Figure I-10), it can be observed that the breaker reaches the instantaneous trip point of the curve at 5 to 7 times its handle (current) rating (see the vertical line portion of the curve). That is, at 250 to 350 amperes for a 50-ampere breaker (depending on the manufacturer's design), the circuit breaker is no longer in a time-delay mode; instead, it trips instantaneously. At currents less than the instantaneous trip level, the circuit breaker is in a time-delay mode. The greater the current that flows through the circuit breaker in this time-delay mode, the faster the breaker trips. The same concept applies to time-delay fuses. Time–current curves are available from the manufacturer of all circuit breakers and fuses. The curves should be consulted to ensure the ground-fault return path has proper characteristics to facilitate the operation of the overcurrent protective device.

4. The path must be *continuous and reliable.* (Note: The word *permanent* was removed from the definition of the term *effective ground-fault current path* during processing of the 2008 *NEC* because of a proposal maintaining that the permanence of the ground-fault path was too subjective. The author has inserted the word *reliable* here to add emphasis.) Implied in these concepts are that the electrical path be continuous for the entire path and that connections are intended to be reliable and thus last for the life of the installation. A significant problem in normally grounded and ungrounded electrical system operation is the lack of a method for monitoring the continuity and integrity of the equipment grounding system. The equipment grounding conductor does not carry current in normal operation; therefore, supplied equipment continues to function normally even though there is a break in the equipment

grounding path. Lights continue to operate and motors continue to run even though the equipment grounding conductor path has been impaired. A break in the equipment grounding path usually is discovered when someone receives a shock by completing a circuit between exposed conductive parts of equipment. The importance of providing a continuous and reliable ground-fault path cannot be overstated.

Path through the Earth. Several previous figures show the parallel path for neutral current that is created by connecting to grounding electrodes at the service and at more than one building on the premises. The interconnection can be completed by connecting the neutral to a grounding electrode at the service and again when the equipment grounding conductor is connected to grounding electrodes at outbuildings.

Figure I-24 shows the hazard created when the neutral, or grounded, conductor that is supplied from a grounded electrical system is not connected to the metal cabinet for the service. Because a relationship with the earth is established at the transformer, a line-to-cabinet fault at the service will simply energize the cabinet at the system voltage and become a shock or electrocution hazard to the unsuspecting person who may contact a metal object. The path through the earth established by the grounding electrode connection is usually very high resistance and will allow system voltage to appear on the metal objects of the service equipment. The solution? Connect the grounded service conductor to the service cabinet. Now, a path or circuit is established from the cabinet back to the utility transformer. Granted, if the fault occurs at the point shown in the service cabinet in Figure I-24, no overcurrent device is in the circuit. Hopefully, enough current will flow in the circuit established to cause the fuse on the supply side of the transformer to open.

A nearly identical shock hazard is created if a feeder or branch circuit that does not include an equipment grounding conductor is extended to a separate building. The only path for a ground fault to the enclosure is through the earth. A ground fault at this point will simply energize the enclosure at system voltage. The solution? Include an equipment grounding conductor with the feeder that provides a fault-return path back to the utility transformer.

As stated in *250.4(A)(5)* for grounded systems and *250.4(B)(4)* for ungrounded systems, the earth is not considered an effective ground-fault current path.

Connections to earth are made from the electrical system and electrical equipment for the purpose

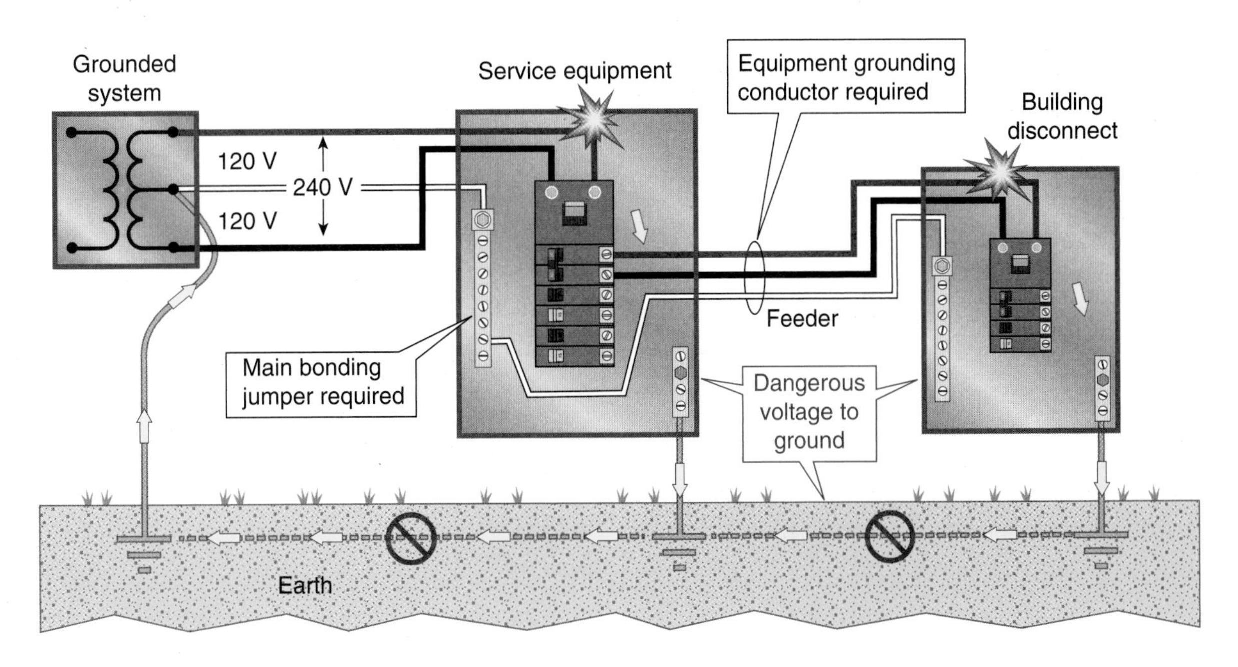

FIGURE I-24 Earth return prohibited. (*© Cengage Learning 2012*)

of dissipating overvoltages and reducing shock hazards, but never for the purpose of carrying fault current. Relief is provided in *250.6* from excessive neutral current flowing through the grounding conductors or system.

Parallel Conductor Installations. The general requirement for installing conductors in parallel is found in *310.10(H)*. Parallel conductor installations consist of two or more conductors connected together at each end to form a set of conductors that have an increased capacity. The decision to install two or more circuit conductors in parallel is usually based on economics as well as on practicality. This is due to the fact that the ampacity of conductors in *Table 310.15(B)(16)* does not increase in direct proportion to the size of the conductors. This is due to the tendency for alternating current to flow on the surface of the conductor rather than through the center. This phenomenon is referred to as "skin effect."

For example, a 250-kcmil copper conductor with insulation rated at 75°C has an allowable ampacity in *Table 310.15(B)(16)* of 255 amperes. Two of these 250-kcmil conductors that are connected in parallel have an allowable ampacity of

$$2 \times 255 = 510 \text{ amperes}$$

A single 500-kcmil copper conductor having the same insulation system rating has twice the circular mil area of a single 250-kcmil conductor but has an allowable ampacity of 380 amperes or 130 amperes less than two 250-kcmil conductors that are connected in parallel. The parallel conductors have a whopping 34% increase in ampacity!

The author has chosen to include a discussion of the requirements for installing conductors in parallel in the *Introduction to Grounding and Bonding* because sizing rules in several sections of *Article 250* refer to this provision. Requirements for the installation of neutral or grounded conductors when installed in parallel are found in *250.24(C)*. Equipment grounding conductors are required to be installed in parallel under the sizing rules in *250.122(F)*.

The requirement in *310.10(H)* is that each set of conductors connected in parallel must be

1. the same length,
2. of the same conductor material,
3. the same size in circular mil area,
4. of the same insulation type, and
5. terminated in the same manner.

Where run in separate raceways or cables, the raceways or cables must have the same electrical characteristics—for example, all of the conduits are to be rigid steel or all of the cables are of interlocking metal-clad (Type MC) cable. Even though rigid metal conduit, intermediate metal conduit, and electrical metallic tubing are all magnetic, they have different electrical characteristics. This can be observed by reviewing the effective length of the various metal circular raceways as an equipment grounding conductor that is provided in the discussion of *250.118* in this book.

Follow the simple rule, make sure all components of the parallel assembly are identical. A parallel installation is shown in Figure I-25.

The same number of conductors must be installed in each parallel conduit or cable. For example, if a 3-phase, 4-wire system is installed, it is required that each conduit contain a phase A, phase B, phase C,

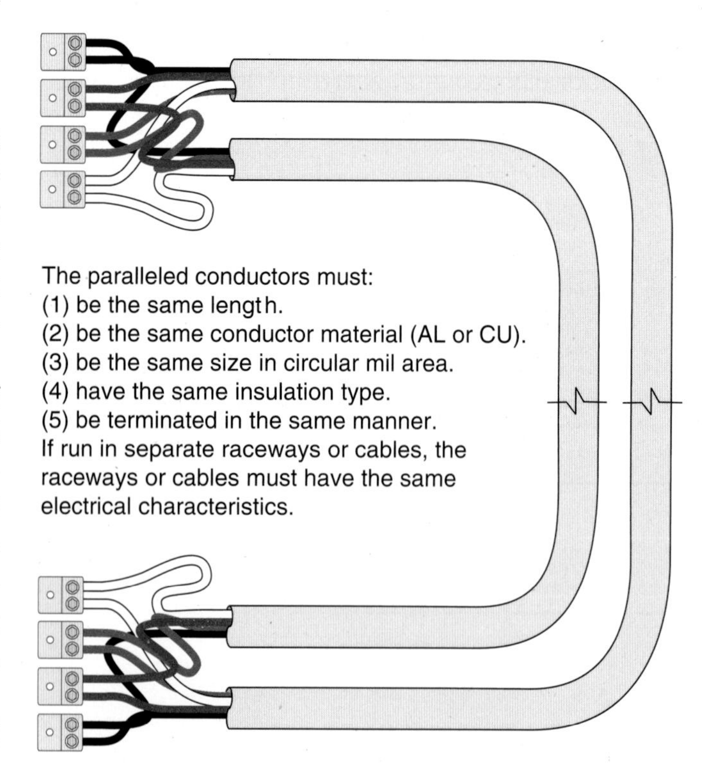

FIGURE I-25 Parallel conductors. (*© Cengage Learning 2012*)

and neutral conductor. It is not permitted to install five conductors of the parallel set in one conduit and three in another.

To keep the impedance of the total circuit as low as possible, each set of conductors installed in parallel must meet the conditions in *310.10(H)*. Generally, each set of parallel conductors must be identical, including all phase (ungrounded) conductors, grounded conductors, neutral conductors, and equipment grounding conductors. The condition most commonly violated is keeping all of the conductors the same length. No range of deviation is permitted in the *NEC*, such as plus or minus 5%.

The purpose of these strict requirements for parallel conductors is to have each set of conductors share the current equally. For example, two 600-kcmil copper conductors are installed in parallel (connected together at each end) between an 800-ampere circuit breaker or fuse and the load. The allowable ampacity of each 600-kcmil conductor in *Table 310.15(B)(16)* is 420 amperes if the conductor has an insulation rated for 75°C. Obviously, neither conductor is large enough to carry all the current that can be supplied through the 800-ampere overcurrent device. If the rules are not followed, the conductor with the lower resistance (impedance in ac circuits) carries more current than the conductor with the higher resistance. The conductor insulation can be damaged by excessive heat in more severe current imbalances.

Underground Parallel Sets. Conductors installed in nonmetallic raceways (typically PVC) that are run underground are permitted to be installed as isolated phase installations. (See *300.3(B)(1), Exception,* and *300.5(I), Exception No. 2.*) This means that all of phase A conductors are in one nonmetallic conduit, all of phase B conductors are in another, and so forth. The benefit to making an installation in this manner is that conductors line up with busbars in switchboards and other large enclosures such as motor control centers. This makes for a neater installation. An example is shown in Figure I-26.

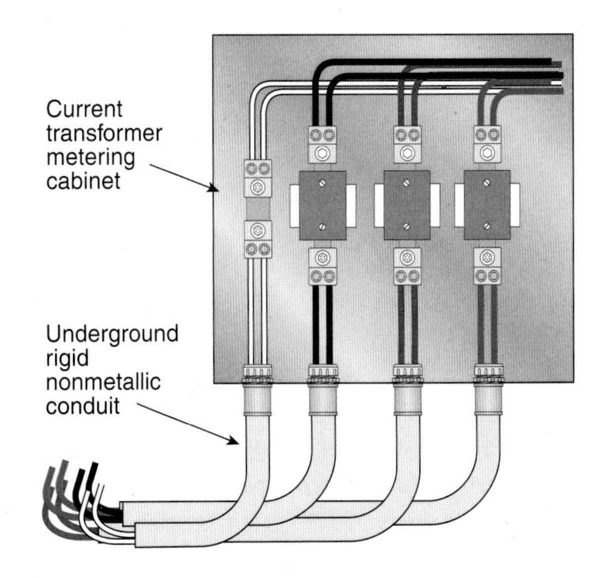

FIGURE I-26 Underground parallel conductor sets. (*© Cengage Learning 2012*)

CAUTION: Adequate wire-bending space must be provided for installing conductors. More wire-bending space is required for straight-in wire connections than for conductors that are bent or deflected 90° to make a connection to a terminal. See *312.6* and the accompanying tables for the rules on deflection of conductors.

Additional requirements for isolated phase installations are that raceways must be installed in close proximity, and the conductors must comply with the provisions of *300.20(B)*. For *close proximity,* no guidance is given in the *NEC* for this term. There should be no problem if the conductors are installed in nonmetallic conduits that are placed in standard duct-bank spacers.

The requirement for complying with *300.20(B)* deals with the avoidance of heating surrounding metal by induction. If isolated phase conductors are connected to a metal enclosure, the metal enclosure acts like a transformer around the conductors. This can create sufficient heat to damage the conduit or the enclosed conductors.

Two solutions to this problem are common. One is to cut a slot between the knockouts (as shown in Figure I-27, *left*) with a hacksaw blade. Although opinions differ on this point, it is best to use nonmetallic or aluminum locknuts to make conduit connections, to prevent closing the slot around the conduit. The other solution is to install a nonmetallic or nonmagnetic plate such as brass or aluminum where the conduits connect to the enclosure as shown in Figure I-27, *right*. In this situation, standard metal locknuts could be used on the conduit connectors.

Let's look at a few other important definitions.

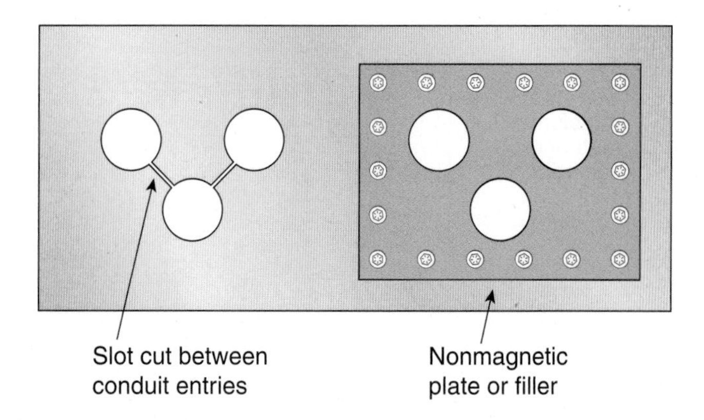

FIGURE I-27 Isolated parallel conductors through metal enclosures. (*© Cengage Learning 2012*)

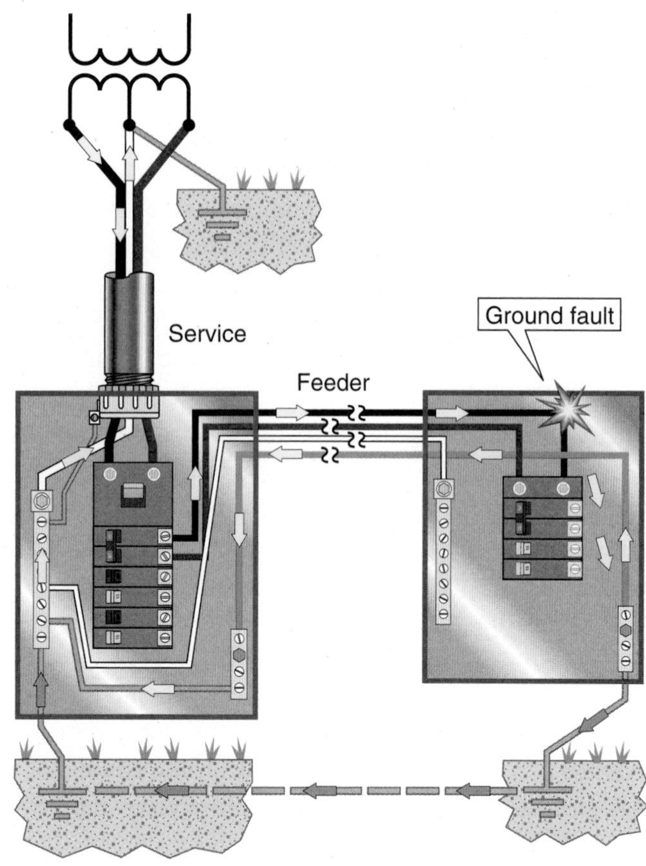

FIGURE I-28 Ground-fault, *Article 100.* (*© Cengage Learning 2012*)

Ground Fault ***(Article 100).*** "An unintentional, electrically conducting connection between an ungrounded conductor of an electrical circuit and the normally non-current-carrying conductors, metallic enclosures, metallic raceways, metallic equipment, or earth."*

Discussion: A ground fault occurs whenever a normally insulated or isolated system conductor contacts another conductive element that is normally not intended to be current carrying. As is indicated in the definition, this other conductive element can be a grounding conductor, a metallic enclosure such as a pull box or panelboard enclosure, metallic raceway, or metallic equipment. The earth is in the ground-fault return path where the system is connected to earth and another earth connection then is made at another point downstream, such as at the service or at the disconnecting means for a building supplied by a feeder from the service (Figure I-28). Because the earth represents a high-impedance path, very little current flows through the earth. As a result, electrical systems and equipment are never connected to the earth for the purpose of carrying fault current.

At first, it may seem unlikely that a ground fault could occur with the grounded conductor. However, grounded conductors originate at the utility transformer, at the service equipment, or at separately derived systems where the system is grounded. After the point where it is intentionally grounded as permitted by the *Code*, the grounded conductor is required to be insulated and isolated from ground. See *110.7* and *250.24(A)(5).* If the grounded conductor is also a current-carrying conductor, such as a neutral, the insulation or isolation prevents the normally non-current-carrying metallic parts or paths such as water pipes and conduits from carrying a portion of the neutral or grounded conductor current.

The following three applications demonstrate the importance of maintaining isolation for the neutral or grounded circuit conductor.

1. Ground-fault circuit interrupters (GFCIs). These safety devices operate on the principle of monitoring current flow on all of the circuit conductors at the same time. This is done by measuring the magnetic field around the conductors by passing them through a torodal coil. The same amount of current passing

*Reprinted with permission from NFPA 70-2011.

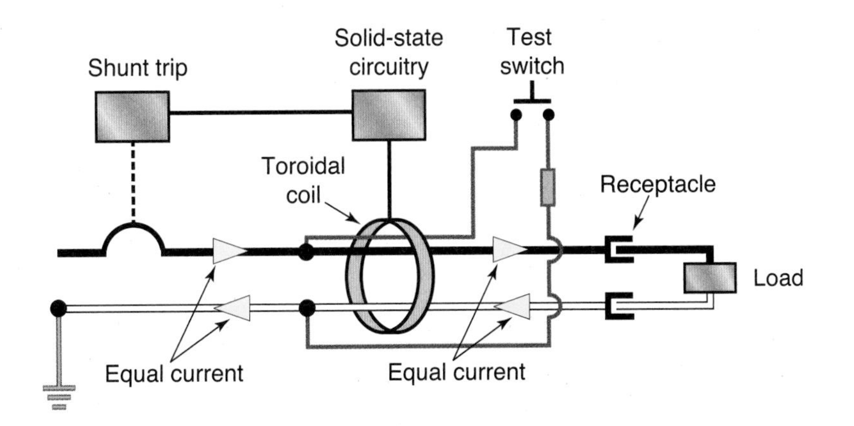

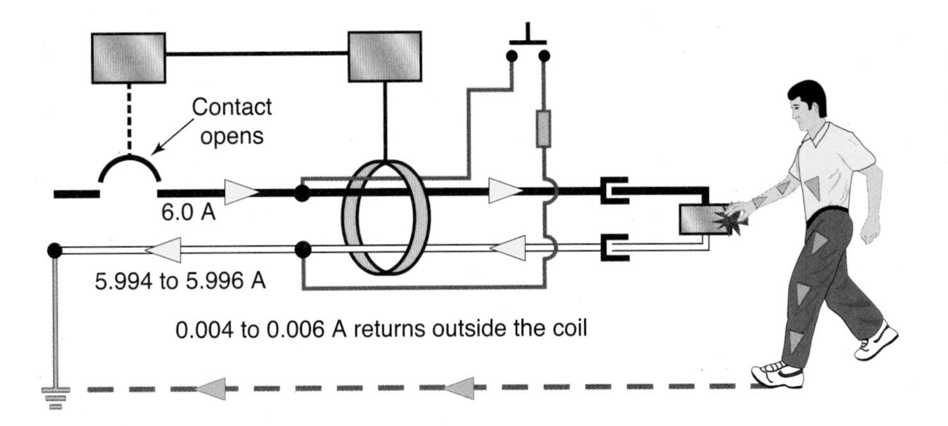

FIGURE I-29 GFCI Principles of operation. (*© Cengage Learning 2012*)

through the coil to the load must return through the coil to the source. This results in magnetic fields that cancel each other in the coil. The principals of operation of a GFCI device are shown in Figure I-29. When a part of the current passing through the coil to the load is returned to the source outside the coil as shown in the test circuit in the device, an imbalance of magnetic fields is created. A Class A GFCI device that is intended for personnel protection is not permitted to open when this imbalance of current is less than 4 mA and must open when the imbalance is 6 mA or more. (1 mA is one-thousandth of an ampere). A fault from the neutral or grounded conductor to a grounded surface or enclosure on the load side of the GFCI device causes some of the grounded conductor current to return over another path. This results in unnecessary or "nuisance" trips of the GFCI device.

2. Ground-fault protection of equipment. This protection is required for grounded systems in which the voltage to ground is greater than 150 volts and the phase-to-phase voltage does not exceed 600 volts if the overcurrent device is rated 1000 amperes or more. See *230.95* for services, *215.10* for feeders, and *240.13* for building disconnecting means. Unintentional ground faults downstream from the ground-fault protection equipment will desensitize the equipment, resulting in loss of protection. Such faults may also cause "nuisance trips" of the equipment ground-fault protection, as a portion of the neutral current flows back on the conduit and thus is outside the current-transformer window(s) through which the circuit conductors pass. Other ground-fault protection equipment monitors return current on the main bonding jumper.
3. Arc-fault circuit interrupters. These circuit protective devices are intended to provide

protection from the effects of arc faults by recognizing characteristics unique to arcing by de-energizing the circuit when an undesirable arc is detected. These circuit protective devices also include ground-fault protection of equipment (GFPE). The leakage or imbalance protection is similar in function to the GFCI device but will cause the device to open the circuit at approximately 50 to 70 mA (milliamperes) of circuit imbalance rather than 5 mA. This imbalance can be caused if the circuit neutral is connected to an equipment grounding conductor or path downstream of the AFCI device. This neutral-to-ground connection can be made unintentionally by a bare equipment grounding conductor in outlet boxes when receptacles are pushed into an outlet box after connections are made. The non-AFCI branch-circuit overcurrent device does not detect this neutral-to-ground connection as a ground fault and does not open the circuit.

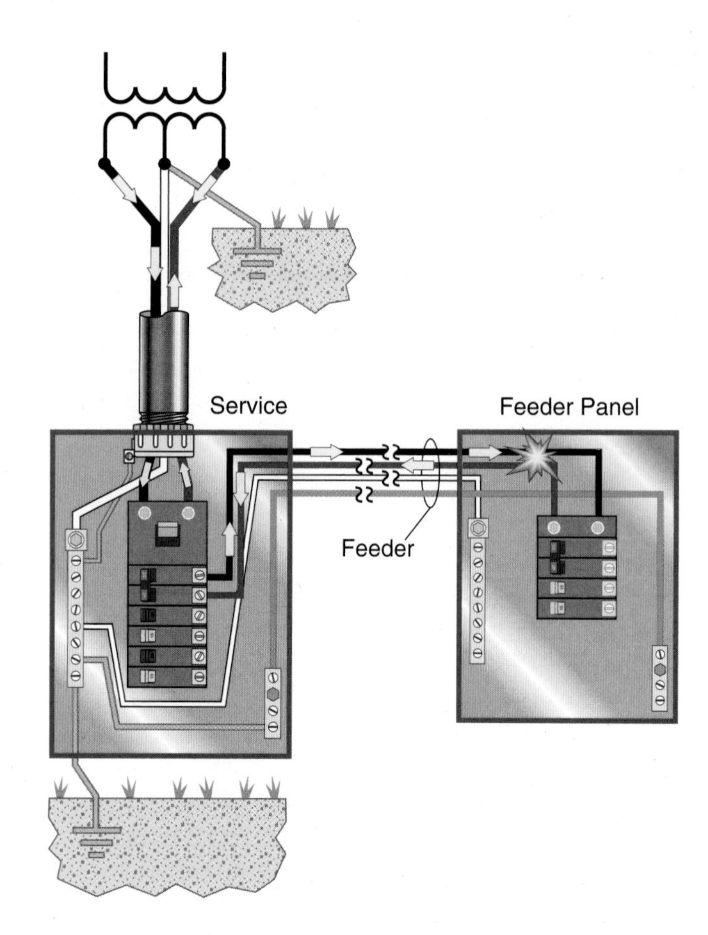

FIGURE I-30 Short circuit. (© *Cengage Learning 2012*)

Short Circuit. An abnormal connection of relatively low impedance, whether made accidentally or intentionally, between two points of different potential on any circuit.

Discussion: The other type of fault that is common in electrical systems is a short circuit, which is not defined in the *NEC.* An example is shown in Figure I-30. These faults are between two ungrounded conductors or between an ungrounded conductor and a grounded or neutral conductor. In simple terms, a short circuit is a circuit whose effective length is reduced by a failure of insulation on the conductors. In other words, one or more of the insulated conductors contacts another before it reaches the end of its original length. It is inappropriate to refer to a short circuit as a ground fault. It is likewise inappropriate to refer to a ground fault as a short circuit. Although statistics proving this point are difficult to obtain, many short circuits are thought to originate as ground faults that escalate to a short circuit or phase-to-phase fault.

Ground-Fault Current Path *(250.2).* "An electrically conductive path from the point of a ground fault on a wiring system through normally non-current-carrying conductors, equipment, or the earth to the electrical supply source."*

> **Informational Note:** "Examples of ground-fault current paths could consist of any combination of equipment grounding conductors, metallic raceways, metallic cable sheaths, electrical equipment, and any other electrically conductive material such as metal water and gas piping, steel framing members, stucco mesh, metal ducting, reinforcing steel, shields of communications cables, and the earth itself."*

Discussion: As can be seen, this definition differs from the definition of *effective ground-fault current path* in *250.2.* This definition, along with the *Informational Note*, demonstrates that ground-fault current will flow on any and all paths that are available to it. As can be seen in Figure I-31 and in the

*Reprinted with permission from NFPA 70-2011.

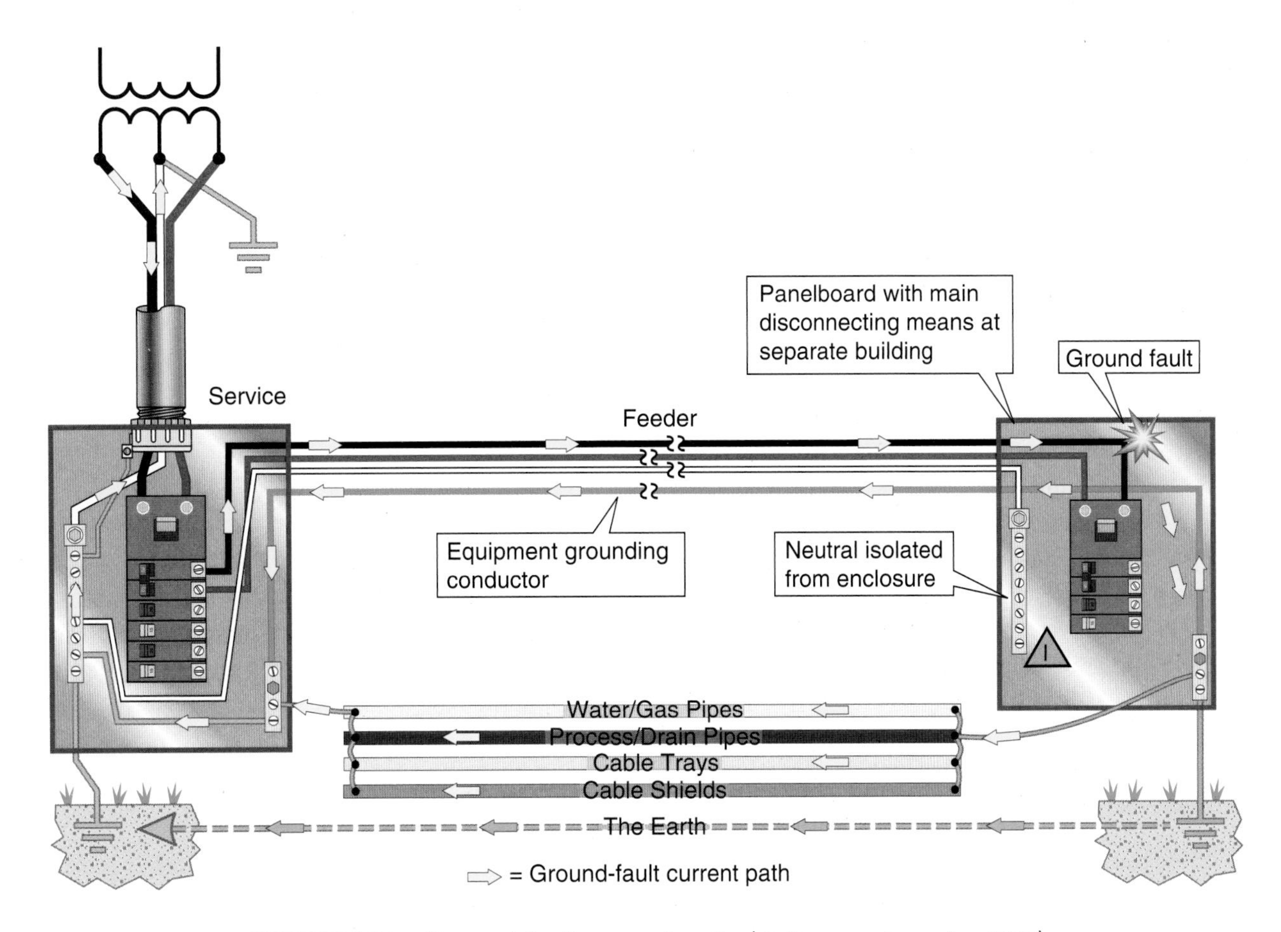

FIGURE I-31 Ground-fault current path. (*© Cengage Learning 2012*)

Informational Note, fault current divides among all the conductive paths that are available and are in the return path to the source. The paths in the illustration show several that are not intended to carry current, such as over pipes, cable trays, cable shields, and, of course, the earth. The amount of current that flows through each path is determined by the impedance of each path. The lower the impedance, the more current flows in the path. Likewise, the higher the impedance, the less current flows in the path.

REVIEW OF OHM'S LAW AND BASIC ELECTRICAL THEORY

Electrical current flowing through any electrical circuit can be compared in some ways with water under pressure flowing through pipes. Water flowing in a water pipe is measured in gallons per minute (gpm), and electricity flowing in a circuit is measured in amperes.

Water flows through a pipe when pressure is exerted on it and a valve is opened. Water pressure is measured in pounds per square inch (psi). An electrical current flows along an electrical conductor when electrical "pressure" is applied to it and a path is provided for the current flow. Just as the water pressure (in psi) causes a flow of water (in gpm), so the electrical pressure (in volts) causes current (in amperes) to flow in the electrical circuit.

It takes more pressure to force the same amount of water through a small pipe than through a larger pipe. A small pipe, with the same pressure applied to it as to a larger pipe, allows less water to flow in a given period of time. It therefore follows that the small pipe offers a greater resistance to the flow of water than does a larger pipe.

In an electrical circuit, a greater electrical pressure (volts) is required to force a given amount of current (amperes) through a small conductor (higher resistance) than the pressure required to force the same amount of current (amperes) through a larger conductor (lower resistance). A smaller conductor allows less current (amperes) to pass than a larger-sized conductor if the same electrical pressure (volts) is applied to each conductor for the same period. The smaller conductor can be assumed to offer greater resistance (ohms) than the larger conductor. Thus, for electrical circuits, we may define *resistance* as the property of a body that resists or limits the flow of electricity through it. This resistance—similar to "friction" in a hose or pipe—is measured in ohms.

Theory Terms and Definitions

The following definitions relate to electrical theory. It is important that installers and inspectors have a working knowledge of electrical theory. Such knowledge is often vital in determining the proper size of conductors for circuits of various loads.

Volt. The unit of electrical pressure. It is commonly accepted that a volt is the pressure or electromotive force required to force 1 ampere of current through a resistance of 1 ohm. The abbreviation for *volt* used in electrical formulas is E (E being the first letter of the term *electromotive force*). The term *volt* is used in honor of Alessandro Volta, who invented the first practical battery.

Ampere. The unit of electrical current. As defined, 1 ampere of current flows through a resistance of 1 ohm under a pressure of 1 volt. The abbreviation for *ampere* for use in electrical formulas is I (I being the first letter of the term *intensity of current*). The term *ampere* is named in honor of André-Marie Ampère. He was a scientist who lived in the late 1700s and early 1800s and is best known for his work with electromagnetism.

Ohm. The unit of electrical resistance in the circuit. As defined, 1 ohm is the resistance through which 1 volt forces 1 ampere. The abbreviation for *ohm* used in electrical formulas is R (for the first letter of the word *resistance*). The symbol used to represent ohms (resistance) is the Greek letter omega (Ω).

Ohm's Law Formula

In its simplest form, Ohm's law states that electrical pressure (volts, or E) equals the product of the current (amperes, or I) and resistance (ohms, or R). The formula is written as

$$E = I \times R$$

In this form, it applies to dc circuits and to ac circuits that consist of resistive loads. It is the practical basis on which most electrical calculations are determined. The formula may be expressed in various forms and by its use, as shown in Figure I-32.

The three forms of the formula are

$$E = I \times R$$
$$I = E \div R$$
$$R = E \div I$$

where

E = electromotive force or voltage
I = intensity of current or amperage
R = resistance or opposition to current flow

If any two values are known, the third can be found by use of the formula. For example, if an electrical circuit has a voltage of 240 and the current is 17 amperes, what is the resistance of the circuit? The following variation of the formula can be used to find the answer:

$$R = E \div I$$
$$R = 240 \div 17$$
$$R = 14.1 \text{ ohms}$$

This formula can be valuable in determining the amount of current that will flow on the circuit, to properly size conductors as well as overcurrent protective devices.

Watt. The unit of measurement of the energy flowing in an electrical circuit at any given moment. It also is the amount of work being performed in the electrical circuit as a result of conversion of electrical energy to some other form of energy, such as heat or mechanical. The watt was named in honor of James Watt, an English scientist. The term *watt* or *kilowatt* has been used more commonly to express the amount of work done in the electrical circuit, rather than the term *joule.* Watts are simply the product of multiplying volts and amperes; today the watt is finding common usage as the volt-ampere. One thousand volt-amperes is referred to as one kilovolt-ampere (kVA).

Watts Wheel

Watts wheel has been developed and published in many manuals, and in several variations, to

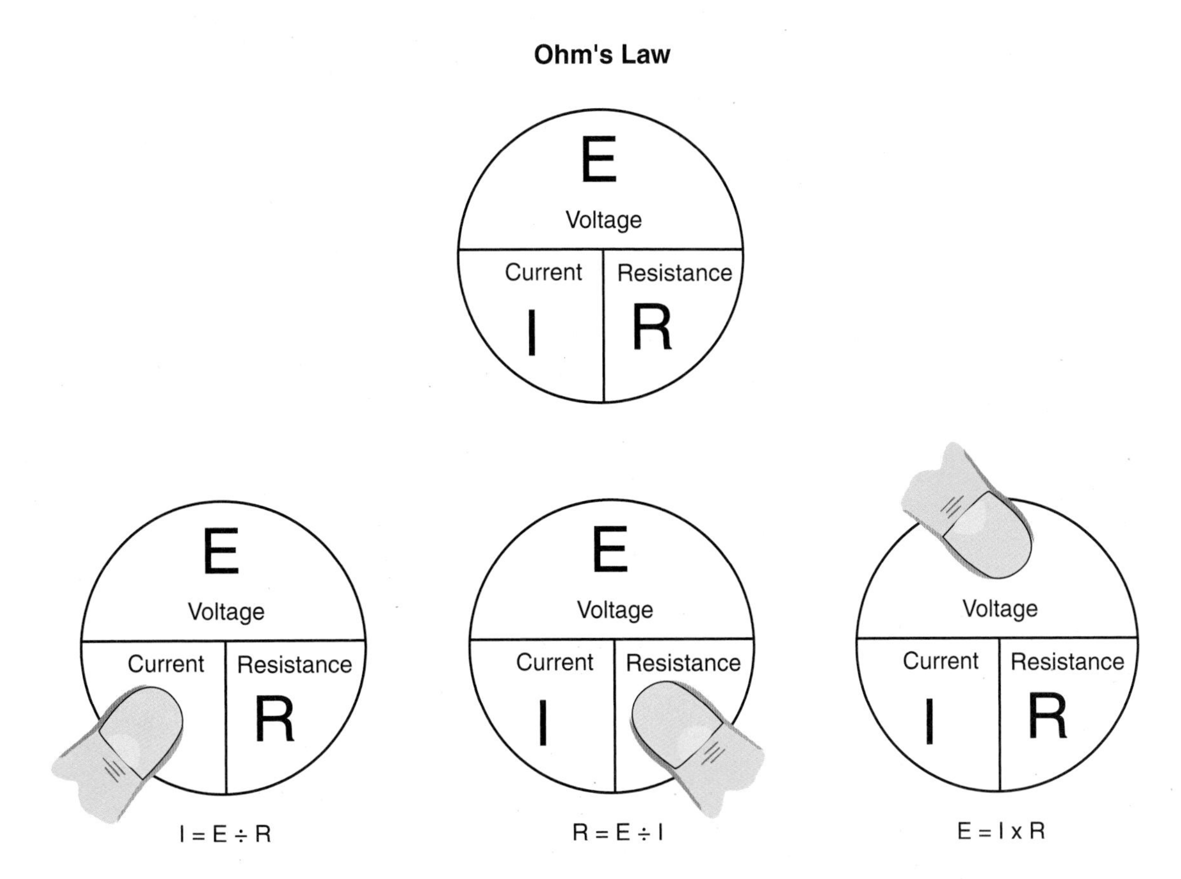

FIGURE I-32 Ohm's law. (*© Cengage Learning 2012*)

illustrate the relationship of the watt to the elements of Ohm's law. "Watts," or "power," is represented on the wheel by the abbreviation "W" (Figure I-33).

As with Ohm's law, if any two values are known, the third value can be determined by use of the formula shown. To use the wheel, first determine what value is desired. The formulas on the outer ring are used to determine the value in the inner ring in the same quadrant. The version presented here is accurate for work with dc circuits and for determining resistive loads of ac circuits where the power factor is near 100%, or unity. The watts wheel cannot be used for determining motor loads, because both power factor and motor efficiency must be factored into the formula.

Impedance

In ac circuits, the term *impedance* is used rather than *ohms* to represent resistance of the circuit. It is represented in electrical formulas by the letter Z. Impedance is the total opposition to current flow in an ac circuit, corresponding to resistance in a dc circuit. Both values are represented and measured in ohms. This concept is important to understand because terms such as *low-impedance path* are used throughout *Article 250* and this book. A low-impedance path is to ac circuits what a low-resistance path is to dc circuits.

Impedance includes resistance, capacitive reactance, and inductive reactance. The last two factors are unique to ac circuits and usually can be ignored in circuits such as incandescent lighting loads and heater circuits consisting of resistive loads.

A detailed explanation of capacitive reactance and inductive reactance is beyond the scope of this text but can be found in many excellent textbooks on electrical theory.

Circuits and Paths

One of the most basic and fundamental laws of electricity is that there must be a complete circuit or path for current to flow in. A break in the circuit due to the operation of a switch or to a loose or broken wire or connection causes current to stop flowing in the circuit.

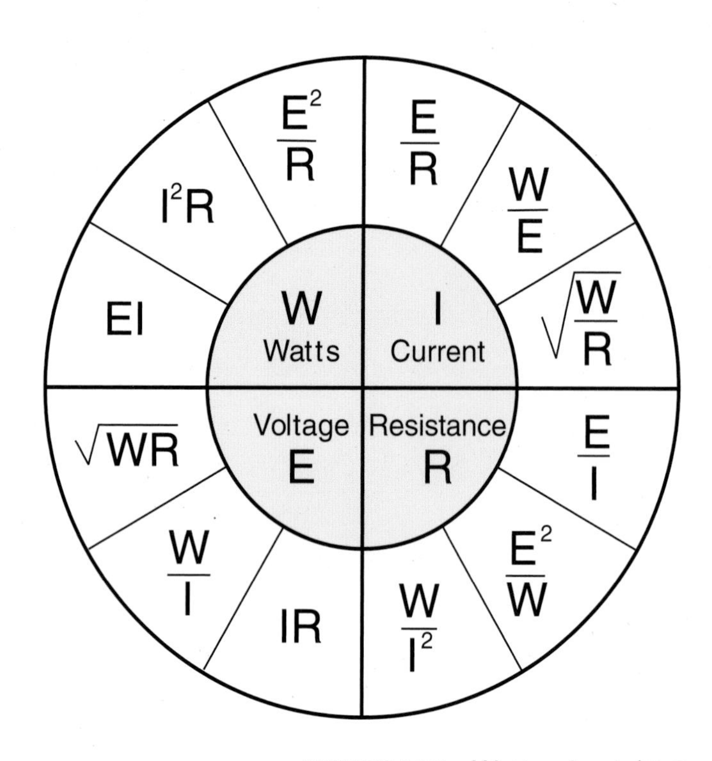

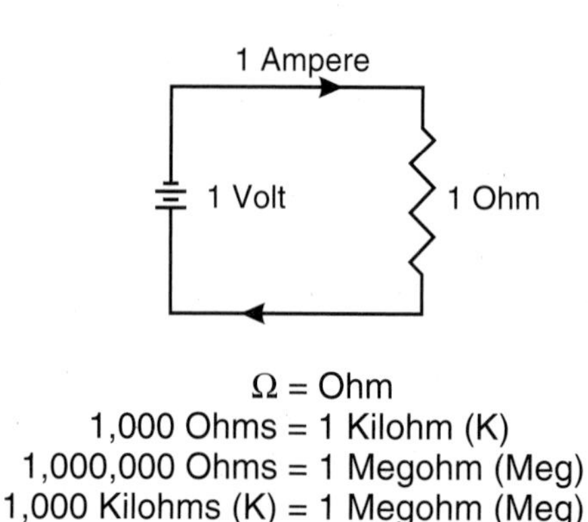

FIGURE I-33 Watts wheel. (*© Cengage Learning 2012*)

The simplest circuit is the series circuit, in which there is only one path for current flow. All work is performed by current passing through only one load, such as a luminaire, a heater, or a motor. As demonstrated in Figure I-34, system voltage can be read across the open point of the circuit, be it at the switch or at the break in the line.

Following is an important concept to our discussion on grounding. As with measurement of system voltage across the open point of the circuit, system voltage can be measured across a break in the ground-fault return path if the ground fault is downstream from the break (Figure I-35). At system

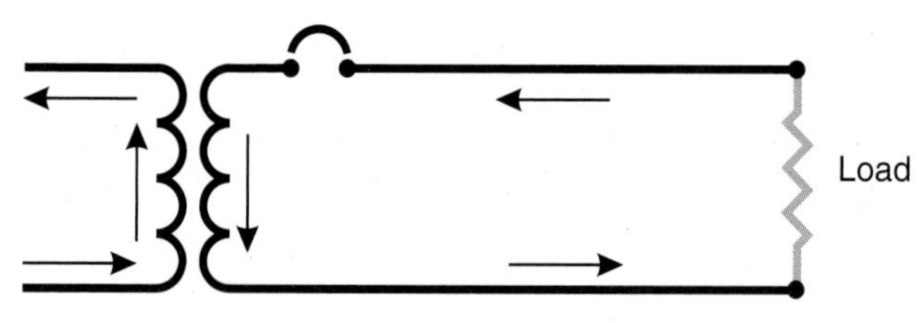

1. No current unless there is a complete path.

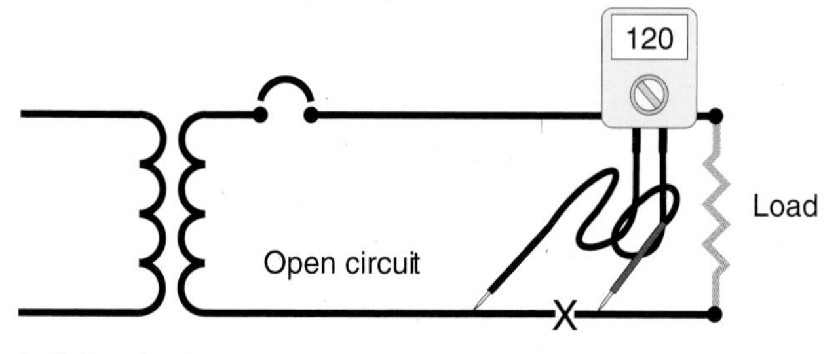

2. If the circuit opens, current ceases to flow.

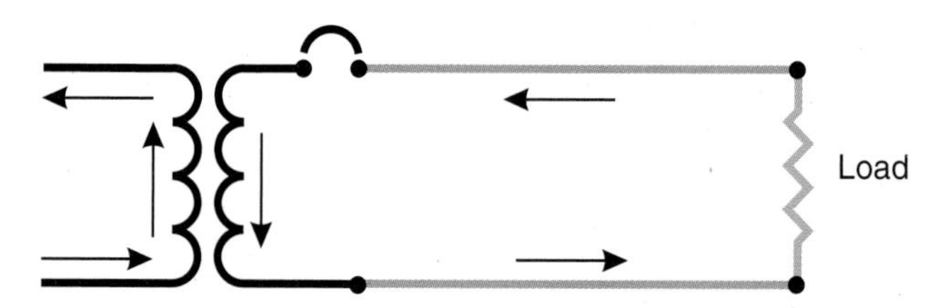

3. Friction (resistance) inside the conductors produces heat when current is flowing in the circuit. The more electrons per second flow, the more heat will be produced.

FIGURE I-34 The basic circuit. (*© Cengage Learning 2012*)

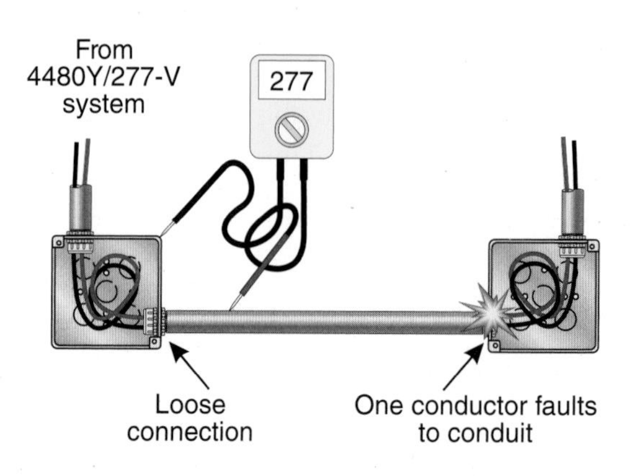

FIGURE I-35 Break in ground-fault path. (*© Cengage Learning 2012*)

voltages greater than 50 volts, this voltage can present a dangerous risk of electric shock or burn.

ELECTRIC SHOCK HAZARDS

Electricity is in many ways a masterful servant. It can be safely said for most Americans that many of the pleasures and comforts we enjoy in our way of life would not be possible without electricity. From the basics of life—such as cooking meals; washing, drying, and ironing clothes; and lighting and heating homes—to creature comforts like radios, stereos, televisions, and computers, many of these tasks are performed regularly and reliably by electricity. Countless offices and factories have electric heating and cooling, lighting, equipment, and machinery. The chaos caused by recent electrical distribution system blackouts demonstrates how much we have come to rely on the electrical distribution system in industry, in commerce, and in our homes.

The greater the use of electricity, however, the greater the opportunity for electrical accidents and incidents. A recent study revealed that 30,000 nonfatal electric shock accidents occur every year. Such a number is difficult or impossible to substantiate, because most electric shock incidents go unreported. The National Safety Council estimates that between 600 and 1000 people die every year from electrocution. Of those killed with voltages below 600, nearly half were working on energized equipment at the time of the electrical accident. Electrocution continues to rank as the fourth highest cause of industrial fatalities, behind traffic accidents, violence or homicide, and construction accidents.

Most people are aware of the hazards associated with the use of electricity. However, few are aware of how little current is required to cause injury or even death. In fact, the current drawn by a 7½ watt, 120-volt lamp, if it passes through the chest (usually hand-to-hand current path) is enough to cause ventricular fibrillation of the heart, potentially resulting in death.

The factors that determine the effects of electrical current on the body can be summarized as follows:

- The amount of current that flows through the body
- The path the current takes through the body
- The length of time the current flows through the body.

The human body can be thought of as a resistor in the electrical circuit. Dry skin is a fairly good resistor, whereas wet skin is a fairly poor resistor. Although different values are used in some studies, the resistance of the extremities (the hands and feet), when dry and without open wounds, generally is accepted to be about 500 ohms. Another 100 ohms is added for the path through the body. Thus, a hand-to-hand shock through the body is estimated to involve a resistance of 1100 ohms (500 + 100 + 500 ohms), as can be seen from Figure I-36. The contact resistance of the hands and feet can decrease with increases in the amount of moisture that is present, through changes in the ambient humidity, sweating, fear, and anxiety.

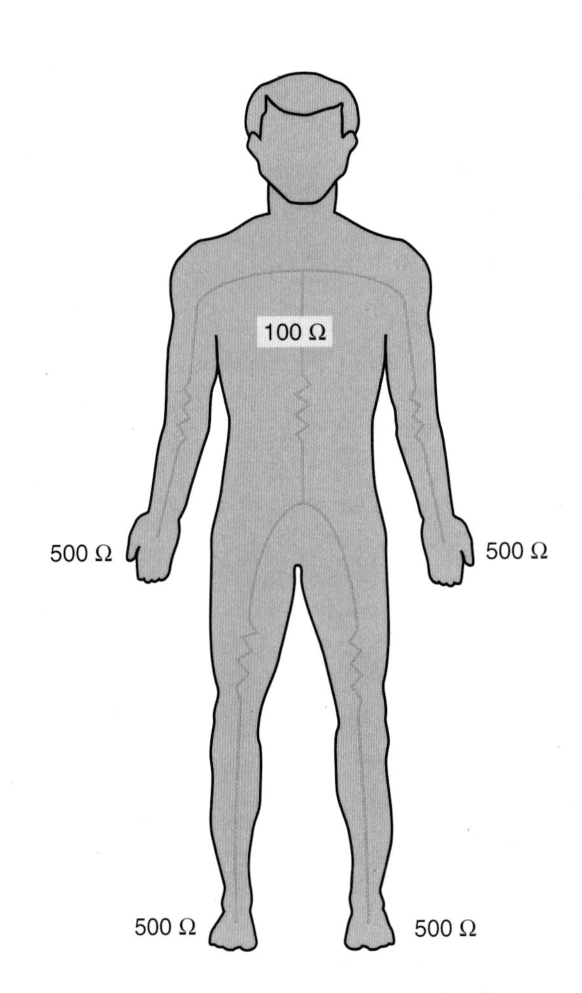

FIGURE I-36 Human body circuit. (© *Cengage Learning 2012*)

Consider the amount of current that flows through a resistance of 1100 ohms for dry skin and 300 ohms for wet skin at various voltages as determined by the Ohm's law formula:

I (amperes) = E (volts) ÷ R (ohms).

	CURRENT (AMPERES OR mA)	
Circuit Voltage	1100 Ohms (Dry Skin)	300 Ohms (Wet Skin)
480	0.436	1.60
277	0.252	0.92
240	0.218	0.80
120	0.109	0.40

Now consider the effects of 60-hertz (60-cycle) ac currents on humans in the study by Charles F. Dalziel ("Dangerous Electric Currents," reported in *AIEE Transactions,* Vol. 65 [1946], p. 579; discussion, p. 1123), presented in the following table. (The effects vary depending on whether the current is dc or ac and on the frequency if it is ac.)

	CURRENT IN mA, 60 HERTZ	
Effect(s)	**Men**	**Women**
Slight sensation on hand	0.4	0.3
Perception of "let go" threshold, median	1.1	0.7
Shock, not painful, and no loss of muscular control	1.8	1.2
Painful shock—muscular control lost by half of participants	9	6
Painful shock—"let go" threshold, median	16	10.5
Painful and severe shock—breathing difficult, muscular control lost	23	15

This study and many others prove the importance of grounding and bonding equipment for keeping the equipment at earth potential or at equal potential (equipotential) and for reducing the shock hazard between pieces of electrical equipment that can be touched by a person who is in contact with the earth or other grounded surface. If there is no potential difference between conductive surfaces, there can be no shock hazard.

A person who receives an electric shock in which the current passes through the chest can experience ventricular fibrillation, immediate heart stoppage, or paralysis of the respiratory system. Ventricular fibrillation is a condition in which the rhythmic pumping action of the heart is interrupted by a disturbance of its electrical signals, which may be caused by the electrical current passing through the chest. In ventricular fibrillation, the heart muscle merely "quivers" and does not pump blood effectively. Death can occur in a few minutes if the condition is not corrected.

Series and Parallel Circuits

For electrical purposes, a "circuit" can be defined as a complete path in which current flows. The simplest circuit typically described in basic electricity classes is a dry cell battery and some copper wire. When the wire is connected from one terminal to the other, current will flow in the circuit. The next experiment that is often done in such classes is to insert a switch and a bell, buzzer, or lamp in the circuit (Figure I-37). This is still a series circuit, as there is only one path for current to flow. The bell, buzzer, or lamp performs work as the current flows through the circuit. Obviously, the switch controls the current flow.

Other, more complex circuits commonly exist in electrical systems. These are parallel circuits and series/parallel circuits. A parallel circuit has more than one path for current to flow. Whenever more than one path for electrical current exists, current flows in every path available. The laws of physics determine how much current flows in each path. One of these laws states: "Current will divide inversely in proportion to the resistance" (or impedance in ac circuits). That is, the greater the resistance or impedance of the path, the less the current. This law can also be stated as follows: "The lower the resistance

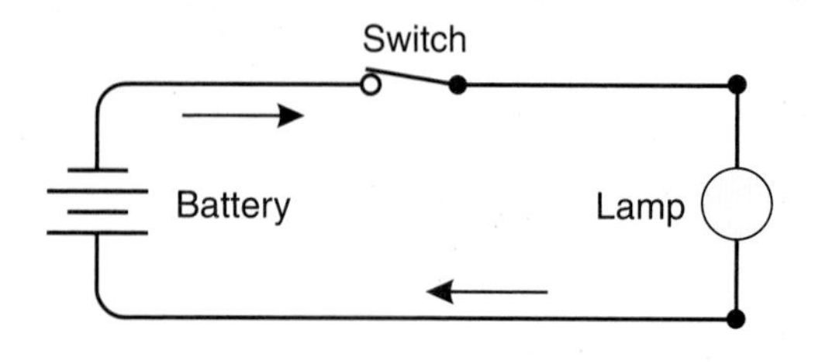

FIGURE I-37 Series circuit.
(*© Cengage Learning 2012*)

or impedance of the path, the greater the current that will flow over that path."

A parallel circuit is one in which there is more than one path for current to flow in. Multiple potential paths are common for branch circuits in homes, businesses, and industrial plants. Loads are connected across the line from phase to phase or from phase to neutral; an example is shown in Figure I-38. In this circuit, the total load on the circuit is a sum of the load on each individual branch of the circuit. This is applicable to our discussion on grounding in at least two ways.

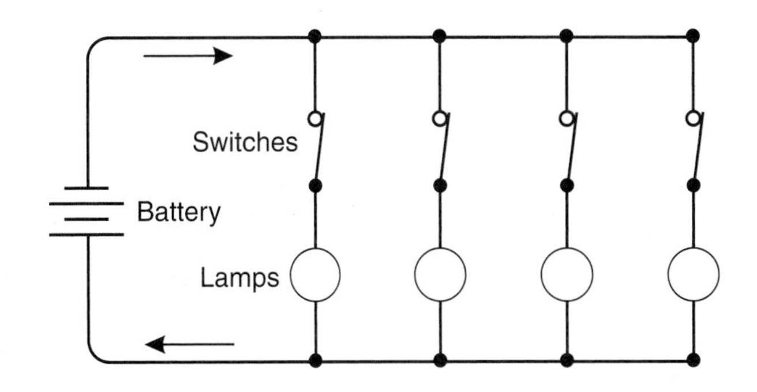

FIGURE I-38 Parallel circuit. (*© Cengage Learning 2012*)

1. Ground-fault current will flow on all the paths that are available to it—both those paths that are intended, such as over conduit or equipment grounding conductors, and those paths that are not intended, such as over cable shields, water pipes, and cable trays. This unintended current flow can cause severe problems in data processing systems, where the ground currents are referred to as "circulating currents."
2. When the neutral is grounded improperly downstream from the service or source of a separately derived system, neutral currents will divide among all of the paths that are available to it. These paths include the intended path over the neutral conductor and any other unintended paths. Unintended paths include current over metal pipes, shields on control cables, process pipes, cable trays, and conduit or cable systems. This undesirable flow of neutral current over unintended paths is often referred to as objectionable current, as covered in *250.6*.

REVIEW QUESTIONS

1. Which of the following statements about the term *electrical system* is *not* true?
   a. The term is defined in the *NEC*, so its usage will be uniform.
   b. The electric utility supplies the system at the service to the building.
   c. The energy supplied by a solar photovoltaic source is a system.
   d. The energy supplied by a generator is a system.
2. Which of the following statements concerning the phrase "conductive body that extends the ground connection" is *not* true?
   a. It is defined in the *NEC*.
   b. It is conduit or wire that extends from the point where an "earth connection" is made.
   c. It is an 8-ft ground rod with a resistance of 25 ohms or less.
   d. It is a grounding electrode consisting of the structural metal frame of a building
3. Which of the following descriptions is the best definition of the term *bonding (bonded)*?
   a. Connected to a grounded panelboard enclosure
   b. A device that makes a connection to the earth
   c. Driving a local ground rod at equipment
   d. The permanent joining of metallic parts to form an electrically conductive path

4. Which of the following locations typically would *not* have areas that have "equipotential bonding" features installed?
   a. Bathrooms in dwelling units
   b. Patient care areas of health care facilities
   c. Electrical equipment associated with swimming pools
   d. Milking parlors of agricultural buildings
5. Which of the following phrases *best* describes the term *grounded conductor*?
   a. A system or circuit conductor that is intentionally grounded
   b. A conductor used to connect the neutral and the equipment grounding conductors
   c. The midpoint of a wye-connected transformer bank
   d. The conductor used to connect exposed conductive surfaces together
6. Which of the following systems does not have a neutral conductor as defined in *Article 100*?
   a. 120-volt, 1-phase, 2-wire
   b. 120/240-volt, 1-phase, 3-wire
   c. 120/240-volt, 3-phase, 4-wire
   d. 208Y/120-volt, 3-phase, 4-wire
7. Which of the following installations accurately describes the term *separately derived system*?
   a. A transformer with a bonding jumper from the primary to the secondary
   b. A generator with a bonding jumper from the neutral to the frame
   c. A 1-phase generator connected to the system by a 3-pole transfer switch
   d. An autotransformer that has a grounded conductor common to both primary and secondary
8. Which of the following statements does *not* accurately describe a function of a ground-fault current path?
   a. A high-impedance grounded neutral system
   b. A low-impedance path to facilitate the operation of an overcurrent protective device
   c. A conductor large enough not to overheat and burn off
   d. All the branch circuit conductors run in the same conduit or cable.
9. Which of the following is *not* a requirement for installing conductors in parallel?
   a. The conductors are of the same length.
   b. All circuit conductors are of the same conductor material.
   c. All circuit conductors are the same size in circular mil area.
   d. All circuit conductors of phase A are black insulation for a 208Y/120-volt system.
10. Which of the following applications does *not* demonstrate the importance of maintaining isolation for the neutral or grounded-circuit conductor?
    a. Ground-fault circuit interrupters
    b. High-impedance grounded neutral systems

c. Ground-fault protection of equipment
d. Isolated systems in health care facilities

11. Which of the following expressions of Ohm's law is *not* correct?
    a. $E = I \times R$
    b. $I = E \div R$
    c. $R = E \div I$
    d. $R = E \times I$
12. Which of the following best describes the definition of *ground*?
    a. Connected to earth
    b. A grounding electrode
    c. The earth
    d. Equipment grounding conductor
13. Soil resistivity is largely determined by its content of _______.
    a. electrolytes
    b. moisture
    c. minerals
    d. dissolved salts
14. Which of the following is *not* a source of electrical energy for the electrical system?
    a. Utility supplied service
    b. Transformer
    c. Fuel cell
    d. Solar photovoltaic system
15. Which of the following terms best describes the definition of *grounded?*
    a. Connected to a ground rod or other grounding electrode
    b. Connected to the equipment grounding terminal bar in a panelboard
    c. Connected to the grounding terminal of a receptacle supplied by an equipment grounding conductor
    d. All of the above
16. In its simplest form, *bonding* means
    a. the same thing as grounding.
    b. connected to a grounding electrode.
    c. providing a path for current flow.
    d. a ground fault.
17. *Equipotential bonding* means
    a. connected to a ground rod on both sides of the building.
    b. providing a low-impedance path to ground.
    c. connecting equipment together to reduce shock hazards.
    d. supplying equipment from two voltage sources.

18. The *system bonding jumper*
    a. is similar in function to the main bonding jumper.
    b. provides a path for fault current to flow.
    c. is installed at a separate derived system.
    d. performs all of the above functions.
19. The *equipment grounding conductor*
    a. connects normally non-current-carrying metal parts together.
    b. connects equipment to the system ground conductor.
    c. connects equipment to the grounding electrode.
    d. performs all of the above functions.
20. Which of the following terms is *not* used to describe an *effective ground-fault current path*?
    a. Intentionally constructed
    b. Low-impedance
    c. 6 AWG or larger
    d. Continuous and reliable
21. *Inverse time* means
    a. the element of time is not an issue so far as operation of the overcurrent device is concerned.
    b. the more current that flows in the circuit, the less time it will take for the overcurrent device to operate.
    c. a time delay is specifically introduced into the overcurrent device so it will operate slower.
    d. circuit breakers open the ground fault faster than fuses.

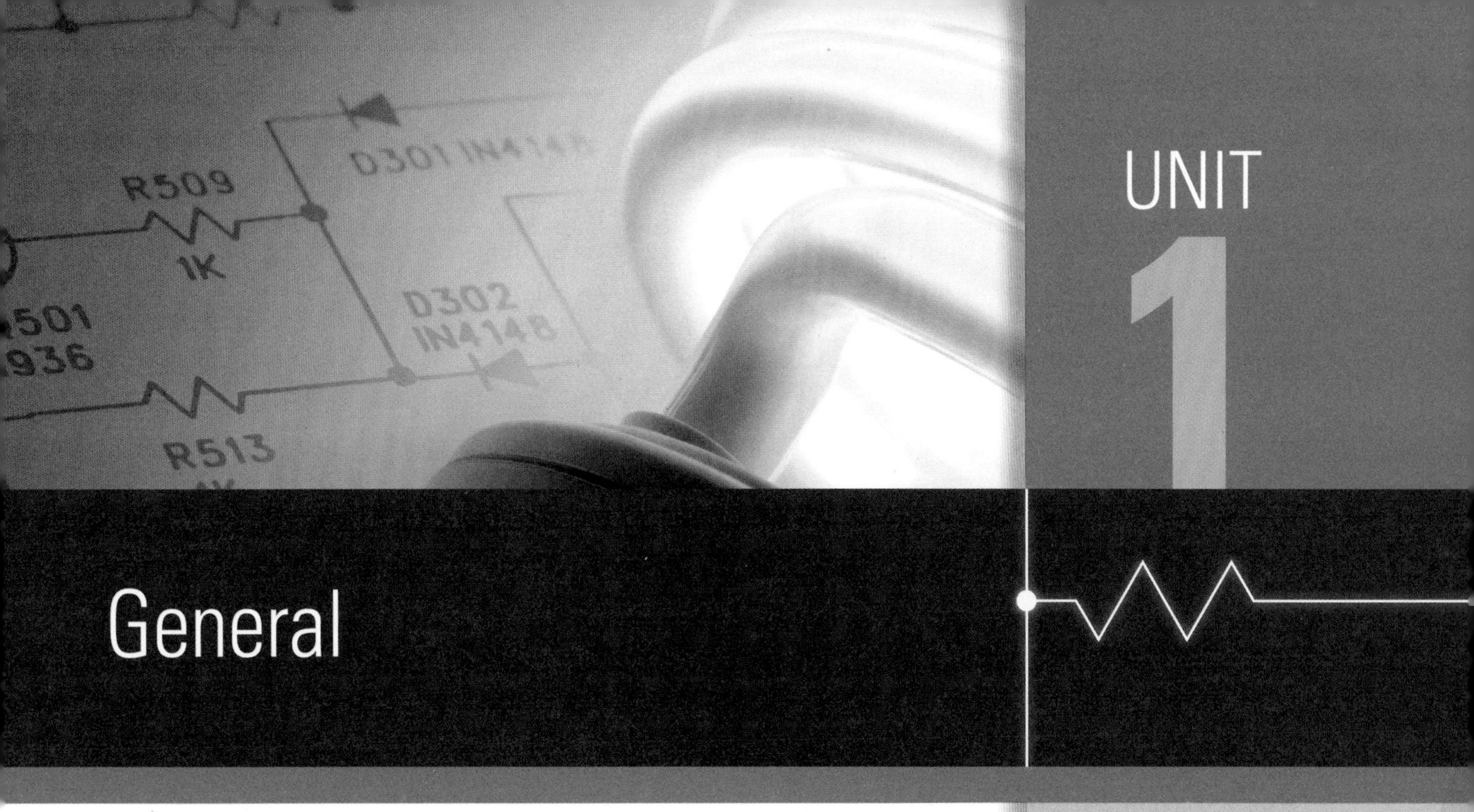

UNIT 1

General

OBJECTIVES

After studying this unit, the reader will understand

- the "performance requirements" of *Article 250*.
- how the "performance requirements" relate to the "prescriptive requirements" in the remainder of *Article 250*.
- objectionable current over grounding conductors.
- connections of grounding and bonding equipment.
- the importance of providing clean surfaces for grounding and bonding connections.

Like all good *NEC* articles, *Article 250* begins with the scope in *250.1*. The scope tells us what is contained in the article. Reading and understanding the scope is important to the student and user of the *NEC*. In addition, *Figure 250.1* is included, which gives guidance about how the article is organized as well as how the parts relate.

250.2 DEFINITIONS

The definitions of several terms important to the understanding of grounding and bonding concepts are included in *250.2*. As they are located in this section, they are applicable only to *Article 250*. We have covered most of them in the Introduction of this text during our review of terms important to grounding and bonding.

Terms located in *250.2* include the following:

Bonding Jumper, Supply Side (New in the 2011 NEC)
Effective Ground-Fault Current Path
Ground-Fault Current Path

250.3 APPLICATION OF OTHER ARTICLES

The Requirement: For other articles that apply to particular cases of installation of conductors and equipment, grounding and bonding requirements are identified in *Table 250.3* that are in addition to, or modifications of, those in *Article 250*.

Discussion: *Table 250.3* covers *Additional Grounding Requirements* that are found in other articles of the *NEC*. These requirements include grounding or bonding rules for such circuits, systems, or occupancies as agricultural buildings in *Article 547*, electric signs and outline lighting in *Article 600*, and health care facilities in *Article 517*.

The rules in articles located in *Chapter 5* through *Chapter 9* of the *NEC* may supplement or amend the rules found in *Chapter 1* through *Chapter 4*. This provision is found in *90.3* regarding the organization of the *NEC*. Examples of such additional rules are those specifying the different methods permitted for equipment grounding conductors in *250.118* and the specific requirements for branch circuits in patient care areas of health care facilities. *NEC 250.118* permits electrical metallic tubing (EMT) to serve as an equipment grounding conductor. However, *517.13* requires two independent equipment grounding paths. Therefore, the path provided by the EMT must be supplemented by an insulated copper equipment grounding conductor. Similar specific grounding or bonding requirements appear in many other articles of the *NEC* and amend or supplement the requirements in *Article 250*.

250.4 GENERAL REQUIREMENTS FOR GROUNDING AND BONDING

The Requirements: The following general requirements identify what grounding and bonding of electrical systems are required to accomplish. The prescriptive methods contained in *Article 250* are required to be followed to comply with the performance requirements of this section.

Discussion: The *NEC* generally is written in "prescriptive style," stating how an installation must be made, rather than in "performance style," stating general goals to be achieved. That is, the *NEC* tells the installer and the inspector what the minimum requirements are to provide an electrical installation that is essentially safe but not necessarily efficient, convenient, or adequate for good service or future expansion of electrical use, *90.1(B)*. *Section 250.4* contains "performance goals" for grounding and bonding of electrical systems. The concepts of grounding and bonding are shown in Figure 1-1.

An example of a performance goal is found in *250.4(A)(1)*. It requires grounded electrical systems be connected to earth in a manner that will limit the voltage imposed by lightning, line surges, or unintentional contact with higher-voltage lines. It also states a purpose of the grounding connection is to stabilize the system voltage to earth during normal operation. Requirements that dictate specific methods for grounding the electrical systems or the minimum size of conductors to be used to achieve the required performance are not included in this section. The remainder of *Article 250* following

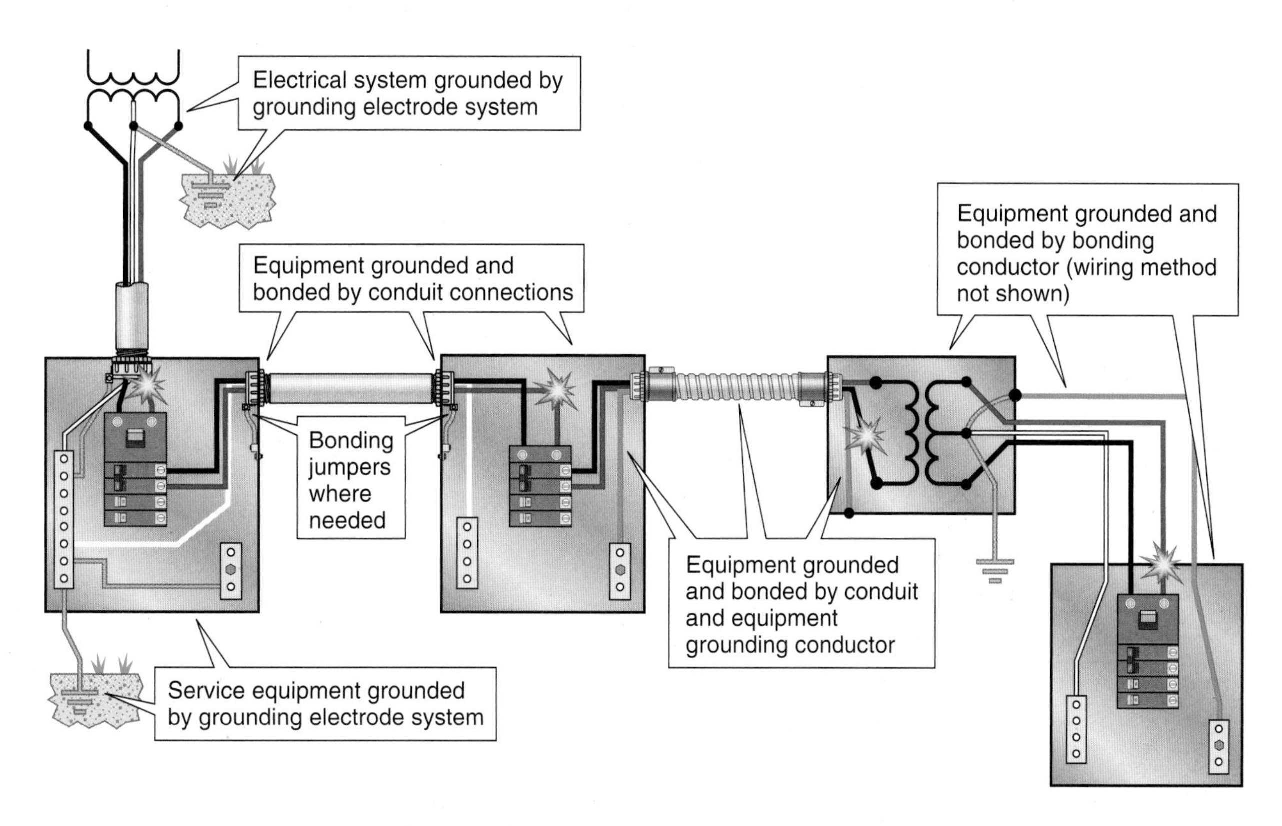

FIGURE 1-1 General requirements for grounding and bonding, *250.4*. (*© Cengage Learning 2012*)

250.4 contains primarily prescriptive requirements. Compliance with these how-to rules in the remainder of *Article 250* will ensure that the performance goals of *250.4* are met.

250.4(A) Grounded Systems

Discussion: *Section 250.4* is divided into two major subsections. Performance requirements for grounded systems are given in *250.4(A)*, whereas *250.4(B)* applies to ungrounded systems. Keep in mind that these requirements in *250.4(A)* refer to the *system* and electrical equipment being grounded, or connected to earth. The equipment associated with the system also is grounded through the connection of the main bonding jumper, bonding conductors or connections and the equipment grounding conductors. For our purposes, the *electrical system* is generally considered to be the source of electrical energy and can consist of the service or a separately derived system. (See the Introduction to Grounding and Bonding in this book for additional discussion on this subject.)

250.4(A)(1) Electrical System Grounding

The Requirement: Electrical systems that are grounded are required to be connected to earth in a way that will limit the voltage imposed by lightning, line surges, or unintentional contact with higher-voltage lines and that will stabilize the voltage of the system to earth during normal operation.

Discussion: This requirement is illustrated in Figure 1-2 and can be thought of as the purpose for grounding electrical systems.

Not all electrical systems are required to be grounded. Rules are given in *250.20* for which systems are required to be grounded. The systems that are permitted to be grounded but are not required to be grounded are given in *250.21*. The circuits that are not permitted to be grounded are identified in *250.22*.

Note that two of the three purposes for grounding the electrical system are to connect the system to earth. This connection provides a path to shunt overvoltages to earth from lightning, line surges, or unintentional contact with higher-voltage lines. Damage

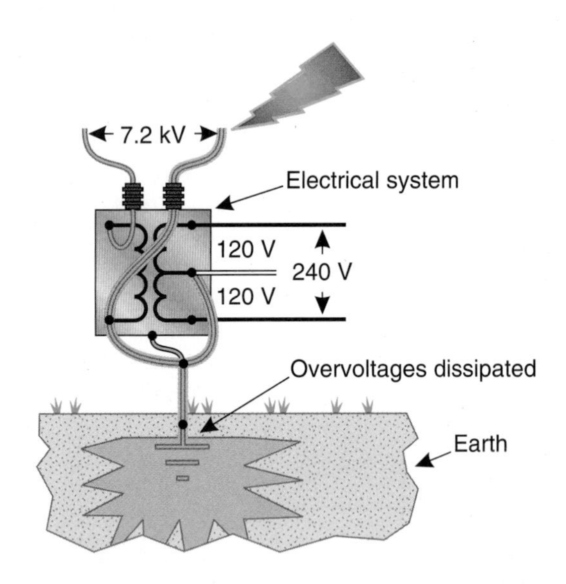

FIGURE 1-2 Grounding the electrical system, *250.4(A)(1)*. (*© Cengage Learning 2012*)

to the electrical equipment or structure from these overvoltages is limited by providing a convenient path for the current to flow to earth, where it will dissipate. Sources of this overvoltage include lightning strikes to the electrical system at either the primary or the secondary voltage level. Other common sources of overvoltage are high-voltage wires that contact the lower-voltage wires that supply the service equipment.

Causes of these high-voltage–to–low-voltage contacts or crosses can be wind, ice, or snowstorms or vehicle accidents in which power poles are damaged. These overvoltages can damage or destroy equipment that is rated for lower voltages.

The third reason for grounding the electrical system is to stabilize the voltage to earth during normal operation. This is best understood by relating grounded systems to ungrounded ones. Systems that are ungrounded establish an artificial voltage and connection to ground by induction or capacitance. This occurs when insulated conductors are installed in metal enclosures such as conduit. A voltage relationship is established to ground by induction. This system voltage-to-ground is not usable for carrying current but can present a shock hazard. A solid connection (in which no impedance such as a resistor is intentionally introduced in the path) or a connection through a resistor or other impedance device stabilizes the voltage to ground.

For solidly grounded systems, a useful line voltage-to-ground connection is possible (Figure 1-3). Typical voltages are 120/240-volt, 1-phase, 3-wire; 208Y/120-volt, 3-phase, 4-wire; and 480Y/277-volt, 3-phase, 4-wire systems. The voltage for the 120/240-volt system is derived from the secondary of a 240-volt, 1-phase transformer that is center-tapped. This center point is then grounded so each ungrounded ("hot") conductor has a voltage-to-ground of 120 volts. This is the most common voltage system used in dwellings and smaller commercial installations. It is shown in the left drawing in Figure 1-3.

The center drawing in Figure 1-3 shows a delta-connected system. This is a 3-phase system and is commonly referred to as 120/240-volt, 3-phase, 4-wire. These systems are used in small to medium-size commercial facilities where it is desirable to have 3-phase power available for motor loads and 120/240 volts for lighting, receptacles, and similar loads. Three 240-volt, 1-phase transformers are connected together as shown. One of the transformers is center-tapped to produce a 120/240-volt system.

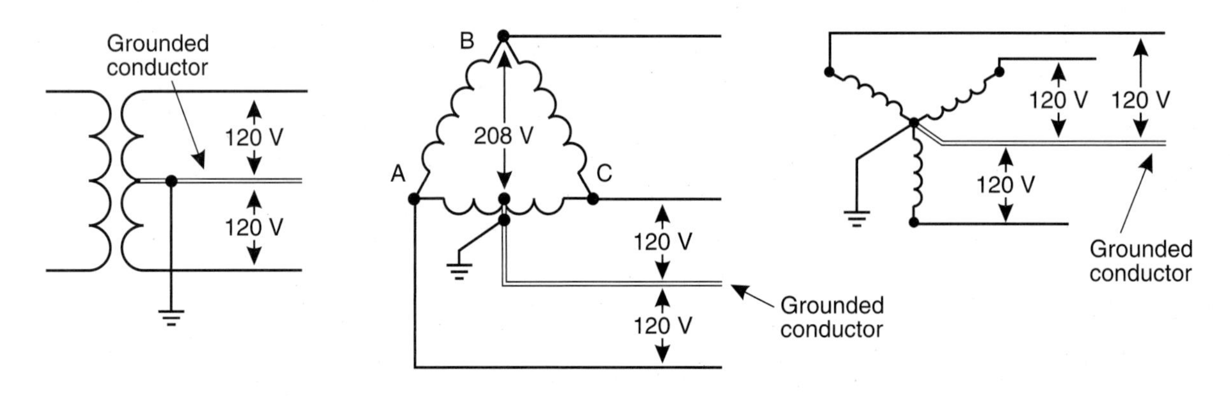

FIGURE 1-3 Typical voltages to ground, *250.4(A)(1)*. (*© Cengage Learning 2012*)

This system is capable of supplying 120-volt loads as well as 240-volt, 1-phase and 240-volt, 3-phase loads. Line-to-neutral connections should be made only to the A and C phases, which result in a 120-volt supply. A line-to-neutral connection from the B phase will result in a 208-volt circuit. The formula used to determine this voltage is ½ of phase-to-phase voltage multiplied by the square root of 3. Inserting the values in the formula results in

$$240 \div 2 = 120 \times 1.732 = 207.84$$
(rounded up to 208 volts in practical usage)

Equipment intended for 120-volt supply that is connected at the 208-volt position will most likely be damaged or destroyed.

Single-phase transformers connected as shown in the right diagram of Figure 1-3 result in a wye connection. As shown, three 120-volt transformers are connected together at a common point where a grounding connection is made. This system is referred to as a 208Y/120-volt, 3-phase, 4-wire system. A 120-volt connection can be made from any ungrounded point on the system to the neutral. The phase-to-phase voltage is

$$120 \times 1.732 = 208$$

This system is commonly used in small to midsize commercial installations. A similar system connected with 277-volt transformers is used in larger commercial and industrial installations. These systems are referred to as 480Y/277-volt, 3-phase, 4-wire.

> **Informational Note:** An important consideration for limiting the imposed voltage is the routing of bonding and grounding conductors. These conductors should not be longer than necessary to complete the connection without disturbing the permanent parts of the installation. Unnecessary bends and loops should be avoided.

Discussion: This *Informational Note* was added during the processing of the 2008 *NEC*. As an *Informational Note*, it contains information considered useful to the user of the *Code* but is not a requirement. The locations of requirements related to the *Informational Note* are found in *NEC Chapter 8, Communications Systems*. *Section 800.100(A)(5)* requires the grounding conductor for communications systems to be run in as straight a line as practicable. A similar requirement is found in *810.21(E)* for radio and television equipment receiving stations and in *820.100(A)(5)* for CATV systems. The reason behind a requirement to run grounding or grounding electrode conductors in a straight line usually is to keep the impedance of the path as low as possible and therefore provide lightning or other high-voltage surges the best path practicable to reach earth.

250.4(A)(2) Grounding of Electrical Equipment

The Requirement: Normally non-current-carrying conductive materials enclosing electrical conductors or equipment, or forming part of such equipment, are required to be connected to earth to limit the voltage to ground on these materials.

Discussion: It should be noted that the phrase "normally non-current-carrying conductive materials" refers to materials that are conductive but are not intended to carry current in normal operation of the electrical system. This phrase refers to equipment such as metal conduits, metal-sheathed cables, junction and pull boxes, panelboard enclosures, switchboard cabinets, and cable tray systems. They are most usually installed to enclose and protect insulated conductors from damage. These conduits and other equipment may be installed as a part of the ground-fault current path in normal operation. (See *250.118* for a list of equipment, including conductors, conduits, and cables, that is permitted to serve as equipment grounding conductors.) The metallic raceways, conductors, and equipment carry fault current when there is a ground fault downstream of the equipment being considered. The equipment then serves as an equipment grounding conductor and provides the ground fault current return path back to the source.

As indicated in this section of *Article 250*, the purpose of connecting the electrical non-current-carrying conductive materials to earth is to *limit the voltage to ground on these materials.* Figure 1-4 illustrates a person in contact with the earth or a grounded surface and in contact with the electrical equipment at the same time. The purpose of this earth connection is to protect the person from electric shock by effectively grounding the electrical equipment.

As discussed in the *Introduction to Grounding and Bonding* in this text, the path through the earth

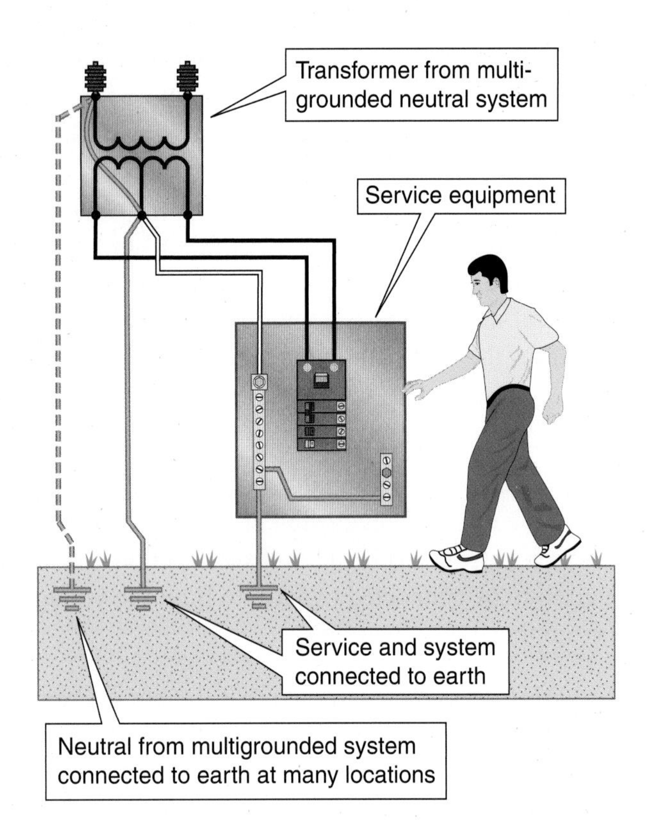

FIGURE 1-4 Grounding electrical equipment, *250.4(A)(2)*. (*© Cengage Learning 2012*)

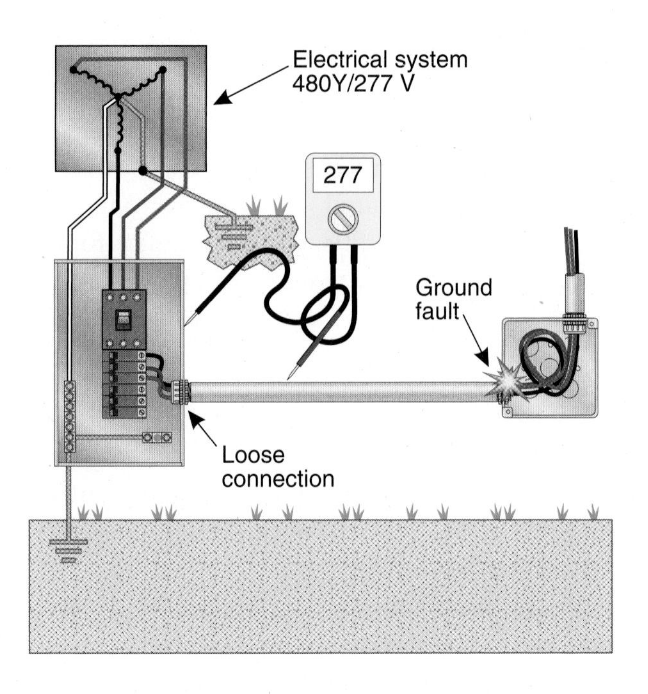

FIGURE 1-5 Break in ground-fault return path, *250.4(A)* (*© Cengage Learning 2012*)

is of high resistance and impedance and is ineffective in carrying fault current. The amount of current flowing through the grounding electrode and into the earth creates a voltage drop across the resistance of the grounding electrode. This voltage rise is limited by the earth connection at the utility transformer and usually multiple locations where the utility makes an earth connection on multi-grounded neutral systems. In actuality, the many, many grounding electrodes in the utility distribution system are bonded together by the utility primary neutral. Extension of the service grounded or neutral conductor to the service equipment and bonding it to the service equipment enclosure will help keep the service equipment at the neutral conductor voltage, which will be at or near earth voltage. However, a break in the neutral conductor at the same time a ground fault occurs could leave the service equipment at a dangerous voltage above ground, depending on the voltage drop across the grounding electrode connection.

As illustrated in Figure 1-5, a break in the ground-fault return path can leave equipment at a dangerous voltage level above ground. A shock or electrocution hazard may result, as the open-circuit voltage will usually be the system voltage. If the grounded system is 480Y/277 volts, the voltage to ground is 277 volts. This would be the open circuit voltage across the open ground-fault current path. This voltage obviously presents a significant shock hazard. Where the system is 120/240 volts, 1-phase, 3-wire, the voltage to ground is 120 volts. Thus, 120 volts would be the open circuit voltage across the impaired or open ground-fault return path.

250.4(A)(3) Bonding of Electrical Equipment

The Requirement: Normally non-current-carrying conductive materials enclosing electrical conductors or equipment, or forming part of such equipment, are required to be connected together and to the electrical supply source in a manner that establishes an effective ground-fault current path.

Discussion: In its simplest form, bonding completes a path through which fault current can flow. Bonding non-current-carrying conductive parts together also keeps the conductive materials at the same electrical potential. This has the effect of eliminating any shock hazard between these bonded enclosures or parts (Figure 1-6). Many specific

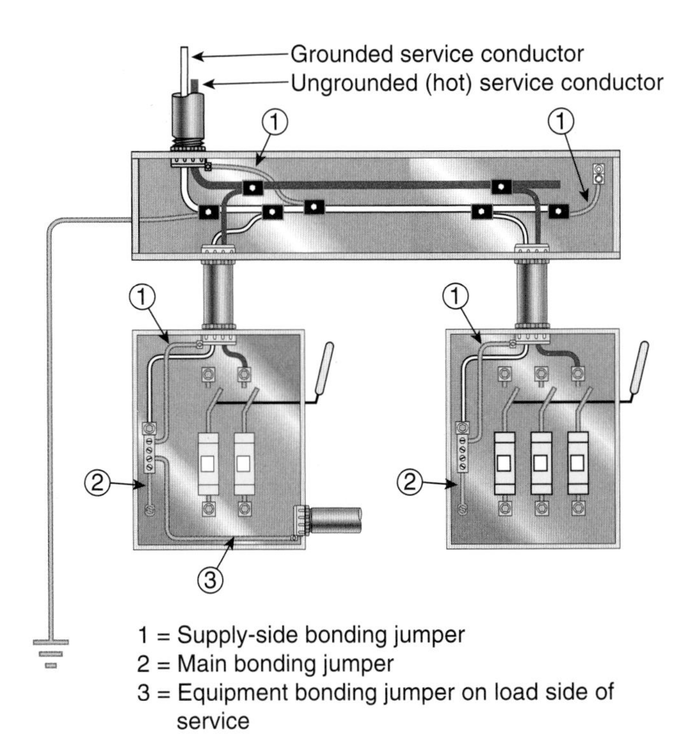

FIGURE 1-6 Bonding of electrical equipment, *250.4(A)(3)*. (*© Cengage Learning 2012*)

rules for bonding are provided in the remaining sections of *Article 250*. These rules concern the following:

- Main bonding jumpers
- System bonding jumpers
- Supply-side bonding jumpers on the supply side of the service or separately derived system
- Equipment bonding jumpers on the load side of the service
- Bonding jumpers to metal piping systems and structural metal frames of buildings.

Bonding must be accomplished around conduit connections whenever the fault-current return path is suspect, as well as when specifically required in the *NEC*. Thus, bonding is required in the following situations:

- When conduit locknuts or bushings are not made up tight
- When it is not certain that locknuts or bushings have adequately penetrated painted electrical enclosures to make a solid connection. (For additional information on the importance of penetrating painted electrical enclosures to provide an effective ground-fault path, see Appendix D, which reproduces the cover page and pages i, 1–5, 11–12, and 24–28 from the Underwriters Laboratories Inc. [UL] report titled "Report of Research on Conduit Fitting Ground-Fault Current Withstand Capability." Permission to reproduce these pages in their entirety has been granted by Underwriters Laboratories Inc.)
- For service equipment enclosures
- Around eccentric or concentric knockouts on enclosures
- Around knockout-reducing washers where the bonding connection is suspect, such as connections to painted pull or junction boxes, wireways or auxiliary gutters, panelboards, motor control centers, and switchboards or other painted enclosures
- For hazardous (classified) locations
- For greater than 250 volts-to-ground systems

Bonding and grounding work together to provide an electrical safety system. The safety system protects people and animals from electric shock. Where installed properly as an effective ground-fault current path, the system also provides a low-impedance path for fault current to return to the source.

250.4(A)(4) Bonding of Electrically Conductive Materials and Other Equipment

The Requirement: Normally non-current-carrying electrically conductive materials that are likely to become energized are required to be connected together and to the electrical supply source in a manner that establishes an effective ground-fault current path.

Discussion: The *Code* does not define the phrase *likely to become energized,* although it is used in several locations. As a result, the phrase is subject to variable interpretation by the authority having jurisdiction (AHJ). The phrase is included in the *NEC Style Manual* as "likely to become energized—failure of insulation on." Usually, metal enclosures for energized conductors such as conduit and pull, outlet, and junction boxes, as well as metal enclosures for appliances and luminaires, all are considered *likely to become energized.*

FIGURE 1-7 Bonding conduits to panelboard cabinet. (*© Cengage Learning 2012*)

Figure 1-7 shows bonding of metal conduit. These wiring methods or enclosures do not carry current while the electrical system is in a normal operational mode. They carry current during abnormal conditions such as when a ground fault occurs and the fault current flows through the wiring methods or enclosures. Other metallic piping, such as that for water, air, gases, or drainage, is considered *likely to become energized* when in the vicinity of electrical conductors or equipment or if they are connected to electrically energized equipment. Also considered *likely to become energized* are metal frames of buildings. Many items related to electrical equipment are simply required to be grounded or bonded without consideration of whether they are *likely to become energized.*

Here again, the *Code* emphasizes the performance requirement of bonding to create an effective ground-fault return path. Specific requirements are addressed later in *Article 250.*

250.4(A)(5) Effective Ground-Fault Current Path

The Requirement: Electrical equipment and wiring and other electrically conductive material likely to become energized are required to be installed in a manner that creates a reliable,[1] low-impedance circuit facilitating the operation of the overcurrent device or ground detector for high-impedance grounded systems. It must be capable of safely carrying the maximum ground-fault current likely to be imposed on it from any point on the wiring system where a ground fault may occur to the electrical supply source. The earth is not to be considered as an effective ground-fault current path.

Discussion: Several important performance concepts are included in this section. Each concept must be complied with for the ground-fault current path to be considered "effective." These concepts include the following:

1. The circuit must be *reliable.* The concept of a *reliable* path includes features such as careful and deliberate construction to be certain that the circuit is reliable for its intended purpose. When conduit is used for the path, this requirement includes installing conduit properly, to include making up all couplings, connectors, locknuts, and bushings wrenchtight. It also includes supporting the conduit properly within the dimensions prescribed in the *NEC* article for the specific conduit or tubing installed.

   Keep in mind that the conduit or tubing becomes a current-carrying conductor while carrying fault current back to the source. Also remember that a single loose connection, such as at a locknut or bushing, or a single loose coupling or connector will interrupt the fault-current path. If this happens, conduit, boxes, receptacles, luminaires, and all other equipment on the load side of the loose connection can be at a dangerous voltage above ground. This elevated voltage can present a shock or electrocution hazard. Figure 1-8 shows a loose connection.

2. The circuit must be of *low impedance.* (Note that no impedance value is given here, or, for that matter, anywhere in *Article 250.*) A low-impedance circuit is one that does not have excessive impedance from improper installation that will unreasonably delay the operation of an overcurrent protective device. Figure 1-9 shows a proper and an improper installation. The low-impedance circuit must allow enough fault current to flow so that the overcurrent device on the supply side of the circuit will open quickly. This

[1]The Code Panel deleted the word "permanent" from the list of requirements for an *effective ground-fault current path.* This author has substituted the word "reliable" in its place to add emphasis to the creation of this path intended to carry fault current.

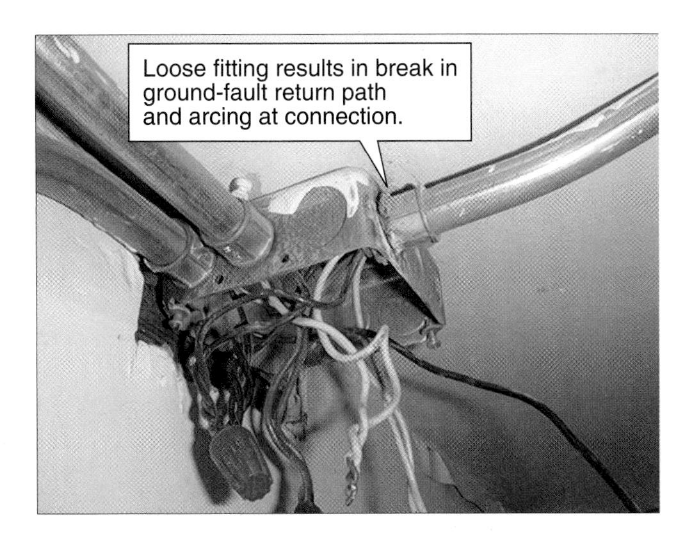

FIGURE 1-8 Loose fitting at box creates shock hazard. (*© Cengage Learning 2012*)

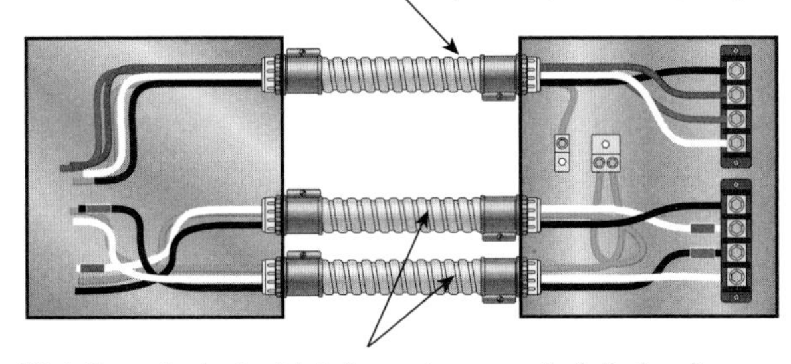

FIGURE 1-9 Low-impedance circuit and violation. (*© Cengage Learning 2012*)

quick operation of the overcurrent device reduces the thermal damage to the conductors and equipment that would be caused by excessive current flowing in the ground-fault return path.

Overcurrent protective devices including molded-case circuit breakers and current-limiting fuses have characteristics of operation that include a concept referred to as "inverse time." This means when the device carries current beyond its rating, the more current that flows in the circuit, the faster the overcurrent device operates. This relation is represented by the curved portion of the time-current curve in Figure 1-10. In normal operation, the overcurrent device will carry up to its rated current indefinitely. (Rules in the product safety standard and in the *NEC* limit the overcurrent device to carrying not more than 80% of its rated current when this load will continue for 3 hours or more. Overcurrent devices rated for continuous duty can carry their rated current indefinitely.)

These overcurrent devices also have an instantaneous trip, or fusing, point. The vertical line in the time-current chart for the circuit breakers shown in Figure 1-10 represents this instantaneous trip, or fusing, point. The width of the curved and the vertical lines represents the range of current and time in which this tripping action can take place. This range reflects manufacturing tolerances or the mechanical action of circuit breakers. For a 50-ampere circuit breaker shown in Figure 1-10, the instantaneous trip point is shown to range from 7 times the rating of the breaker or 350 amperes to as much as 18 times the rating or 900 amperes.

The time for the breaker to trip is shown up the left and right sides of the chart, and the values at the top and bottom are multiples of rated current. This means the 50-ampere breaker will trip without any delay if from 350 to 900 amperes is flowing in the short circuit or ground fault.

The time-current chart for a current-limiting fuse is shown in Figure 1-11. The curve has a different shape than a circuit breaker due to the melting characteristics of the fuses. In this chart, the time values are shown at the left, and current in amperes is at the bottom. For example, if 400 amperes is flowing in a faulted circuit, a 60-ampere fuse will clear in about 0.1 seconds or in 6 cycles (at 60 Hz). At the same current, a 30-ampere fuse will clear in less than 0.02 seconds.

The impedance of the fault-current path should be low enough to allow sufficient current to flow in the path to reach the vertical, or "instantaneous," portion of the time-current curve. As a result, clearing of the faulted circuit will occur as rapidly as possible, with the least amount of thermal damage to the circuit components.

Two *NEC* sections have specific requirements related to this low-impedance path, although

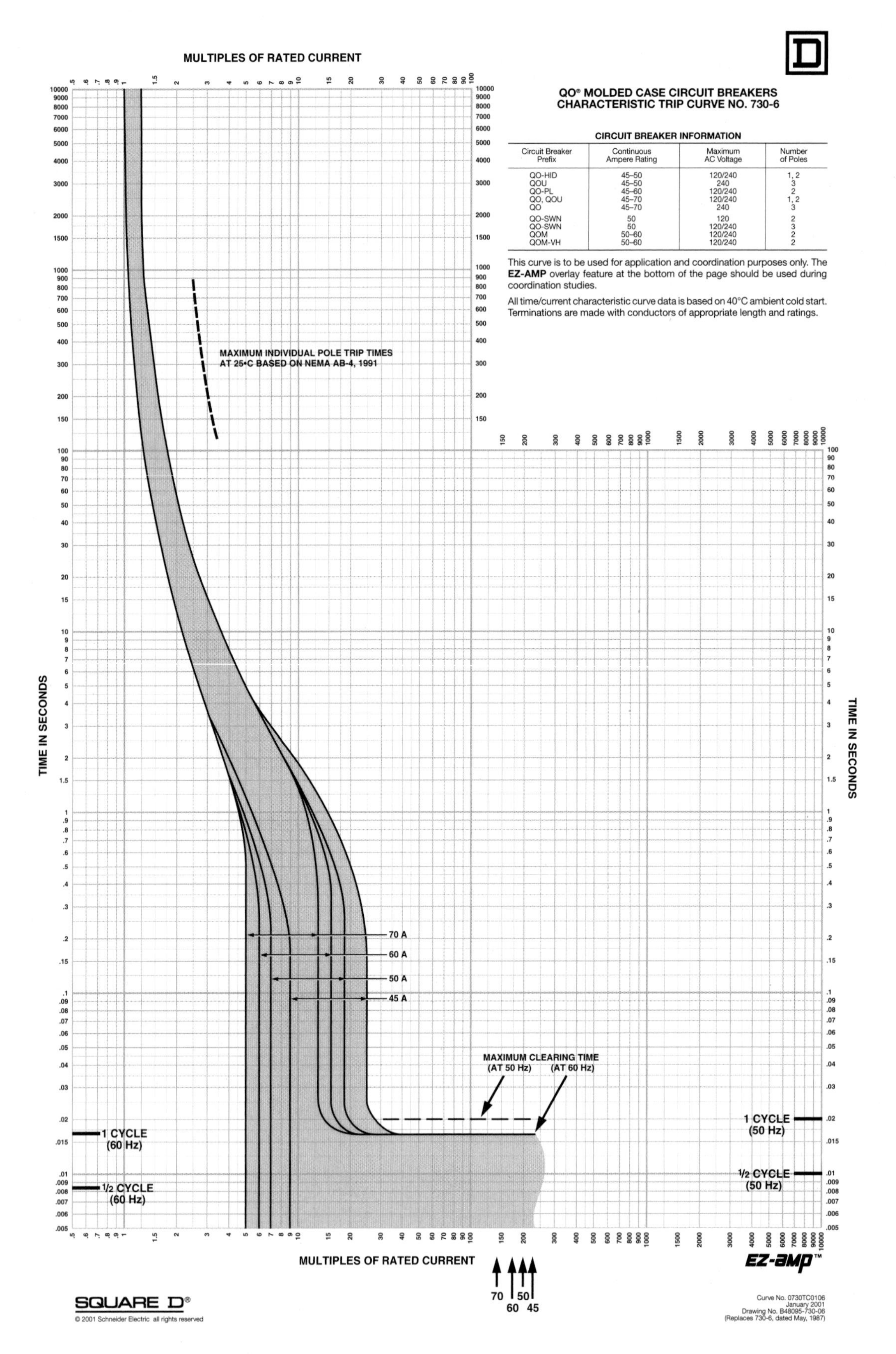

Circuit Breaker Prefix	Continuous Ampere Rating	Maximum AC Voltage	Number of Poles
QO-HID	45–50	120/240	1, 2
QOU	45–50	240	3
QO-PL	45–60	120/240	2
QO, QOU	45–70	120/240	1, 2
QO	45–70	240	3
QO-SWN	50	120	2
QO-SWN	50	120/240	3
QOM	50–60	120/240	2
QOM-VH	50–60	120/240	2

FIGURE 1-10 Time-current chart for circuit breakers. (*Courtesy of Square D/Schneider Electric*)

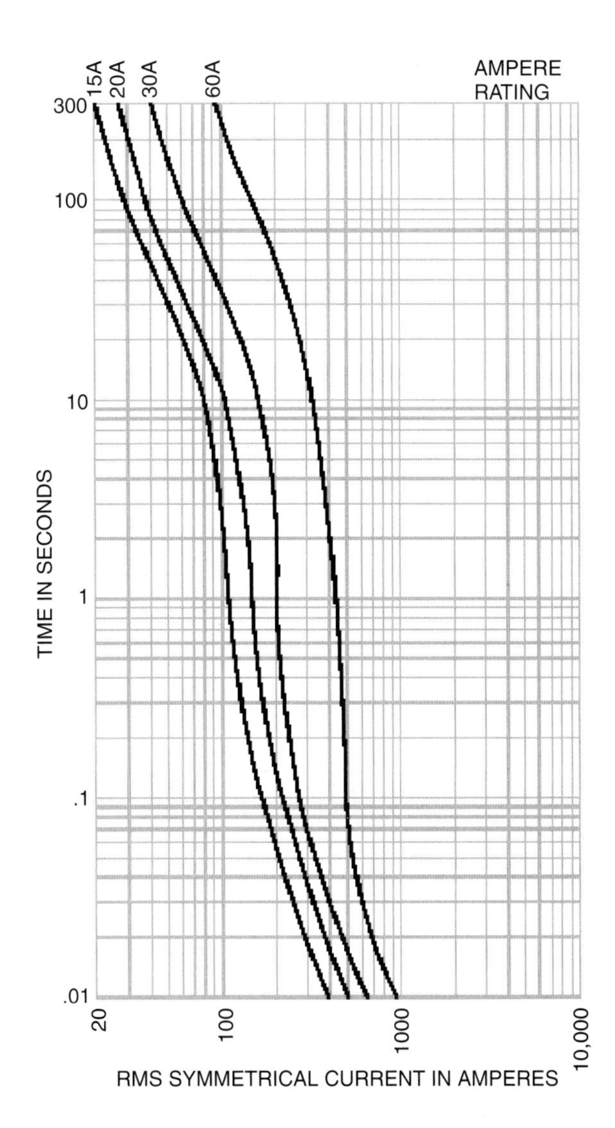

FIGURE 1-11 Time-current chart for fuses. *(Courtesy of Cooper Bussmann)*

neither section states the purpose for the requirement. This lack of explanation is not unusual for construction of *NEC* rules. The scope of the *NEC* is to provide an electrical installation that is "essentially free from electrical hazards." Most of the requirements in the *NEC* state the rule but do not give the reason behind the rule. The reason behind the requirement in the *NEC* can usually be found in the two documents published during the processing of each edition of the *NEC.* These documents are the *Report on Proposals* and the *Report on Comments.* These documents are available from NFPA and can also be downloaded in Portable Document Format (PDF) from NFPA's Web site at http://www.nfpa.org.

NEC 250.134(B) is in *Part VII, Methods of Equipment Grounding,* and contains a requirement that when an equipment grounding conductor is installed, it must be contained within the same raceway or cable or must be otherwise run with the circuit conductors. *NEC 300.3(B)* has similar requirements for the installation of all conductors of a circuit. It states, "All conductors of the same circuit and, where used, the grounded conductor and all equipment grounding conductors and bonding conductors shall be contained within the same raceway, auxiliary gutter, cable tray, cablebus assembly, trench, cable, or cord, unless otherwise permitted in accordance with *300.3(B)(1)* through *(4)*." Two exceptions to this rule are found following *250.134(B).*

As indicated in these rules, it is critical for ac circuits that all of the circuit conductors, including neutral and equipment grounding conductors, be installed in close proximity to each other. This keeps the impedance as low as possible. Furthermore, because impedance is the total opposition to current flow for an ac circuit (as resistance is the opposition to current flow in dc circuits), it is imperative that the impedance be as low as possible so enough current will flow in the circuit to cause the overcurrent device to clear the fault rapidly.

A study of the effect of circuit conductor arrangement and its relationship to the impedance of a circuit was performed many years ago by R. K. Kaufmann, an electrical engineer with General Electric. The results of his findings were presented in a "white paper" at the American Institute of Electrical Engineering (AIEE) (now IEEE) Summer and Pacific General Meeting (held in Los Angeles, California, on June 24 and 25, 1954).[2] The paper is reproduced in Appendix A of this book. The results of the test proved conclusively that the lowest-impedance path for ac circuits is achieved when the equipment grounding conductor either is the metal conduit itself or is contained within the conduit. Locating the

[2]The AIEE was the predecessor of the Institute of Electrical and Electronic Engineers (IEEE).

equipment grounding conductor only 1 ft (304 mm) away from the supply conductor significantly increases the impedance of the circuit.

3. The conductor must be capable of safely carrying the maximum ground-fault current likely to be imposed on it. That is, the conductor must be large enough to carry the current without overheating excessively or melting. Overheating of equipment grounding conductors or bonding conductors to the point of damaging circuit conductors would be a failure in circuit design. See the discussion at *250.122* in this book on sizing equipment grounding conductors based on the short-time heating characteristics of the conductor. Although this section does not give any specific size requirements for the conductor, several other sections of *Article 250* do. (We shall address those sizing requirements when we get to the applicable section.)

As stated in *250.4(A)(5)*, the earth is not permitted to be the only *(sole)* path for fault current. Where there is a grounding electrode connection at the service equipment and another grounding electrode connection at the utility transformer, it is inevitable that a parallel path for neutral current is established through the earth.

One of the laws of physics is that current will flow over every complete path or circuit that is available to it. Accordingly, if there is a return path from the service to the utility transformer over the neutral and the earth, current will flow over each path.

The laws of physics also dictate that the amount of current that will flow over the parallel paths is determined by the impedance of each path. Therefore, the path with the lowest impedance will carry the most current and the path with the highest impedance will carry the least current.

It usually can safely be assumed the neutral conductor will provide the lowest-impedance path and will carry the most current. Here is an important safety rule that should be emphasized: *Article 250* does not intend that the path through the earth be used to carry current. As discussed earlier, the earth (ground) connection is made for other purposes that are related to safety—but carrying fault current is not one of them.

250.4(B) Ungrounded Systems

Discussion: The requirements in this section deal with performance requirements for grounding and bonding enclosures in which the system is not grounded. These rules are not a contradiction in terms or requirements. Even though the system is not grounded, metallic enclosures for the system conductors must be grounded and bonded to protect against shock hazards and to provide a path for overvoltages to flow to earth.

As we shall see when we look at *250.21,* some electrical systems are not required to be grounded. The ungrounded systems are typically delta-connected, although some wye-connected systems are permitted to be ungrounded as well. Industry practice usually is to use delta-connected 3-phase, 3-wire systems when it is desired to have an ungrounded system.

250.4(B)(1) Grounding Electrical Equipment

The Requirement: Non-current-carrying conductive materials enclosing electrical conductors or equipment, or forming part of such equipment, are required to be connected to earth in a manner that will limit the voltage imposed by lightning or unintentional contact with higher-voltage lines and limit the voltage to ground on these materials.

Discussion: This requirement is identical in scope to that for grounded systems in *250.4(A)(1).* Please refer to that section for discussion of this topic.

250.4(B)(2) Bonding of Electrical Equipment

The Requirements: Non-current-carrying conductive materials enclosing electrical conductors or equipment, or forming part of such equipment, are required to be connected together and to the supply system grounded equipment in a manner that creates a low-impedance path for ground-fault current that is capable of carrying the maximum fault current likely to be imposed on it.

Discussion: Electrical systems that are not grounded have no direct connection to a grounding electrode.

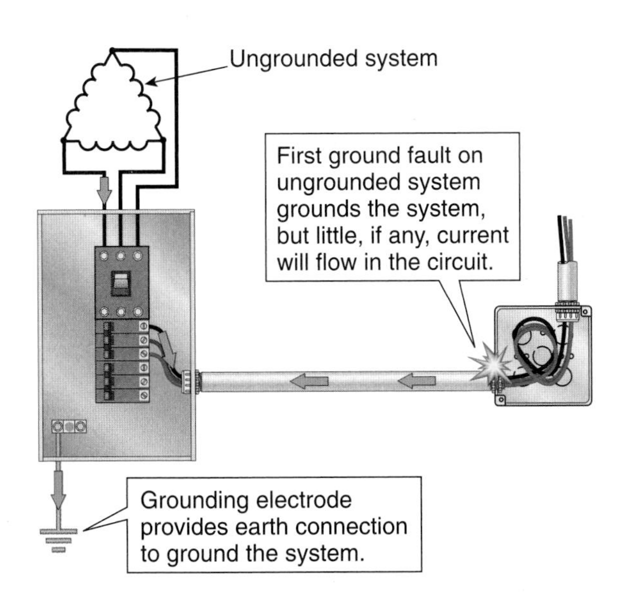

FIGURE 1-12 First ground fault on ungrounded system, *250.4(B)(2)*. (*© Cengage Learning 2012*)

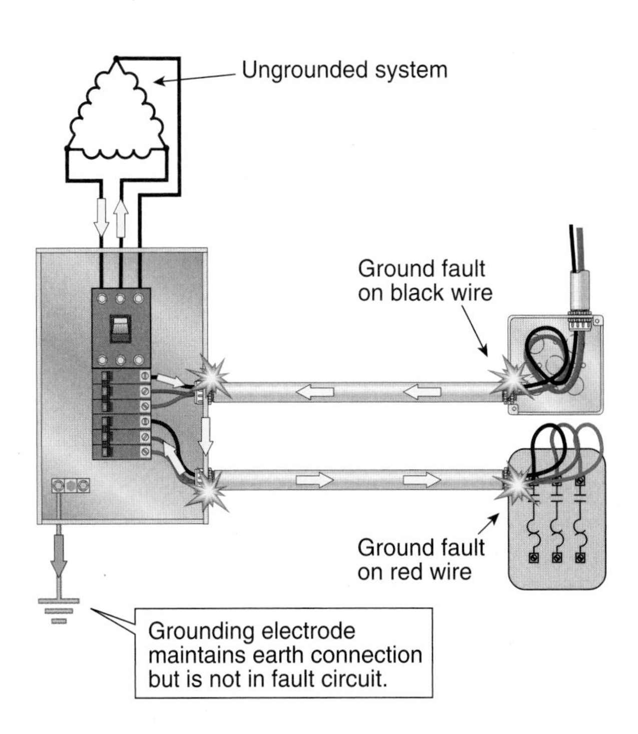

FIGURE 1-13 Two ground faults on different phases of ungrounded system, *250.4(B)(2)*. (*© Cengage Learning 2012*)

However, these systems are capacitively coupled to ground if the conductors are installed in metal raceways and the raceways are grounded. This charging voltage establishes a voltage reference to ground, but little current flows in the first ground fault. As illustrated in Figure 1-12, the first ground fault on an ungrounded system simply grounds the system. If the system is from a 3-phase delta-connected transformer bank, it becomes a corner-grounded 3-phase system if the faulted conductor is solidly connected at the point of the fault. This can occur when the insulation of a conductor is damaged and is solidly connected to a conductive path such as a metal conduit, metal cable armor, or metal enclosure. These first faults on an ungrounded system are often intermittent or sputtering faults rather than solid or bolted faults.

As illustrated in Figure 1-13, the second ground fault on an ungrounded system, if the fault is on a different phase from the first fault, becomes a phase-to-phase fault. In this situation, a lot of current can flow through the one or more paths between the faults. The fault flows over the equipment grounding conductor path provided by the wiring method, as well as through enclosures such as panelboards, wireways, and conduit. Every conduit locknut and bushing connection, as well as every coupling and connector, becomes a current-carrying connection. Any loose connection can cause arcing and sparking as current tries to flow through the connection. Every connection must be made wrenchtight to prevent arcing and sparking of components. A good connection also is necessary to ensure the integrity and performance of the fault return path. Arcing and sparking from loose connections can throw sparks that are sufficient to ignite combustible objects in the vicinity. One or more overcurrent devices open, provided that the overcurrent devices and equipment grounding conductors are sized properly.

250.4(B)(3) Bonding of Electrically Conductive Materials and Other Equipment

The Requirements: Electrically conductive materials that are likely to become energized are required to be connected together and to the supply system grounded equipment in a manner that creates a low-impedance path for ground-fault current that is capable of carrying the maximum fault current likely to be imposed on it.

Discussion: Providing a low-impedance path for fault current is just as important for ungrounded systems as it is for grounded systems. As previously mentioned, the first ground fault on an ungrounded system, if it is a solid connection, simply grounds the system, but little if any current flows in this path because there is no circuit established. The second ground fault, if on another phase conductor, establishes a phase-to-phase fault. Current flows in this fault as determined by the available fault current and the impedance of the fault path. See the illustration and discussion of this subject in *250.4(B)(2)*.

250.4(B)(4) Path for Fault Current

The Requirement: Electrical equipment, wiring, and other electrically conductive material likely to become energized are required to be installed in a manner that creates a low-impedance circuit from any point on the wiring system to the electrical supply source to facilitate the operation of overcurrent devices should a second fault occur on the wiring system. The earth is not permitted to be used as the sole or only equipment grounding conductor and is not considered to be an effective fault-current path.

Discussion: The requirements for providing a low-impedance fault return path in this section parallel those for grounded systems in *250.4(A)(5)*. See the discussions in that section. This section includes an *Informational Note* to clarify that a second fault that occurs through the equipment enclosures and bonding means or conductors is considered a ground fault.

250.6 OBJECTIONABLE CURRENT

Discussion: The requirements in this section have been subject to various interpretations because the *NEC* does not identify or define what is meant by *objectionable current.* Grounding electrode conductors and equipment grounding conductors are not intended to carry current in normal electrical system operation. As shown in Figure 1-14, *objectionable current* occurs when the system neutral is grounded improperly downstream from the service equipment. This allows normal system neutral current to flow on the equipment grounding or bonding conductor, which would be considered *objectionable current.*

In some cases, these grounding conductors are sized according to their short-time current rating rather than their continuous-current rating. For

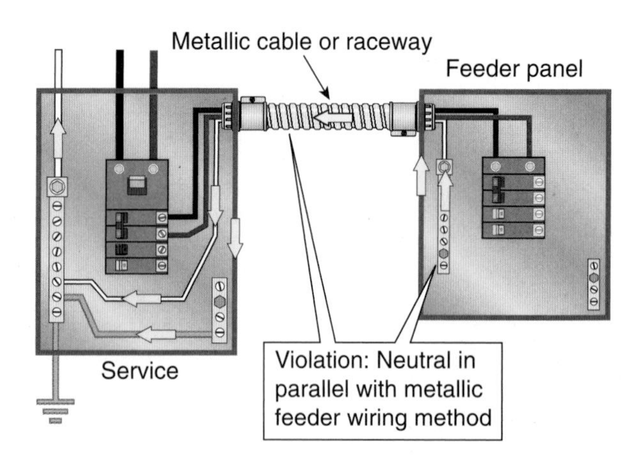

FIGURE 1-14 Objectionable current over grounding conductors from improper neutral connection, *250.6.* (*© Cengage Learning 2012*)

example, *Table 250.122* requires an equipment grounding conductor not smaller than 2 AWG copper be installed for a circuit having a 500-ampere overcurrent device. A 2-AWG copper conductor with 75°C-rated insulation has an allowable ampacity of 115 amperes in *Table 310.15(B)(16)*. The allowable ampacity of the conductor represents 23% of the rating of the overcurrent device. It therefore follows that the construction of *Table 250.122* intends that the overcurrent device will open the faulted circuit before the temperature of the conductor exceeds the thermal damage level of the insulation. This equipment grounding conductor is not intended to carry current on a normal or long-time basis.

250.6(A) Arrangement to Prevent Objectionable Current

The Requirement: The grounding of electrical systems, circuit conductors, surge arresters, surge-protective devices, and conductive normally non-current-carrying metal parts of equipment is required to be installed and arranged in a manner that will prevent objectionable current.

Discussion: As mentioned earlier, the *NEC* is silent on how much *objectionable* or *unintended* current flowing over grounding conductors is too much. If the system neutral is grounded again beyond the service or source of a separately derived system, then neutral current flows over the equipment grounding

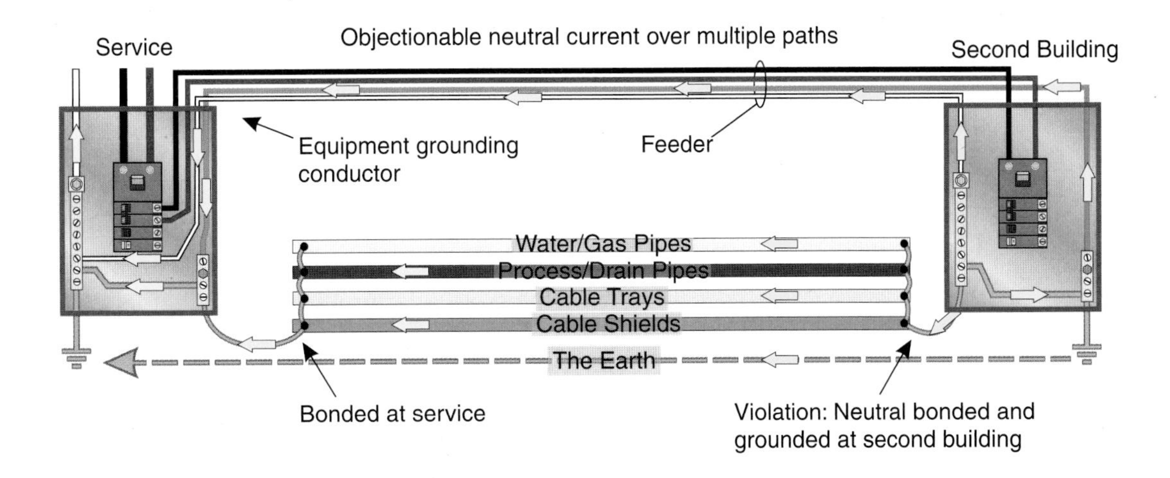

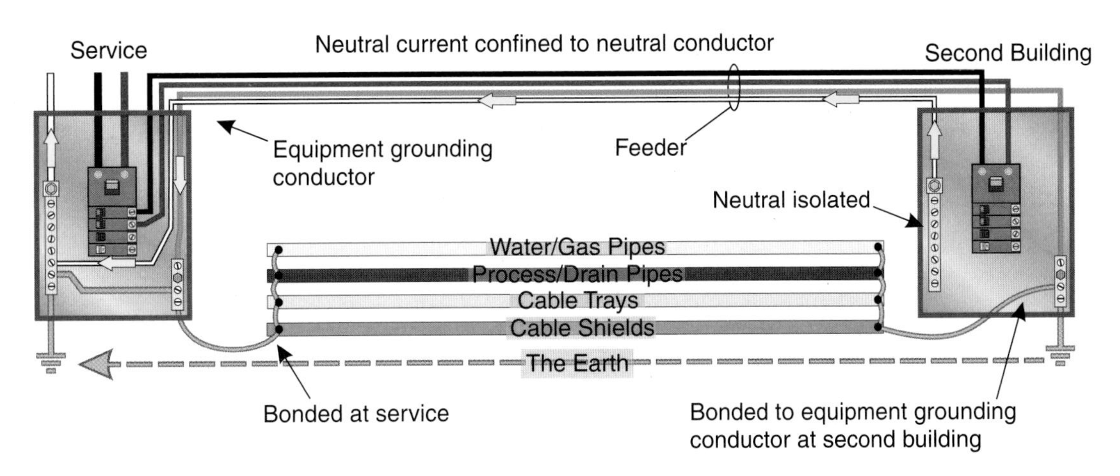

FIGURE 1-15 Arrangement to prevent objectionable current, *250.6(A)*. (*© Cengage Learning 2012*)

conductors and over any other path that is in parallel with the neutral conductor, as shown in Figure 1-15. These other or parallel paths include metal piping systems such as water, sprinkler, gas, or process piping and cable tray or metallic structural members. The equipment grounding conductor(s) are intended to carry current only when a ground fault is present on the system. Neutral current flowing on the equipment grounding conductor(s) should be considered *objectionable.*

Another source of *objectionable current* can be from grounding the system neutral conductor at several places on the premises. This places the earth in parallel with the neutral conductor. Although the earth presents a high resistance and impedance path compared with the neutral conductor, some neutral current flows in this path through the earth just because the path exists. As discussed earlier in the Introduction to Grounding and Bonding in this book, the laws of physics require that current flows through every complete circuit or path that is available. The grounding electrode conductors carry the neutral current to the grounding electrodes for this neutral current that flows through the earth. This neutral current flowing over grounding electrodes can be considered *objectionable* when it becomes excessive.

250.6(B) Alterations to Stop Objectionable Current

The Requirements: If the use of multiple grounding connections results in objectionable current, this section permits one or more of the following alterations to be made, provided that the requirements of *250.4(A)(5)* or *250.4(B)(4)* are met:

1. Discontinue one or more but not all of such grounding connections.

2. Change the locations of the grounding connections.
3. Interrupt the continuity of the conductor or conductive path causing the objectionable current.
4. Take other suitable remedial and approved action.

Discussion: As shown in Figure 1-16, neutral current can flow over equipment grounding conductors to grounding electrodes and back to the source even though the neutral conductor is not connected to earth on the load side of the service. Also, as can be seen in the figure, a parallel path is created every time another earth or grounding electrode connection is made. Because current will flow in every path or circuit that is available, neutral current will flow to and through the earth in multiple circuits.

This figure is simplified from what really happens in electrical systems. It is common for the electric utility to connect its primary neutral conductor to one or more grounding electrodes at every transformer and to connect a bonding jumper from primary to secondary at its transformer. Thus, the primary neutral connects together all the grounding electrodes in the neighborhood. So, the neutral current can flow from your installation through the earth to the transformer at the neighbors and from there back to the windings of the transformer that supplies your service.

These scenarios demonstrate that current flow, if significant enough, can be considered *objectionable current.*

As can be seen, any alterations to the grounding connections in an effort to reduce the current that flows over the grounding conductors cannot reduce the *effective ground-fault return path* required by *250.4(A)(5)* for grounded systems or *250.4(B)(4)* for ungrounded systems. Therefore, connections can be eliminated or changed so long as the effectiveness of the path itself is not reduced. Corrections to reduce the undesired current flow should include the following:

1. Do not reground the neutral at buildings or structures supplied by a feeder from the service equipment. Regrounding the neutral at buildings or structures on the premises is permitted to continue by *250.32(B)(1), Exception,* only for existing installations that were in compliance with the *Code* rules when installed. Regrounding the neutral at additional buildings or structures causes neutral current to flow through the earth. Although the neutral current through the earth path will be limited by the resistance of the grounding electrode system, as well as the resistance and impedance of the path through the earth, neutral current through the earth is

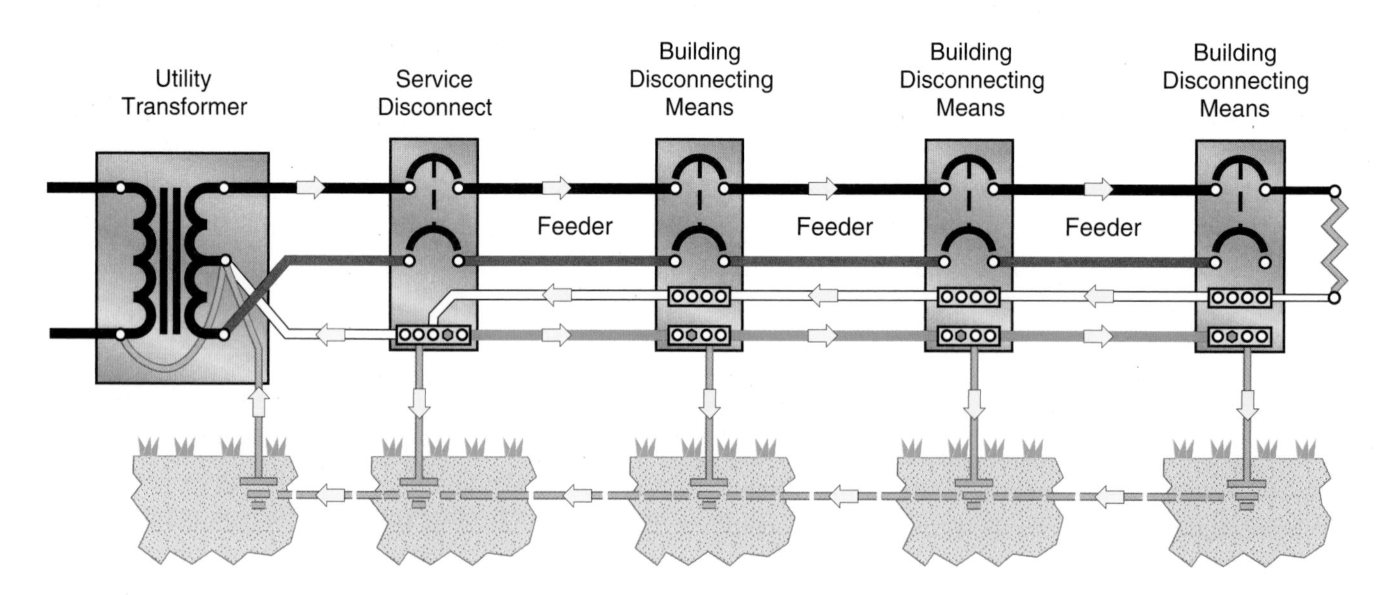

FIGURE 1-16 Parallel path for neutral current created by multiple grounding electrode connections. (*© Cengage Learning 2012*)

enhanced by the usually low earth connections at the electric utility's transformers and poles. The electric utility grounds the system neutral conductor for multigrounded neutral systems not less than 4 times per mile and at each transformer connection. The primary neutral bonds all these grounding electrodes together. This creates an extensive grounding electrode system of very low earth-connection resistance. As a result, it is unavoidable that some neutral current will be carried by the earth.

To avoid regrounding the neutral, install an equipment grounding conductor with the feeder to the building or structure, and ground the equipment grounding conductor at the building or structure disconnecting means.

2. Do not reground the neutral in a feeder panelboard supplied from the service in the same building. An exception for which grounding is required is where a new system is created by a transformer or generator or another separately derived system.
3. Other solutions can be approved by the AHJ. The AHJ may require that the solution be developed under engineering supervision.

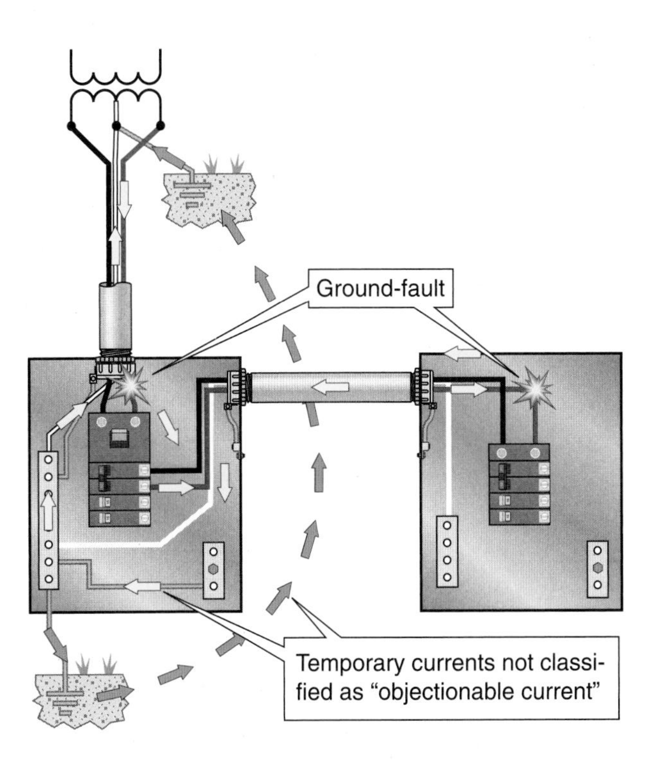

FIGURE 1-17 Temporary currents not classified as objectionable currents, *250.6(C)*. (*© Cengage Learning 2012*)

250.6(C) Temporary Currents Not Classified as Objectionable Currents

The Requirements: Temporary currents resulting from accidental conditions, such as ground-fault currents, that occur only while the grounding conductors are performing their intended protective functions are not to be classified as objectionable current for the purposes specified in *250.6(A)* and *250.6(B)*.

Discussion: For a grounded system, the first ground fault, regardless of whether it occurs on the line side or the load side of the service, causes current to flow over metal conduits, metal tubing, and metal-clad cables (see Figure 1-17). To complete a circuit, the current flows to the main bonding jumper in the service or the system bonding jumper at the separately derived system and returns to the source. This current that flows until an overcurrent device opens the circuit is not classified as the *objectionable* current mentioned in this section. For ungrounded systems, the first ground fault does not cause current to flow but simply grounds the system so long as a solid connection is established. As discussed earlier, the second ground fault on the previously ungrounded system, if it occurs on a different phase and is another solid or arcing connection, causes current to flow between the points of the faults over the metallic paths that are available. Like current flowing in the grounded system, the current that flows on the metallic parts of the ungrounded system is not classified as the *objectionable* current mentioned in this section.

250.6(D) Limitations to Permissible Alterations

The Requirements: The provisions of this section are not to be considered as permitting electronic equipment from being operated on ac systems or branch circuits that are not connected to an equipment grounding conductor as required by *Article 250*. Currents that introduce noise or data errors in electronic equipment are not to be considered the *objectionable currents* addressed in this section.

Discussion: It has often been reported that some manufacturers of data processing equipment have instructed electricians to avoid connecting the equipment grounding conductor from the branch circuit to their equipment. Rather, these manufacturers have instructed the electricians to install a local grounding electrode such as a ground rod in the vicinity of the data processing equipment and then connect the equipment to the grounding electrode. This type of installation is, of course, a violation of *NEC 250.50, 250.110,* and *250.134* and can present a serious shock hazard as well. Connecting a remote grounding electrode in this manner creates a ground-fault return path through the earth, which represents both a high-resistance and a high-impedance path. Owing to the resistance of the grounding electrode connections to the earth and the resistance of the earth itself, it is nearly impossible to establish a connection in this manner that will carry enough current (at 600 volts and below) to clear even a 15-ampere fuse or circuit breaker.

Some manufacturers of data processing equipment refer to the conduit or branch-circuit ground as a "dirty ground" because of the electrical interference or noise that travels on this path. It also is sometimes referred to as the "safety ground." These manufacturers often request a "clean ground" for their equipment. See the discussion in *250.146(D).* The correct method of installing a local grounding electrode for data processing equipment is shown in Figure 1-18. This local grounding electrode, if installed to supplement the equipment grounding conductor in the branch circuit, is considered an auxiliary grounding electrode as provided in *250.54.*

- Install equipment grounding conductor with branch circuit conductors.
- Local auxiliary grounding electrode permitted to supplement but not replace equipment grounding conductor.

FIGURE 1-18 Limitations to permissible alterations for data-processing systems, *250.6(D).* (*© Cengage Learning 2012*)

250.8 CONNECTION OF GROUNDING AND BONDING EQUIPMENT

250.8(A) Permitted Methods

The Requirement: Equipment grounding conductors, grounding electrode conductors, and bonding jumpers must be connected by one of the following means:

1. Listed pressure connectors
2. Terminal bars
3. Pressure connectors listed as grounding and bonding equipment
4. Exothermic welding process
5. Machine screw-type fasteners that engage not less than two threads or are secured with a nut
6. Thread-forming machine screws that engage not less than two threads in the enclosure
7. Connections that are part of a listed assembly
8. Other listed means

Discussion: This section provides general as well as specific requirements regarding connection of equipment grounding conductors, grounding electrode conductors, and bonding jumpers. Most of the connectors used to make these connections are listed by a recognized electrical products testing laboratory. Exothermic welding is not required to be listed as it is a field applied process and the quality of the connections is determined by several variables, including using proper graphite forms and ensuring the conductors to be connected are clean and dry.

Listed pressure connectors, listed clamps, or other listed means are permitted or required to be used for making connections of grounding conductors. A sample of these connectors is shown in Figure 1-19. These connectors are specifically designed for either copper or aluminum conductors. Many specific requirements are found in *NEC 110.14* for making electrical connections. These requirements can be summarized as follows:

- Connectors must be identified for the material of the conductor, such as "CU" for copper and

1. Listed pressure connectors

2. Terminal bars

3. Pressure connectors listed as grounding and bonding equipment

4. Exothermic welding process

5. Machine screw-type fasteners that engage not less than two threads or are secured with a nut

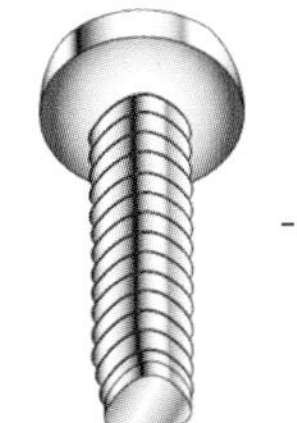
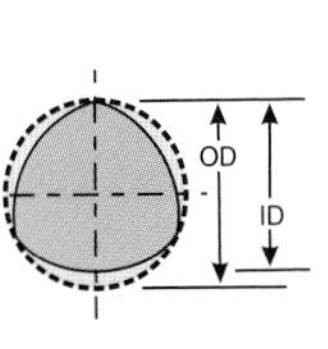

6. Thread-forming screws that engage not less than two threads in the enclosure

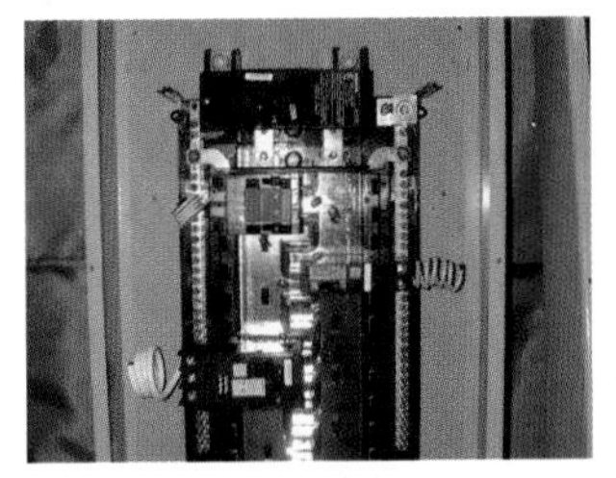
7. Connections that are a part of a listed assembly

8. Other listed means

FIGURE 1-19 Methods permitted for connection of grounding and bonding equipment. (*Photos 1, 2, 3, 5, and 8 courtesy of Thomas and Betts. Photos 4, 6, and 7 © Cengage Learning 2012*)

"AL" for aluminum. Many connectors are dual rated and marked "AL/CU."

- Conductors of dissimilar metal are not to be intermixed in a terminal or splicing connector so that physical contact occurs between dissimilar conductors, such as between copper and aluminum conductors, unless the device is identified for such use. It is important to carefully follow the manufacturer's installation instructions. Some connectors are specifically listed for connecting copper and aluminum conductors when physical contact occurs between the conductors only in dry locations. Other connectors permit making connection of dissimilar materials only in a separate part of the connector such as a multibarrel lug or splicer. Split-bolt connectors used for connecting copper to aluminum wires have a spacer that separates the conductors.
- Materials such as solder, fluxes, inhibitors, and compounds are required to be suitable for the use.
- Terminals are suitable for connecting only one conductor unless otherwise identified.
- Terminals for connecting aluminum are so identified. Connectors without a marking are suitable for connecting copper only.

Most pressure wire connectors in use today are of the set-screw type, wire-binding screw type, or compression or twist-on type. In every case, the appropriate type of connector for the size and conductor material must be selected to ensure a safe and proper connection.

Sheet metal screws are not permitted to be used to connect grounding conductors or connection devices to enclosures. Generally, machine screw threads are required for electrical connections. Both standard and fine threads are used. When a screw is used in connecting a terminal to an enclosure such as to a box, the product safety standard generally requires not less than two full threads to be engaged. Accordingly, fine-thread machine screws such as 10-32 often are used for ground screws used in outlet boxes.

As shown in Figure 1-19, exothermic welding involves the use of a form (made from graphite to withstand the high temperatures involved) that is designed for the specific size of wire or busbar to be welded, as well as for the material to which the connection is to be made, such as a ground rod, ground wire, reinforcing steel, or structural steel column. It is very important that all manufacturer's instructions be carefully followed. These instructions typically include using the proper form or mold, being certain the materials to be welded are clean and dry, and using the correct amount of welding compound. The material reaches a temperature of approximately 4000°F (2204°C) when it is ignited. This high temperature fuses the materials to be joined together. Proper safety procedures must be followed, such as wearing safety glasses and wearing gloves before the compound is ignited.

Exothermic-welded connections are not required to be listed or labeled because these connections involve a field process. Good workmanship is critical to ensure a proper connection, particularly being certain that the proper form is used and that the conductors are clean and dry. Some engineering specifications require these completed welded connections to be examined by the use of X-ray equipment. To ensure that a good-quality connection has been made, less critical connections should be inspected visually for excessive voids and sags, and be struck with a hammer to make certain the components are securely joined.

Compression wire connectors are designed for either copper or aluminum and are marked either on the smallest shipping carton or on the connector itself. Some connectors that are designed for aluminum wire termination come prefilled with an inhibitor compound. Manufacturers of other connectors may recommend or require the use of an inhibitor compound for aluminum splices or terminations. A sample of a compression tool and connectors is shown in Figure 1-20. All compression connectors are designed for connecting a specific range of wire sizes. Some accept a wide range of conductor sizes, whereas others have a limited range. The importance of selecting the appropriate size of connector cannot be overemphasized.

Of equal importance is selecting the proper compression tool. As a rule, the compression tool and the connector must be from the same manufacturer. Both mechanical and hydraulic tools are available.

Compression Wire Connectors

FIGURE 1-20 Compression wire connectors and compression tool. (*Courtesy of Thomas and Betts*)

Some manufacturers use a color-coded system with exchangeable dies. For example, a connector with a red dot requires that a die with a corresponding red mark be placed in the compression tool.

Specific requirements can be found in *250.70* for connection of grounding electrode conductors to grounding electrodes.

250.8(B) Methods Not Permitted

The Requirement: Connection devices or fittings that depend solely on solder are not permitted to be used.

Discussion: Although not generally used today, solder connections are permitted where the conductors are held together mechanically and then soldered. Smaller wires can be twisted together to make the secure connection and then soldered. Because of the extra time and labor costs involved as well as the difficulty of taking connections apart for circuit testing or modification, soldering has fallen out of favor in the electrical industry.

250.12 CLEAN SURFACES

The Requirements: Nonconductive coatings (such as paint, lacquer, and enamel) on equipment to be grounded are required to be removed from threads and other contact surfaces to provide good electrical continuity or be connected by means or fittings designed so as to make such removal unnecessary.

Discussion: Paint used on electrical enclosures can be thought of as an insulator. Metal enclosures that are commonly painted include switchboards, panelboards, motor control centers, motor controllers and operators, wireways, auxiliary gutters, and pull and junction boxes. When conduit, electrical metallic tubing, or metal cables are connected to these enclosures, the paint should be removed, at least inside, to ensure a good electrical connection.

This section includes the phrase *or be connected by means or fittings designed so as to make such removal (paint or other insulating coatings) unnecessary.* Underwriters Laboratories has verified that it has not evaluated a connector or fitting for such a

FIGURE 1-21 Locknuts for connection of conduit or fittings to enclosures. (*© Cengage Learning 2012*)

purpose. Therefore, every connection in the ground return path—be it locknuts, bushings, connectors, couplings, or fittings—should be looked at with the following question in mind: "Have I made a good, solid connection that is capable of carrying fault current?"

As shown in Figure 1-21, locknuts generally are made in one of two ways:

1. They may be cut (punched) from flat sheet steel, bent into a slight crown, punched or drilled for the size of pipe thread, and threaded, with the tabs (or ears) bent and left with a fairly sharp edge that contacts the enclosure when tightened.
2. They may be cast from iron, aluminum, zinc, or other material. These cast locknuts may have a fairly flat contact surface against the enclosure or a sharper tang.

When a locknut is used to make a connection to a painted enclosure, be certain it has scraped the paint while being tightened to make a good electrical connection. If a good electrical connection has not been made, the grounding and bonding path is interrupted, which may leave equipment at a shock hazard if a ground fault occurs downstream from the isolated connection. If a low-impedance and secure connection cannot be verified visually, a fitting should be installed such as a bonding bushing and bonding jumper. If no eccentric or concentric rings exist, a bonding-type locknut can be used.

Review the information in Appendix D, Report of Research on Conduit Fitting Ground-Fault Current Withstand Capability, on testing conduit fittings. A portion of the report concludes failures of several connections were the result of the locknut not penetrating the enclosure paint.

REVIEW QUESTIONS

1. Which of the following *NEC* sections provides the requirement for the concept that the requirements for grounding in *Chapter 5* through *Chapter 7* can modify or supplement the rules in *Chapter 1* through *Chapter 4*?

   a. *517.17*

   b. *250.118*

   c. *250.3*

   d. *90.3*

2. Which of the following statements is correct about the style in which the *NEC* is written?

   a. It is generally written in "performance" style that states the safety goals to be achieved.

   b. It is generally written in a "legal" style so that the installer will be more likely to comply with the requirements.

   c. It is generally written in "prescriptive" style that states how the installation must be made.

   d. It is written in neither "performance" nor "prescriptive" style.

3. Which of the following statements about the reasons for grounding an electrical system is *not* true?

   a. To limit the overvoltage imposed by lightning

   b. To limit the overvoltage from contact with utility high-voltage lines

   c. To provide a low-impedance ground-fault return path

   d. To stabilize the voltage of the system to earth during normal operation

4. Which of the following statements about bonding is *not* true?

   a. Bonding completes a path for fault current to flow through.

   b. Bonding conductive parts together keeps the conductive materials at the same electrical potential.

   c. Bonding's primary purpose is to make an earth connection.

   d. Bonding has the effect of eliminating any shock hazard between bonded enclosures or parts.

5. Which of the following statements about providing an *effective ground-fault current path* is *not* true?

   a. It must be a reliable rather than a temporary path.

   b. It must provide a low-impedance circuit.

   c. It must be capable of safely carrying the maximum ground-fault current likely to be imposed.

   d. It permits the earth to be considered an effective ground-fault current path.

6. Which of the following is *not* a purpose for grounding equipment associated with an ungrounded system?
   a. Even though the system is not grounded, metallic enclosures for the system conductors must be grounded and bonded to protect against shock hazards.
   b. If the system is not required to be grounded, there is no requirement for the equipment associated with the ungrounded system to be grounded.
   c. Even though the system is not grounded, grounding the enclosure provides a path for overvoltages to flow to the earth.
   d. Even though the system is not grounded, bonding the enclosures together provides a path for fault current to flow over.
7. Which of the following statements does *not* describe *objectionable current*?
   a. Neutral current flowing over the neutral grounding conductor
   b. Neutral current flowing over the equipment grounding conductor
   c. Neutral current flowing over cable shields
   d. Neutral current flowing over metal conduit
8. Which of the following connection methods is *not* permitted to be used to connect grounding conductors to enclosures or equipment?
   a. Exothermic welding
   b. Listed pressure connectors
   c. Listed clamps
   d. Sheet metal screws
9. Which of the following statements about wire connectors or wire connections is *not* correct?
   a. The marking "AL/CU" on the connector means that it is suitable for both copper and aluminum wire connections.
   b. The marking "AL/CU" on the connector means that it is acceptable to connect copper and aluminum in direct contact.
   c. A wire connector that is not marked for the type of conductor material is suitable for only copper connections.
   d. Connectors that are suitable for connecting more than one conductor are marked to indicate the suitability.
10. Which of the following statements about painted electrical enclosures is *not* true?
    a. Paint used on electrical enclosures can be thought of as an insulator.
    b. Where conduit, electrical metallic tubing, or metal cables are connected to these enclosures, the paint should be removed.
    c. Fittings and locknuts are readily available that are listed to make removal of paint unnecessary.
    d. Every connection in the ground return path should be evaluated to ensure it is a good, solid connection that is capable of carrying fault current.
11. The typical voltages for a 120/240 volt, 3-phase, 4-wire delta-connected system are:
    a. 120 volts phase A to phase B
    b. 208 volts phase B to ground

c. 208 volts phase B to phase C

d. 240 volts phase C to ground

12. A fault-return path of ________ is essential for proper action of the overcurrent device on the supply side.

a. low-impedance

b. low-resistance

c. low- inductance

d. low-capacitance

13. The curved portion of the circuit breaker time-current chart is best described as the:

a. one-cycle tripping time.

b. time-delay component.

c. instantaneous-trip component.

d. total clearing time.

14. Keeping all circuit conductors, including the equipment grounding conductor, in the same raceway, cable or similar enclosure is essential to keep the ________ as low as practicable:

a. impedance

b. resistance

c. inductance

d. capacitance

15. If more than one path exists for neutral current to flow, current will flow:

a. in only the path of lowest resistance.

b. in only the paths that are in series.

c. in all paths that are available.

d. only when there is a ground-fault.

16. Temporary currents flowing in the circuit until the overcurrent device opens are considered "objectionable" and are not permitted.

a. True

b. False

17. Due to data processing equipment being sensitive to electrical noise on the equipment grounding conductor, it is permitted to isolate the equipment from the "dirty ground" and connect the equipment to only a local ground rod.

a. True

b. False

18. A sheet metal enclosure must be not less than ________ thick if a machine screw type fastener of the 10-32 size is used to connect an equipment grounding conductor to the enclosure.

a. 1⁄32 in.

b. 1⁄16 in.

c. 3⁄32 in.

d. 1⁄8 in.

UNIT 2

System Grounding

OBJECTIVES

After studying this unit, the reader will understand

- systems that are required to be grounded, those that are permitted but not required to be grounded, and those circuits that are not permitted to be grounded.
- grounding requirements for grounded systems.
- installing and sizing grounded service and system conductors.
- system conductors that are required to be grounded.
- function and sizing of main and system bonding jumpers.
- grounding requirements for separately derived systems.
- grounding for two or more buildings supplied by a feeder or branch circuit.
- requirements for grounding portable and vehicle-mounted generators.
- requirements for high-impedance grounded neutral systems.

250.20 ALTERNATING-CURRENT CIRCUITS AND SYSTEMS TO BE GROUNDED

The Requirement: Alternating-current circuits and systems are required to be grounded (connected to earth) if the system meets any of the conditions in *250.20(A), (B), (C), or (D).*

Discussion: In considering whether the electrical systems are required to be grounded, the voltage supplied by the system is a major factor. In some cases, the requirement for grounding depends on whether the system is 3-phase and whether the neutral conductor is used as a system (current-carrying) conductor. Most circuits and systems that are not required to be grounded are permitted to be grounded. Only a few systems or circuits are not permitted to be grounded. When systems that are not required to be grounded are grounded, they must be grounded in accordance with the applicable provisions of *Article 250.*

> **Informational Note:** An example of a system permitted but not required to be grounded is a corner-grounded delta transformer connection. See *250.26(4)* for the conductor to be grounded.

Discussion: Although this *Informational Note* refers to a corner-grounded delta-connected transformer as a system permitted but not required to be grounded, such a system must meet the 150-volt test of *250.20(A)(1).* If the delta-connected transformer bank has 120-volt transformer connections, the system would be required to be grounded, because if one corner of the delta-connected system is grounded, the other phases of the delta connections would be less than 150 volts to ground. Many of these optionally grounded systems are 240-volt or 480-volt, 3-phase, 3-wire, delta-connected systems. As shown in Figure 2-1, if one transformer phase is grounded, the voltage to ground on the other phases would be greater than 150 volts, and thus this system would not be required to be grounded. See the discussion following *250.20(A)(1)* for additional information.

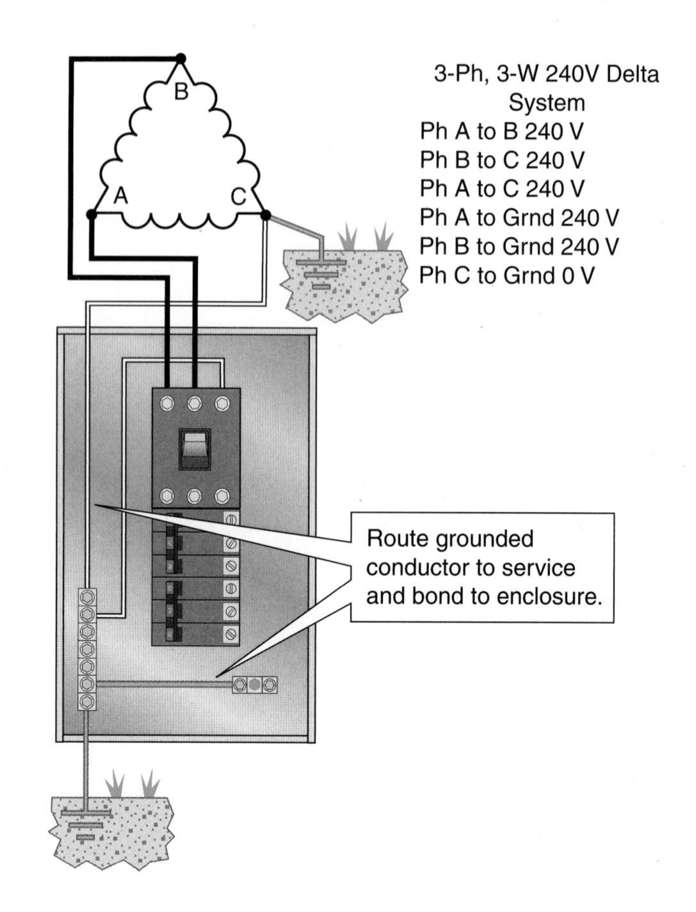

FIGURE 2-1 Systems permitted but not required to be grounded, *250.20 Informational Note.* (*© Cengage Learning 2012*)

250.20(A) Alternating-Current Systems of Less Than 50 Volts

The Requirements: Alternating-current systems of less than 50 volts are required to be grounded under any of the following conditions:

1. Where supplied by transformers, if the transformer supply system exceeds 150 volts to ground.
2. Where supplied by transformers, if the transformer supply system is ungrounded.
3. Where installed outside as overhead conductors.

Figure 2-2 illustrates these three conditions.

Discussion: *Voltage to ground* is defined in *Article 100.* The main principles can be summarized as follows: For grounded circuits, the voltage to ground is the difference in potential between the ungrounded conductor or portion of the circuit and the conductor

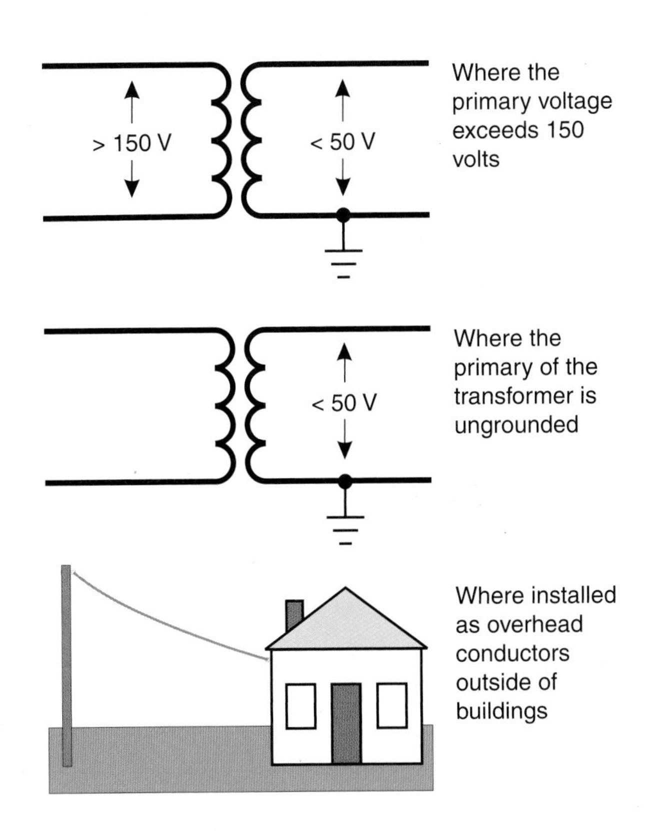

FIGURE 2-2 Alternating-current systems less than 50 volts required to be grounded, *250.20(A)*. (© *Cengage Learning 2012*)

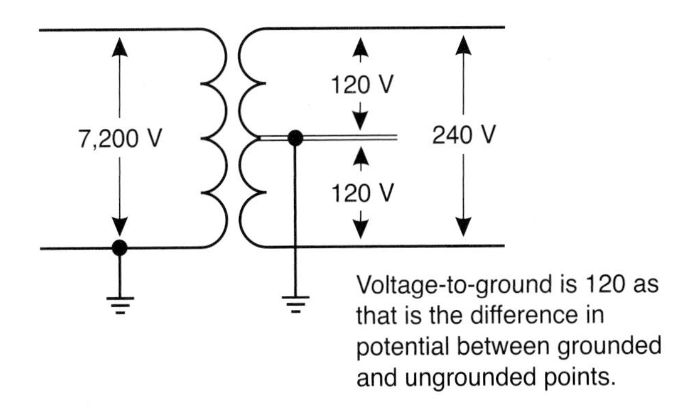

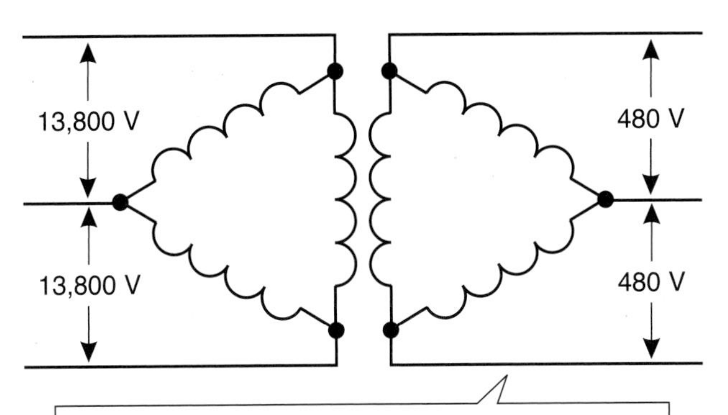

FIGURE 2-3 Definition of *Voltage to Ground, Article 100*. (© *Cengage Learning 2012*)

or portion of the circuit that is grounded. For ungrounded circuits, the voltage to ground is considered the greatest voltage between the given conductor and any other conductor of the circuit. This rule applies because if one of the conductors of the ungrounded system becomes grounded, the other side of the circuit or system has a potential to ground that equals the system voltage. See Figure 2-3.

250.20(B) Alternating-Current Systems of 50 Volts to 1000 Volts

The Requirement: Alternating-current systems of 50 volts to less than 1000 volts that supply premises wiring and premises wiring systems are required to be grounded under any of the following conditions:

Discussion: An understanding of the term *premises wiring systems* as it is defined in *Article 100* is essential because it determines the systems that are required to be grounded. As defined, *premises wiring* begins at the *service point* (another term defined in *Article 100*) and includes "interior and exterior wiring, including power, lighting, control, and signal circuit wiring, together with all of their associated hardware, fittings, and wiring devices, both permanently and temporarily installed."* Premises wiring includes the source of power from the electric utility (the service[s]) and other sources of power such as a battery, a solar photovoltaic system, or a generator, transformer, or converter windings. *Premises wiring* extends from the source of power to the outlet(s) and includes the service(s) and all feeders and branch circuits. The definition of *premises wiring* also includes the wiring from the source of power, such as a separately derived system to the outlets if there is no service point. Such wiring does not include wiring internal to appliances, luminaires, motors, controllers, motor control centers, and similar equipment.

*Reprinted with permission from NFPA 70-2011.

The Requirement: *250.20(B)(1).* The system is required to be grounded if it can be grounded so the maximum voltage to ground on the ungrounded conductors does not exceed 150 volts.

Discussion: When one of the *system* conductors is grounded, the other system conductors rise to a specific voltage above ground. For systems supplied by transformers, this voltage to ground is determined by the voltage supplied to the primary and the ratio of the primary to the secondary windings.

Common systems that meet the conditions of *250.20(B)(1)* and therefore must be grounded include the following:

120-volt, 1-phase, 2-wire systems
120/240-volt, 1-phase, 3-wire systems
208Y/120-volt, 3-phase, 4-wire systems

These systems are shown in Figure 2-4. It is uncommon for the electric utility to supply a 120-volt service. Almost all services in this class are 120/240-volt, 1 phase. This system consists of a transformer with a 240-volt secondary that is center tapped. This allows the occupant to use equipment rated for 240 volts as well as 120 volts. The system is single phase in spite of the multiple voltages that are available.

The 208Y/120 volt, 3-phase, 4-wire system consists of three 120-volt transformers that are connected together at one end. That point is the common point where the system is grounded. Of course, the primary is also 3-phase, with the phases 120 electrical degrees apart.

If you use a volt meter and measure from phase to phase, it will read approximately 208 volts. This can be determined mathematically as

$$120 \text{ volts} \times 1.732 = 207.84 \text{ volts}$$

In common usage, this is rounded up to 208. The multiplier 1.732 is used because it is the square root of 3 and this is a 3-phase system.

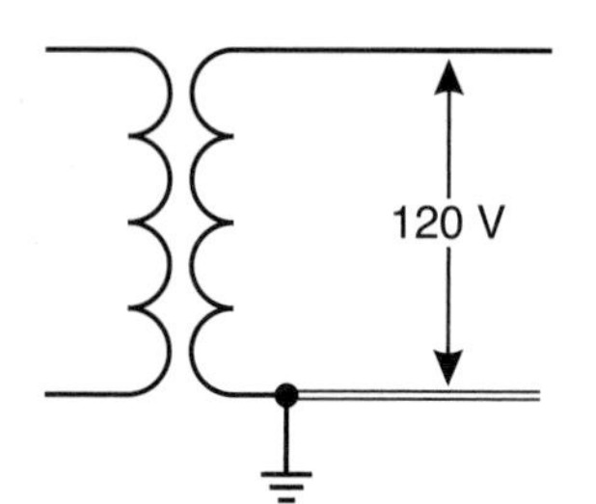

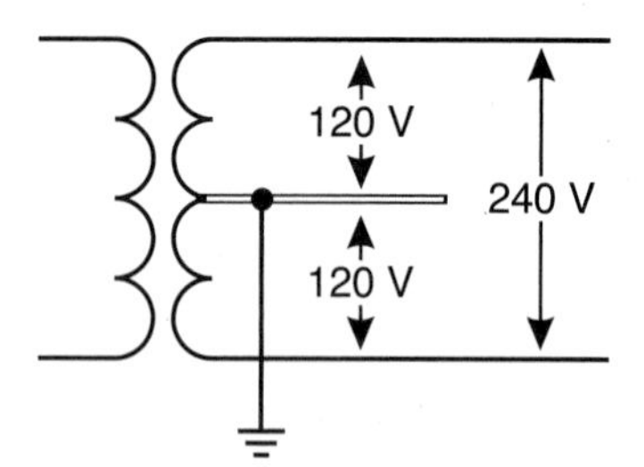

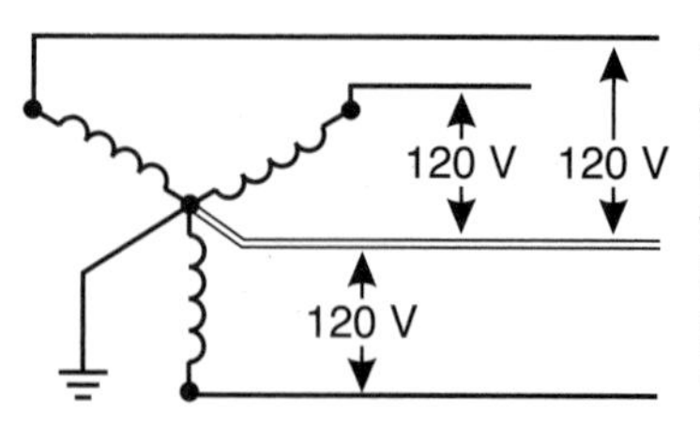

FIGURE 2-4 Systems 50 to 1000 volts required to be grounded, *250.20(B)(1).* *(© Cengage Learning 2012)*

The Requirement: *250.20(B)(2).* The system is required to be grounded if it is a 3-phase, 4-wire, wye-connected system in which the neutral is used as a circuit conductor.

Discussion: Three-phase, 4-wire, wye-connected systems are created by connecting three single-phase transformers or three windings in a "Y" (wye) configuration. The center points of the three transformers are connected together. As can be seen in Figure 2-5, this system can easily supply single-phase loads by connecting a load from the center terminal of the transformers to one of the outer terminals of a transformer. The center terminal is usually identified as "XO" and the ungrounded terminals are identified as "X1," "X2," or "X3." It can also supply a higher-voltage "single-phase" load by connecting the load between two of the outer terminals of the transformers. For 208Y/120-volt transformers, the voltage between phases is 208 volts, and for 480Y/277-volt transformers the voltage is 480 volts. The transformer bank can also supply 3-phase loads by connecting loads between each of the ungrounded terminals of the transformers.

Although not required by the *Code,* these wye-connected transformers usually are delta connected on the primary side. This configuration allows for

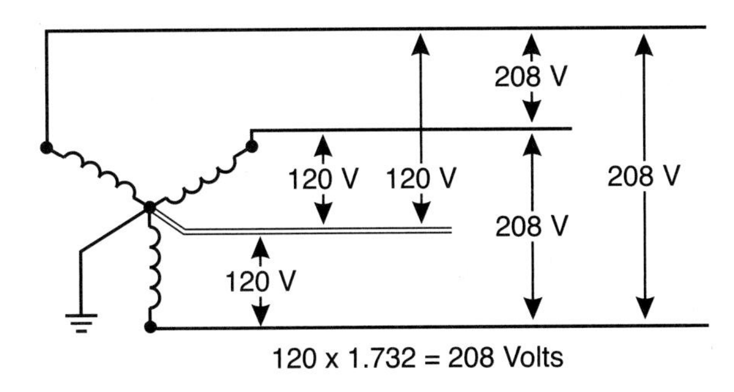

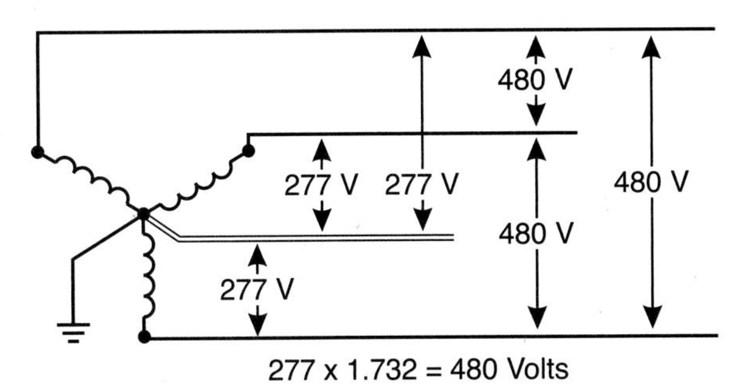

FIGURE 2-5 Wye-connected systems required to be grounded, *250.20(B)(2)*. (© *Cengage Learning 2012*)

efficient operation, with a 3-wire supply to the primary and a 4-wire secondary.

The wording *in which the neutral is used as a circuit conductor* indicates that a line-to-neutral load is supplied. For 480Y/277-volt systems, these loads would be rated at 277 volts, such as 277-volt lighting systems. Therefore, a system supplied by a 3-phase, wye-connected transformer where the neutral is not grounded is not required to have a neutral extended to a disconnect, such as a service disconnecting means, if there are no line-to-neutral loads supplied. (Recall the 150-volts-to-ground rules discussed under *250.20(B)(1)*. These 208Y/120-volt wye-connected systems are required to be grounded even though line-to-neutral loads are not supplied.)

Systems that meet the conditions of *250.20(B)(2)* and therefore must be grounded include the following:

208Y/120-volt, 3-phase, 4-wire

480Y/277-volt, 3-phase, 4-wire, in which the neutral is used as a circuit conductor

575Y/332-volt, 3-phase, 4-wire (used in some industrial plants; may not have a grounded system conductor when there are no line-to-neutral loads)

600Y/346-volt, 3-phase, 4-wire (used in some industrial plants; may not have a grounded system conductor when there are no line-to-neutral loads)

These systems are represented in Figure 2-5.

The Requirement: *250.20(B)(3)*. The system is required to be grounded if it is 3-phase, 4-wire, delta connected, in which the midpoint of one phase winding is used as a circuit conductor.

Discussion: These systems are commonly installed when it is desired or required to supply both 3-phase loads at the higher delta-connected system voltages and single-phase loads for lighting and/or receptacles. The common voltages are 120/240-volt, 3-phase, 4-wire. The midpoint of one of the transformers is tapped to create a portion of the system that is 120/240-volt, single-phase. As a result, the system is required to be grounded.

The *Code* refers to such a system as one that has a *conductor or busbar having the higher phase voltage to ground*. This conductor having the higher voltage to ground is required to be identified durably and permanently by an outer finish that is orange in color or by other effective means. This identification is required to be placed at each point on the system where a connection is made if the grounded conductor also is present. See the requirements for identification in *110.15*, for phase arrangement in *408.3(E)*, and for field marking requirements in *408.3(F)*.

The higher voltage to ground is approximately 208 volts. See Figure 2-6. The voltage of the

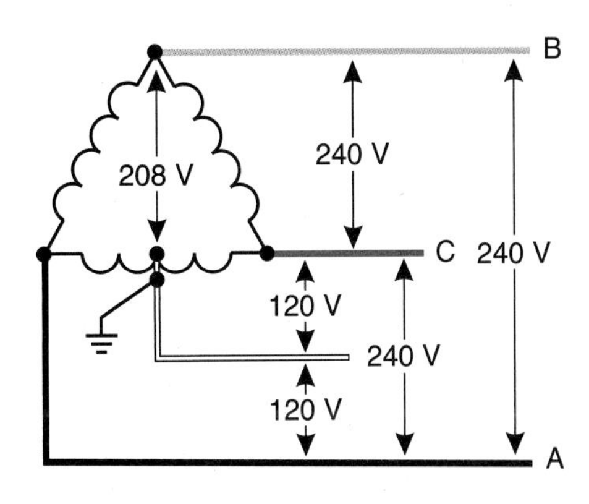

FIGURE 2-6 Delta-connected systems required to be grounded, *250.20(B)(3)*. (© *Cengage Learning 2012*)

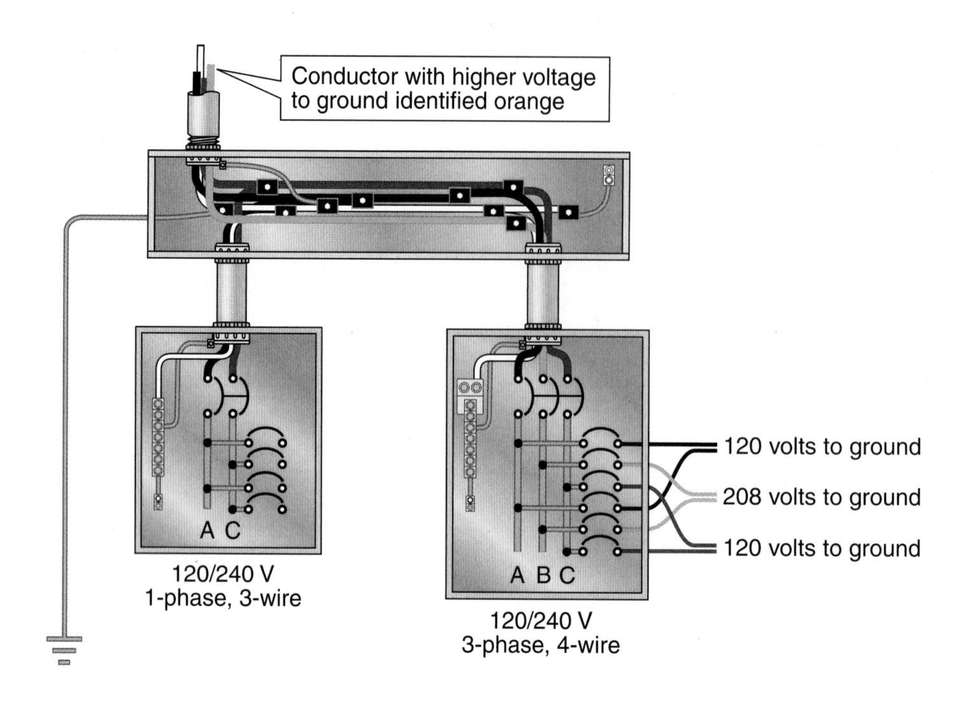

FIGURE 2-7 Conductor with higher voltage to ground, *250.20(B)(3)*. (*© Cengage Learning 2012*)

conductor with the higher voltage to ground is determined by the following formula:

$$\frac{1}{2} \text{ the system voltage} \times \sqrt{3}$$

For a 120/240-volt, 3-phase, 4-wire system, the voltage of the conductor with the higher voltage to ground is

$$(240/2) = 120 \times 1.732 = 207.8 \text{ volts}$$

As can be seen from Figure 2-7, it is important for electricians to be aware of these higher voltages when connecting branch circuits from panelboards supplied by 120/240-volt, 3-phase, 4-wire systems. For 208Y/120-volt systems, it is common to install a multiwire branch circuit with a conductor connected to phases A, B, and C and the neutral. On this circuit, each ungrounded conductor has a voltage to ground of 120 volts. If a similar set of conductors is installed from a 120/240-volt, 3-phase, 4-wire system, two of the ungrounded conductors—phase A and phase C—will have a voltage to ground of 120 volts, and the B phase conductor will have a voltage to ground of 208 volts. Obviously, this higher voltage can have disastrous consequences if the equipment supplied is intended to be supplied at 120 volts.

In an attempt to avoid improper connections, some utilities require two separate service panelboards to be installed. The first one will be 120/240 volt, 3 phase, 4 wire, and the second will be 120/240 volt, 1 phase, 3 wire. Electricians are encouraged to connect 3-phase-only loads to the 3-phase panelboard and not to connect any line-to-neutral loads to it. Line-to-neutral loads should be supplied only from the 1-phase panelboard.

CAUTION: If the electrical system is grounded by the utility, be aware that *250.24(C)* requires the grounded conductor to be installed as far as the service disconnecting means and to be bonded to the enclosure. This requirement applies for the service installed for 3-phase power loads, even though all loads will be connected 3 phase. This grounded system conductor serves to carry any line-to-ground-fault current that develops at the service equipment or downstream from the service. See the comment in *250.24(C)* for the sizing rules.

Depending on the connected loads, it is not uncommon for electric utilities to install a larger transformer for the single-phase loads (A- to C-phase transformer with a grounded midpoint) and smaller transformer for the 3-phase loads. It is also not uncommon for the electric utility to install two transformers connected in open delta configuration (the A- to C-transformer with a grounded midpoint plus one other transformer).

250.20(D) Impedance Grounded Neutral Systems

The Requirement: Impedance grounded neutral systems are required to be grounded in accordance with *250.36* or *250.186.*

Discussion: Grounding of high-impedance grounded neutral systems is discussed later in this book, in *250.36.* Obviously, these systems are grounded, but through an impedance device rather than through a solid connection.

250.21 ALTERNATING-CURRENT SYSTEMS OF 50 VOLTS TO 1000 VOLTS NOT REQUIRED TO BE GROUNDED

250.21(A) General

The Requirements: The following ac systems of 50 volts to 1000 volts are permitted to be grounded but are not be required to be grounded:

1. Electric systems used exclusively to supply industrial electric furnaces for melting, refining, tempering, and the like.
2. Separately derived systems used exclusively for rectifiers that supply only adjustable-speed industrial drives.
3. Separately derived systems supplied by transformers that have a primary voltage rating less than 1000 volts, provided that all the following conditions are met:
   a. The system is used exclusively for control circuits.
   b. The conditions of maintenance and supervision ensure that only qualified persons service the installation.
   c. Continuity of control power is required.
4. Other systems that are not required to be grounded in accordance with the requirements of *250.20(B).*

Discussion: This section contains rules on electrical systems that are permitted but are not required to be grounded. As shown in the list, many of these systems are not grounded and most of these systems are installed in industrial occupancies. The most common reason electrical systems are not grounded, when there is an option to do so, is to try to achieve operational reliability. Figure 2-8 shows these systems.

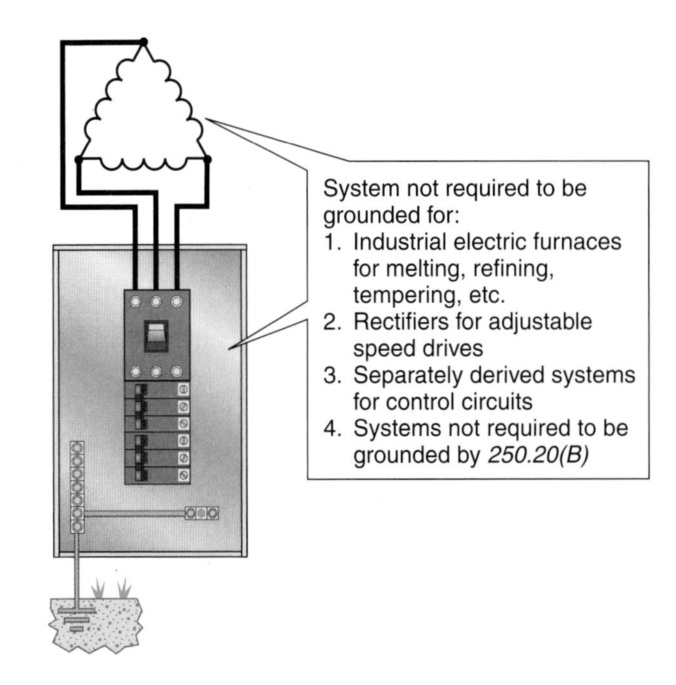

FIGURE 2-8 Systems permitted but not required to be grounded, *250.21.* (*© Cengage Learning 2012*)

As discussed extensively in Unit 1, ungrounded systems are not affected by the first ground fault other that the ground fault simply grounds the system. Little, if any, current will flow in the fault path. Certainly, not enough current will flow to cause an overcurrent device to operate. See Figure 1-12 and the accompanying discussion on *250.4(B)(2).*

If the first ground fault is not located and the insulation integrity restored before the second ground fault occurs on a different phase, a short circuit or phase-to-phase fault is established. The impedance of the system and the available fault current determine the amount of current that will flow in the short circuit. Conductive paths between the faults will carry current. The path includes the metallic wiring method that usually includes conduits, motor starters, disconnects, wireways, locknuts, and fittings. Arcing and sparking occurs where these connections are not tight, which can be the source of ignition in the presence of combustible material. See Figure 1-13 and the accompanying discussion on *250.4(B)(2).*

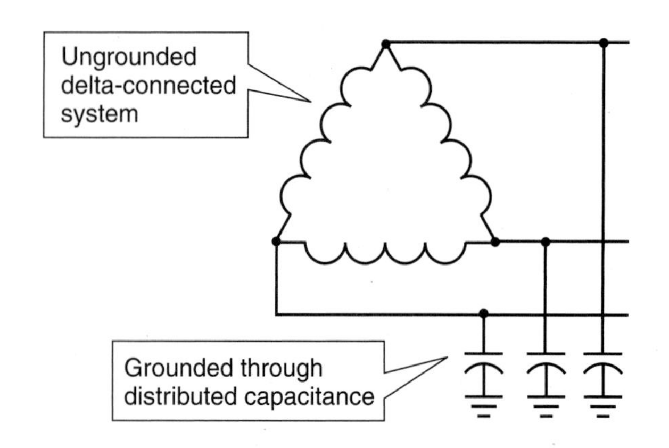

FIGURE 2-9 Systems grounded through distributed capacitance. (*© Cengage Learning 2012*)

These systems are ungrounded in name only. In reality, the systems are inductively and capacitively coupled to ground, as shown in Figure 2-9. This occurs when the energized conductor is insulated and installed inside metal conduits and other raceways that are grounded. The insulation acts like the dielectric between conductive plates of a capacitor.

An induced voltage to ground builds up on the conductor, which commonly will be about one half of the system voltage so long as there are no sputtering or intermittent ground faults on the system. For a 480-volt ungrounded system, the "half-voltage" of approximately 240 volts presents a severe shock hazard to maintenance personnel, who may not be aware of the operational nature of these systems. Sputtering or intermittent faults occur when the insulation breaks down, often as a result of damage to the conductors or where chafing occurs, such as at vibrating motors or other equipment. Because the system is not grounded and little current flows in the fault, not enough current is present either to weld the conductor to the metal conduit or other enclosure through a high-temperature arc or to burn itself clear. These sputtering faults can cause transient overvoltages on the system that are several times the system voltage. This elevated voltage can easily exceed the dielectric withstand rating of conductor insulation as well as that of motor and transformer windings. Premature failure of these system components can often be traced back to these transient overvoltages.

250.21(B) Ground Detectors

The Requirement: Ground detectors (for ungrounded systems) are required to be installed in accordance with the following:

1. Ungrounded alternating current systems as permitted in *250.21(A)(1)* through *(A)(4)* operating at not less than 120 volts and not exceeding 1000 volts are required to have ground detectors installed on the system.
2. The ground detection–sensing equipment is required to be connected as close as practicable to where the system receives its supply.

Discussion: Ground detection on ungrounded systems has direct operational benefits to industrial plants because the first ground fault will be detected and indicated. The type of system ground-fault indication varies by the sophistication of the equipment and can range from light bulbs that extinguish when a ground-fault occurs on that phase to both a light and horn or bell and local and remote annunciation.

Ungrounded electrical systems are usually installed to provide operational continuity because the first ground fault does not cause an overcurrent device to open but simply grounds the system. Ideally, maintenance electricians will locate the ground fault and make repairs before a second ground fault occurs. A second ground fault on a different phase from the first fault can result in the opening of one or more overcurrent devices provided the current flowing in the fault path is above the tripping point of the circuit breaker or the melting point of a fuse. This can result in disruption of the operations in the plant.

Ground-detection systems are available ranging from the fairly simple to the sophisticated. The *Code* does not specify the level or type of ground-detection systems to be installed. Some systems, consisting of light bulb arrangements, are made on site. Manufacturers of busway systems often have a ground-fault–indicating plug available as an accessory that plugs into the busway. This equipment often includes a resistor that is connected to ground and has voltage taps for connecting instrument lamps. This equipment has the effect of grounding the ungrounded system through the resistors, which

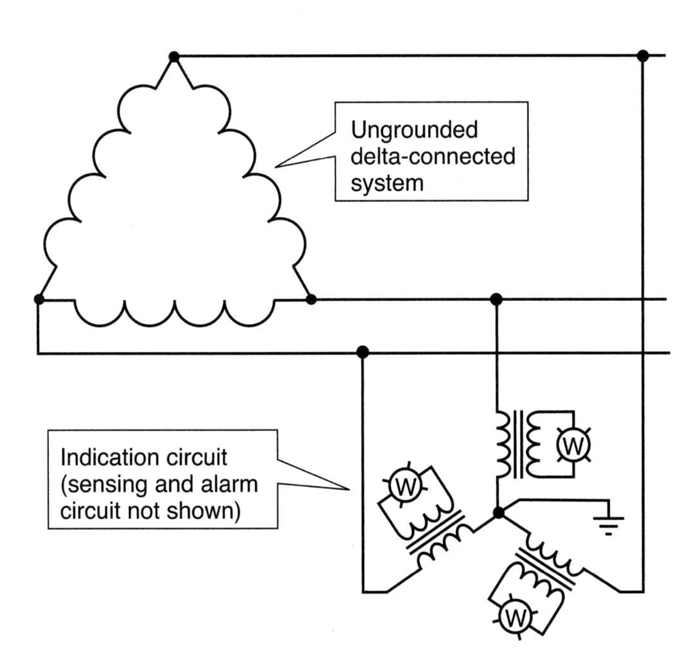

FIGURE 2-10 Ground-detection system.
(© Cengage Learning 2012)

should carry enough current to prevent transient overvoltages.

Ground-detection equipment is also available that is listed by a nationally recognized electrical testing laboratory. It features a sophisticated design and construction, including a sensing circuit, alarm circuit, and an indication circuit. The indication circuit (shown in Figure 2-10) includes three indicator lamps, an alarm light, and an alarm bell. Remote annunciators are also available.

250.22 CIRCUITS NOT TO BE GROUNDED

The Requirements: As illustrated in Figure 2-11, the following circuits are not permitted to be grounded:

1. Circuits for electric cranes operating over combustible fibers in Class III locations, as provided in *503.155.*

Discussion: According to *500.5(D),* Class III locations are those that are hazardous because of easily ignitible fibers or flyings but in which such fibers or flyings are not likely to be in suspension in the air in quantities sufficient to produce ignitible mixtures.

Hot particles dropping from ground faults that might occur on cranes that operate over these easily

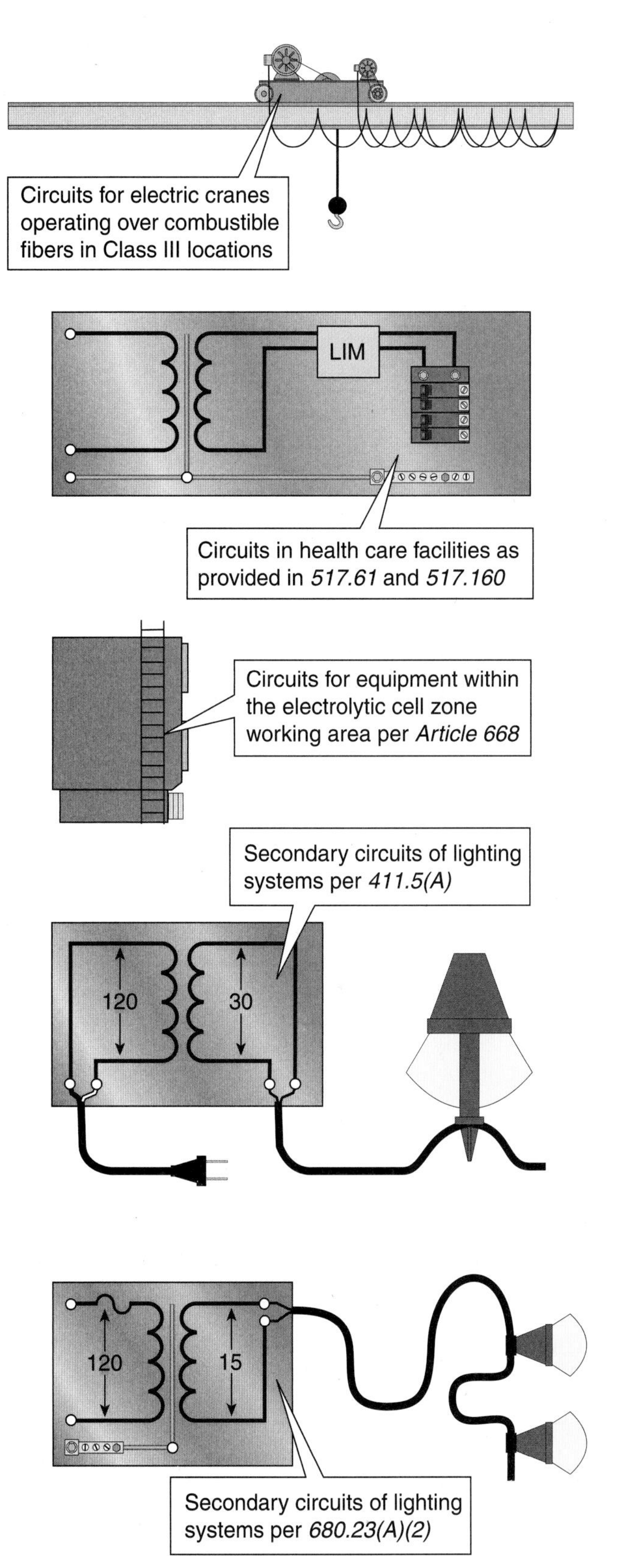

FIGURE 2-11 Circuits not to be grounded, *250.22.*
(© Cengage Learning 2012)

ignitible fibers or flyings might be a source of ignition if the temperature of the particles is above the ignition temperature of the fibers or flyings. As a result, the circuits for these cranes are required to be ungrounded. The ungrounded circuit is safer, because the first ground fault will not produce an arc or cause an overcurrent protective device to operate. A ground detector must be installed that gives an alarm and automatically de-energizes the contact conductors in case of a fault to ground or gives a visual and audible alarm so long as power is supplied to the contact conductors and the ground fault remains.

More Requirements:

2. Circuits in health care facilities as provided in *517.61* and *517.160* are not permitted to be grounded.

Discussion: The circuits referred to here are from isolated (ungrounded) transformers for equipment used in flammable inhalation anesthetizing locations in health care facilities. Most health care facilities have stopped using flammable inhalation anesthetics such as cyclopropane in favor of more modern, nonflammable anesthetics. However, isolated (ungrounded) power systems are sometimes used in surgeries or special procedure rooms that are defined as "wet procedure locations" by the health care facility administration. All receptacles and fixed equipment within the area of the wet procedure location are required by *517.20(A)* to be provided with special protection against electric shock by one of the following means:

1. "A power distribution system that inherently limits the possible ground-fault current due to a first fault to a low value, without interrupting the power supply."*

This requirement equates to an ungrounded secondary type system. These systems are referred to as "isolated power systems." See *517.20(B)* and *517.160.*

2. "A power distribution system in which the power supply is interrupted if the ground-fault current does, in fact, exceed a value of 6 mA."*

Although this section does not specifically require a Class A ground-fault circuit interrupter for personnel protection, the operating characteristics are identical to those of the GFCI.

This section no longer relates to a determination of whether interruption of power under fault conditions can be tolerated regarding procedures being performed on patients. The administration of the health care facility has the responsibility for determining the type of shock protection to be employed as well as for designating wet procedure locations.

More Requirements:

3. Circuits for equipment within the electrolytic cell working zone as provided in *Article 668* are not permitted to be grounded.

Discussion: *Article 668* applies to the installation of the electrical components and accessory equipment of electrolytic cells, electrolytic cell lines, and process power supply for the production of aluminum, cadmium, chlorine, copper, fluorine, hydrogen peroxide, magnesium, sodium, sodium chlorate, and zinc.

Power to the cell lines is usually dc and is not required to be grounded. Alternating-current systems supplying fixed and portable electrical equipment within the cell line working zone also are not required to be grounded.

More Requirements:

4. Secondary circuits of lighting systems as provided in *411.5(A)* are not permitted to be grounded.

Discussion: *Article 411* covers lighting systems that operate at 30 volts or less. A transformer reduces the source voltage, which is usually 120 volts to 30 volts. These systems are not permitted to be grounded because the secondary conductors are permitted to be bare.

More Requirements:

5. Secondary circuits of lighting systems as provided in *680.23(A)(2)* are not permitted to be grounded.

Discussion: Transformers used for the supply of underwater luminaires, together with the transformer enclosure, are required to be listed by a qualified electrical testing laboratory as a swimming pool and spa transformer. The transformer is required to be an isolated winding type with an ungrounded secondary that has a grounded metal barrier between the primary and secondary windings.

Swimming pool and spa transformers are covered under the product category WDGV in the Underwriters Laboratories *White Book* (*Guide Information for Electrical Equipment*).

The Guide Card information reads in part, "This category covers swimming pool and spa transformers of the isolated two-winding type having a grounded metal barrier between the primary and secondary windings, and intended to supply swimming pool, spa or submersible (fountain) luminaires in accordance with *Article 680* of ANSI/NFPA 70, *National Electrical Code*. The primary rating is 120 V and the maximum secondary ratings are 15 V rms and 1 kVA. The transformers are provided with integral overload protection.

"These products are provided with a power supply cord or have provisions for conduit connection to the branch circuit supply. Transformers not provided with a power supply cord are provided with leads or with studs or terminal pads to which Listed pressure wire connectors can be factory or field installed to accommodate field wiring. Wire binding screws or studs with cupped washers should be used for copper wire 10 AWG max.

"Transformers provided with a power supply cord are intended for supplying low-voltage submersible (fountain) luminaires as indicated by marking on the transformer. They are not intended for use with a swimming pool or spa luminaires.

"Unless marked otherwise, these transformers are not suitable for connection to a conduit which extends directly to a wet-niche or no-niche luminaire."[U]

250.24 GROUNDING SERVICE-SUPPLIED ALTERNATING-CURRENT SYSTEMS

250.24(A) System Grounding Connections

The Requirement: A premises wiring system supplied by a grounded ac service is required to have a grounding electrode conductor connected to the grounded service conductor, at each service, in accordance with *250.24(A)(1)* through *(A)(5)*.

[U]"Reprinted from the White Book with permission from Underwriters Laboratories Inc.®, Copyright © 2010 Underwriters Laboratories Inc.®."

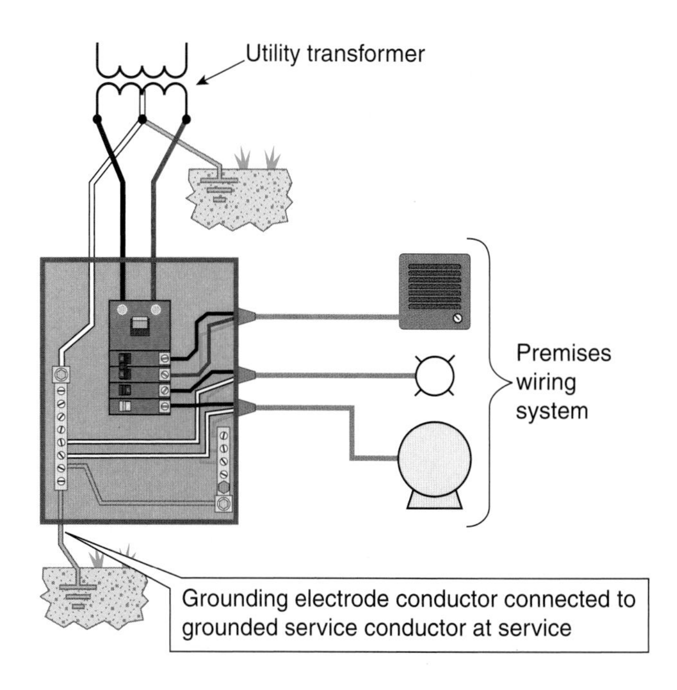

FIGURE 2-12 Grounding service-supplied ac systems, *250.24(A)*. (*© Cengage Learning 2012*)

Discussion: This connection is normally made to the bus for the neutral or grounded service conductor in the service equipment (see Figure 2-12). The phrase *at each service* is both narrow and broad in meaning. As can be seen in *230.71,* the service disconnecting means can consist of a single enclosure or up to six enclosures grouped at one location. Additional locations permitted for the grounding electrode connection can be found in *250.24(A)(1)* and *250.64(D).*

250.24(A)(1) General

The Requirement: The grounding electrode conductor connection is required to be made at any accessible point from the load end of the service drop or service lateral to and including the terminal or bus to which the grounded service conductor is connected at the service disconnecting means.

> **Informational Note:** See the definitions of Service Drop, Service Conductors, Underground and Service Lateral in *Article 100.*

Author's Note: Several changes were made to the definitions of conductors related to services in *Article 100* of the 2011 *NEC*. For example, if the underground supply conductors are installed by the electric utility, they are defined as the *service lateral*—if the conductors are installed by the

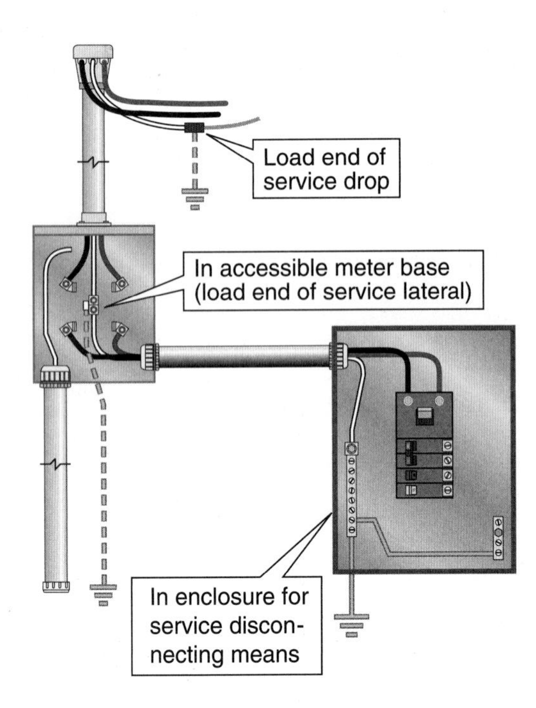

FIGURE 2-13 Grounding electrode conductor connection at accessible location, *250.24(A)(1)*. (*© Cengage Learning 2012*)

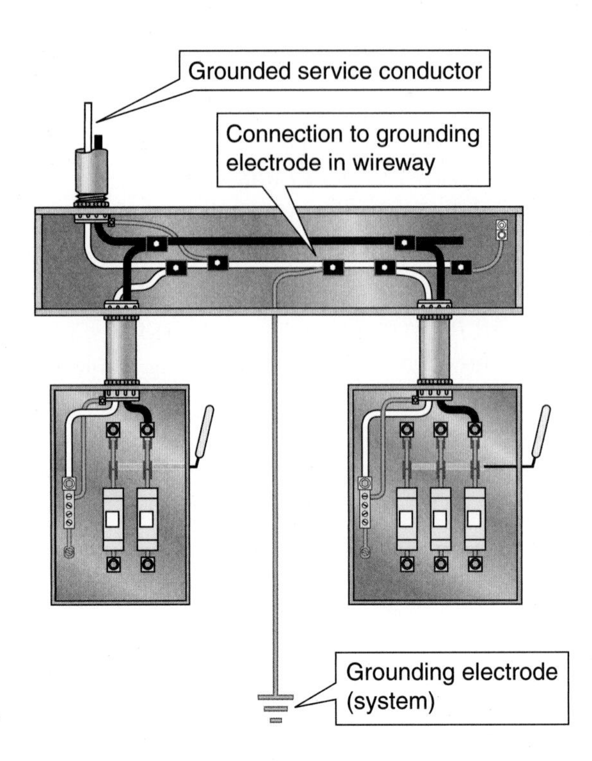

FIGURE 2-14 Grounding electrode conductor connection in wireway, *250.24(A)(1)*. (*© Cengage Learning 2012*)

owner, they are defined as *service conductors-underground.* The *service drop* is usually installed by the electric utility and connects to *service-entrance conductors* at the building or structure. Corresponding changes were made to the terminology in *Article 230.*

Discussion: Note that this rule refers to making the connection at any *accessible* point (see Figure 2-13). The term *accessible* is defined in *Article 100* and means, so far as equipment is concerned, *admitting close approach; not guarded by locked doors, elevation, or other effective means.** Locations for making the connection to the service grounded conductor that are generally considered accessible include the following:

1. At the weatherhead for overhead services.
2. At the meter socket or current transformer enclosure. This location is sometimes not permitted by the electrical inspector, because the connection is protected by the seal installed on the metering enclosure by the serving utility. Check with the local AHJ.
3. At a wireway or auxiliary gutter on the line side of the service equipment. See Figure 2-14. This is perhaps the most desirable location when the service consists of more than one service disconnecting means, as permitted by *230.71,* because a single connection can be made for all of the service disconnecting means rather than multiple connections, including tap connections, permitted in *250.64(D).*
4. Within the service equipment enclosure. This location is almost universally accepted by the AHJ. It is illustrated in Figure 2-13.

250.24(A)(2) Outdoor Transformer

The Requirement: Where the transformer supplying the service is located outside the building, at least one additional grounding connection is required to be made from the grounded service conductor to a grounding electrode, either at the transformer or elsewhere outside the building.

> ***Exception:*** *The additional grounding connection is not permitted to be made on high-impedance grounded neutral systems. The system is required to comply with 250.36.**

*Reprinted with permission from NFPA 70-2011.

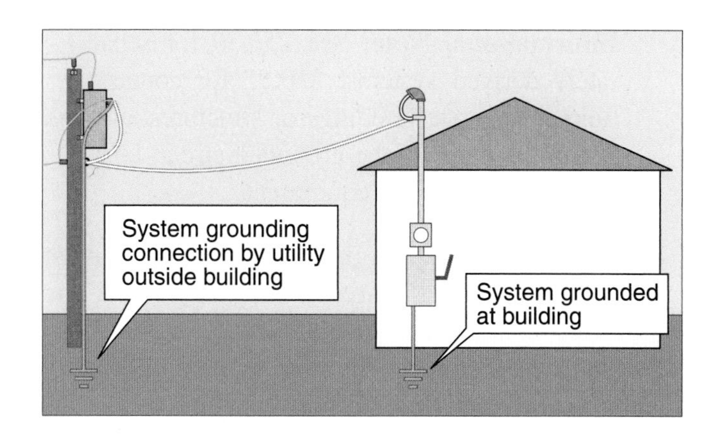

FIGURE 2-15 Grounding connections if outdoor transformer, *250.24(A)(2).* (© *Cengage Learning 2012*)

Discussion: This requirement ensures that the electrical system is grounded twice: at the source transformer and again at the service disconnecting means (see Figure 2-15). Note that this rule does not apply to serving utilities, as described in *90.2(B)(5).* Despite the utility exemption in *90.2(B)(5),* service utilities almost always ground their systems in compliance with *NEC* rules but for a different reason: They ground their electrical systems in compliance with the *National Electrical Safety Code ANSI C-2.* This standard is produced by the Institute of Electrical and Electronic Engineers.

250.24(A)(3) Dual-Fed Services

The Requirement: For services that are dual fed (double-ended) in a common enclosure or grouped together in separate enclosures and employing a secondary tie, a single grounding electrode connection to the tie point of the grounded conductor(s) from each power source is permitted.

Discussion: This requirement really amounts to an exception to the general rule, although it is stated separately. It simply permits a single grounding electrode connection to the tie point for the double-ended service, rather than a double connection to this equipment (see Figure 2-16).

250.24(A)(4) Main Bonding Jumper as Wire or Busbar

The Requirement: Where the main bonding jumper specified in *250.28* is a wire or busbar and is

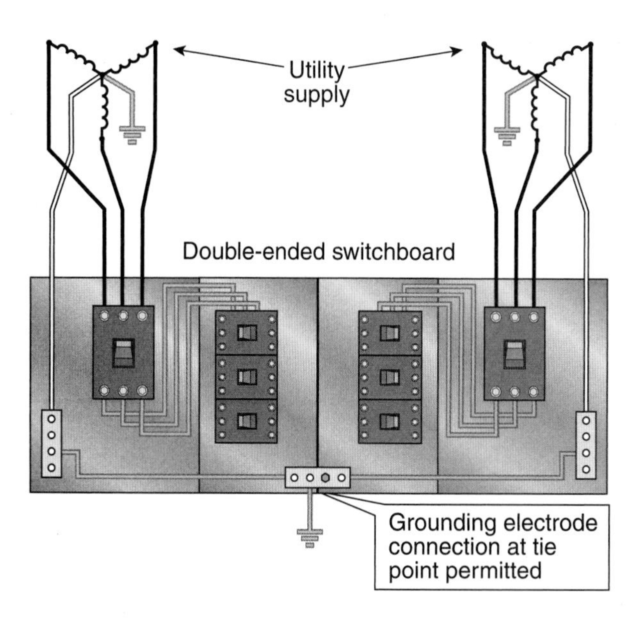

FIGURE 2-16 Grounding electrode connection for dual-fed services, *250.24(A)(3).* (© *Cengage Learning 2012*)

installed from the grounded conductor terminal bar or bus to the equipment grounding terminal bar or bus in the service equipment, the grounding electrode conductor is permitted to be connected to the equipment grounding terminal, bar, or bus to which the main bonding jumper is connected rather than to the neutral bus.

Discussion: This requirement is illustrated in Figure 2-17. Normally, and to comply with the rules in *250.24(A)(1),* the grounding electrode conductor is connected directly to the neutral terminal bar in the service equipment. Electrical inspectors often interpret a connection of the grounding electrode conductor to other than the neutral terminal bar as creating a splice in the grounding electrode conductor, in violation of the rules in *250.64(C). Section 250.25(A)(4)* permits the connection of the grounding electrode conductor to be made to the equipment grounding bus for a special reason. Making the connection to the equipment grounding terminal bar rather than at the neutral terminal bar allows a residual-type ground-fault protection system to function. (A residual-type ground-fault protection system is sometimes referred to as a "ground-strap" type.)

For the sake of simplicity, a residual-type equipment ground-fault protection system can be

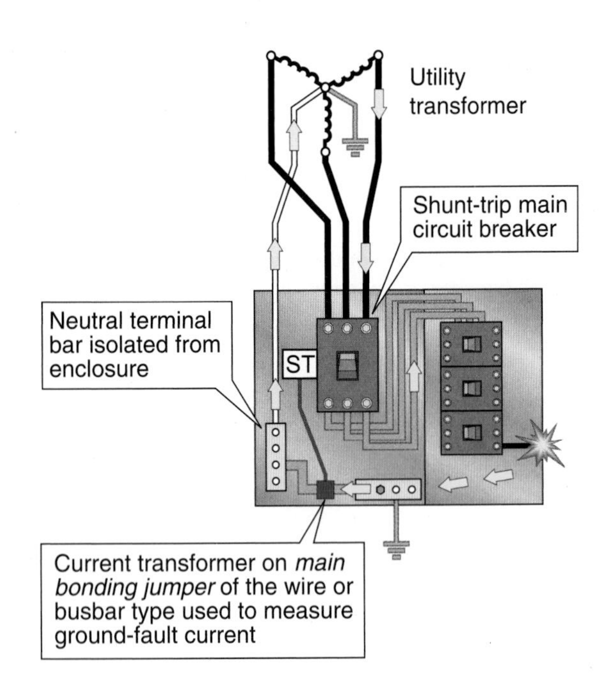

FIGURE 2-17 Grounding electrode connection if main bonding jumper is a wire or busbar, *250.24(A)(4)*. (© Cengage Learning 2012)

considered to operate by forcing all of the return current from a ground fault to flow through the main bonding jumper. A current transformer is placed on the main bonding jumper to measure the current flowing on it. A controller can be connected to set various pickup and time-delay points or periods. At the predetermined amount of fault current, a signal is sent to open a shunt-trip circuit breaker or contactor (relay), which opens the faulted circuit. By making the grounding electrode connection on the load side of the main bonding jumper, a single return path for ground-fault current is established. This single path allows the residual or ground-strap type of ground-fault equipment to work correctly.

250.24(A)(5) Load-Side Grounding Connections

The Requirements: A grounded conductor is not permitted to be connected to normally non-current-carrying metal parts of equipment or to equipment grounding conductor(s), or to be reconnected to ground on the load side of the service disconnecting means, except as otherwise permitted in *Article 250.*

Informational Note: See *250.30* for separately derived systems, *250.32* for connections at separate buildings or structures, and *250.142* for use of the grounded-circuit conductor for grounding equipment.

Discussion: This section provides some of the most basic and important requirements in *Article 250.* The grounded conductor is not generally permitted to be grounded again or to be used to ground or bond equipment beyond the point of grounding at the service equipment. The *Informational Note* following this section gives an example of the three locations at which a grounding connection of the neutral conductor downstream of the service is acceptable. In some of these locations, specific conditions must be met before regrounding of the neutral is acceptable. Those requirements are discussed later in those sections.

Figure 2-18 shows the basic service disconnecting means where the neutral of a 120/240-volt,

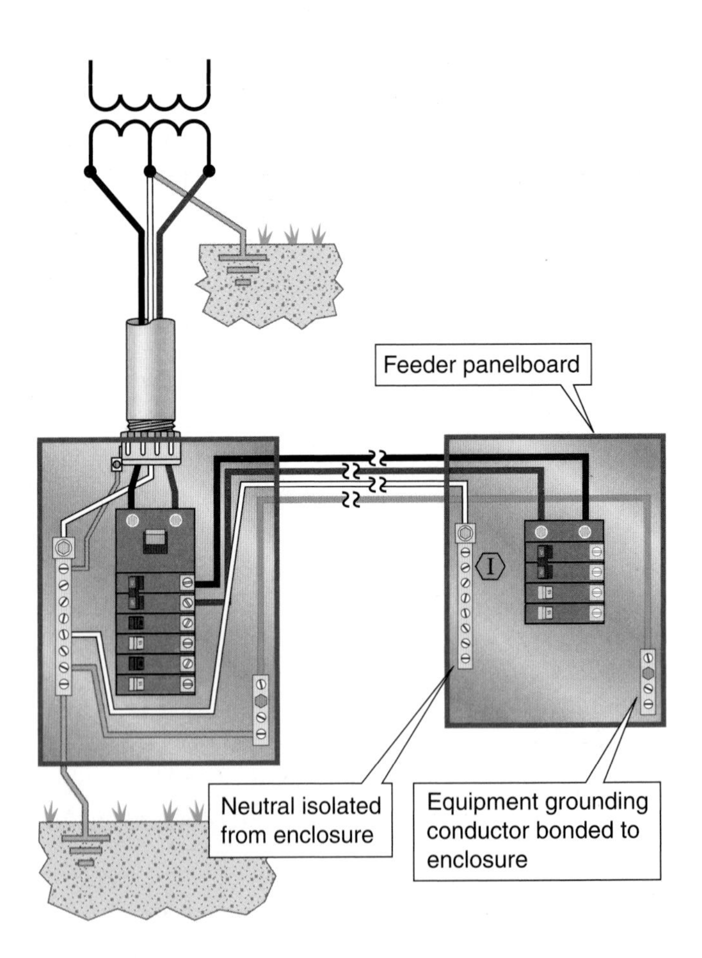

FIGURE 2-18 Neutral isolated in panelboard, *250.24(A)(5)*. (© Cengage Learning 2012)

FIGURE 2-19 Load-side grounding connections, *250.24(A)(5)*. (*© Cengage Learning 2012*)

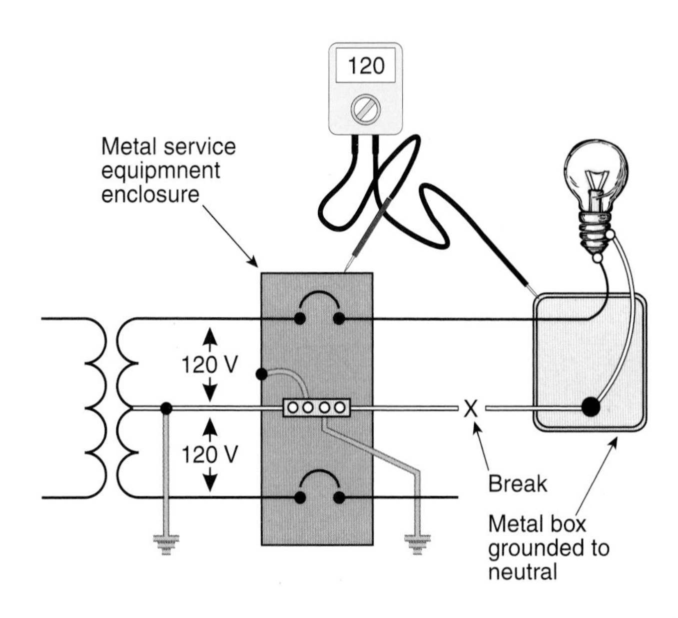

FIGURE 2-20 Hazard of using neutral to ground equipment. (*© Cengage Learning 2012*)

1-phase, 3-wire service is grounded. A feeder consisting of two ungrounded conductors, one neutral conductor, and one equipment grounding conductor supplies a distribution panelboard. At the distribution panelboard (shown in Figure 2-19), the bonding jumper between the neutral bar and the enclosure is not installed. Sometimes this is referred to in the electrical trade as "floating the neutral." The neutral conductor and terminal bar are electrically isolated from the enclosure. The equipment grounding conductor with the feeder is installed to an equipment grounding terminal bar that has been installed in good electrical contact with the enclosure.

Connecting the neutral to a metallic enclosure on the load side of the service or point of bonding for a separately derived system is the leading cause of power quality problems. These improper connections allow neutral current to flow on multiple unintended paths, including conduit, EMT, and cable armor or sheaths, as well as over shields for data processing cables.

The neutral is permitted or required to be grounded beyond the service for separately derived systems (see *250.30*), under specific conditions for existing separate buildings or structures [see *250.32(B)*], and under specific conditions for grounding equipment, in *250.142*.

Isolating the neutral on the load side of the service or point of grounding for a separately derived system prevents a shock hazard that could occur if the neutral were used to ground equipment improperly. Figure 2-20 illustrates how a ground fault "downstream" of the service could create a shock hazard if the equipment is grounded to the neutral. Actually, two events must occur for the shock hazard to be present. The neutral connection to the panelboard must be loose somewhere on its line or supply side. A line-to-neutral load, such as a single incandescent lamp (or other load-connected line to neutral), will then energize the panelboard at a dangerous elevated voltage above ground. The same hazard would be present at any other equipment with a line-to-neutral connection.

250.24(B) Main Bonding Jumper

The Requirement: For a grounded system, an unspliced main bonding jumper is required to be used to connect the equipment grounding conductor(s) and the service-disconnect enclosure to the grounded conductor within the enclosure for each service disconnect, in accordance with *250.28*.

Discussion: The main bonding jumper provides the essential connection between the grounded system conductor (often a neutral) and the enclosure. Equipment grounding conductors such as conduit, cable sheaths, and wire-type conductors are connected to the service metal enclosure such as the cabinet. The main bonding jumper connects the equipment grounding conductors to the system grounded conductor, which is often a neutral. Recall that for current to flow, a complete circuit must exist. As can be seen in Figure 2-21, the main bonding jumper provides the critically important link to complete the circuit from the fault to the system grounded conductor, which in turn returns fault current back to the utility transformer or other source of energy. Additional discussion on the requirements for main and system bonding jumpers takes place in *250.28* in this book.

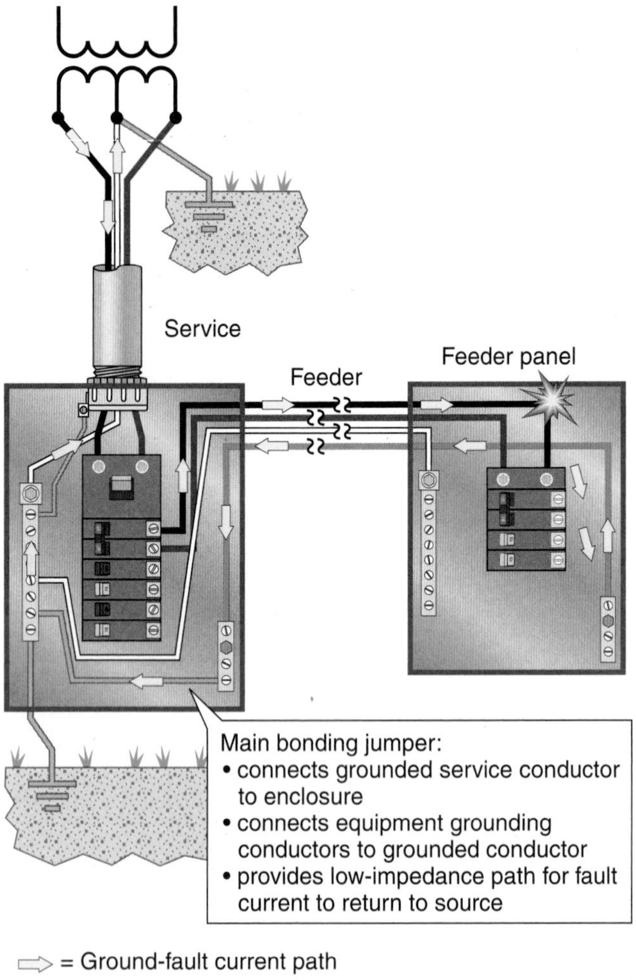

Service
Feeder
Feeder panel
Main bonding jumper:
• connects grounded service conductor to enclosure
• connects equipment grounding conductors to grounded conductor
• provides low-impedance path for fault current to return to source

⇨ = Ground-fault current path

FIGURE 2-21 Main bonding jumper, *250.24(B)*. (*© Cengage Learning 2012*)

Exception: Where more than one service disconnecting means is located in an assembly listed for use as service equipment such as a switchboard, an unspliced main bonding jumper is only required to be located in one of the sections of the assembly enclosure.

Discussion: Switchboards are generally large pieces of equipment that may be installed as the service disconnecting means. Typically, this is floor-mounted equipment that consists of several sections that are bolted together on the job site before the service is energized. The sections are thereby connected together mechanically and electrically. In addition, an equipment grounding bus is installed in contact with each section. Because of electrical continuity of the equipment, a single main bonding jumper is permitted to be installed between the neutral bar and the enclosure (see Figure 2-22). Usually, this main bonding jumper is furnished by the manufacturer and is the proper size for the ampere rating of the

FIGURE 2-22 Main bonding jumper for service-disconnecting means in listed assembly, *250.24(B), Exception No. 1*. (*© Cengage Learning 2012*)

switchboard. The main bonding jumper for switchboards usually consists of a busbar or copper conductor. The bonding jumper is identified as such by the manufacturer.

250.24(C) Grounded Conductor Brought to Service Equipment

The Requirement: If an ac system operating at less than 1000 volts is grounded at any point, the grounded conductor(s) is required to be routed with the ungrounded conductors to each service disconnecting means and to be connected to each disconnecting means grounded conductor(s) terminal or bus. A main bonding jumper is required to connect the grounded conductor(s) to each service disconnecting means enclosure. The grounded conductor(s) must be installed in accordance with *250.24(C)(1)* through *(C)(4)*.

Discussion: This *Code* section provides another critically important requirement in *Article 250.* If a system supplying the service is grounded, usually by the electric utility, the grounded system conductor (usually a neutral) must be run to each service disconnect and be bonded to each service disconnecting means. As discussed in *250.24(B),* this grounded system conductor provides the critically important path from a ground fault back to the source to complete the circuit. This complete circuit allows current to flow through the overcurrent device so it will clear the faulted circuit off the line. (See Figures 2-23 to 2-25.)

The requirement to route the grounded system conductor with the phase (ungrounded) conductors is intended to keep all of the circuit conductors in close proximity (see Figures 2-23 through 2-25). As discussed earlier, keeping all of the circuit conductors together in an ac system is essential to maintain the impedance as low as possible. This is important because the grounded system conductor serves the critical role of providing the path for fault current to return to the source.

Discussion: As illustrated in Figure 2-26, it is important that the grounded service conductor be run to each service enclosure and be connected to the enclosure either directly or through a main bonding jumper even though the grounded conductor (often a neutral) is not used beyond the service. This is often the case when a 1-phase service disconnect and a 3-phase service disconnect are installed from a 3-phase, 4-wire system. Often, the 1-phase service supplies lighting and receptacle loads, and the 3-phase service supplies motor and other 3-phase loads. The grounded conductor from the 3-phase service provides the low-impedance path back to the source even if line-to-neutral loads are not supplied by the service.

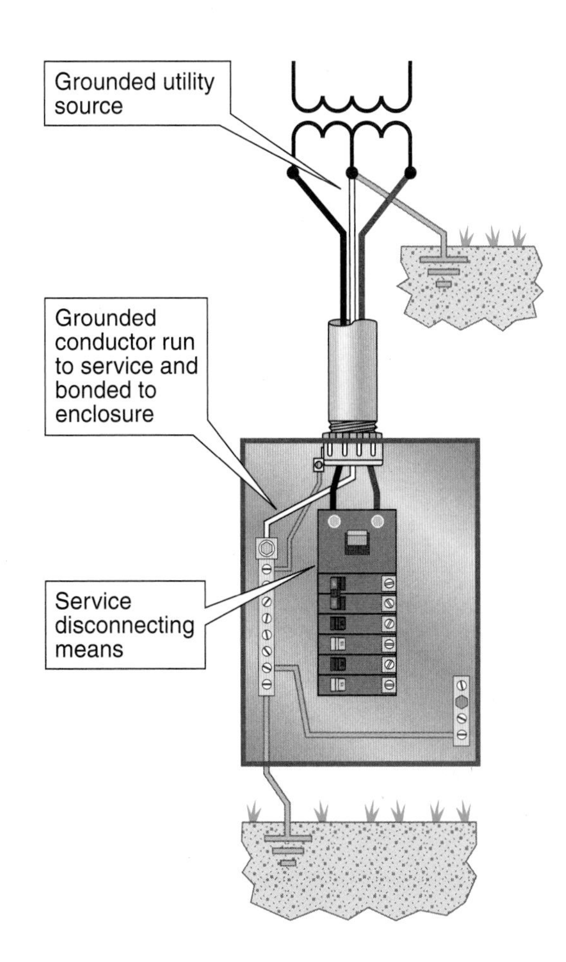

FIGURE 2-23 Grounded conductor to service equipment routed with phase conductors, *250.24(C).* (*© Cengage Learning 2012*)

> **Exception:** Where more than one service disconnecting means are located in a single assembly listed for use as service equipment, it is permitted to run the grounded conductor(s) to the assembly common grounded conductor(s) terminal or bus. The assembly must include a main bonding jumper for connecting the grounded conductor(s) to the assembly enclosure.

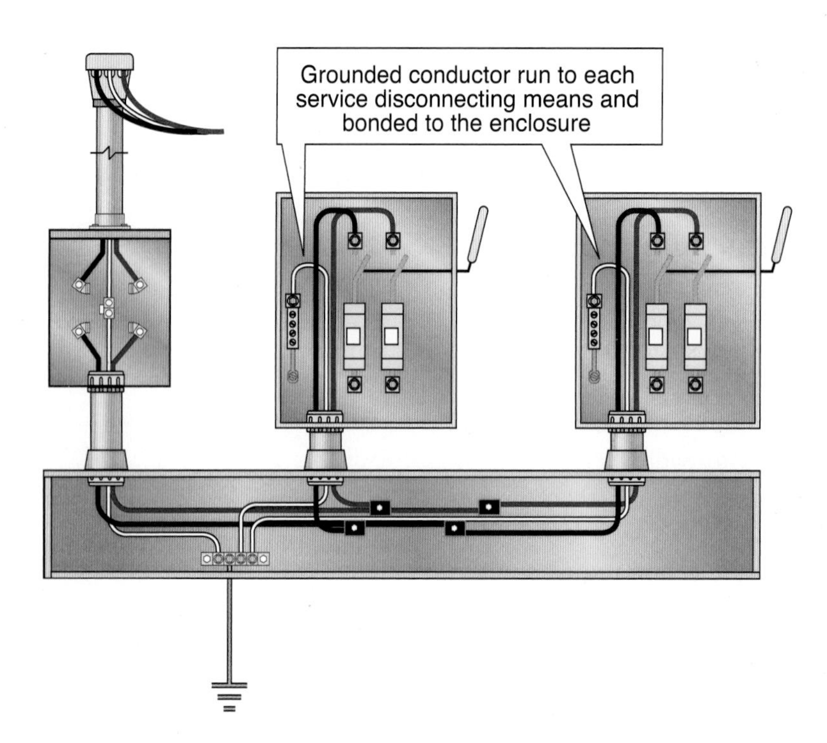

FIGURE 2-24 Grounded conductor brought to service equipment and bonded to enclosure, *250.24(C)*. (*© Cengage Learning 2012*)

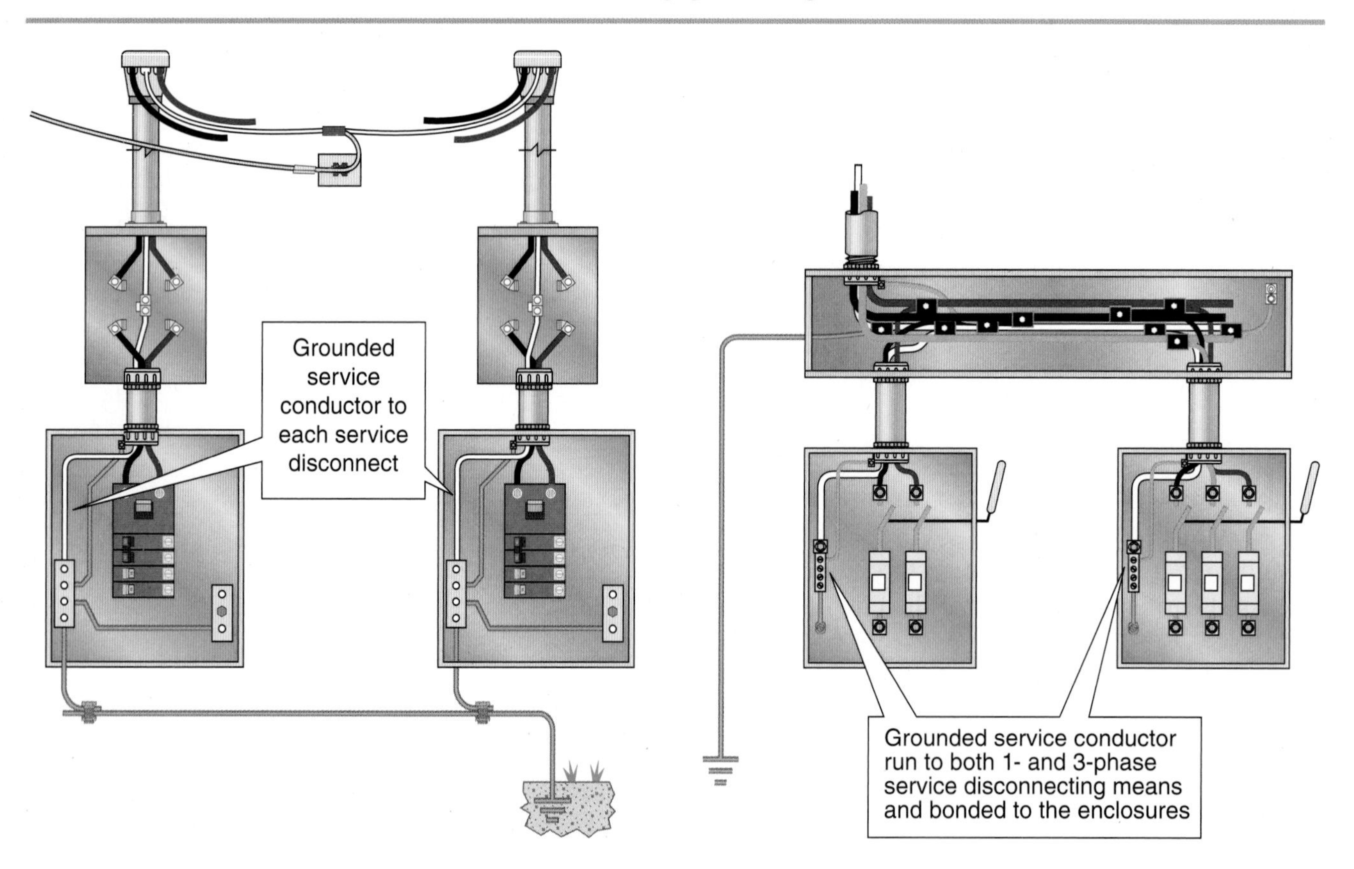

FIGURE 2-25 Grounded service conductor to two or more service-disconnecting means, *250.24(C)*. (*© Cengage Learning 2012*)

FIGURE 2-26 Grounded service conductor to 1-phase and 3-phase service-disconnecting means, *250.24(C)*. (*© Cengage Learning 2012*)

FIGURE 2-27 Grounded conductor to listed assembly, *250.24(C) Exception.* (*© Cengage Learning 2012*)

Discussion: This exception mirrors *Exception No. 1* to *250.24(B)* and recognizes that multisection switchboards or motor control centers used as service equipment are bolted together and are equipped with an equipment grounding bus that also connects the sections together (see Figure 2-27).

250.24(C)(1) Sizing for a Single Raceway

The Requirements: The grounded conductor is required to be not smaller than the required grounding electrode conductor specified in *Table 250.66* but is not required to be larger than the largest ungrounded service-entrance conductor(s).

Discussion: The requirements for sizing the grounded conductor in this section are intended to ensure that it is large enough to carry fault current back to the utility transformer. The conductor size determined by reference to *Table 250.66* satisfies this requirement. Even though the title of *Table 250.66* is *Grounding Electrode Conductor for Alternating-Current Systems*, the table is used for sizing many other conductors and this is one of them. The reference in the title of *250.24(C)(1)* indicates these are sizing requirements that apply if all the service conductors are in a single raceway. The raceway might be conduit, EMT, or a wireway. Figure 2-26 is an example of this requirement, where a single conduit supplies a single wireway. Another example is Figure 2-28, which shows an individual set of service-entrance conductors from a current-transformer enclosure to individual service disconnecting means. For each of these and similar installations, use the size of the service conductor in the raceway in the first or second column of *Table 250.66*, based on whether the service conductors are copper or aluminum. The third or fourth column is used to determine the minimum-size conductor permitted to be used for the grounded (neutral) service conductor.

The rule would also apply to Type MC cables that are installed for service conductors. As shown in Figure 2-28, parallel underground service conductors supply a current transformer cabinet for utility metering. Individual sets of service-entrance conductors supply each service disconnecting means.

The grounded conductor for the underground service conductor must be sized on the basis of the larger of the calculated load or from *Table 250.66*, using the size of the ungrounded ("hot") conductors in the raceway.

Routing and sizing of grounding system conductors for parallel service conductors that are installed in a single wireway are shown in Figure 2-29.

Basically, three steps are required for properly sizing the grounded system conductor (often the neutral). First, perform a load calculation in accordance with *220.61.* This gives the minimum-size conductor required to carry the maximum unbalanced load from the ungrounded service conductor. Second, determine the size of the ungrounded service conductor and refer to *Table 250.66.* Note that even though the table title is *Grounding Electrode Conductor for Alternating-Current Systems,* the table is used for several other purposes, and this is one of them. Third, install the largest conductor.

Here's How: For example, if 350-kcmil ungrounded aluminum service-lateral conductors are installed in each conduit, *Table 250.66* requires a 2 AWG copper or 1/0 AWG aluminum grounded conductor. This becomes the minimum size required to carry fault current back to the source from the service equipment.

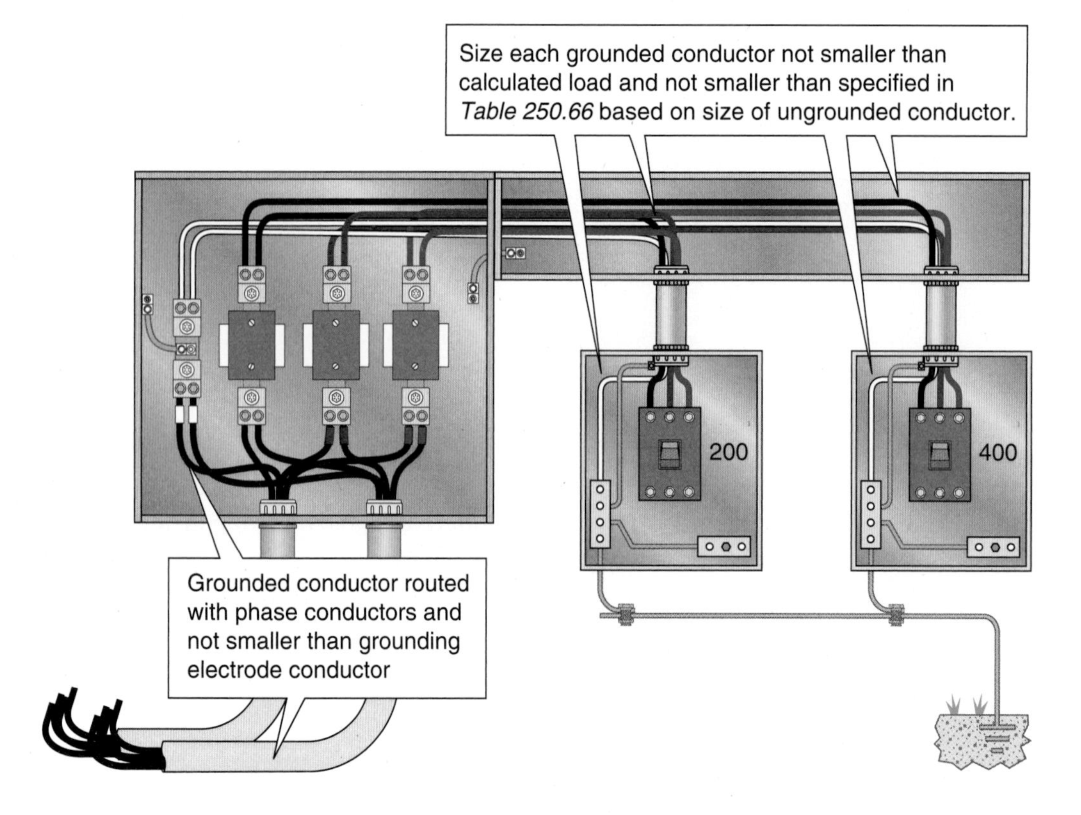

FIGURE 2-28 Routing and sizing of grounded conductor for individual service disconnects, *250.24(C)(1)*. (*© Cengage Learning 2012*)

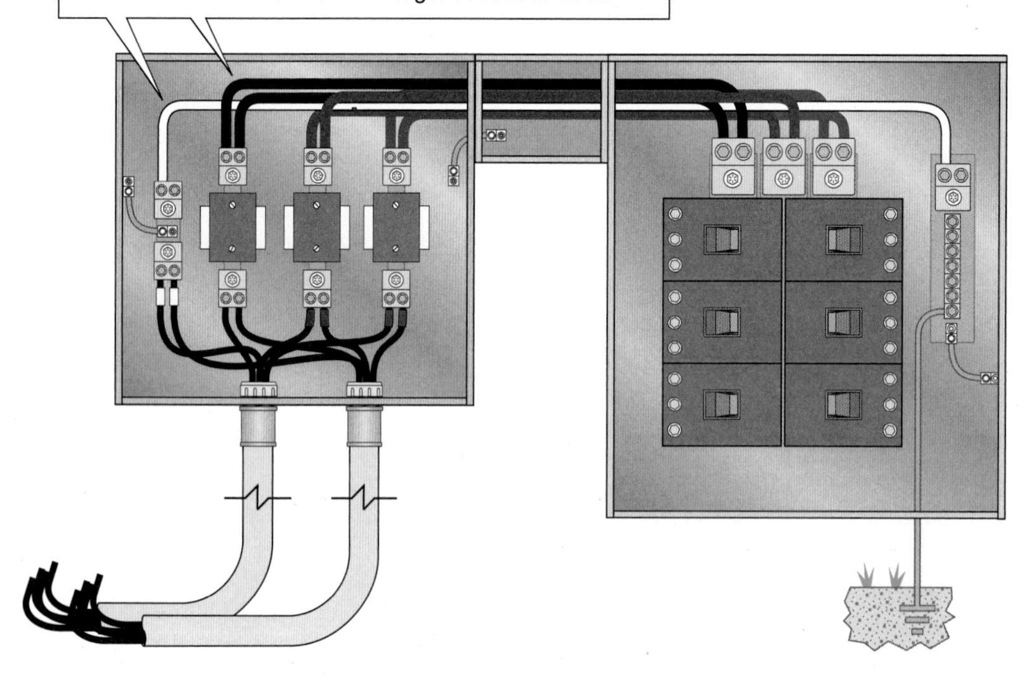

FIGURE 2-29 Routing and sizing of grounded conductor for parallel service conductors installed in a single wireway or auxiliary gutter, *250.24(C)(1)*. (*© Cengage Learning 2012*)

Compare the size of conductor required to carry the unbalanced current from the load calculation in *220.61* with the size of conductor from *Table 250.66* and install the larger one. The requirements of *310.10(H)* on installing parallel conductors must be complied with as well. The smallest conductor generally permitted to be installed in parallel is 1/0 AWG. This rule has the effect of requiring the same size copper or aluminum wire as used in our example.

If 3/0 AWG copper conductors supply the 200-ampere service disconnect from the current transformer cabinet, *Table 250.66* requires a 4 AWG copper or 2 AWG aluminum grounded conductor. A larger conductor may be required as a result of a load calculation.

For the 400-ampere service disconnect, select a 1/0 AWG copper or 3/0 AWG aluminum grounded conductor, based on a 500-kcmil copper service-entrance conductor. Again, a larger conductor may be required as a result of a load calculation.

More Requirements: In addition, for sets of ungrounded service-entrance conductors larger than 1100-kcmil copper or 1750-kcmil aluminum, the grounded conductor must not be smaller than 12½% of the area of the largest service-entrance ungrounded conductor.

Discussion: This requirement covers installations in which the ungrounded service conductors are larger than specified in *Table 250.66*. Note that *Table 250.66* covers ungrounded service conductors up to 1100-kcmil copper and 1750-kcmil aluminum. For installations with service conductors larger than this, the minimum size of the grounded conductor must be calculated using a simple formula. A typical installation of a service with service-entrance conductors in parallel is shown in Figure 2-29. Keep in mind this formula determines the minimum-size conductor for carrying fault current back to the source and does not take neutral load into account. The formula is applied as follows:

- Add together the circular mil area of one set of the ungrounded service-entrance conductors.
- Multiply this value by 0.125 (12½%).
- This gives the minimum size of grounded (neutral) conductor.

If the calculation does not result in a standard conductor size or in a size less than 250 kcmil, refer to *Table 8* of *Chapter 9* and round up to the next standard size.

For some standard combination of service conductors, refer to Table 2-1 for the minimum size permitted for a single copper grounded (neutral) service conductor.

This requirement and discussion contemplates that all of the ungrounded conductors are to be installed in one raceway, such as a metal wireway or metal auxiliary gutter. One ungrounded or neutral conductor can then be installed to serve as the return fault-current conductor for all of the ungrounded service conductors. These larger installations utilize installing two or more conductors in parallel (electrically joined at each end to create a set of conductors that have adequate ampacity).

It is common for design electrical engineers to specify that all service-entrance conductors, including ungrounded and grounded (neutral), be the same size.

Note the rules in *310.10(H)* for installing conductors in parallel.

Note: Applying an adjustment factor for more than three current-carrying conductors in the raceway

TABLE 2-1

Minimum Size of Grounded Conductor

Service Ampere Rating	Total Area of Ungrounded Conductors	Minimum Area of Grounded Conductor	Minimum Size of Grounded Conductor
1000	(3) 400 kcmil = 1200 kcmil	150,000 cm	3/0
1200	(4) 350 kcmil = 1400 kcmil	175,000 cm	4/0
1600	(5) 400 kcmil = 2000 kcmil	250,000 cm	250 kcmil
2000	(6) 400 kcmil = 2400 kcmil	300,000 cm	300 kcmil

1 kcmil = 1000 cm; cm = circular mil area.

is not required in a metal wireway or auxiliary gutters in which the number of conductors at a cross section does not exceed 30. See *366.23(A)* for metal auxiliary gutters and *376.22(B)* for metal wireways. These sections act like exceptions to the rules in *310.15(B)(3)(a)* for adjustment factors when there are more than three current-carrying conductors in a raceway. The rationale is that if the maximum conductor fill in the auxiliary wireway or wireway does not exceed 20%, the conductors can readily dissipate heat.

250.24(C)(2) Parallel Conductors in Two or More Raceways

The Requirement: If the ungrounded service-entrance conductors are installed in parallel in two or more raceways, the grounded conductor is also required to be installed in parallel. The size of the grounded conductor in each raceway is required to be based on the total circular mil area of the parallel ungrounded conductors in the raceway as indicated in *(C)(1)*, but not smaller than 1/0 AWG. See Figure 2-30.

Discussion: This is the second method for installing conductors in parallel. [The first method is in *250.24(C)(1)*.] Here, identical sets of conductors are installed in parallel in two or more conduits or cables. All the rules discussed previously for installing conductors in parallel apply. With this method, the set of ungrounded conductors connected in parallel is not normally larger than 1100-kcmil copper or 1750-kcmil aluminum. Thus, *Table 250.66* can usually be used to determine the minimum size of grounded conductor for each raceway, as follows:

- Determine the maximum-size ungrounded conductor in the raceway or cable.
- Refer to *Table 250.66.*
- Determine the minimum-size copper or aluminum grounded conductor.

This grounded conductor cannot be smaller than 1/0 AWG to comply with the rules in *310.10(H)* and must be installed in each raceway or cable.

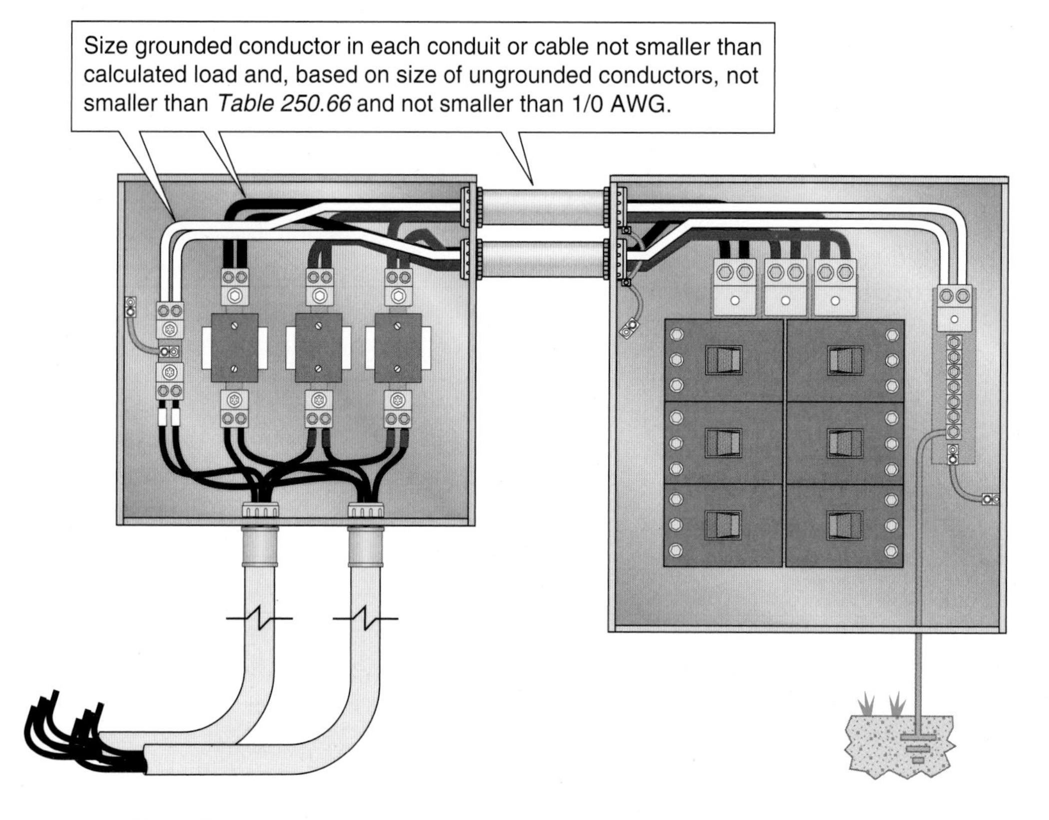

FIGURE 2-30 Size of parallel grounded service conductor in individual conduits, *250.24(C)(2)*. (*© Cengage Learning 2012*)

Often, electrical design engineers will require a full-size grounded (often a neutral) conductor in each raceway or cable. To comply, the installer must select a grounded conductor(s) of appropriate size:

- As required for the load calculated by *220.61*
- As required for carrying fault current as discussed previously
- As required by the project electrical engineer or designer

250.24(C)(3) **Delta Connected Service** The grounded conductor of a 3-phase, 3-wire delta service is required to have an ampacity not less than the ungrounded conductors.

Discussion: This requirement deals with a corner-grounded supply system in which the transformers are connected in a delta configuration, as illustrated in Figure 2-31. In a system connected in this manner, loads are typically connected only 3 phase, such as for motors. As a result, the current on each conductor is usually equal. So, the size of all conductors, including the grounded conductor, is required to be the same. These delta-connected systems often are not required to be grounded by *250.20(B)* but are grounded by choice to avoid problems associated with ungrounded systems.

Note that this rule clarifies that the grounded conductor of a 3-phase, 3-wire, corner-grounded delta system is required to be directly connected to the neutral terminal bar of the enclosure. In the past, some have connected the grounded conductor to a circuit breaker or a fuse rather than to the neutral terminal bar or enclosure.

As previously mentioned, this corner-grounded system provides 3-phase power, typically for motor loads, but does not have the disadvantages of transient overvoltages inherent in ungrounded systems. One disadvantage of this system is a single ground fault on one of the ungrounded circuit conductors will cause an overcurrent device to open if enough current flows in the circuit.

These systems can best be understood if they are thought of as single-phase systems. The similarities are summarized in Table 2-2.

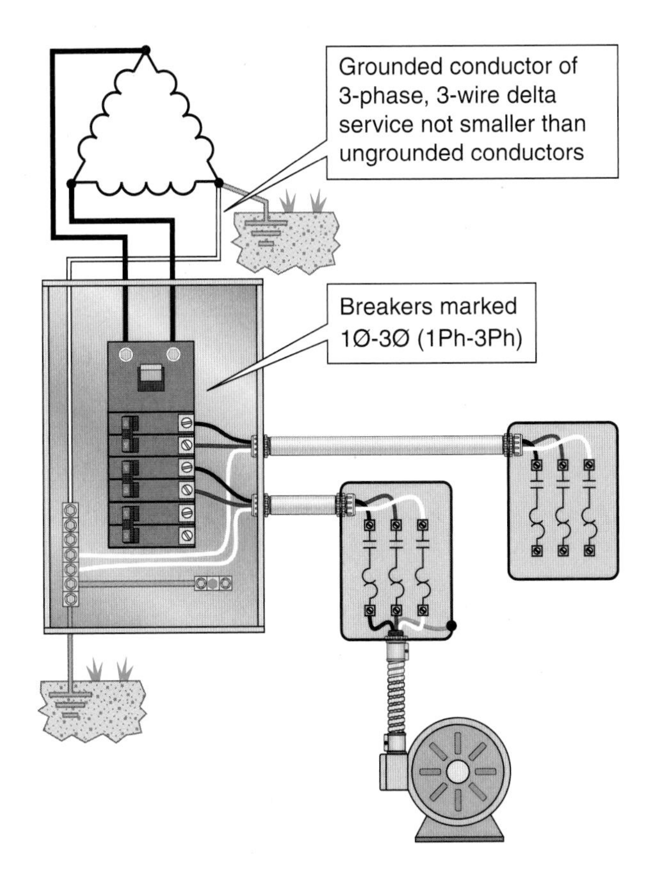

FIGURE 2-31 Grounded conductor for 3-phase, 3-wire delta-connected service, *250.24(C)(3)*. (*© Cengage Learning 2012*)

250.24(D) Grounding Electrode Conductor

The Requirement: A grounding electrode conductor is required to be used to connect the equipment grounding conductors, the service-equipment enclosures, and, where the system is grounded, the grounded service conductor to the grounding electrode(s) required by *Part III* of *Article 250*. The grounding electrode conductor is required to be sized according to *250.66* and high-impedance systems grounded according to *250.36*. See Figure 2-32.

Discussion: We reviewed the requirements of making the grounding electrode connection in *250.21* and *250.21(A)(1)*. The requirements for a grounding electrode system are described in *250.50* in this book; the grounding electrode, in *250.52;* the methods of installing the grounding electrodes, in *250.53;* and installation rules for the grounding electrode conductors, in *250.64*. In addition, the

TABLE 2-2

Comparison of Requirements for 1-Phase and 3-Phase Systems

	120/240 V, 1-∅, 3-W	120/240 V, 3-∅, 3-W	Section
Two ungrounded conductors and one grounded conductor	Yes	Yes	N/A
Color code of grounded conductors	Yes	Yes	*200.6*
Use of grounded conductor	Yes	Yes	*200.7*
Grounded system conductor permitted to be bare to the service	Yes	Yes	*230.41*
Grounded system conductor required to be grounded at each service	Yes	Yes	*250.24(A)(1)*
Grounded system conductor required to be connected to service equipment enclosure	Yes	Yes	*250.24(C)*
Grounded conductor required to be insulated on the load side of the service disconnecting mean	Yes	Yes	*310.106(D)*
Two-pole circuit breaker required to be rated	N/A	Yes	*240.85*

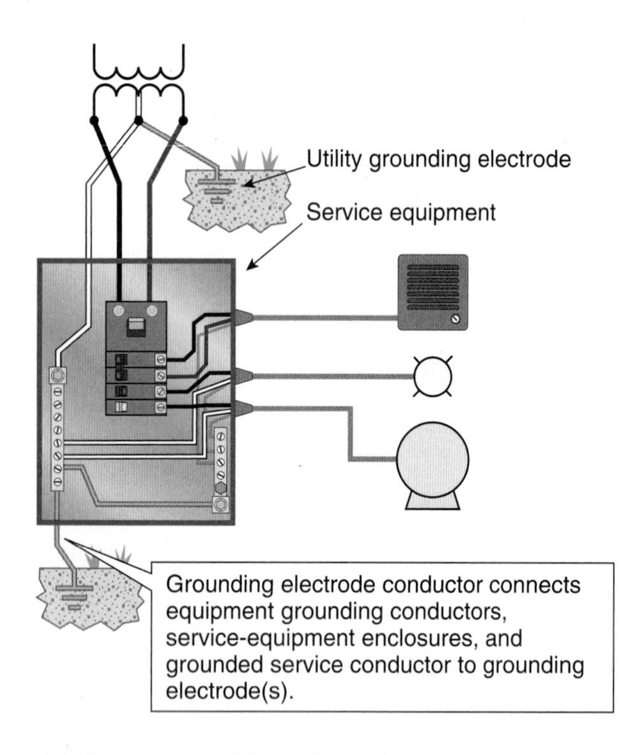

FIGURE 2-32 Grounding electrode conductor connection to service, *250.24(D)*. (*© Cengage Learning 2012*)

requirements for installing high-resistance grounded systems are covered in *250.36* in this book and the requirements for sizing grounding electrode conductors in *250.66.*

250.24(E) Ungrounded System Grounding Connections

The Requirements: A premises wiring system that is supplied by an alternating-current service that is ungrounded is required to have, at each service, a grounding electrode conductor connected to the grounding electrode(s) required by *Part III* of *Article 250.* The grounding electrode conductor must be connected to a metal enclosure of the service conductors at any accessible point from the load end of the service drop or service lateral to the service disconnecting means.

Discussion: The title of this section—"Ungrounded System Grounding Connection"—may appear self-contradictory. How can an ungrounded system have grounding connections? The requirements for grounding an ungrounded system result in a "case ground" or a "metal enclosure ground" connection to the grounding electrode system (see Figure 2-33). The source system is left ungrounded. The reasons for grounding the metal enclosures for the service equipment are covered in *250.4(B).*

The *NEC* is silent in this section on the minimum sizing of the grounding electrode conductor to be used to connect the enclosure(s) to the grounding electrode(s). In *250.66,* however, the rules for sizing the grounding electrode conductor are the same for

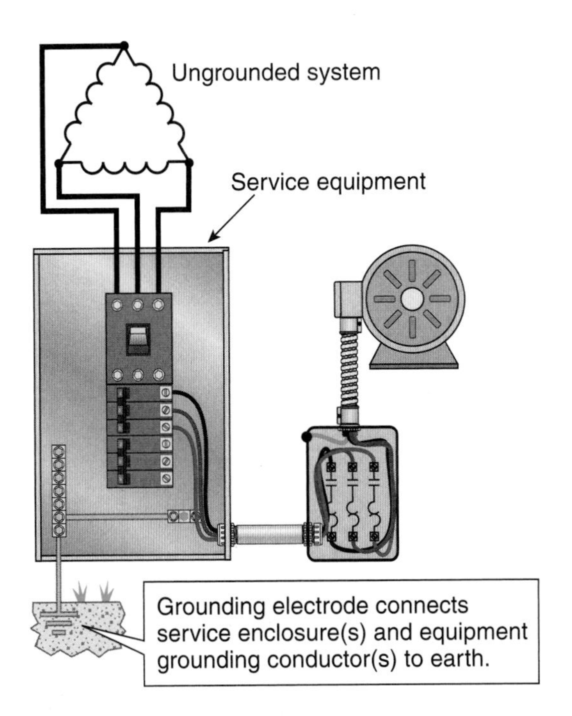

FIGURE 2-33 Ungrounded system grounding connections, *250.24(E)*. (*© Cengage Learning 2012*)

grounded and for ungrounded systems. This finding supports the stated rationale for installation of the grounding electrode and conductor—which is not to clear ground faults.

250.26 CONDUCTOR TO BE GROUNDED—ALTERNATING-CURRENT SYSTEMS

The Requirement: For ac premises wiring systems, the conductor to be grounded is as specified in the following:

1. Single-phase, 2-wire—one conductor

Discussion: Either side of the system can be grounded. These commonly are 120-volt, single-phase systems (see Figure 2-34).

More Requirements:

2. Single-phase, 3-wire—the neutral conductor

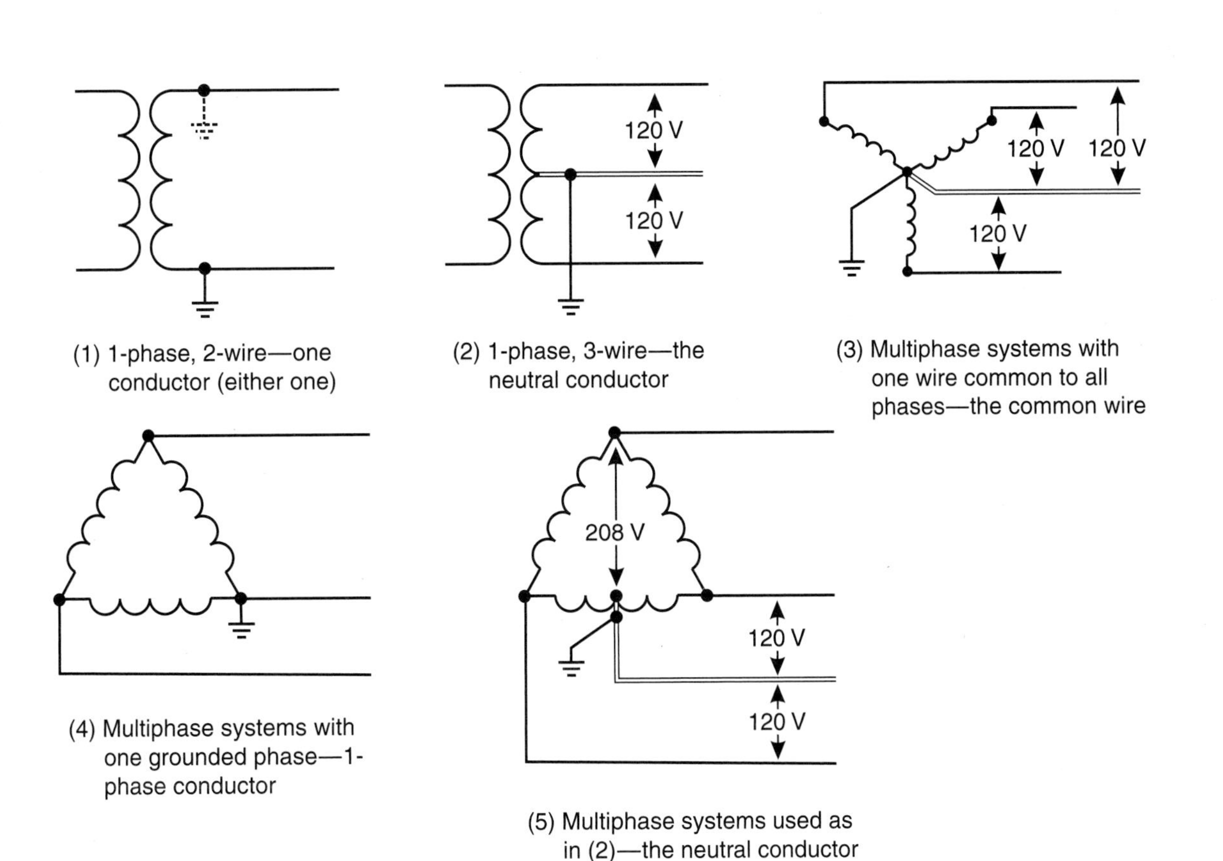

FIGURE 2-34 Conductor to be grounded: ac systems, *250.26*. (*© Cengage Learning 2012*)

Discussion: These typically are 120/240-volt, 1-phase, 3-wire systems and are installed in dwellings and for installations not requiring 3-phase motors or power.

More Requirements:

3. Multiphase systems having one wire common to all phases—the common conductor

Discussion: These systems are commonly 3-phase, 4-wire, wye-connected. Common voltages are 208Y/120, 480Y/277, and, in some industrial plants, 600Y/346.

More Requirements:

4. Multiphase systems where one phase is grounded—1-phase conductor

Discussion: These systems typically are 3-phase, 3-wire, delta-connected in which one corner of the system is grounded. For obvious reasons, the systems are referred to as "corner grounded."

More Requirements:

5. Multiphase systems in which one phase is used, as in (2)—the neutral conductor

Discussion: These systems typically are 120/240-volt, 3-phase, 4-wire. They are often installed when the higher phase-to-phase voltage of 240 volts is desired and 3-phase power is desired, often for motors. In addition, 120-volt power is available for lighting and receptacles. One of the three 240-volt transformers is center-point grounded to create this system.

The requirements in this *NEC* section ensure standardized system grounding connections.

250.28 MAIN BONDING JUMPER AND SYSTEM BONDING JUMPER

The Requirement: For a grounded system, main bonding jumpers and system bonding jumpers are required to be installed as follows.

Discussion: Main bonding jumpers are installed at the service equipment, and system bonding jumpers are installed at separately derived systems (see Figure 2-35). The definition of *main bonding jumper* and *system bonding jumper* are in *Article 100* of the *NEC*. The main and system bonding jumpers perform identical functions; they bond the grounded system conductor to the metal enclosure and serve as the fault-current return path. Thus, they serve a vital and important function in the safety system.

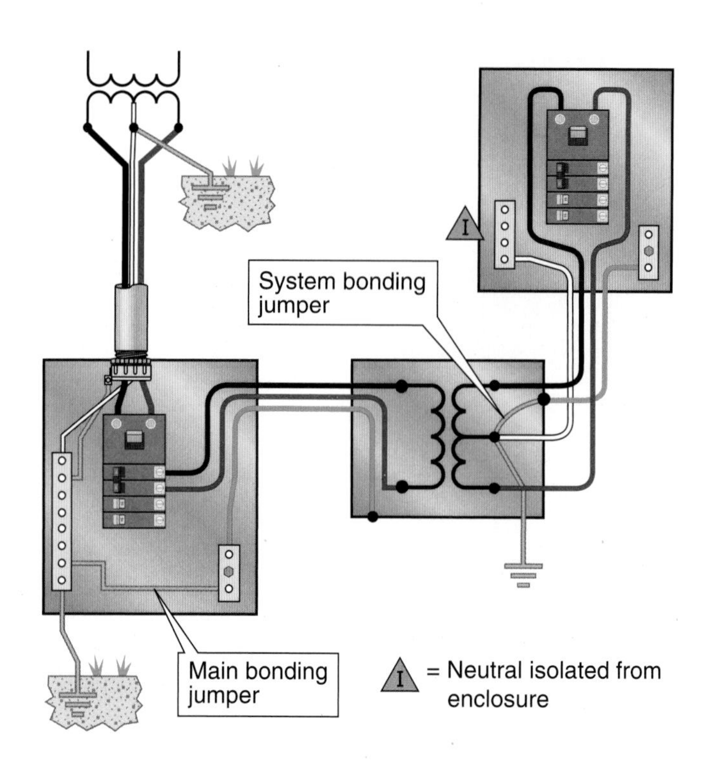

FIGURE 2-35 Main and system bonding jumpers, *250.28*. (*© Cengage Learning 2012*)

250.28(A) Material of Main and System Bonding Jumpers

The Requirements: Main bonding jumpers and system bonding jumpers are required to be of copper or other corrosion-resistant material. A main bonding jumper and a system bonding jumper can consist of a wire, bus, screw, or similar suitable conductor.

Discussion: Main bonding jumpers are typically furnished by the electrical equipment manufacturer for the listed equipment intended to be used as service equipment. The equipment is most often listed by an electrical products testing laboratory in compliance with a product safety standard. For the U.S. safety system, Underwriters Laboratories (UL) produces the safety standards for testing and manufacture of most electrical equipment.

FIGURE 2-36 Main bonding jumper in listed panelboard, *250.28(A)*. (*© Cengage Learning 2012*)

FIGURE 2-37 "Suitable only for use as service equipment" designation, *250.28*. (*© Cengage Learning 2012*)

The panelboard shown in Figure 2-36 is furnished with a bonding jumper that is sized in compliance with the product safety standard. It can then be used without examination to ensure proper size. See *90.7*. A partial list follows of equipment commonly used as service equipment and the product safety standards that apply:

Panelboards	UL67, Panelboards
Switchboards	UL891, Dead-front Switchboards
Power outlets	UL231, Power Outlets
Motor control	UL845, Electric Motor Control Centers
Enclosed switches	UL98, Enclosed Switches

Larger equipment probably will have a bonding jumper that consists of one or more conductors or busbars. Smaller equipment may have a bonding jumper consisting of a screw or strap.

CAUTION: Some service equipment may be marked "Suitable Only for Use as Service Equipment"(see Figure 2-37). Such equipment has the neutral terminal bar bonded to the enclosure at the factory. It is not suitable for use at the feeder position or "downstream" from the service position. Locating the equipment where it would be supplied by a feeder would generally violate the requirement in *250.24(A)(5)* that a grounding connection is not permitted to be made to any grounded conductor on the load side of the service disconnecting means. The UL *White Book* contains the following information at several locations where the phrase "Suitable Only for Use as Service Equipment" appears: "A section marked for use at services may also be used to provide the main control and disconnecting means for a separately derived system." It should be noted that *250.30(A)(1)* generally permits a system bonding jumper at only one location or point. So, this equipment marked to indicate that the neutral is bonded to the enclosure would be considered to be the point where the system bonding jumper is connected. Another connection of the neutral of the separately derived system on the supply side of the equipment marked to indicate that the neutral is bonded to the enclosure is not generally permitted.

250.28(B) Construction of Main and System Bonding Jumpers

The Requirement: Where a main bonding jumper or a system bonding jumper is a screw only, the screw must be identified with a green finish that is required to be visible with the screw installed.

Discussion: The main bonding jumper is permitted to be a screw but must be identifiable after installation (see Figure 2-38). This identification is intended to distinguish the bonding jumper screw from other mounting screws for the neutral terminal bar.

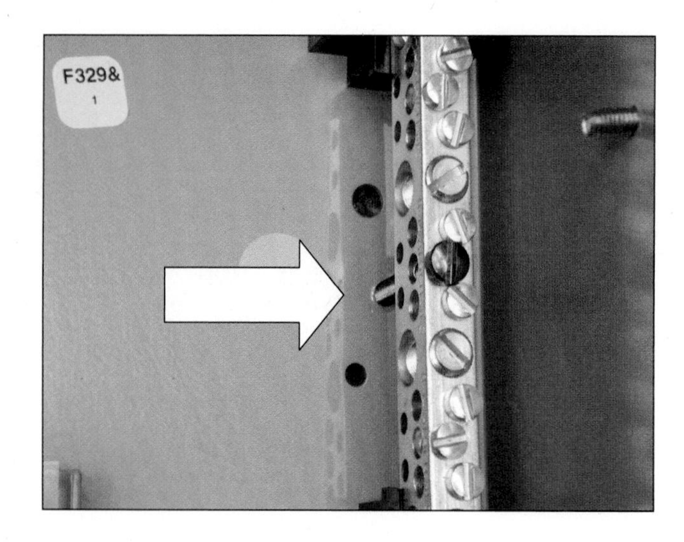

FIGURE 2-38 Screw as main bonding jumper, *250.28(B)*. (© Cengage Learning 2012)

Electrical product safety standards dictate the size of screw required for various panelboards and similar equipment. The size of the screw can vary, depending on the ampere rating of the equipment such as a panelboard. Nothing in the *NEC* describes the minimum size of the bonding screw. Again, the product safety standard dictates the appropriate size, which then can be used without calculation.

250.28(C) Attachment of Main and System Bonding Jumpers

The Requirement: Main bonding jumpers and system bonding jumpers are required to be connected in the manner specified by the applicable provisions of *250.8*.

Discussion: Although this section refers to *250.8* for the general rules for connecting the bonding jumper, the connection also must be made in compliance with the manufacturer's instructions. These installation instructions often dictate the location for installation of the bonding jumper, as well as providing a requirement that the fasteners be tightened to a specific torque.

250.28(D) Size of Main and System Bonding Jumpers

250.28(D)(1) General

The Requirement: Main bonding jumpers and system bonding jumpers must not be smaller than the sizes shown in *Table 250.66*. If the supply conductors are larger than 1100-kcmil copper or 1750-kcmil aluminum, the bonding jumper is required to have an area that is not less than 12½% of the area of the largest phase conductor.

Discussion: As discussed previously, main and system bonding jumpers in listed equipment such as switchboards and panelboards can be used without the need to calculate the appropriate size. The manufacturer of the listed equipment is required to comply with the sizing rules in the appropriate product safety standard. The bonding jumper can be used on the basis of the listing, as covered in *90.7* of the *NEC*.

Proceed as follows when it is necessary to determine or calculate the minimum size of the main or system bonding jumper:

Here's How: For ungrounded copper system conductors of 1100 kcmil or smaller or ungrounded aluminum system conductors of 1750 kcmil or smaller, simply determine the largest circular mil area of the ungrounded conductors; then refer to *Table 250.66* for the minimum size of main or system bonding jumper. As discussed previously, *Table 250.66* is used for several purposes other than that indicated by the title, which is selection of the grounding electrode conductor.

For example, consider an 800-ampere service or separately derived system that is being installed. Two sets of 500-kcmil Type XHHW conductors are installed in parallel.

$$500 \text{ kcmil} \times 2 = 1000 \text{ kcmil}$$

Table 250.66 requires a 2/0 AWG copper main or system bonding conductor.

Here's How: For ungrounded copper system conductors larger than 1100 kcmil or aluminum system conductors larger than 1750 kcmil, proceed as follows:

1. Determine the total circular mil area of the largest set of conductors.
2. Multiply this total by 12½% (0.125).
3. Refer to *Table 8* of *Chapter 9* if the result is less than 250 kcmil.
4. Round up to the next standard size if necessary.

For example, a 1200-ampere service is being installed. Four 350-kcmil copper conductors are installed in parallel for the ungrounded system conductors.

$$4 \times 350 \text{ kcmil} = 1400 \text{ kcmil}$$

$$1400 \text{ kcmil} \times 1000 = 1{,}400{,}000 \text{ cm}$$

$$1{,}400{,}000 \text{ cm} \times 0.125 = 175{,}000 \text{ cm}$$

The next larger standard size after 175,000 cm is 4/0 AWG (211,600 cm; *Table 8, Chapter 9*), which must be installed if the main bonding jumper supplied by the manufacturer of listed service equipment is not installed.

250.28(D)(2) Main Bonding Jumper for Service with More Than One Enclosure

The Requirement: If a service consists of more than a single enclosure as permitted in *230.71(A),* the main bonding jumper for each enclosure is required to be sized in accordance with *250.28(D)(1),* based on the largest ungrounded service conductor serving that enclosure.

Discussion: This requirement was added during the processing of the 2008 *NEC* and adds clarity for installations of services consisting of more than a single service disconnecting means enclosure. *Section 230.71(A)* permits up to six separate enclosures to be used for the service disconnecting means.

The rule for sizing the main bonding jumper for each individual enclosure is identical to the rule for sizing a main bonding jumper for a single enclosure. The size of the service-entrance conductor that is connected to the supply equipment terminals is used to determine the minimum size of the main bonding jumper for each enclosure. This was reviewed in the discussion on *250.28(D)(1), General.* Once again, if listed service equipment is installed, the manufacturer will furnish the proper size of main bonding jumper. It can be installed without calculation as to its proper size.

250.28(D)(3) Separately Derived System with More Than One Enclosure

The Requirement: If a separately derived system supplies more than a single enclosure, the system bonding jumper for each enclosure is required to be sized in accordance with *250.28(D)(1),* based on the largest ungrounded feeder conductor serving that enclosure, or a single system bonding jumper must be installed at the source and sized in accordance with *250.28(D)(1),* based on the equivalent size of the largest supply conductor, determined by the largest sum of the areas of the corresponding conductors of each set.

Discussion: This requirement was added during the processing of the 2008 *NEC* and adds clarity for installations of separately derived systems consisting of more than a single disconnecting means enclosure.

The system bonding jumper is generally permitted to be installed at any single point from the source of the separately derived system to the first system disconnecting means. See *250.30(A)(1).* The system bonding jumper is required to be at the source if it is located outdoors. If multiple disconnecting means are installed from a separately derived system, the system bonding jumper must be installed at either the source (a single point) or at the individual enclosures (a single point in each). The "single point rule" in *250.30(A)(1)* would not permit the system bonding jumper to be installed at the source for supply to one disconnecting means and at the disconnecting means for another feeder.

If a system bonding jumper is to be installed in each disconnecting means enclosure, the size of the derived conductor that is connected to the system disconnecting means terminals is used to determine the minimum size of the system bonding jumper for each enclosure. This was reviewed in the discussion on *250.28(D)(1), General.* Once again, if the disconnecting means being installed is listed as service equipment (though in this instance it is being used as a feeder panelboard and not as a service disconnecting means), the manufacturer will furnish the proper size of system bonding jumper. The UL *Guide Information for Electrical Equipment (White Book)* states under Panelboards (QEUY), "Panelboards marked for use at services may also be used to provide the main control and means of cutoff for a separately derived system."

If the system bonding jumper is installed at the source of the separately derived system, it must be sized for the total circular mil area of the derived conductors.

Here's How: As shown in Figure 2-39, a separately derived system supplies two feeders. One

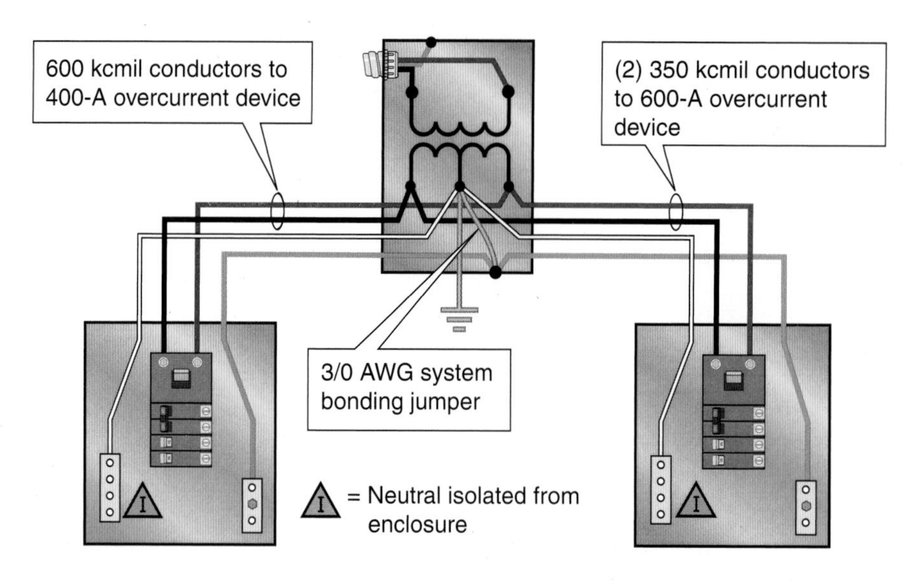

FIGURE 2-39 Separately derived system with more than one enclosure, *250.28(D)(3).* (*© Cengage Learning 2012*)

feeder is to a 400-ampere panelboard, and the derived conductors are 600-kcmil copper. The second feeder is to a 600-ampere panelboard, and the supply conductors are two 350 kcmil copper conductors in parallel. A single system bonding jumper is to be installed from a bonding terminal bar installed in the transformer to the XO terminal in the transformer.

The sizing of the system bonding conductor must comply with *250.28(D)(1).* That section requires the system bonding jumper be not smaller than a grounding electrode conductor from *Table 250.66,* based on the size of the derived conductors. If the derived conductors are larger than 1100-kcmil copper, the system bonding jumper is not permitted to be smaller than 12½% of the circular mil area of the derived conductors. The following table shows how to calculate the system bonding jumper for the installation shown in Figure 2-39.

Feeder	Derived Conductors	Number per Phase	Circular Mil Area
400 Ampere	600 kcmil	1	600 kcmil
600 Ampere	350 kcmil	2	700 kcmil
		Total Area	1300 kcmil
System Bonding Jumper 12½% of 1300 kcmil			162,500 cm

The next standard size of conductor from *Table 8* of *Chapter 9* is 3/0 AWG with a circular mil area of 167,800, which must be installed.

250.30 GROUNDING SEPARATELY DERIVED ALTERNATING-CURRENT SYSTEMS

The Requirements: In addition to complying with *250.30(A)* for grounded systems or as provided in *250.30(B)* for ungrounded systems, separately derived systems shall comply with *250.20, 250.21, 250.22* and *250.26.*

Discussion: This is a new introductory paragraph for *250.30.* It requires separately derived system to comply with the rules that apply as follows:

- *250.20* gives the rules on those systems that are required to be grounded
- *250.21* states the systems that are permitted but not required to be grounded
- *250.22* provides a list of the circuits or systems that are not permitted to be grounded
- *250.26* provides rules on which conductor is required to be grounded for those systems where grounding is required or performed if system grounding is optional.

- *250.30(A)* contains the rules for grounding and bonding separately derived systems that are grounded
- *250.30(B)* provides the requirements for grounding equipment related to ungrounded separately derived systems.

Two *Informational Notes* provide helpful information:

> **Informational Note No. 1:** An alternate ac power source such as an on-site generator is not a separately derived system if the grounded conductor is solidly interconnected to a service-supplied-system grounded conductor. An example of such situations is where alternate source transfer equipment does not include a switching action in the grounded conductor and allows it to remain solidly connected to the service-supplied grounded conductor when the alternate source is operational and supplying the load served.
>
> **Informational Note No. 2:** See 445.13 for the minimum size of conductors that must carry fault current.

Discussion: As mentioned in the Introduction chapter in this book, the term *separately derived system* is defined in *NEC Article 100* as "A premises wiring system whose power is derived from a source of electric energy or equipment other than a service. Such systems have no direct connection from circuit conductors of one system to circuit conductors of another system, other than connections through the earth, metal enclosures, metallic raceways, or equipment grounding conductors."*

The second sentence of the definition is a significant revision from that in earlier editions of the *NEC*. As shown in Figure I–19, a circuit is established from the system supplied by the utility to the separately derived system through equipment grounding conductors and through the earth. These paths connect the grounded conductors of the two systems together. The ungrounded conductors, referred to in the definition as *circuit conductors*, are not interconnected.

In reality, based on the previous definition of separately derived system, almost no systems were *separately derived,* as a path from the grounded conductor of one system to the grounded conductor of the other system is commonplace. See Figure I–19 for a graphic representation of the definition of separately derived systems.

A separately derived system can be recognized by determining how the grounded (often a neutral) conductor is treated. The phrase "no direct connection from circuit conductors of one system to circuit conductors of another system, other than connections through the earth, metal enclosures, metallic raceways, or equipment grounding conductors" is the key.

To qualify as a separately derived system, the system must meet the following tests:

1. The system is a premises wiring system (see the definition of this term in *Article 100*).
2. The power is derived from a source of electric energy or equipment other than a service.
3. No direct connection from circuit conductors of one system to circuit conductors of another system exists, other than connections through the earth, metal enclosures, metallic raceways, or equipment grounding conductors.

Sources of electrical energy include a battery, a solar photovoltaic system, and a fuel cell. These sources of energy produce dc, which has to be converted to ac through an inverter. Generators are most commonly of the ac type, which are used to supply power into the electrical system. Separately derived systems also can be supplied from transformer or converter windings. In the strictest sense, a transformer is not a *source of electric energy*. Transformers simply modify the source voltage either higher or lower, depending on the ratio between the number of turns in the primary and the secondary windings. However, the *NEC* has long recognized transformers as separately derived systems, depending on how they are connected. A fairly simple test can determine whether a transformer or a generator is a separately derived system.

Here's How: For transformers, check to be sure there is not a bonding jumper from primary to secondary. The primary may be of any voltage, and the secondary of any voltage. For the transformer in

*Reprinted with permission from NFPA 70-2011.

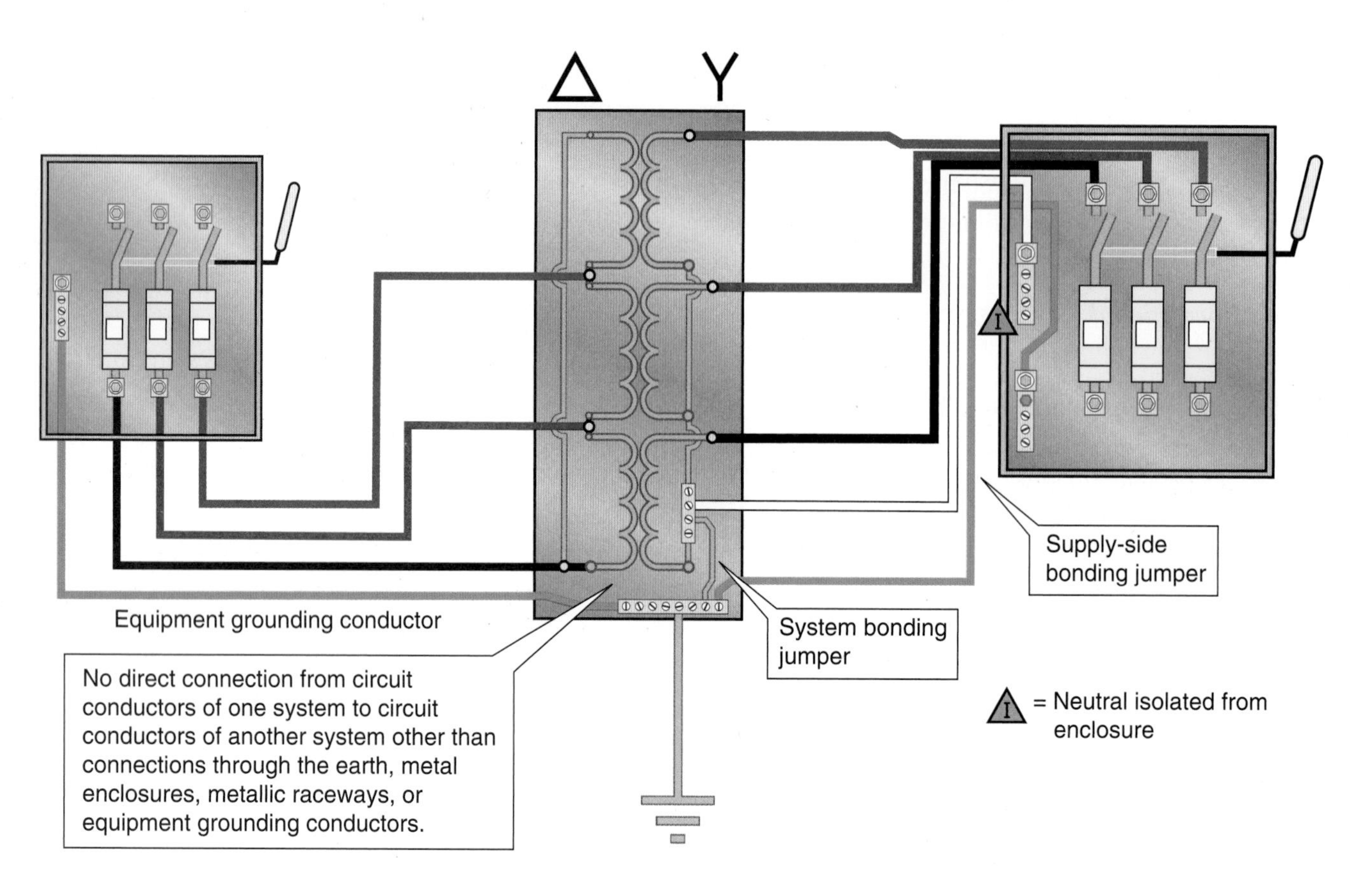

FIGURE 2-40 Transformer-type separately derived system, *250.30*. (*© Cengage Learning 2012*)

Figure 2-40, assume the supply voltage is 480 volts and the secondary voltage is 208Y/120. So long as a bonding jumper is not installed from primary to secondary, the system is separately derived. Connection of the equipment grounding conductor to the transformer enclosure on the primary and the installation of the system bonding jumper and the supply-side bonding jumper on the secondary are permitted within the definition. These connections do not constitute a *direct connection from circuit conductors of one system to circuit conductors of another system* as included in the definition of the *separately derived system.*

Here's How: For generators, the key to determining whether the system is separately derived is to look at the transfer switch to see how the neutral or grounded conductor is handled. If the grounded conductor (often a neutral) is switched, the system is separately derived (see Figure 2-41). If the grounded conductor (often a neutral) is not switched, the system is not separately derived.

Separately derived systems are required to be grounded and bonded in accordance with *250.30* if the system produced is required to be grounded in *250.20.*

Systems, such as from transformers or generators, that are not separately derived are required to comply with other sections of *Article 250.* See *250.35* for specific rules that apply to generators that are permanently installed. Other systems are grounded and bonded as any other electrical equipment. The electrical system is grounded by the electric utility and again at the service. Equipment grounding and bonding conductors are installed to provide a ground-fault return path back to the source, which is the utility transformer. The neutral or grounded conductor is generally required to be electrically isolated from electrical enclosures throughout the system beyond the service.

Autotransformers by their very nature are not separately derived systems, because one conductor is common to both primary and secondary

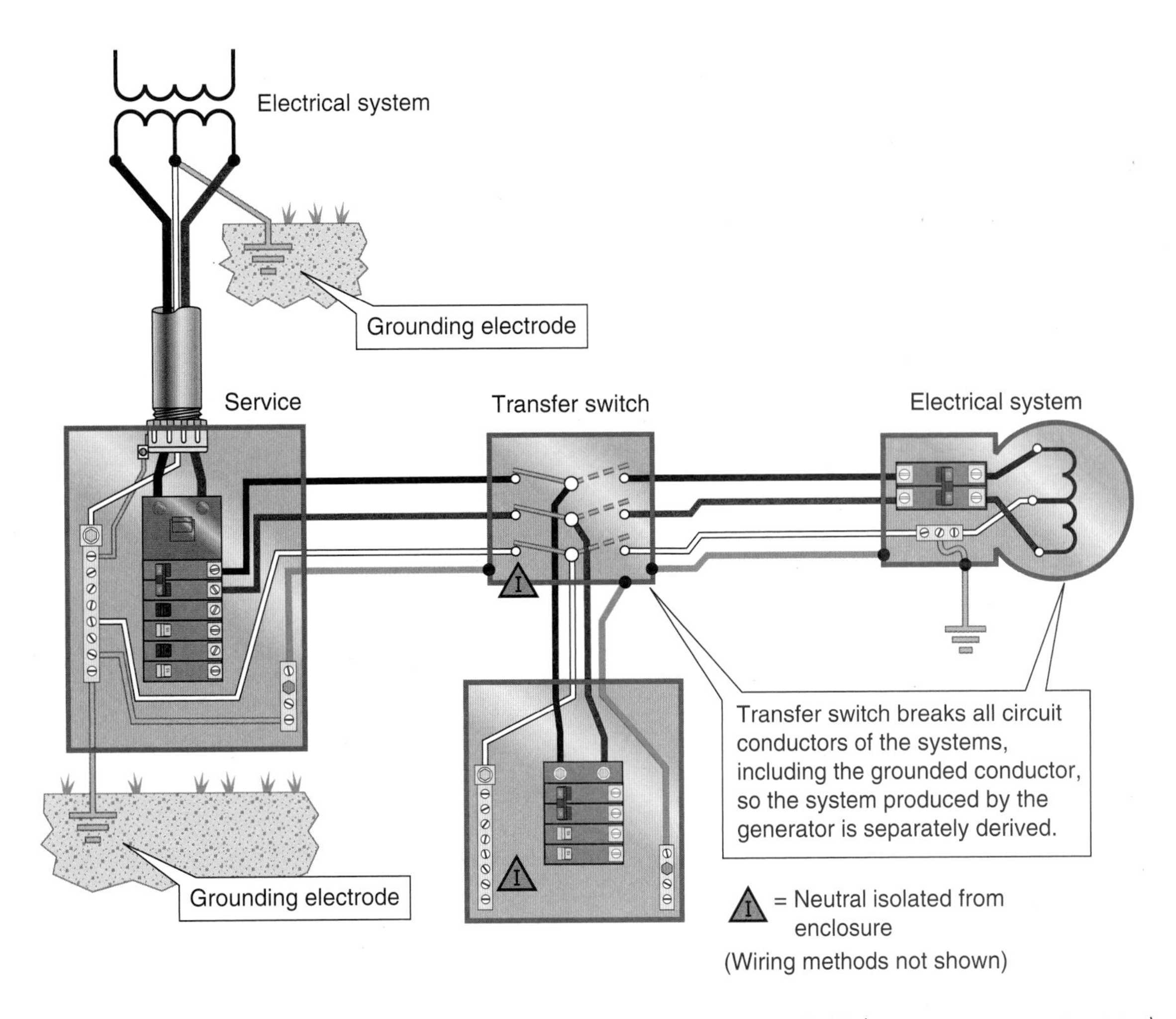

FIGURE 2-41 Generator-type separately derived ac systems, *250.30.* (*© Cengage Learning 2012*)

windings. See Figure 2-42. Autotransformers are available in many sizes and configurations. They are often referred to as "buck/boost" transformers. Often, the same transformer can be connected to either "buck" or "boost" the voltage. If connected in the "buck" orientation, the voltage is reduced. If connected in the "boost" orientation, the voltage is increased. If the autotransformer has a grounded conductor, it must be the common conductor. As a result, it is not permitted to ground the grounded conductor at or beyond the autotransformer if the autotransformer has a common grounded conductor.

NEC 250.30 requires that the system produced by a separately derived system be considered in light of the requirements for grounding in *250.20(A)* and *(B)*. If the separately derived system meets the voltage, phase, or usage rules in *250.20(A)* or *(B)*, it must be grounded. If the system does not meet those rules, grounding of the system generally is optional. Recall that there are a few circuits that are not permitted to be grounded, as provided in *250.22*. For separately derived systems that are required to be grounded, grounding is accomplished in accordance with the rules in *250.30(A)*.

250.30(A) Grounded Systems

The Requirements: A separately derived ac system that is grounded is required to comply with *(A)(1)* through *(A)(8)*. Except as otherwise permitted in *Article 250,* a grounded conductor is not permitted to be connected to normally non-current-carrying metal parts of equipment, to equipment grounding conductors or be reconnected to ground on the load side of the system bonding jumper.

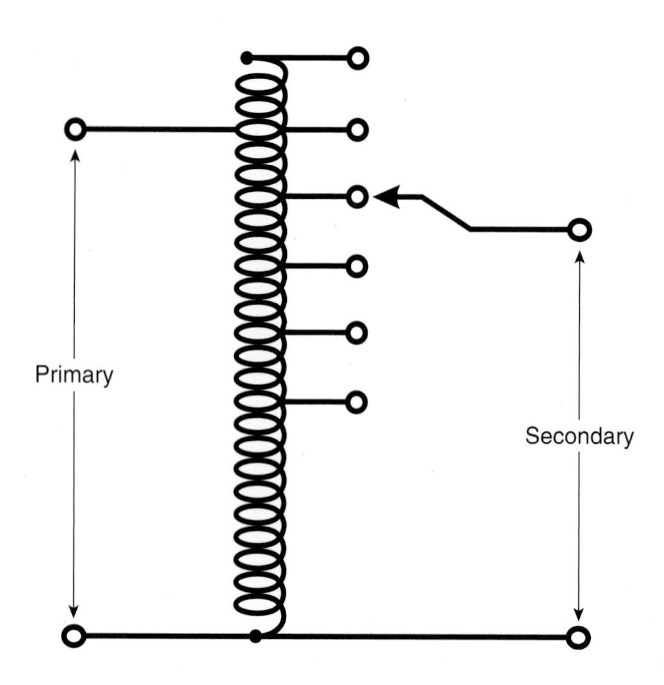

FIGURE 2-42 Autotransformer has a conductor common to primary and secondary. (© Cengage Learning 2012)

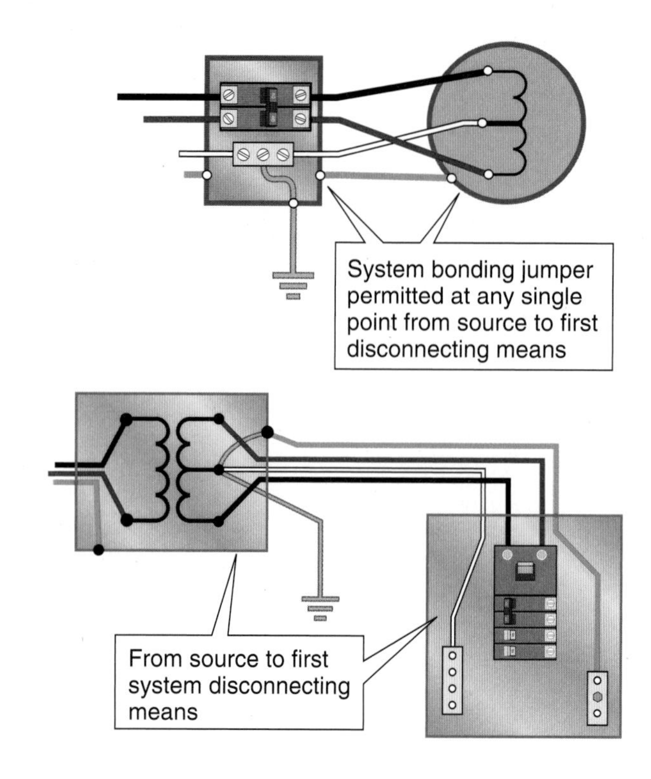

FIGURE 2-43 System bonding jumper, *250.30(A)(1)*. (© *Cengage Learning 2012*)

Discussion: This rule mirrors the requirement that the neutral not be regrounded on the load side of the service disconnecting means in *250.24(A)(5)*. The grounded conductor is permitted to be grounded again under specific conditions in *250.30(A)(1), Exception No. 2,* and for existing buildings or structures that were wired in that manner when previous editions of the *NEC* permitted doing so. See *250.32(B)(1), Exception*. It is also permitted to use the grounded conductor for grounding certain equipment under specific conditions provided in *250.142*.

250.30(A)(1) System Bonding Jumper

The Requirements: An unspliced system bonding jumper shall comply with *250.28(A)* through *(D)*. This connection must be made at any single point on the separately derived system from the source to the first system disconnecting means or overcurrent device, or it must be made at the source of a separately derived system that has no disconnecting means or overcurrent devices, in accordance with *250.30(A)(1)(a)* or *(b)*. The system bonding jumper is required to remain within the enclosure where it originates. If the source is located outside the building or structure supplied, a system bonding jumper is required to be installed at the grounding electrode connection in compliance with *250.30(C)*.

Discussion: The system bonding jumper in a separately derived system performs a function identical to that of the main bonding jumper in a service. It provides the essential link between the equipment grounding conductor(s) or the metal enclosures for equipment to the source of the separately derived system. These metal enclosures include the transformer and cabinets for panelboards. Without the system bonding jumper, no path exists for ground-fault current to return to the source. As a result, overcurrent devices will not operate to clear line-to-enclosure faults. (See Figure 2-43.)

The system bonding jumper is required to be composed of and sized in an identical manner as the main bonding jumper for the service equipment. As can be seen, all of the rules in *250.28* for the main bonding jumper apply to the system bonding jumper for separately derived systems (Figure 2-44). The size of the derived conductors is used in *Table 250.66* to determine the minimum size of the system bonding jumper. As discussed in *250.28,* a main bonding jumper furnished as a part of listed service

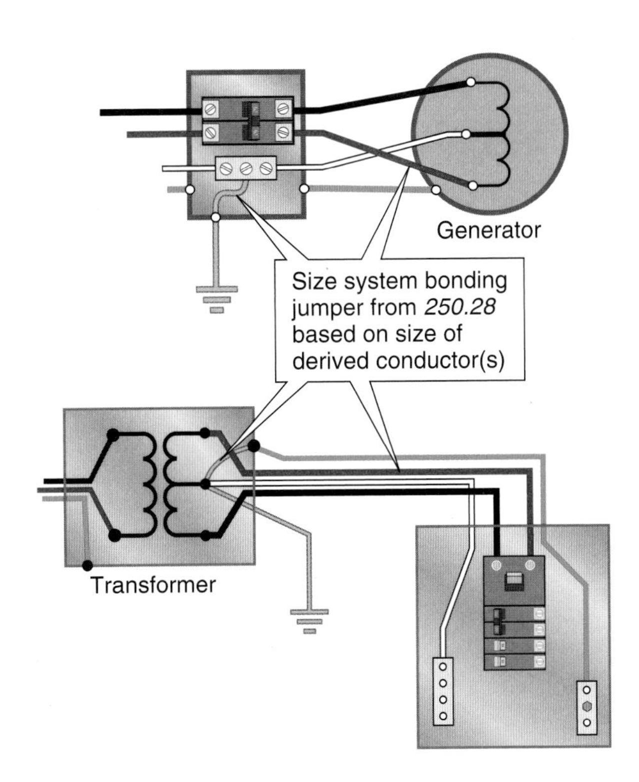

FIGURE 2-44 Size of system bonding jumper, *250.30(A)(1)*. (*© Cengage Learning 2012*)

equipment also can be used without calculation as the system bonding jumper for separately derived systems. If a wire or busbar system bonding jumper is field-installed in a transformer or generator, the appropriate size must be determined by the methods discussed in *250.28(D)* in this book.

The general rule is that a system bonding jumper is required and permitted to be installed at only one location for the separately derived system (see Figures 2-40, 2-41, 2-43, and 2-44). This embodies the concept of single-point grounding. A single point is established for the grounding electrode and system bonding conductor connection. As will be seen later, the single connection point can be located at one of several locations. This single point for grounding and bonding connection reduces the likelihood of parallel paths for neutral current.

> **Exception No. 1:** For systems installed in accordance with *450.6,* a single system bonding jumper connection to the tie point of the grounded circuit conductors from each power source is permitted.

Discussion: This exception is identical in scope and function to the exception in *250.24(A)(3)* for services. For larger separately derived systems, two transformers from different separately derived systems may be tied together on the secondary to provide extra capacity or system operational redundancy. A single system bonding jumper that is installed at the tie point is permitted in this case.

As can be seen in Figure 2-45, the two *separately derived systems* that have a common neutral connection no longer meet the definition of separately derived systems in *NEC Article 100,* as the neutral conductor is a *circuit conductor.* When the two previously separately derived systems cease to meet the definition, in reality the rules in *250.30(A)* no longer apply. Were it not for this exception, the other pertinent rules in *Article 250* would be applicable.

> **Exception No. 2:** A system bonding jumper at both the source and the first disconnecting means is permitted where installing them does not establish a parallel path for the grounded conductor. If a grounded conductor is used in this manner, it must not be smaller than the size specified for the system bonding jumper in *250.28* but is not required to be larger than the ungrounded conductor(s). For the purposes of this exception, connection through the earth is not considered as providing a parallel path.

Figure 2-46 shows a system bonding jumper installed at the source and the first system disconnecting means.

Discussion: This exception allows the grounded conductor from the separately derived system to be used to bond the enclosure for the first overcurrent device or disconnecting means to the enclosure for the separately derived system, provided that installing a bonding jumper at both locations does not result in providing a parallel path for current to flow. If the grounded system conductor (often a neutral) is used as the bonding conductor between the enclosures, it must be sized not smaller than a supply-side bonding jumper. As such, the grounded conductor is serving several roles, including providing the path for unbalanced current to return to the source, bonding the enclosures together, and providing the return path for ground-fault currents. Sizing the bonding conductor requires going to *250.102(C),* which sends the installer to *Table 250.66* for ungrounded conductors

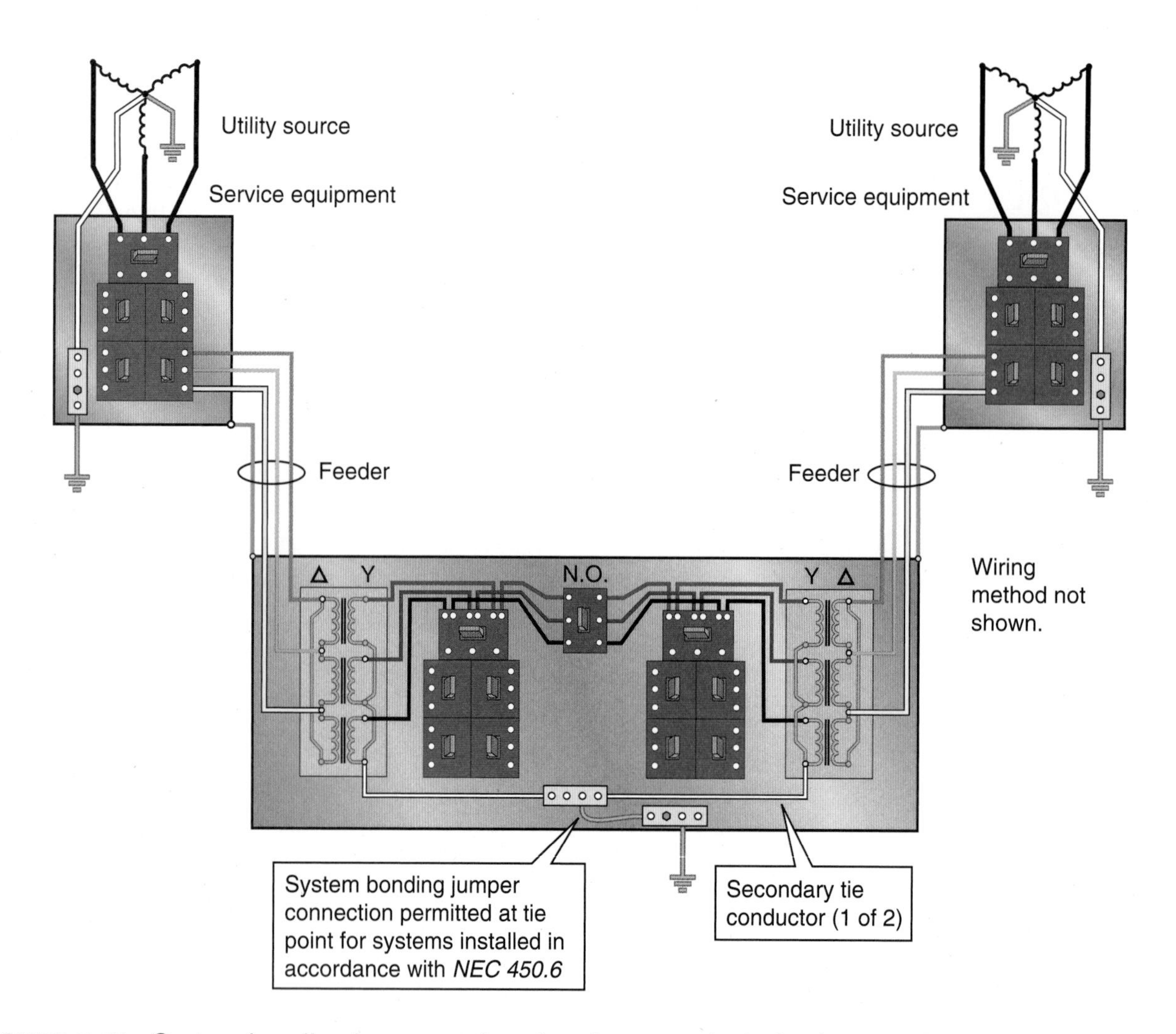

FIGURE 2-45 System bonding jumper at tie point of separately derived system installed in accordance with *450.6. 250.30(A)(1), Exception No. 1.* (*© Cengage Learning 2012*)

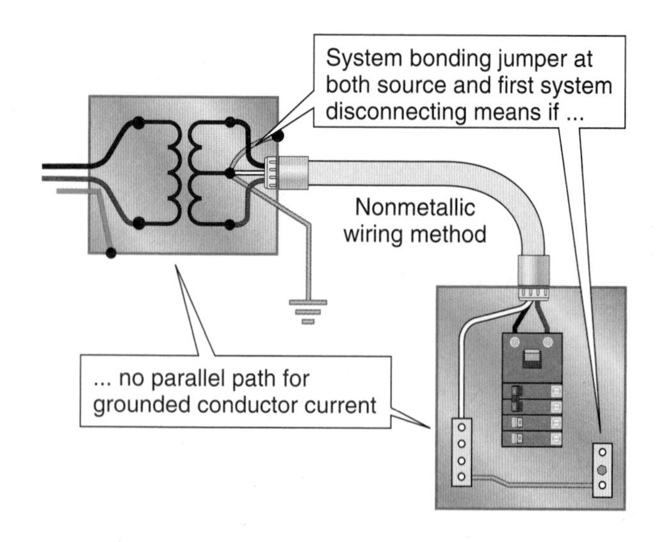

FIGURE 2-46 System bonding jumper at both source and first system-disconnecting means, *250.30(A)(1), Exception No. 2.* (*© Cengage Learning 2012*)

within the scope of the table. For larger installations that do not fit in *Table 250.66,* a calculation using the 12½% rule must be used. This procedure is explained in detail in *250.28(D)* of this book.

As discussed earlier, when a parallel path exists, current will divide among all the paths that are available. There is no sound engineering principle for allowing neutral current to flow on other than the grounded-circuit conductor for separately derived systems. If the grounded conductor is installed improperly, a parallel path for neutral current can be established by one of the following methods:

1. Using metal wiring methods between the enclosure for the separately derived system and the disconnecting means
2. Mounting the metal enclosure for the separately derived system and the disconnecting

means on common conductive materials such as steel columns and conduit or pipe racks

3. Using metal piping systems or metal structural members for the grounding electrode for the separately derived system and equipment grounding conductors that subsequently have contact with the grounded surfaces

Many other possible interconnections exist in industrial and commercial facilities.

Granted, a parallel path for neutral current exists at the service, depending on how the equipment is installed and what the wiring method is. This is discussed later, in *250.102* in this book. However, the reason for permitting the parallel path at the service does not exist at a separately derived system.

It appears a conflict was introduced in processing the 2011 *NEC* that affects this exception. A new rule in *250.30(A)(2)* requires a supply-side bonding jumper to be installed if the source and the first disconnecting means are installed in separate enclosures. So, when the supply-side bonding jumper is installed to comply with the new rule, the exception to *250.30(A)(1)* can no longer be used because bonding the neutral at each end of the circuit would create a parallel path with the supply-side bonding jumper.

It is unknown at the time of this writing if changes will be made to coordinate these rules before the processing of the 2014 *NEC*. For additional information, you can check at www.nfpa.org or www.necplus.org.

Exception No. 3: The size of the system bonding jumper for a system that supplies a Class 1, Class 2, or Class 3 circuit, and is derived from a transformer rated not more than 1000 volt-amperes, must not be smaller than the derived phase conductors and not be smaller than 14 AWG copper or 12 AWG aluminum.

Discussion: These small transformers are often used for control circuits or systems. The enclosure for the transformer is permitted to be used as the grounding electrode for the system. If the system is required to be grounded by *250.20,* a bonding jumper is permitted to be installed from the secondary of the transformer to the enclosure. This bonding jumper also serves as the grounding electrode for the system.

250.30(A)(1)(a) System Bonding Jumper Installed at the Source

The Requirement: The system bonding jumper is required to connect the grounded conductor to the supply-side bonding jumper and the normally non-current-carrying metal enclosure.

Discussion: The system bonding jumper is permitted to be installed at any single point from the source to the first system disconnecting means. As indicated in this section, if the system bonding jumper is installed at the source, usually at the transformer or generator, it serves to connect the grounded conductor (most often a neutral) to the supply-side bonding jumper and the normally non-current-carrying metal enclosure. The ground-fault current return path with the supply-side bonding jumper located at the source is shown in the upper drawing in Figure 2-47.

250.30(A)(1)(b) System Bonding Jumper Installed at the First Disconnecting Means

The Requirement: The system bonding jumper is required to connect the grounded conductor to the supply-side bonding jumper, the disconnecting means enclosure, and the equipment grounding conductor(s).

Discussion: As indicated in this section, if the *system bonding jumper* is installed at the first disconnecting means, usually at the panelboard or fusible switch, it serves to connect the grounded conductor (most often a neutral) to the supply-side bonding jumper, the disconnecting means enclosure, and the equipment grounding conductor(s).

The ground-fault current return path with the supply-side bonding jumper located at the first disconnecting means is shown in the lower drawing of Figure 2-47. As can be seen, the ground-fault current return path to the source is now using the grounded conductor. This requires the neutral conductor to be sized not smaller than a supply-side bonding jumper or system bonding jumper. This subject is discussed in greater detail in *250.30(A)(3)* in this text.

250.30(A)(2) Supply-Side Bonding Jumper

The Requirement: "If the source of a separately derived system and the first disconnecting means are located in separate enclosures, a supply-side

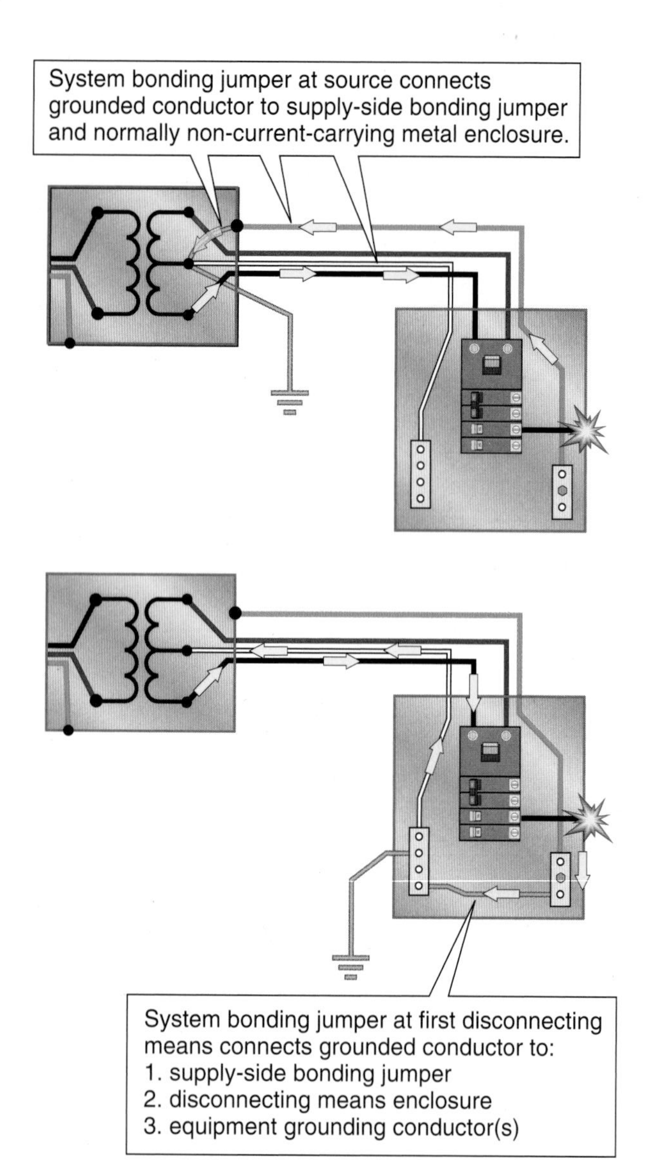

FIGURE 2-47 Connection of system bonding jumper at source and first disconnecting means with ground-fault return paths, *250.30(A)(1)(a)* and *(b)*. (*© Cengage Learning 2012*)

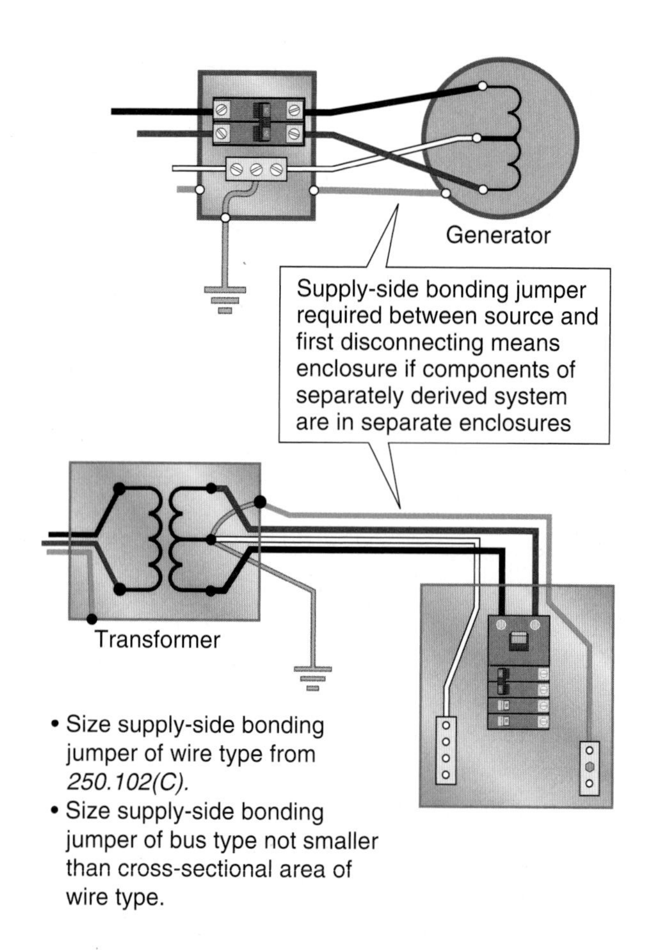

FIGURE 2-48 Requirements for supply-side bonding jumper, *250.30(A)(2)*. (*© Cengage Learning 2012*)

bonding jumper is required to be installed with the circuit conductors from the source enclosure to the first disconnecting means. A supply-side bonding jumper is not required to be larger than the derived ungrounded conductors. The supply-side bonding jumper is permitted to be of the non-flexible metal raceway type or of the wire or bus type as follows:

"a. A supply-side bonding jumper of the wire type is required to comply with *250.102(C)*, based on the size of the derived ungrounded conductors.

"b. A supply-side bonding jumper of the bus type is required to have a cross-sectional area not smaller than a supply-side bonding jumper of the wire type as determined in *250.102(C)*."

Discussion: As illustrated in Figure 2-48, a supply-side bonding jumper is required from the source of the separately derived system to the metal enclosure for the disconnecting means if the components are in separate enclosures. This is necessary whether the system bonding jumper is located at the source or at the first disconnecting means. An effective ground-fault return path must be provided from the enclosure for the source of the separately derived system to the first disconnecting means, to comply with the rules in *250.4(A)(5)* and *250.4(B)(4)*. This return path is necessary to enable enough current to return to the source to clear a ground fault regardless of where it occurs.

As can be seen, sizing of the bonding jumper, if of the wire type, is determined by reference to

250.102(C). The title of that section indicates that the rules apply to the size of the supply-side bonding jumper. Because of similarities between wiring on the supply side of the service and the conductors between the source of the separately derived system and the overcurrent protection, the same rules apply. The reason is conductors from the utility source to the service do not have overcurrent protection. Overload protection only (not overcurrent protection) is provided by connecting the conductors to one or more overcurrent protective devices at the load end of the service conductors. Similarly, overcurrent protection of conductors from the source of the separately derived system is provided at their load end but not at their line end.

In keeping with similar rules for sizing the service grounded or neutral conductor as well as the main and system bonding conductors, the supply-side bonding jumper from the enclosure for the separately derived system and the first disconnecting means is sized from *Table 250.66* according to the size of the ungrounded derived conductor. When ungrounded phase conductors are larger than 1100-kcmil copper or 1750-kcmil aluminum, the bonding jumper has to comply with the 12½% rule.

Here's How: The method for determining the minimum size of the supply-side bonding jumper is covered in detail in the earlier discussion of *250.24(C)* as well as in *250.28(D)* for main and system bonding jumpers. Simply determine the circular mil area of the largest phase conductor and use this size in *Table 250.66* to determine the minimum size of the bonding jumper. If the ungrounded conductors from the source to the first disconnecting means and overcurrent protection are paralleled in two or more raceways or cables, the equipment bonding jumper, if routed with the raceways or cables, is required to be run in parallel. The size of the bonding jumper for each raceway or cable is required to be based on the size of the ungrounded conductors in each raceway or cable.

The fault return path from the source of the separately derived system may consist of the wiring method if it is adequate. For example, electrical metallic tubing and wireways are permitted to be used without regard to length and must simply be sized for the contained conductors.

CAUTION: Flexible metal conduit and liquidtight flexible metal conduit have fairly severe restrictions on their use as equipment grounding conductors. Section *250.30(A)(2)* permits *nonflexible metal raceways* to be used for the supply-side bonding conductor or jumper. This language permits rigid metal conduit, intermediate metal conduit, and electrical metallic tubing but excludes flexible metal conduit (FMC) and liquidtight metal conduit (LFMC). Although specifically excluded by this section from use as the supply-side bonding jumper, FMC and LFMC could be used as the wiring method if a properly sized supply-side bonding jumper of the wire type is installed through these wiring methods. Refer to Table 2-3 and to *250.118(5)* and *(6)* for their permitted use.

The supply-side bonding jumper between the separately derived system and the first overcurrent protective device serves as the means to clear a ground fault on the load or downstream side of the source, be it a transformer, generator, or other type of separately derived system. Figure 2-49 shows two scenarios, one where the system bonding jumper is located at the source and another where the system bonding jumper is located at the first overcurrent protective device.

TABLE 2-3

Limitations of Flexible Metal and Liquidtight Flexible Metal Conduits

Wiring Method	Size	Maximum Overcurrent Protection	Length in Ground Return Path	*NEC* Reference
Flexible metal conduit	All sizes	20 amperes	6 ft (1.8 m)	*250.118(5)*
Liquidtight flexible metal conduit	⅜ and ½	20 amperes	6 ft (1.8 m)	*250.118(6)*
	¾ through 1¼	60 amperes	6 ft (1.8 m)	*250.118(6)*
	Larger than 1¼	Not acceptable	Not acceptable	*250.118(6)*

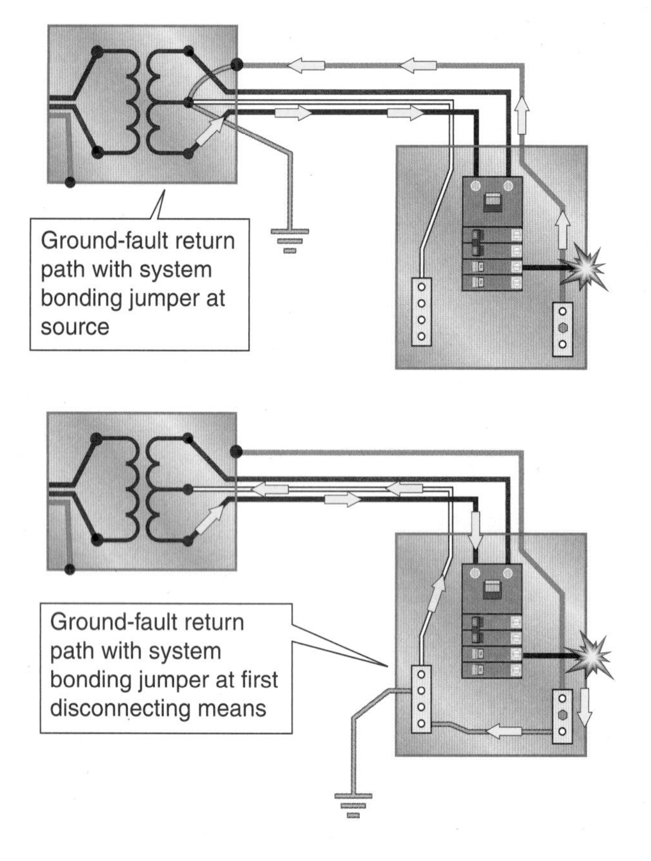

FIGURE 2-49 Ground-fault return path over supply-side bonding jumper and over neutral. (*© Cengage Learning 2012*)

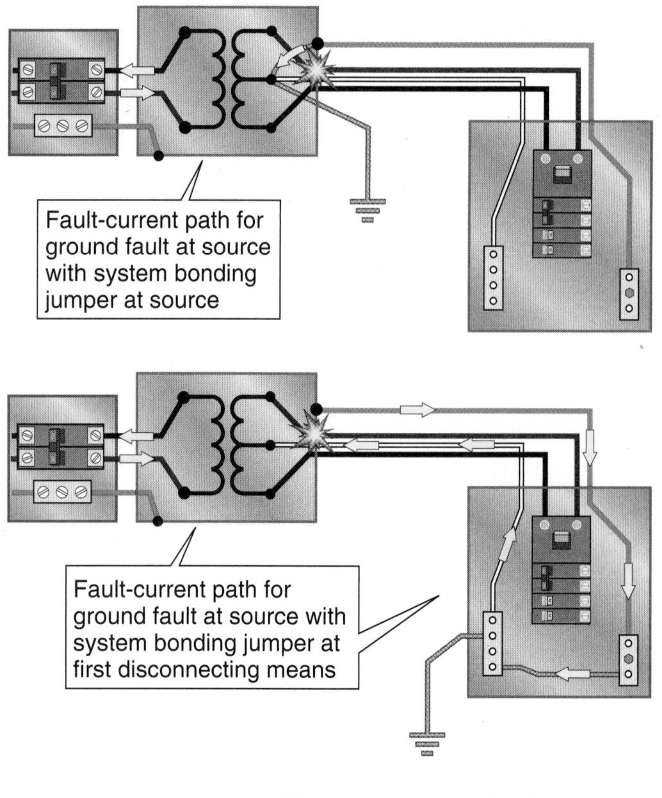

FIGURE 2-50 Clearing ground faults at source of separately derived system. (*© Cengage Learning 2012*)

As shown in the upper drawing of Figure 2-49 with the system bonding jumper in the source of the separately derived system, fault current flows from the source to the panelboard, through an overcurrent device, and returns to the source over the supply-side bonding conductor and the system bonding jumper. In the lower drawing with the system bonding jumper in the panelboard, fault current flows from the source to the panelboard, through an overcurrent device, over the system bonding jumper, and returns to the source over the grounded system conductor. In both installations, there should be sufficient fault current to cause the overcurrent protective device in series with the fault to open the faulted circuit.

Figure 2-50 shows separately derived systems where the fault is located at the source of the system. In the upper drawing, the current from the ground fault in the enclosure has a relatively short path over the system bonding jumper. At this point, the conductors do not have overcurrent protection. The current flowing in the faulted circuit on the secondary of the transformer is reflected in the primary, based on the ratio of the windings of the transformer. The overcurrent protective devices on the primary will open the circuit, provided adequate current is flowing through them. As can be seen, the fault current path on the secondary does not pass through an overcurrent device.

The lower drawing in Figure 2-50 illustrates the current path where the ground fault is in the source enclosure and the system bonding jumper is in the panelboard. Fault current must travel over the supply-side bonding jumper between enclosures, over the system bonding jumper, and return to the source over the grounded system conductor. Like the upper drawing, current flowing in the faulted circuit on the secondary of the transformer is reflected in the primary, based on the ratio of the windings of the transformer. The overcurrent protective devices on the primary will open the circuit, provided adequate current is flowing through them.

250.30(A)(3) Grounded Conductor

The Requirements: If a grounded conductor is installed and the system bonding jumper is *not* located at the source, *250.30(A)(3)(a)* through *(A)(3)(d)* must be complied with.

250.30(A)(3)(a) Sizing for a Single Raceway

The Requirement: The grounded conductor is required to be not smaller than the required grounding electrode conductor specified in *Table 250.66,* but is not required to be larger than the largest ungrounded derived conductor(s). For sets of ungrounded conductors larger than 1100-kcmil copper or 1750-kcmil aluminum, the grounded conductor must not be smaller than 12½% of the area of the largest set of ungrounded derived conductors.

Discussion: This requirement is identical to that for sizing grounded conductors for the service and ensures that the grounded conductor is large enough to carry fault current back to the source (see Figure 2-49). See the discussion in *250.24(C)(1)* in this book.

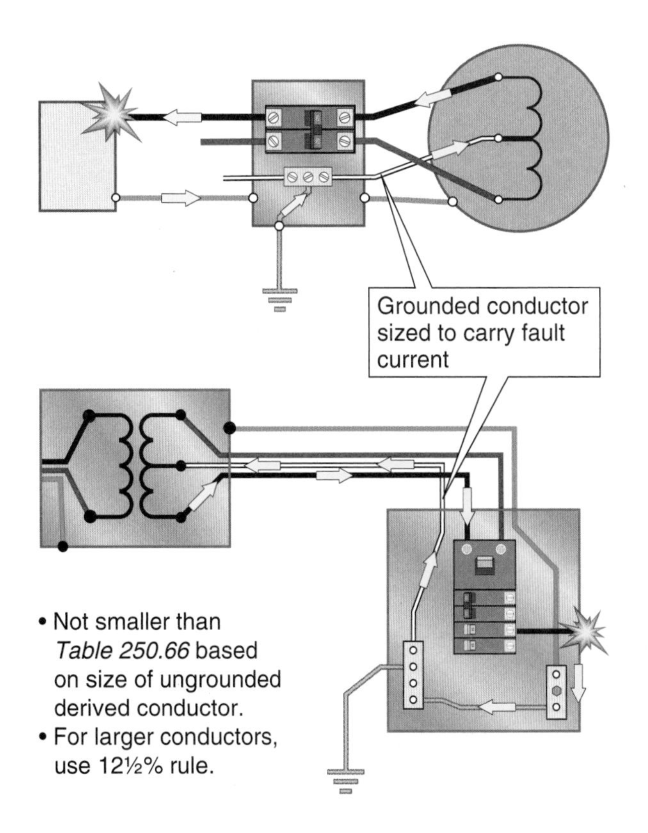

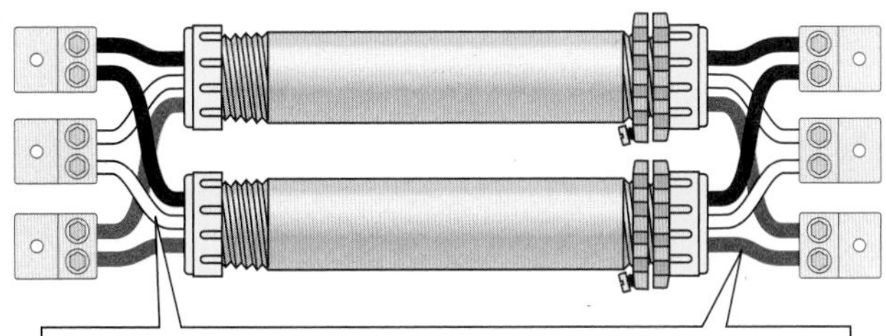

FIGURE 2-51 Minimum size of grounded conductor to serve as ground-fault return path when system bonding jumper is not located at source, *250.30(A)(3)*. (*© Cengage Learning 2012*)

250.30(A)(3)(b) Parallel Conductors in Two or More Raceways

The Requirements: If the ungrounded conductors are installed in parallel in two or more raceways, the grounded conductor is also required to be installed in parallel. The size of the grounded conductor in each raceway is required to be based on the total circular mil area of the parallel derived ungrounded conductors in the raceway as indicated in *(a)* but not smaller than 1/0 AWG.

Discussion: This requirement is identical to that for sizing grounded conductors for the service and ensures that the grounded conductor is large enough to carry fault current back to the source. See Figure 2-51 and the discussion in *250.24(C)(2)* in this book.

250.30(A)(3)(c) **Delta-Connected System** "The grounded conductor of a 3-phase, 3-wire delta system is required to have an ampacity not less than that of the ungrounded conductors."

Discussion: This requirement is identical to that for sizing the grounded conductor for a delta-connected service and ensures that the grounded conductor is large enough to carry fault current back to the source. The grounded conductor is required to be the same size as the ungrounded conductors. See Figure 2-51 and the discussion in *250.24(C)(1)* in this book.

250.30(A)(4) Grounding Electrode

The Requirements: The grounding electrode for the separately derived system is required to be as near as practicable to and preferably in the same area as the grounding electrode conductor connection to the system. The grounding electrode must be the nearest one of the following:

1. A metal water pipe grounding electrode, as specified in *250.52(A)(1)*
2. A structural metal grounding electrode, as specified in *250.52(A)(2)*

Discussion: Although the general purpose for grounding electrical systems contained in *250.4(A)(1)* indicates that the system be connected to earth in a manner that will limit the voltage imposed by lightning, line surges, or unintentional contact with higher-voltage lines and that will stabilize the voltage to earth during normal operation, the primary reason for grounding separately derived systems is indicated by the grounding electrodes required to be used. Connecting the separately derived systems to the grounding electrodes mentioned is primarily for the purpose of bonding electrical structures or equipment that could become energized, and to equalize the potential of equipment supplied by the separately derived system to that of equipment that is supplied from the service. See Figure 2-52.

The grounding electrode is required to be the nearest effectively grounded metal water pipe or effectively grounded structural metal grounding electrode. Both the water pipe grounding electrode and the structural metal member grounding electrode are required to be bonded and become a part of the grounding electrode system, so they would be connected to the service neutral or grounded conductor. The separately derived system then uses the same water pipe or structural metal member as the grounding electrode. Potential differences are thereby equalized.

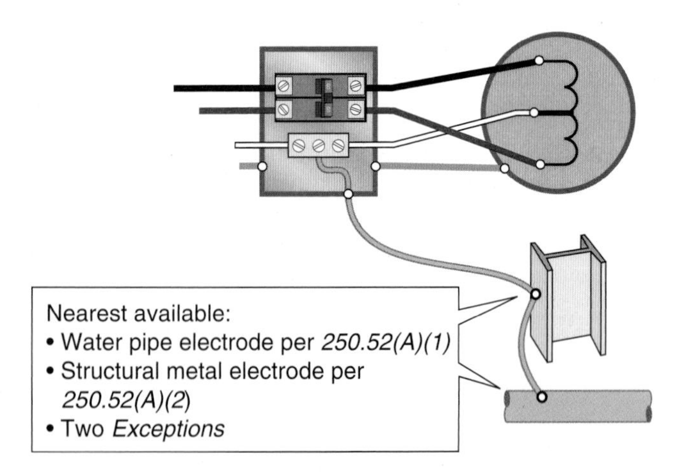

FIGURE 2-52 Grounding electrodes for separately derived systems, *250.30(A)(4)*. (*© Cengage Learning 2012*)

The previous *250.52(A)(1)* requiring the grounding electrode connection to water pipe be made within 5 ft (1.52 m) from the point of entrance to the building has been relocated to *250.68(C)(1)*. An exception is included for industrial and commercial buildings or structures in which conditions of maintenance and supervision ensure that only qualified persons service the installation. At locations in compliance with this rule, interior metal water piping located more than 5 ft (1.52 m) from the point of entrance is permitted to be used as a conductor to connect to the water pipe grounding electrode. It is required that the entire length of the interior metal water pipe being used for the grounding electrode conductor be exposed, other than for short sections passing perpendicularly through walls, floors, and ceilings. Note that the definition of *exposed* in *Article 100* does not require that the interior metal piping be visible for its entire length.

The reference to *250.52(A)(2)* deals with ways to determine whether the metal frame of the building or structure is connected to earth and thus becomes a grounding electrode. Bonding the structural metal to a concrete-encased grounding electrode is recognized to establish the structural metal as a grounding electrode.

Note that if it is determined that the metal water piping system and/or the metal structure of the building in the vicinity of the separately derived system is not effectively grounded and is not required to be used as a grounding electrode for the separately derived system, requirements are in *250.104(D)* for bonding these appurtenances. Interior metal water piping and structural metal members of the building are required either to be used as the grounding electrode for the separately derived system or to be bonded to the separately derived system.

> **Exception No. 1:** Any of the other electrodes identified in *250.52(A)* are required to be used where the electrodes specified above in *250.30(A)(4)* are not available.

Discussion: The other electrodes referred to in *250.52(A)* include concrete-encased electrodes, ground rings, rod and pipe electrodes, plate electrodes, other listed electrodes such as chemically charged rods, and other local grounding electrodes such as underground tanks or piping systems. These other electrodes typically are outside of the building or structure and may be at a greater distance from the separately derived system.

CAUTION: Installing another grounding electrode for any purpose, including for a separately derived system, does not relieve the installer from the requirement to bond all of the grounding electrodes together. See *250.50* and *250.58.*

> **Exception No. 2 to (1) and (2):** If a separately derived system originates in listed equipment suitable for use as service equipment such as a unit substation, the grounding electrode used for the service or feeder equipment is permitted to be used as the grounding electrode for the separately derived system.

250.30(A)(5) Grounding Electrode Conductor, Single Separately Derived System

The Requirement: A grounding electrode conductor for a single separately derived system is required to be sized in accordance with *250.66* for the derived ungrounded conductors. It must be used to connect the grounded conductor of the derived system to the grounding electrode, as specified in *250.30(A)(4).* This connection must be made at the same point on the separately derived system where the system bonding jumper is installed. Figure 2-53 shows this configuration schematically.

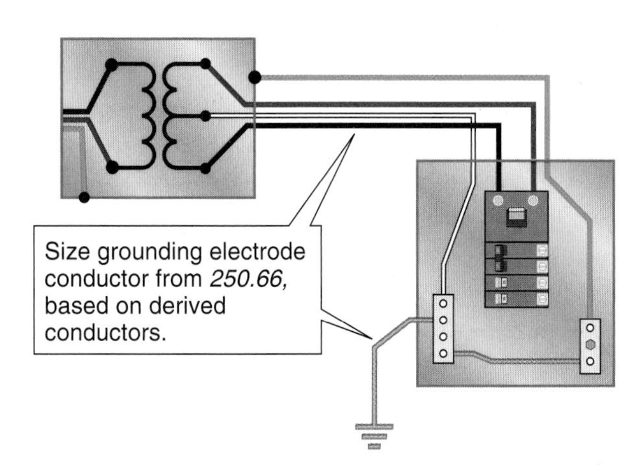

FIGURE 2-53 Grounding electrode conductor, single separately derived system, *250.30(A)(5).* (*© Cengage Learning 2012*)

Discussion: By now, the requirements are looking familiar and almost repetitious. The grounding electrode conductor connects the system conductor that is required to be grounded to the grounding electrode. The grounding electrode conductor is sized from *250.66* in accordance with the size of the supply ungrounded conductors.

Here's How: Assume that a separately derived system supplies a 600-ampere panelboard. Two copper conductors with 75°C insulation are selected on the basis of *Table 310.15(B)(16)* and are to be connected in parallel.

$$350 \text{ kcmil THWN at } 310 \text{ amperes} \times 2 = 620 \text{ amperes}$$

$$350 \text{ kcmil} \times 2 = 700 \text{ kcmil}$$

Table 250.66 requires a 2/0 AWG or larger grounding electrode conductor for this separately derived system. The connection of the grounding electrode conductor to the system must be made at the same location where the system bonding jumper is connected. This continues the concept of single-point grounding.

> **Exception No. 1:** Where the system bonding jumper specified in *250.30(A)(1)* is a wire or busbar, it is permitted to connect the grounding electrode conductor to the equipment grounding terminal bar or bus provided the equipment grounding terminal bar or bus is of sufficient size for the separately derived system.

Discussion: This exception is identical to that for making a grounding electrode connection at the service of identical configuration. Generally, the grounding electrode conductor is connected to the neutral conductor bus or terminal bar. The exception permits the connection to be made at the equipment grounding terminal bar for the purpose of facilitating the operation of a residual-type equipment ground-fault protection system. These are sometimes called ground strap–type equipment ground-fault protection systems. For additional information, see the discussion in *250.24(A)(4).*

> **Exception No. 2:** If a separately derived system originates in listed equipment that is suitable as service equipment, the grounding electrode conductor from the service or

feeder equipment to the grounding electrode is permitted to serve as the grounding electrode conductor for the separately derived system, provided the grounding electrode conductor is of sufficient size for the separately derived system.

Discussion: This equipment usually is listed as "unit substation." It typically consists of three sections. The first section is often a medium-voltage section with supply conductors over 600 volts for premises wiring systems that have over 600-volt distribution. Unit substations are also available with the primary voltage under 600 volts, typically 480 volts. The center section usually is a transformer-type separately derived system. The third section contains distribution equipment for the derived system. The sections are bolted together and have an equipment grounding bus connected to the three sections. (See Figure 2-54.)

The exception is intended to permit the grounding electrode conductor for the service or high-voltage section as well as the equipment grounding bus to serve as the grounding electrode system for the separately derived system, provided that the conductor and bus are large enough.

Here's How: Determine the minimum-size grounding electrode conductor required for the separately derived system. This is done by determining the circular mil area of the derived conductors, referring to *250.66,* and selecting the minimum-size grounding electrode conductor. To use the grounding electrode conductor installed for the service equipment or high-voltage section of the unit substation, it must be not smaller than that required for the separately derived system. The same is true for the size of the equipment grounding bus in the unit substation.

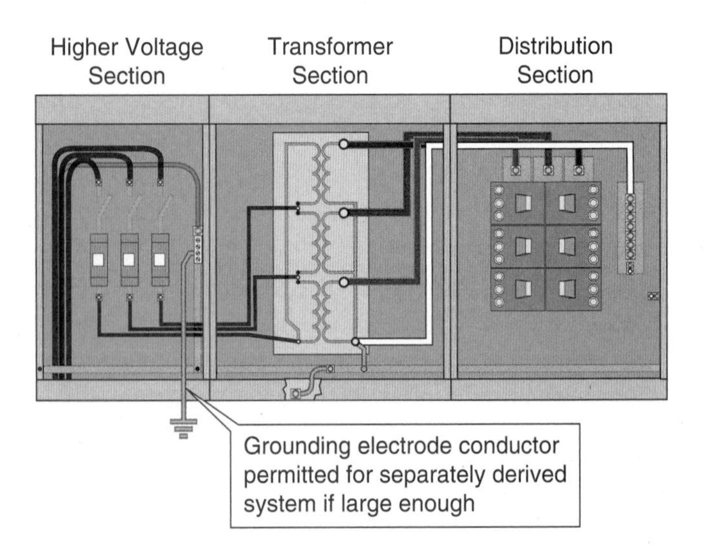

FIGURE 2-54 Grounding electrode for separately derived system as part of unit substation, *250.30(A)(5), Exception.* (*© Cengage Learning 2012*)

The grounding electrode to be used for the separately derived system determines the size of grounding electrode conductor required. If a water pipe or structural metal grounding electrode is used as provided in *250.30(A)(7),* the size of the grounding electrode conductor is determined from *Table 250.66* on the basis of the size of the derived conductors. If these grounding electrodes are not available and other grounding electrodes are used as permitted in *Exception No. 1* to *250.30(A)(5),* the requirements for sizing the grounding electrode conductor in *250.66(A), (B),* and *(C)* may be applied as appropriate.

CAUTION: The conductors derived from a separately derived system often require a grounding electrode conductor larger than the grounding electrode conductor required for the higher-voltage primary supply.

Exception No. 3: A grounding electrode conductor is not required for a system that supplies a Class 1, Class 2, or Class 3 circuit and is derived from a transformer rated not more than 1000 volt-amperes, if the grounded conductor is bonded to the transformer frame or enclosure by a jumper sized in accordance with *250.30(A)(1), Exception No. 3,* and the transformer frame or enclosure is grounded by one of the means specified in *250.134.*

Discussion: This exception is identical to *Exception No. 3* to *250.30(A)(1)* except that it applies to the sizing of the grounding electrode conductor rather than to the sizing of the system bonding jumper. In essence, a grounding electrode conductor is not required when a properly sized system bonding conductor is installed. This connects the secondary of the Class 1, Class 2, or Class 3 separately derived system to the enclosure.

250.30(A)(6) Grounding Electrode Conductor, Multiple Separately Derived Systems

The Requirement: A common grounding electrode conductor for multiple separately derived systems is permitted. If installed, the common grounding electrode conductor is required to be used to connect the grounded conductor of the separately derived system to the grounding electrode, as specified in *250.30(A)(4)*. A grounding electrode conductor tap shall then be installed from each separately derived system to the common grounding electrode conductor. Each tap conductor is required to connect the grounded conductor of the separately derived system to the common grounding electrode conductor. This connection is required to be made at the same point on the separately derived system where the system bonding jumper is installed.

Discussion: The concept of installing a common grounding electrode conductor and a grounding electrode conductor tap to two or more separately derived systems must be clearly understood as being an exception to the general rule of connecting each separately derived system individually to the grounding electrodes specified in *250.30(A)(5)*. As can be seen in that section, the grounding electrode for the separately derived system is required to be the *nearest* effectively grounded structural metal member of the building or the *nearest* effectively grounded metal water pipe. See Figure 2-55.

If these grounding electrodes do not exist in the area served by a separately derived system, the installation of a common grounding electrode and grounding electrode taps is permitted.

> **Exception No. 1:** If the system bonding jumper specified in *250.30(A)(1)* is a wire or busbar, it is permitted to connect the grounding electrode conductor to the equipment grounding terminal bar or bus provided the equipment grounding terminal bar or bus is of sufficient size for the separately derived system.

Discussion: This exception is identical to that for making a grounding electrode connection at the service if the installation is identical. Generally, the grounding electrode conductor is connected to the neutral conductor bus or terminal bar in the service.

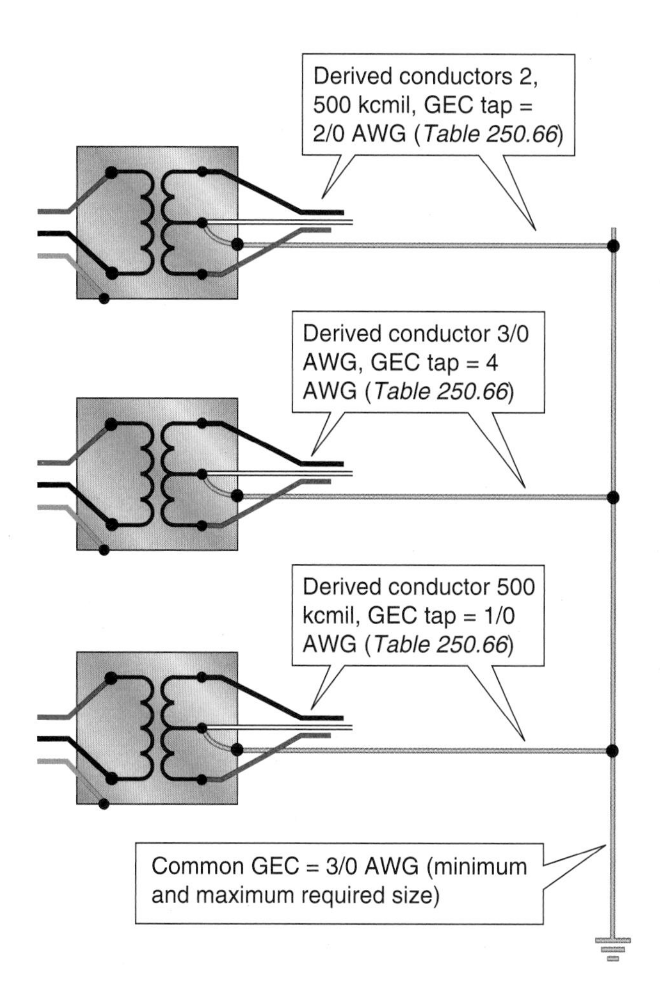

FIGURE 2-55 Grounding electrode conductor (GEC), multiple separately derived systems, *250.30(A)(6)*. (*© Cengage Learning 2012*)

The exception permits the connection to be made at the equipment grounding terminal bar for the purpose of facilitating the operation of a residual-type equipment ground-fault protection system. These are sometimes called ground strap-type equipment ground-fault protection systems. For additional information, see the discussion in *250.24(A)(4)*.

> **Exception No. 2:** A grounding electrode conductor is not required for a system that supplies a Class 1, Class 2, or Class 3 circuit and is derived from a transformer rated not more than 1000 volt-amperes, provided the system grounded conductor is bonded to the transformer frame or enclosure by a jumper sized in accordance with *250.30(A)(1), Exception No. 3,* and the transformer frame or enclosure is grounded by one of the means specified in *250.134*.

Discussion: This exception is identical to *Exception No. 3* to *250.30(A)(1)* except that it applies to the sizing of the grounding electrode conductor for the tap grounding electrode conductor rather than to the sizing of the system bonding jumper. In essence, a grounding electrode conductor is not required when a properly sized system bonding conductor is installed. This connects the secondary of the Class 1, Class 2, or Class 3 separately derived system to the enclosure.

250.30(A)(6)(a) Common Grounding Electrode Conductor Size

The Requirement: The common grounding electrode conductor is permitted to be one of the following:

1. A conductor of the wire-type not smaller than 3/0 AWG copper or 250-kcmil aluminum
2. The metal frame of the building or structure that complies with *250.52(A)(2)* or is connected to a grounding electrode system by a conductor not smaller than 3/0 AWG copper or 250-kcmil aluminum

Discussion: As can be seen in this section, two methods are provided to serve as a common grounding electrode conductor. The first option is new to the 2011 *NEC*. It allows the metal frame of a building that is recognized as a grounding electrode in *250.52(A)(2)* to serve as the grounding electrode conductor for the separately derived system. See the discussion in *250.52(A)(2)* for the methods provided for creating a grounding electrode from the metal building structure.

The other method provided is to install a 3/0 AWG copper or 250-kcmil aluminum common grounding electrode conductor. Note that these sizes are also the maximum size for the grounding electrode conductor required in *Table 250.66*. When smaller separately derived systems are installed, this maximum-sized grounding electrode conductor may be larger than was required in previous editions of the *NEC*. Because these common grounding electrode conductors often are installed in concealed spaces, installing the maximum-size conductor required at the time of construction accommodates the addition of future separately derived systems.

250.30(A)(6)(b) Tap Conductor Size

The Requirement: Each tap grounding electrode conductor is required to be sized in accordance with *250.66*, based on the derived phase conductors of the separately derived system it serves.

Discussion: This tap conductor is a grounding electrode conductor tap. When we think of a *tap conductor*, we generally expect the tap conductor to be smaller than the common conductor. No requirement is in the *NEC* for this concept. Each conductor must be sized for the component it serves.

Here's How: The grounding electrode conductor tap conductor is sized using *Table 250.66* for the size of the derived ungrounded conductors from the individual separately derived system and extends to the common grounding electrode conductor. Figure 2-53 illustrates sizing grounding electrode conductor tap conductors for three separately derived systems. In each case, determine the size of the derived ungrounded conductors. Then go to *Table 250.66* to determine the minimum size of grounding electrode tap conductor.

Discussion: An exception is provided for this tap conductor sizing, identical to that for sizing the grounding electrode conductor for the separately derived system originating in listed unit substations.

250.30(A)(6)(c) Connections

The Requirements: All tap connections to the common grounding electrode conductor are required to be made at an accessible location by one of the following methods:

1. A connector listed as grounding and bonding equipment
2. Listed connections to aluminum or copper busbars not less than ¼ in. × 2 in. (6 mm × 50 mm). Where aluminum busbars are used, the installation must comply with *250.64(A)*
3. By the exothermic welding process

Tap conductors are required to be connected to the common grounding electrode conductor in such a manner that the common grounding electrode conductor remains without a splice or joint. See Figure 2-56.

Discussion: This section describes the methods permitted to be used to make connections from the tap grounding electrode conductors to the common grounding electrode conductor. In addition to the

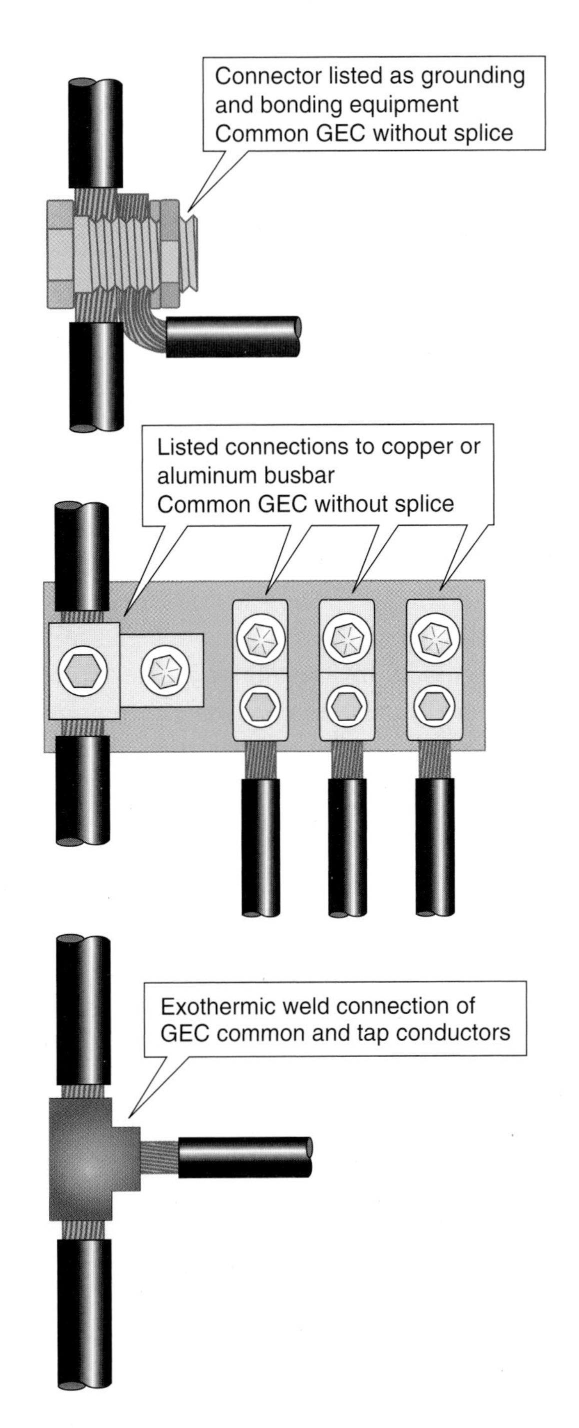

FIGURE 2-56 Connection of grounding electrode conductor taps to common grounding electrode conductor (GEC), *250.30(A)(6)(c)*. (*© Cengage Learning 2012*)

specific connection requirements in this section, the general rules for connecting grounding and bonding equipment in *250.8* apply.

The phrase *connector listed as grounding and bonding equipment* in this section means a connector that has passed a short-time, high-current test as prescribed in the UL Standard 467. These connectors are identified as grounding and bonding equipment, usually on the shipping carton. Included in this category are some split bolts and some compression connectors. See Figure 2-56 for an illustration of a connection made with a split bolt.

The phrase *listed connections to aluminum or copper busbars* allows any listed connector to be used. It is the responsibility of the installer to select a connector that is suitable for the connection being made, including the size of conductor, and for connecting the conductor material, either copper or aluminum, to other conductors or to the busbar provided for in (2). Figure 2-56 shows connections being made to a busbar with listed connectors and bolts, nuts, and washers.

Although not specifically mentioned in the list of connection methods in *250.30(A)(6)(c)*, use of irreversible compression connectors is an ideal way to make these connections from conductor to conductor or from conductor to busbar. Many of these compression connectors are *listed as grounding and bonding equipment*. Be certain to select the appropriate compression connector as well as the appropriate tool to make these connections. Carefully follow all manufacturer's instructions.

It is important that the rules in *110.14* be followed, as well as any installation instructions from the manufacturer of the wire connector. Extensive information is available in the UL General Information Directory under the heading "Wire Connectors and Soldering Lugs (ZMVV)."

When an aluminum or copper busbar is used to make the connection between the common and tap grounding electrode conductors, specific installation rules must be followed. The busbar must be not smaller than ¼ in. × 2 in. (6 mm × 50 mm). Although specific rules for mounting the busbar are not provided in this section, other general rules for mounting electrical equipment in *Article 110* must be followed. These rules include *Neat and workmanlike* requirements in *110.12* and *Mounting electrical equipment* in *110.13(A)*. If aluminum busbars are used, the installation must comply with *250.64(A)* to avoid problems with corrosion. Aluminum busbars should not be mounted directly on masonry walls that are subject to dampness or wetness but should be mounted on suitable standoffs.

250.30(A)(7) Installation

The Requirements: The installation of all grounding electrode conductors is required to comply with *250.64(A), (B), (C),* and *(E).*

Discussion: *NEC 250.64(A)* applies to installing aluminum and copper-clad aluminum grounding electrode conductors; *(B)* provides requirements on securing and protecting grounding electrode conductors; *(C)* has requirements for maintaining the grounding electrode in one continuous length but includes acceptable methods for splicing them; and *(E)* provides rules on bonding ferrous (magnetic) wiring methods used to protect the grounding electrode from physical damage. These installation rules are discussed in Unit 3 of this book.

250.30(A)(8) Bonding

The Requirement: Structural steel and metal piping are required to be connected to the grounded conductor of a separately derived system in accordance with *250.104(D).*

Discussion: As seen in *250.30(A)(4),* effectively grounded metal water pipes and metal structural steel are required to be used as the grounding electrode for separately derived systems when they are in the vicinity of the separately derived system. *NEC 250.104(D)* contains requirements for bonding structural steel and metal piping that are in the vicinity of the separately derived system but are not being used as a grounding electrode.

250.30(B) Ungrounded Systems

The Requirements: The equipment of an ungrounded separately derived system must be grounded as specified in *(B)(1)* through *(B)(3).*

1. **Grounding Electrode Conductor.** A grounding electrode conductor, sized in accordance with *250.66* for the derived phase conductors, is required to be used to connect the metal enclosures of the derived system to the grounding electrode, as specified in *250.30(A)(5)* or *(A)(6)* as applicable. This connection is permitted to be made at any point on the separately derived system from the source to the first system disconnecting means. If the source is located outside the building or structure supplied, a grounding electrode connection is required to be made in compliance with *(C).*
2. **Grounding Electrode.** Except as permitted by *250.34* for portable and vehicle-mounted generators, the grounding electrode is required to comply with *250.30(A)(4).*
3. **Bonding Path and Conductor.** A supply-side bonding jumper is required to be installed from the source of a separately derived system to the first disconnecting means in compliance with *250.30(A)(2).*

Discussion: The metal enclosures for ungrounded separately derived systems are required to be grounded in similar fashion to that for the enclosures for ungrounded service equipment. As previously discussed, this results in a case ground, in which the enclosures are connected to the grounding electrode system, but the electrical system itself is not grounded (see Figure 2-57). Grounding the

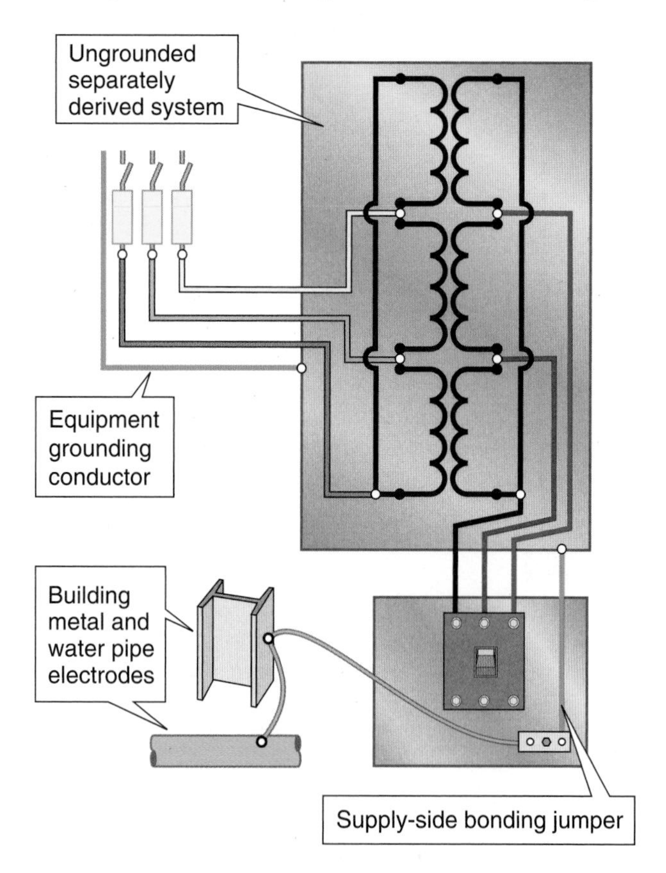

FIGURE 2-57 Grounding electrode and grounding electrode conductor for ungrounded separately derived system, *250.30(B).* (*© Cengage Learning 2012*)

equipment for an ungrounded system protects people from shock hazards, because the equipment that contains or is supplied from the separately derived system is at or near earth potential.

It is required that a supply-side bonding jumper be installed from the enclosure for the source to the enclosure for the first disconnecting means. The supply-side bonding jumper is required to be sized as though it were from a grounded system as a result of the pointer to *250.30(A)(2)*. Determine the size of the derived conductors from the separately derived system. Then, take this size to *Table 250.66* to determine the minimum size of the supply-side bonding jumper.

For separately derived systems permitted to be ungrounded, see *250.20(B)*.

250.30(C) Outdoors Source

"If the source of the separately derived system is located outside the building or structure supplied, a grounding electrode connection shall be made at the source location to one or more grounding electrodes in compliance with *250.50*. In addition, the installation shall comply with *250.30(A)* for grounded systems or with *250.30(B)* for ungrounded systems."*

> ***Exception:*** *The grounding electrode conductor connection for impedance grounded neutral systems shall comply with 250.36 or 250.186 as applicable.*

Discussion: This is a new section in the 2011 *NEC*. Similar to the rules for services in *250.24(A)(2)*, a grounding electrode conductor connection is required at the source of a separately derived system that is located outside.

Figure 2-58(A) shows a generator that is located outdoors. Connection to a grounding electrode is shown at the generator. The generator qualifies as a separately derived system because all system and circuit conductors are switched in the transfer switch. In Figure 2-58(B), a unit substation is located outdoors. The transformer section also qualifies as a separately derived system because the circuit conductors are isolated electrically from the primary conductors other than through equipment grounding conductors that are internal to the equipment.

As can be seen in both Figures 2–58(A) and (B), the system bonding jumper is located at or near the same point where the grounding electrode conductor connection is made. This is done to comply with *250.30(A)(5)*.

As pointed out in the exception to this section, the rules for grounding electrode connections are modified for impedance-grounded systems in *250.36* and *250.186* as applicable.

The sizing of the system bonding jumper and grounding electrode conductor are covered previously in *250.30(A)*.

For ungrounded systems, refer to Figure 2-59. The ungrounded system originates outside so a grounding electrode connection to the metal equipment is required at that location. As shown in Figure 2-59(A), an equipment grounding conductor is installed from the source of the separately derived system to the enclosure for the building disconnecting means.

In Figure 2-59(B), the source (transformer) of the separately derived system is located remotely from the unit substation. As a result, a grounding electrode and grounding electrode conductor are required at each of the locations for the electrical equipment. In addition, because there is no overcurrent protection provided where the feeder from the transformer to the building originates, the overcurrent protection for the conductors must comply with an applicable section of *240.21(C)*.

250.32 BUILDINGS OR STRUCTURES SUPPLIED BY FEEDER(S) OR BRANCH CIRCUIT(S)

250.32(A) Grounding Electrode

The Requirement: Building(s) or structure(s) supplied by feeder(s) or branch circuit(s) are required to have a grounding electrode or grounding electrode system installed in accordance with Part III of *Article 250*. The grounding electrode conductor(s) must be connected in accordance with *250.32(B)* or *(C)*. Where there is no existing grounding electrode, the grounding electrode(s) required in *250.50* are required to be installed.

*Reprinted with permission from NFPA 70-2011.

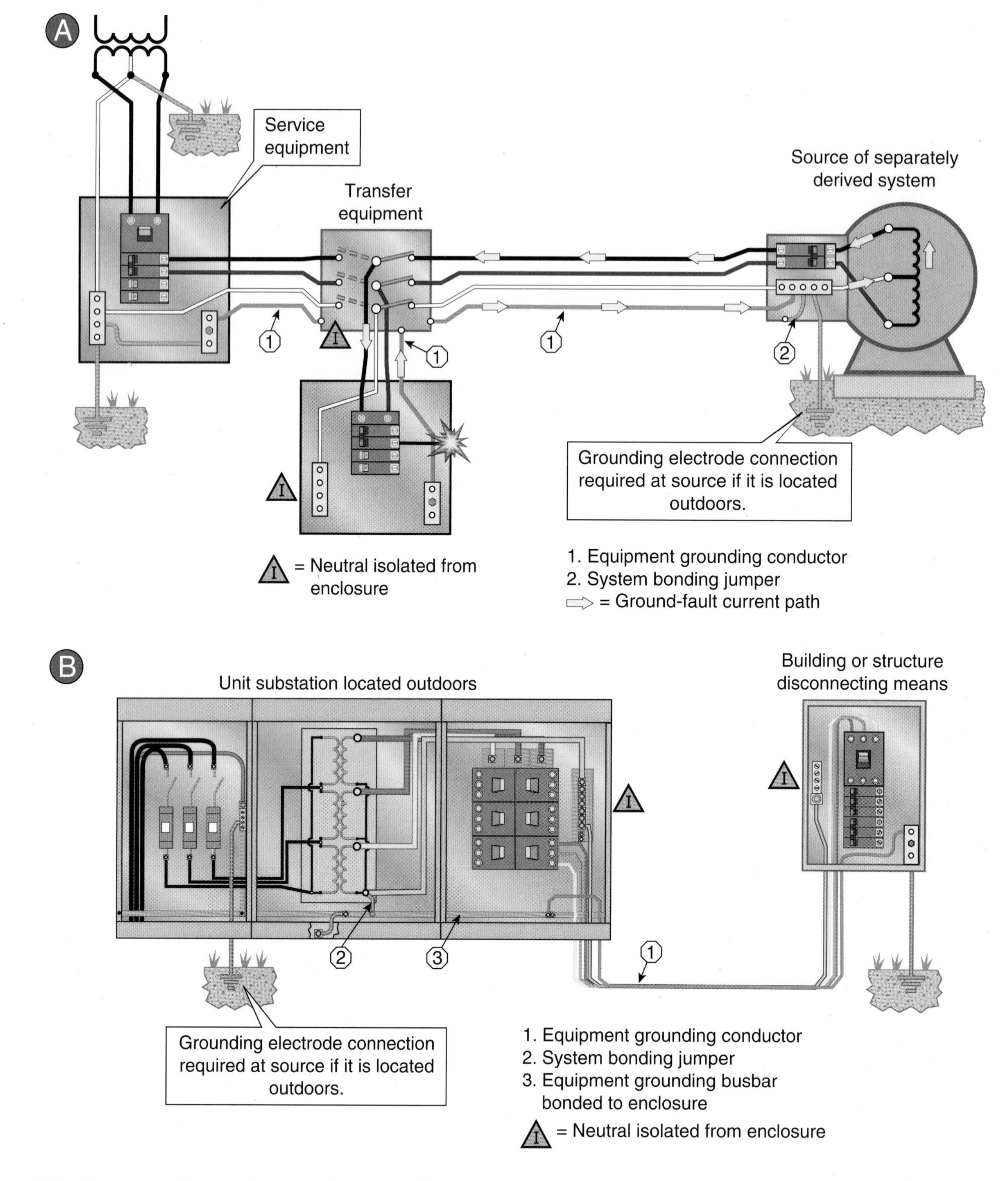

FIGURE 2-58 Grounding requirements for grounded separately derived systems having the source located outdoors, *250.30(C)*. (*© Cengage Learning 2012*)

Discussion: As a result of the reference to *Part III* of *Article 250,* all of the grounding electrodes described in *250.52(A)(1)* through *(A)(4)* that are present at the building or structure are required to be bonded together to form a grounding electrode system. Note that the *Code* does not require these electrodes to be installed if they are not present, but to be bonded together if they are present.

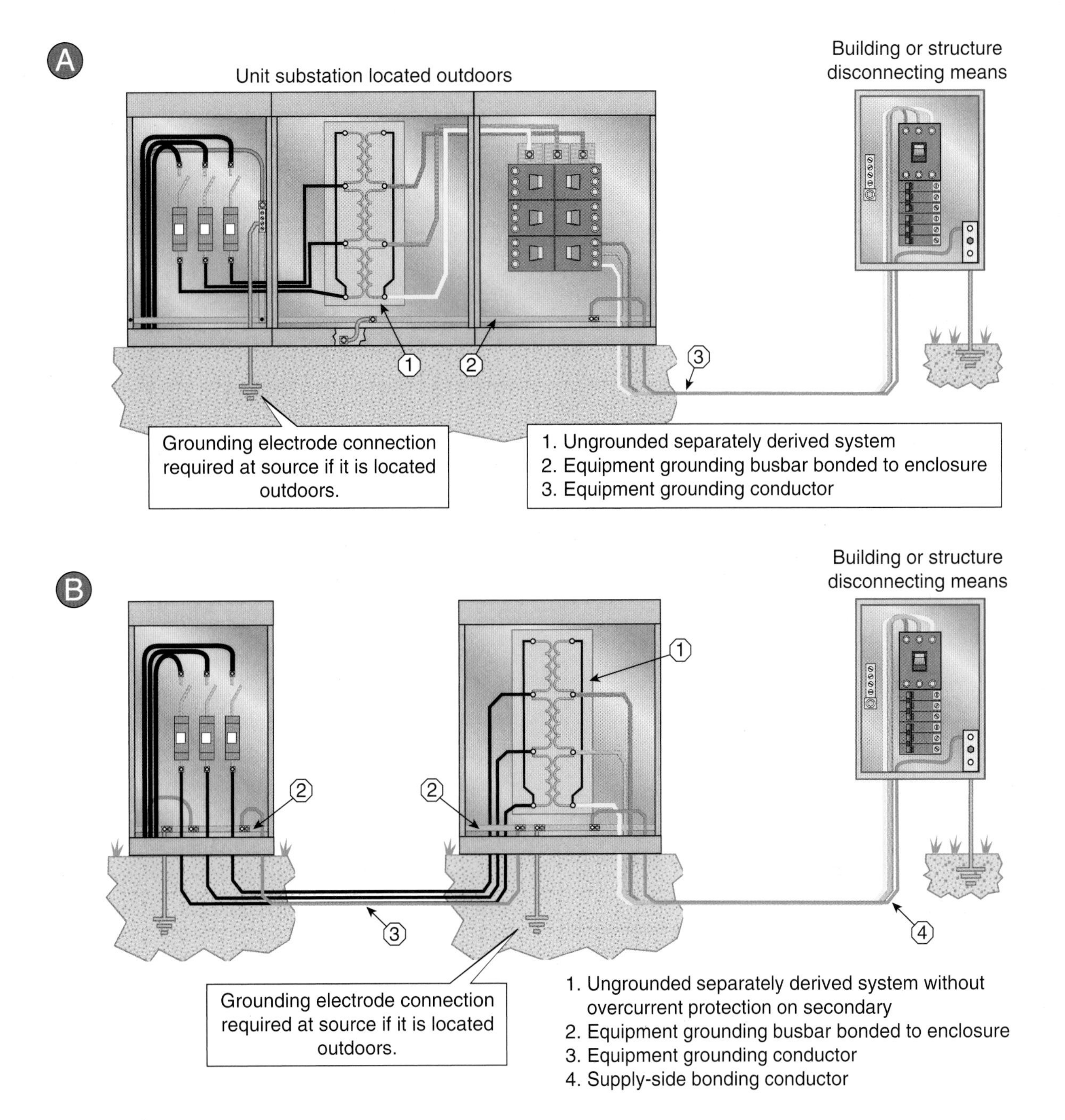

FIGURE 2-59 Grounding requirements for ungrounded separately derived systems having the source located outdoors. *250.30(B)(1)* and *250.30(C)*. (*© Cengage Learning 2012*)

Figure 2-60 shows the general requirements for the grounding system related to supplying a building or structure by a feeder or branch circuit. The phrase *where available* was changed to *if present* in *250.50* in the 2005 *NEC* to require the use of concrete-encased grounding electrodes. An exception is provided for concrete-encased electrodes at *existing* buildings so concrete is not required to be damaged to connect to the concrete-encased electrode.

The grounding electrodes described in *250.52 (A)(1)* through *(A)(4)* are as follows:

- *(A)(1)* Metal underground water pipe
- *(A)(2)* Metal frame of the building or structure
- *(A)(3)* Concrete-encased electrode
- *(A)(4)* Ground ring

If the above grounding electrodes are not present at the building or structure supplied by a feeder

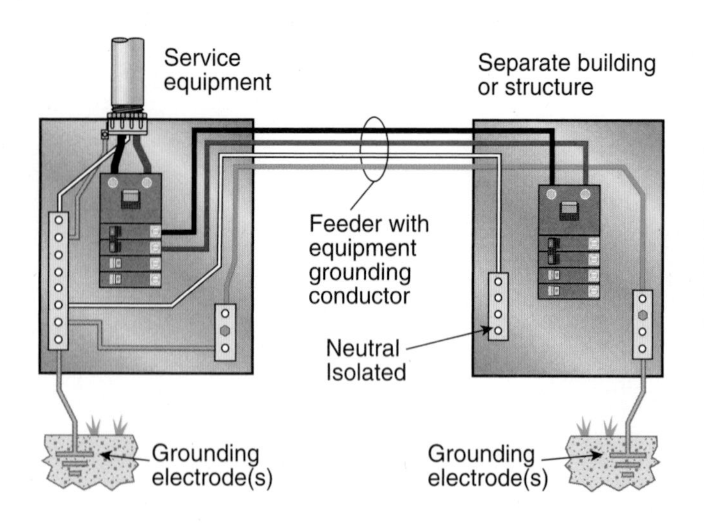

FIGURE 2-60 Grounding requirements for buildings or structures supplied by feeder(s) or branch circuit(s), *250.32(A)*. (*© Cengage Learning 2012*)

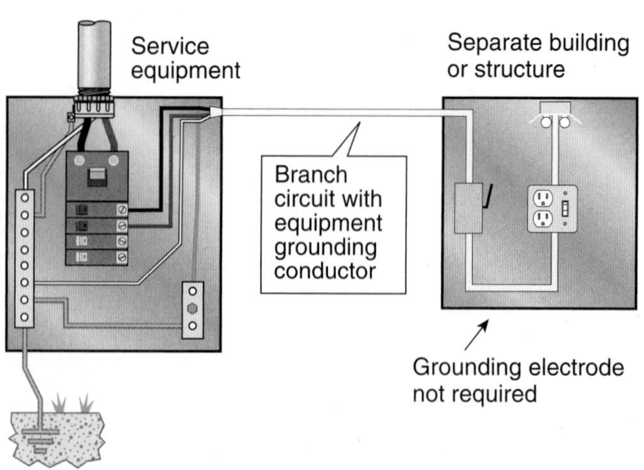

FIGURE 2-61 Exception for buildings or structures supplied by only one branch circuit, *250.32(A), Exception*. (*© Cengage Learning 2012*)

or branch circuit, one or more of the grounding electrodes identified in *250.50(A)(5)* through *(A)(8)* must be installed. The grounding electrodes described in *250.52(A)(5)* through *(A)(8)* include the following:

- *(A)(5)* Rod and pipe electrodes
- *(A)(6)* Other listed electrodes
- *(A)(7)* Plate electrodes
- *(A)(8)* Other local metal underground systems or structures

These grounding electrodes are required to be bonded together to form the grounding electrode system and are required to be connected as specified in *250.32(B)* or *(C)* (see Figure 2-60).

An Exception: *A grounding electrode is not required where only a single branch circuit, including a multiwire branch circuit, supplies the building or structure and the branch circuit includes an equipment grounding conductor for grounding the normally non-current-carrying metal parts of equipment.*

Discussion: Note that this exception from the requirement to have a grounding electrode or grounding electrode system at the second or additional building or structure applies only if the building or structure is supplied by a single branch circuit. In addition, the branch circuit must include an equipment grounding conductor for grounding equipment at the additional building or structure (see Figure 2-61).

The branch circuit can be any of the branch circuits recognized or permitted in the *Code*. This includes a 2-wire or a multiwire branch circuit. An equipment grounding conductor recognized in *250.118* must either enclose the circuit conductors or be run with the branch circuit conductors. Table 2-4 summarizes some of the branch circuits permitted to supply an additional building or structure.

Conductors at higher voltages, such as 480 volts, typically are installed as feeders to a second or additional building rather than as an individual branch circuit.

250.32(B) Grounded Systems

The Requirement: An equipment grounding conductor as described in *250.118* is required to be run with the supply conductors and be connected to the

TABLE 2-4

Permitted Circuit Configurations

Voltage	Number of Current-Carrying Conductors	Common Designation (All with Ground)
120	2	120-V, 2-wire
120/240*	2	120/240-V, 3-wire
208Y/120*	3 or 4	208Y/120-V, 3-phase, 4-wire

*Considered a multiwire branch circuit.

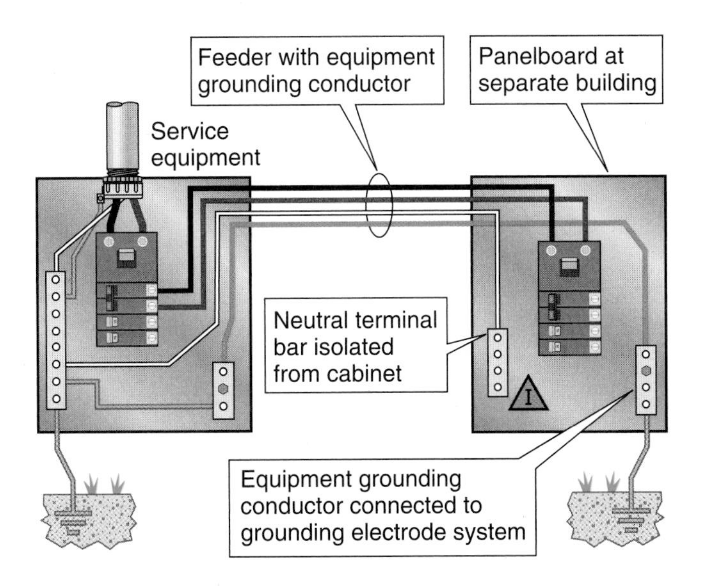

FIGURE 2-62 Grounding at separate building or structure supplied with equipment grounding conductor, *250.32(B)*. (*© Cengage Learning 2012*)

building or structure disconnecting means and to the grounding electrode(s). The equipment grounding conductor must then be used for grounding or bonding of equipment, structures, or frames required to be grounded or bonded. The equipment grounding conductor must also be sized in accordance with *250.122*. Any installed grounded conductor (often a neutral) is not permitted to be connected to the equipment grounding conductor or to the grounding electrode(s).

Discussion: The equipment grounding conductor that is run with the branch circuit or feeder is required to be bonded to the equipment grounding terminal bar in the panelboard or building disconnecting means at the additional building or structure served. See Figure 2-62. The terminal bar for the grounded conductor—which is often a neutral—is isolated electrically from the enclosure. To accomplish this, the main bonding jumper furnished with the panelboard or disconnecting means is not installed to connect the neutral terminal bar to the enclosure. In fact, it is often a good idea to discard the main bonding jumper to prevent it from being installed inappropriately at some later time.

The feeder or branch circuit supplying the building or structure may be installed either overhead or underground. Obviously, the applicable *NEC* rules for installing the feeder or branch circuit must be complied with. The equipment grounding conductor is permitted to be any of the appropriate ones from *250.118*. Permitted are rigid and intermediate steel conduits. Some AHJs and electrical engineers will require these conduits to be coated with a protective tape or paint to protect them from corrosion. Electrical metallic tubing is not often accepted for underground installations, as the EMT is not designed or intended for that purpose.

The equipment grounding conductor of the wire type is required to be sized from *250.122*. The rating of the overcurrent device at the origination of the feeder or branch circuit is used in *Table 250.122* to determine the appropriate size of the equipment grounding conductor. Conductors or cables that are listed for direct burial are often installed.

250.32(B) Exception

The Requirements: *For installations made in compliance with previous editions of this Code that permitted such connection, the grounded conductor run with the supply to the building or structure is permitted to serve as the ground-fault return path if all the following requirements continue to be met:*

1. *an equipment grounding conductor is not run with the supply to the building or structure,*
2. *there are no continuous metallic paths bonded to the grounding system in both buildings or structures involved, and*
3. *ground-fault protection of equipment has not been installed on the supply side of the feeder(s).*

If the grounded conductor is used for grounding as provided in this Exception, the size of the grounded conductor to the building or structure must be the largest of either of the following:

1. *The calculated load as determined by 220.61*
2. *An equipment grounding conductor as required by 250.122.*

Discussion: This *Exception* was previously *Section 250.32(B)(2)* and permitted using the grounded system conductor (often a neutral) for two functions. The section was changed to an exception to the general rule in *250.32(B)* and applies to only existing electrical installations. All installations made after the adoption of the 2008 *NEC* are not permitted to use this exception. The first function served by

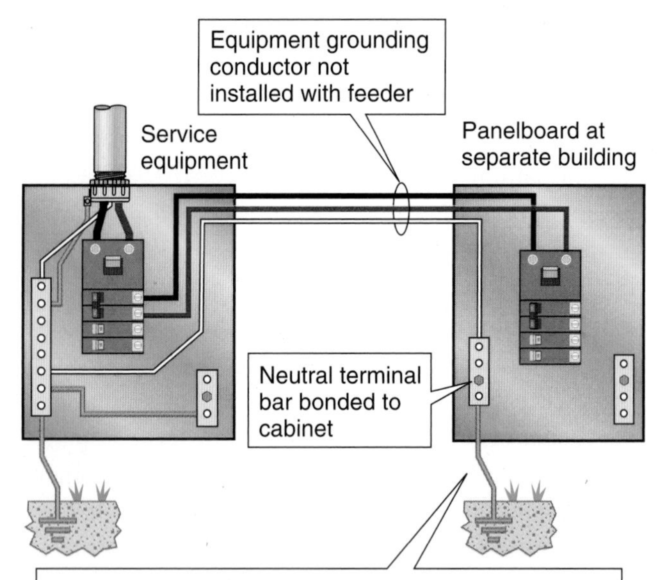

FIGURE 2-63 Exception for grounding at existing separate building or structure previously using the grounded system conductor for the ground-fault return path, *250.32(B), Exception.* (*© Cengage Learning 2012*)

the grounded (neutral) conductor is that of returning unbalanced current from phase conductors back to the source. The second function is that of serving as an equipment grounding conductor to return line-to-ground-fault currents back to the source (see Figure 2-63).

Strict conditions are in place in this exception to prevent neutral current from returning to the source over unintended paths such as metal piping or conduit systems. The intended path, of course, is over the neutral conductor that is installed with the feeder conductors to the second or additional building(s) or structure(s).

The three conditions identified in this exception must continue to be satisfied before the grounded conductor is permitted to be used for the ground-fault return path. See Figure 2-63. The language in the exception now requires that the conditions that permitted the exception to be used originally be verified for the neutral to continue to be used as the ground-fault return path. The conditions are as follows:

1. An equipment grounding conductor is not run with the supply to the building or structure. The equipment grounding conductor can be any of those identified in *250.118.* This includes a properly sized conductor or the metallic wiring method such as conduit or cable that might enclose the feeder conductors between the buildings. If any equipment grounding conductor is run between the buildings or structures, it is not permitted to reground the grounded system conductor at the separate building or structure.
2. There are no continuous metallic paths bonded to the grounding system in both buildings or structures involved. Continuous metallic paths between the buildings can consist of one or many objects. A partial list follows:
   a. Metallic wiring methods such as conduit, tubing, and cable sheaths
   b. Any metal piping system such as water, sprinkler system, compressed air, steam, gas, process system, drainage, and so on
   c. Cable tray systems
   d. Shields for cable systems such as for telephone, communications, data, alarm, control, and so on
   e. Conveyors or other process transportation systems

It is difficult to bond and ground all the wiring methods, equipment, communications cables, and process equipment at each building or structure on the premises and not to have an electrically conductive interconnection between buildings. The more complex the systems, the more difficult it is to maintain electrical isolation. The simple solution is to install an equipment grounding conductor between buildings and not attempt to use the grounded (neutral) conductor for that purpose.

3. Ground-fault protection of equipment has not been installed on the supply side of the feeder.

This third condition intends to prevent a common problem for equipment ground-fault protection systems. Systems for ground-fault protection of equipment will not function properly if the system grounded conductor is grounded again on the load side or downstream from the protection equipment.

Only if all three conditions were met, the system grounding conductor (often a neutral) was permitted to be grounded again at the second or additional building.

As indicated, when the system grounding conductor (often a neutral) is grounded again at the second or additional building, the conductor serves two functions: It carries the unbalanced current from the phase conductors back to the source and serves as the equipment grounding conductor. As a result, the grounded conductor is required to be sized not smaller than required for load calculations of the feeder or service neutral load in *220.61* or as an equipment grounding conductor in *250.122*.

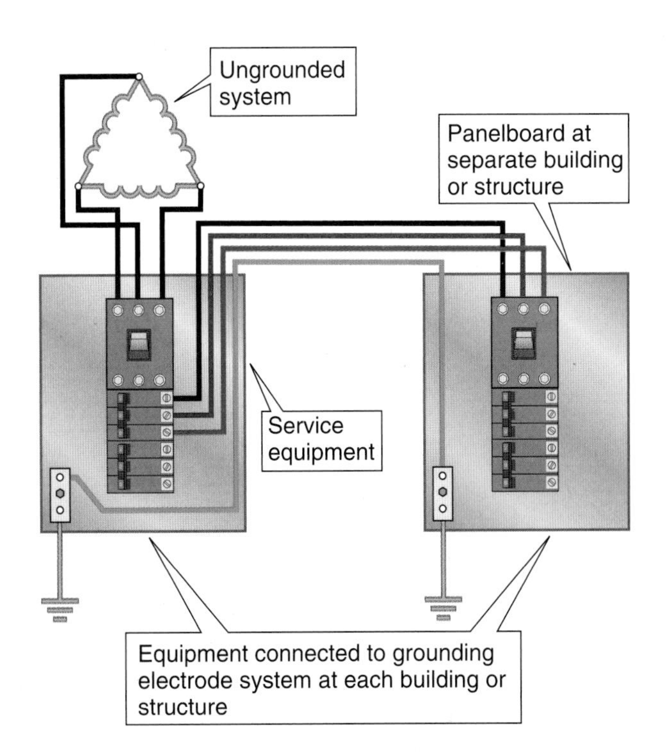

FIGURE 2-64 Grounding of equipment for ungrounded systems at separate building or structures, *250.32(C)*. (*© Cengage Learning 2012*)

250.32(C) Ungrounded Systems

250.32(C)(1) Supplied by a Feeder or Branch Circuit

The Requirement: An equipment grounding conductor as described in *250.118* shall be installed with the supply conductors and be connected to the building or structure disconnecting means and to the grounding electrode(s). The grounding electrode(s) is also required to be connected to the building or structure disconnecting means.

Discussion: This section changed significantly during the processing of the 2011 *NEC*. As can be seen, an equipment grounding conductor is now required to be installed between buildings or structures supplied by a feeder from an ungrounded system. Though not mentioned in this section, equipment grounding conductors of the wire type are required to be sized from *NEC Table 250.122*.

For grounding equipment for an ungrounded system at the second or additional building, the grounding electrodes at the second building are required to be bonded together to form the grounding electrode system (see Figure 2-64). A grounding electrode conductor then connects the grounding electrode system to the building or structure disconnecting means. This is often referred to as a "case ground." The equipment grounding conductor that is installed with the feeder or branch circuit must be connected to the metal enclosure for the building disconnecting means and to the grounding electrode conductor.

250.32(C)(2) Supplied by a Separately Derived System

The Requirements:

(a) Having Overcurrent Protection. If overcurrent protection is provided where the conductors originate, the installation shall comply with *(C)(1)*.

(b) Without Overcurrent Protection. If overcurrent protection is not provided where the conductors originate, the installation shall comply with *250.30(B)*. If installed, the supply-side bonding jumper shall be connected to the building or structure disconnecting means and to the grounding electrode(s).

Discussion: Section *250.32(C)(2)(a)*, by referring to *250.30(C)(1)*, requires an equipment grounding conductor be installed with the feeder or branch circuit installed from the source of the ungrounded system. This equipment grounding conductor is

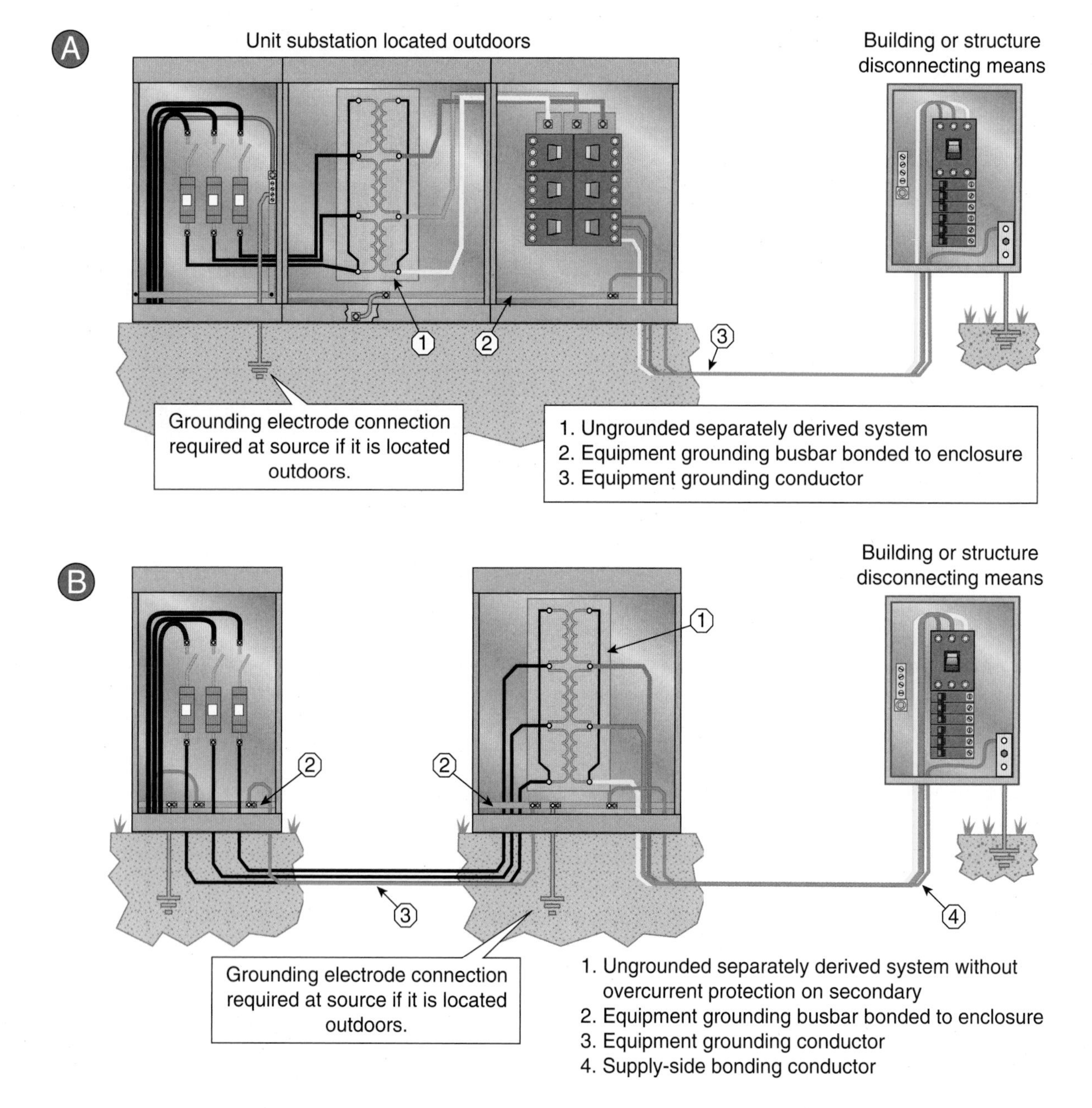

FIGURE 2-65 Grounding and bonding requirements for ungrounded separately derived system having the source located outdoors. (*© Cengage Learning 2012*)

required to be sized in accordance with *250.122*. See Figure 2-65(A).

Section *250.32(C)(2)(b)* provides the rules for the installation of a feeder from an ungrounded system that does not have overcurrent protection where the conductors originate. See Figure 2-65(B). By reference to *250.30(B)*, a grounding electrode conductor connection to the metallic enclosures of the separately derived system is required. The grounding electrode conductor is sized by going to *Table 250.66,* using the size of the derived ungrounded conductor. In addition, a supply-side bonding jumper is required from the source of the separately derived system to the first disconnecting means. Because an overcurrent protective device is not located where the conductors originate, the size of the supply-side bonding jumper is also determined by reference to *Table 250.66*. The size of the supply-side bonding jumper must not be smaller than 12.5% of the circular mil area of the derived conductors if they are larger than 1100-kcmil copper or 1750-kcmil aluminum.

Though not covered in *Article 250*, be sure you comply with the overcurrent protection rule for transformers in *Article 450* and for conductors in *Article 240*.

250.32(E) Grounding Electrode Conductor

The Requirement: The size of the grounding electrode conductor to the grounding electrode(s) must not be smaller than given in *250.66,* based on the largest ungrounded supply conductor to the second or additional building or structure. The installation of the grounding electrode must comply with *Part III* of *Article 250.*

Discussion: As indicated, the size of the grounding electrode conductor is based on *250.66* rather than on *250.122*. The rule for using *250.66* is that the size of the grounding electrode conductor is based on the size of the feeder conductors that supply the additional building or structure, not on the rating of the overcurrent protective device on the supply side of the feeder. See Figure 2-66.

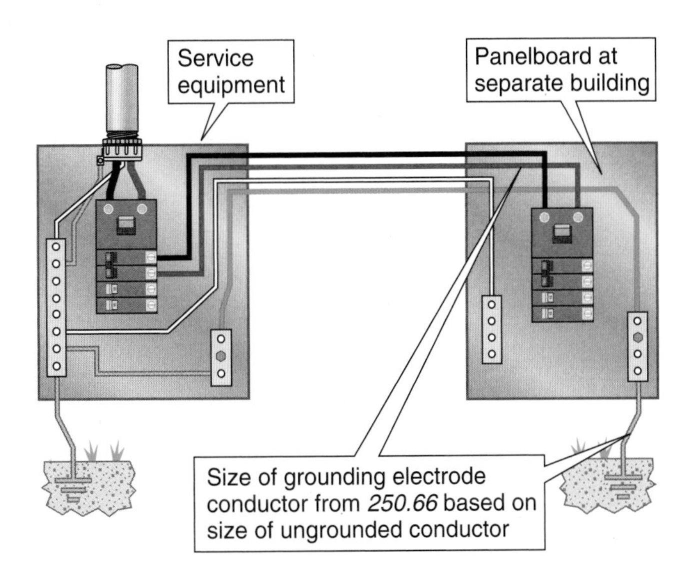

FIGURE 2-66 Size of grounding electrode conductor at separate building or structure, *250.32(E)*. (© Cengage Learning 2012)

250.34 PORTABLE AND VEHICLE-MOUNTED GENERATORS

250.34(A) Portable Generators

The Requirements: The frame of a portable generator is not required to be connected to a grounding electrode, as defined in *250.52* for a system supplied by the generator under the following conditions:

1. The generator supplies only equipment mounted on the generator, cord-and-plug-connected equipment through receptacles mounted on the generator, or both, and
2. The normally non-current-carrying metal parts of equipment and the receptacle terminal(s) for the connection to the equipment grounding conductor are connected to the generator frame.

Discussion: As illustrated in Figure 2-67, the system neutral and the equipment grounding conductor of a

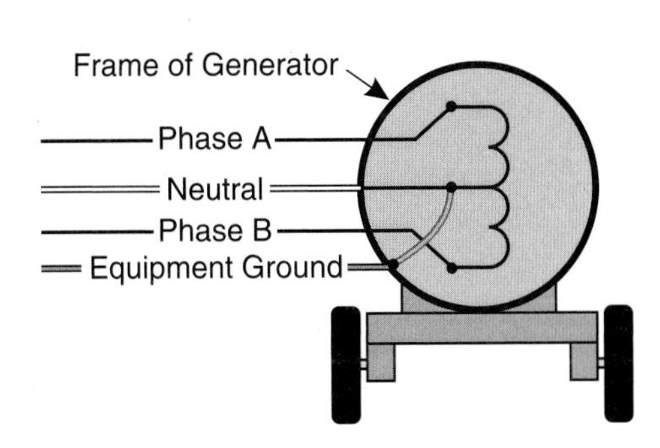

FIGURE 2-67 Grounding and bonding requirements for portable generators, *250.34(A)*. (© Cengage Learning 2012)

portable generator are required to be bonded to the frame of the generator. Without the connection between the neutral and the equipment grounding conductor, a circuit would not exist for current from a line-to-ground fault to return to its source. Keep in mind that all equipment must comply with *250.4(A)* by providing an effective ground-fault return path.

The *NEC* no longer refers to the frame of the generator as serving as the grounding electrode for the system produced by the generator. This section previously stated, "The frame of a portable generator shall not be required to be grounded and shall be permitted to serve as the grounding electrode for a system supplied by the generator."* The present language is a technical correction to the *Code* because the frame of the generator is not grounded unless an earth connection is made through one or more grounding electrodes. Grounding-type receptacles and ground-fault circuit-interrupter receptacles are typically installed on portable generators for supply of equipment through flexible cords.

UL Safety Standard 2201 for portable generators requires that portable generators for connection to the wiring system of a building or structure be considered a separately derived system. The guide card (FTCN) reads, "This category covers internal-combustion-engine-driven generators rated 15 kW or less, 250 V or less, which are provided only with receptacle out lets for the ac output circuits. The generators may incorporate alternating- or direct-current generator sections for supplying energy to battery-charging circuits.

"When a portable generator is used to supply a building or structure wiring system:

"1. The generator is considered a separately derived system in accordance with ANSI/NFPA 70, *National Electrical Code (NEC).*

"2. The generator is intended to be connected through permanently installed listed transfer equipment that switches all conductors other than the equipment grounding conductor.

"3. The frame of a listed generator is connected to the equipment-grounding conductor and the grounded (neutral) conductor of the generator. When properly connected to a premises or structure wiring system, the portable generator will be connected to the premises or structure grounding electrode for its ground reference.

"4. Portable generators used other than to power building or structure wiring systems are intended to be connected to ground if required by the NEC.[U]

The designation as a separately derived system requires the neutral and equipment grounding conductor be bonded to the frame of the generator and the transfer equipment to switch all conductors other than the equipment grounding conductor. This designation and switching arrangement prevents the neutral from being in parallel, as it is required to be switched in the transfer equipment.

The neutral or frame of the portable generator is not required to be connected to a grounding electrode so long as all power loads are supplied from receptacles mounted on the generator. Because the neutral and equipment grounding conductor are connected to the frame as required in *250.34(C)*, overcurrent protection provided on the generator will function correctly.

Figure 2-68 shows a type of transfer equipment that has often been installed to provide for the supply of a portion of a premises wiring system by a flexible cord from a portable or vehicle-mounted generator. The flanged inlet receptacle provides for the flexible cord connection. Circuits to be supplied have been wired through the transfer equipment and are selected by the rocker-type switches.

The drawing in the middle of Figure 2-68 shows the typical circuit where the portable or vehicle-mounted generator is connected to the premises wiring system through this transfer equipment. Because the equipment grounding conductor and the neutral are connected together at the service equipment and at the generator, neutral current will be shared by the equipment grounding conductor. The lower photo shows the inside of the transfer equipment shown in the upper photo. As can be seen, the neutral conductor is not switched. The solid connection in the transfer equipment is noncompliant with the UL 2201 Safety Standard.

*Reprinted with permission from NFPA 70-2011.

[U]Reprinted from the *White Book* with permission from Underwriters Laboratories Inc.® Copyright © 2010 Underwriters Laboratories Inc.®.

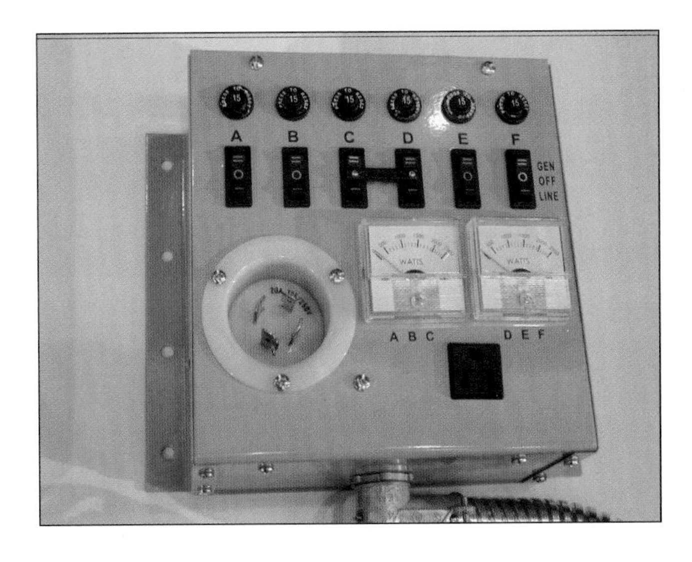

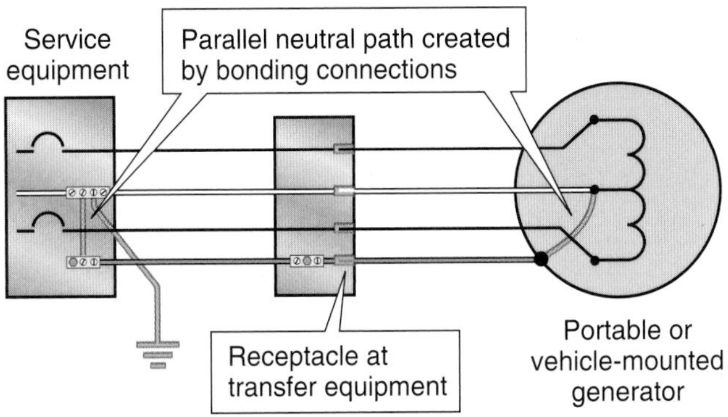

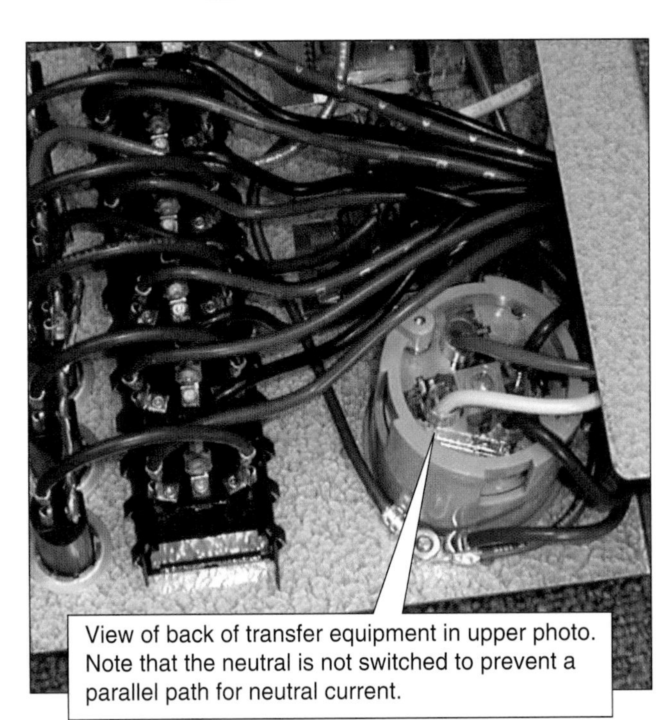

FIGURE 2-68 Connection of portable or vehicle-mounted generator to premises wiring system. (© *Cengage Learning 2012*)

250.34(B) Vehicle-Mounted Generators

The Requirements: The frame of a vehicle and the generator frame are not required to be connected to a grounding electrode as defined in *250.52* for a system supplied by a generator located on the vehicle under the following conditions:

1. The frame of the generator is bonded to the vehicle frame, and
2. The generator supplies only equipment located on the vehicle or cord-and-plug-connected equipment through receptacles mounted on the vehicle, or both equipment located on the vehicle and cord-and-plug-connected equipment through receptacles mounted on the vehicle or on the generator, and
3. The normally non-current-carrying metal parts of equipment and the receptacle terminal(s) for the connection to the equipment grounding conductor are connected to the generator frame.

Discussion: The rules for using the electrical system produced by a vehicle-mounted generator without connecting it to a grounding electrode driven into the earth are similar to those for a portable generator (see Figure 2-69). As with the portable generator, it is important that the equipment grounding conductor and the system neutral be connected together and bonded

FIGURE 2-69 Grounding of vehicle-mounted generators, 250.34(B). (© *Cengage Learning 2012*)

to the generator frame. The generator frame is required to be bonded to the vehicle frame. Note that these rules apply only where the generator supplies only equipment located on the vehicle or cord-and-plug-connected equipment through receptacles mounted on the vehicle, or both equipment located on the vehicle and cord-and-plug-connected equipment through receptacles mounted on the vehicle or on the generator. These rules do not apply where the generator is permanently installed and supplies premises wiring systems through a transfer switch or transfer equipment. See *250.35* for the rules for grounding and bonding generators that are permanently installed. These generators are required to be grounded and bonded under the rules in *250.30* if they are a separately derived system.

250.34(C) Grounded Conductor Bonding

The Requirement: A system conductor that is required to be grounded by *250.26* must be connected to the generator frame if the generator is a component of a separately derived system.

Discussion: This rule applies to both portable and vehicle-mounted generators. It ensures that a bonding jumper is connected between the system conductor that is required to be grounded by *250.26* and the generator frame when the generator is used as a component of a separately derived system. Without the bonding jumper in place, a return path for ground-fault current is not established.

If the generator is used as a component of a separately derived system, compliance with all of the rules set forth in *250.30* is required.

250.35 PERMANENTLY INSTALLED GENERATORS

The Requirement: A conductor that provides an effective ground-fault current path is required to be installed with the supply conductors from a permanently installed generator(s) to the first disconnecting mean(s) in accordance with (A) or (B).

"(A) **Separately Derived System.** If the generator is installed as a separately derived system, the requirements in 250.30 must be complied with.

"(B) **Nonseparately Derived System.** If the generator is installed as a nonseparately derived system, and overcurrent protection is not integral with the generator assembly, a supply-side bonding jumper shall be installed between the generator equipment grounding terminal and the equipment grounding terminal, bar, or bus of the disconnecting mean(s). It shall be sized in accordance with *250.102(C)* based on the size of the conductors supplied by the generator."*

See Figure 2-69.

Discussion: These requirements ensure an effective ground-fault return path is provided for permanent generator installations. Some manufacturers produce generators that have the overcurrent devices mounted on the generator. Others supply the generator with overcurrent protection as an accessory that is mounted remote from the generator. In some cases, the installer provides the overcurrent protection that may be mounted adjacent to or some distance from the generator.

The rule in *250.35(A)* requires compliance with *250.30* if the system provided by the permanently installed generator is a separately derived system. Separately derived systems are covered in the Introduction to Grounding and Bonding and in *250.30* of this book. As indicated in Figure 2-70, the ground-fault return path (circuit) when the transfer switch is in the normal position is over the equipment grounding conductor(s) to the service and returns to the utility source over the grounded system conductor.

Important rules are provided in *250.35(B)* if the system supplied by the permanently installed generator is not separately derived. This means that the neutral is not switched in transfer equipment. The rules change depending on where the overcurrent device for the supplied conductors is located. If the overcurrent device is located remote from the generator, a properly sized supply-side bonding jumper is required. The size of the supply-side bonding jumper is determined by reference to *Table 250.66,* based on the size of the ungrounded conductors.

If the overcurrent device is located at the generator, an equipment grounding conductor or equipment grounding jumper is required that is determined from *Table 250.122,* based on the rating of the overcurrent device.

Section 250.35(B)(1) covers bonding between the generator and the first overcurrent device. As

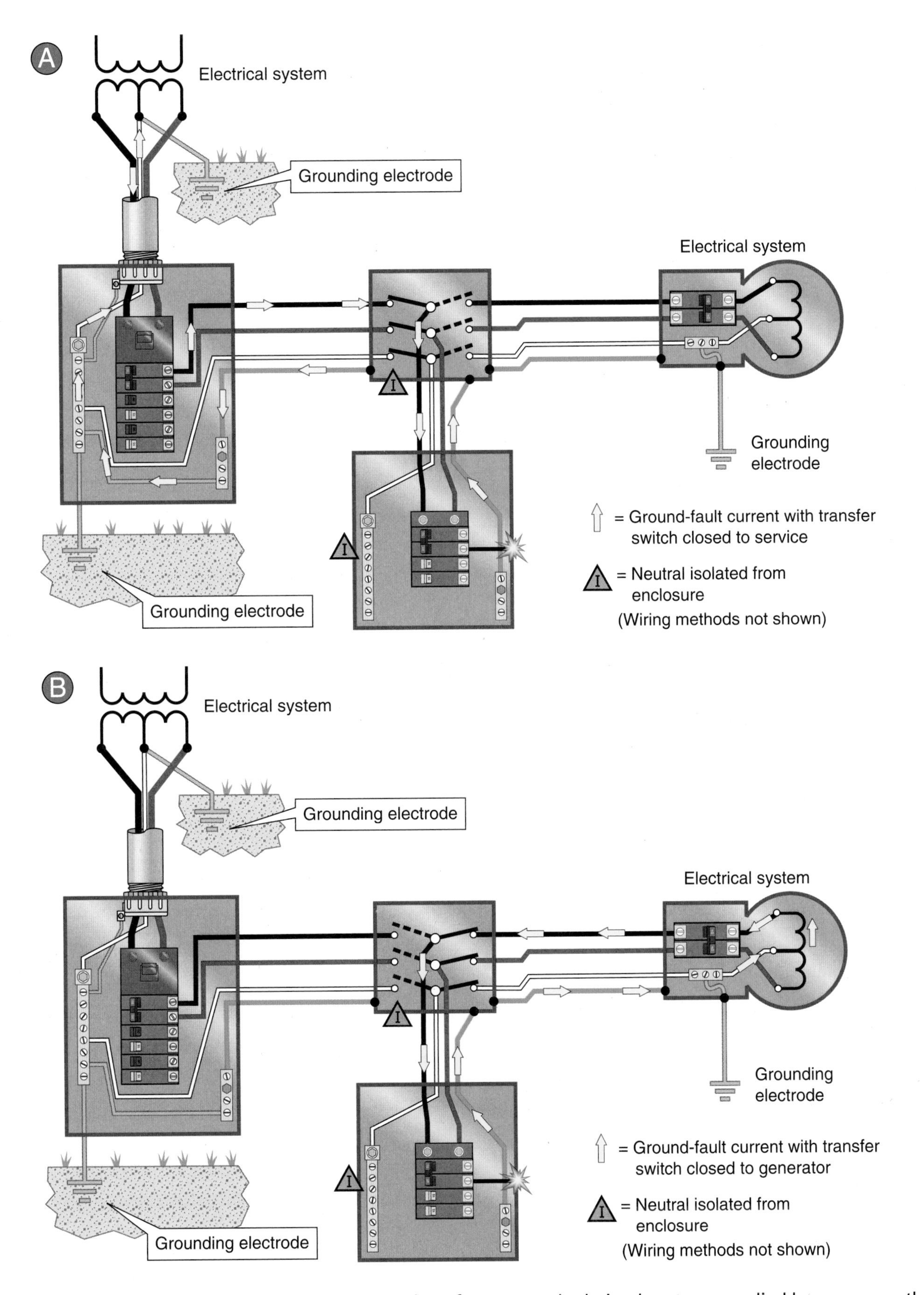

FIGURE 2-70 Grounding and bonding connections for separately derived system supplied by permanently installed generators with the system bonding jumper at the source, *250.35(B)(1)*. (*© Cengage Learning 2012*)

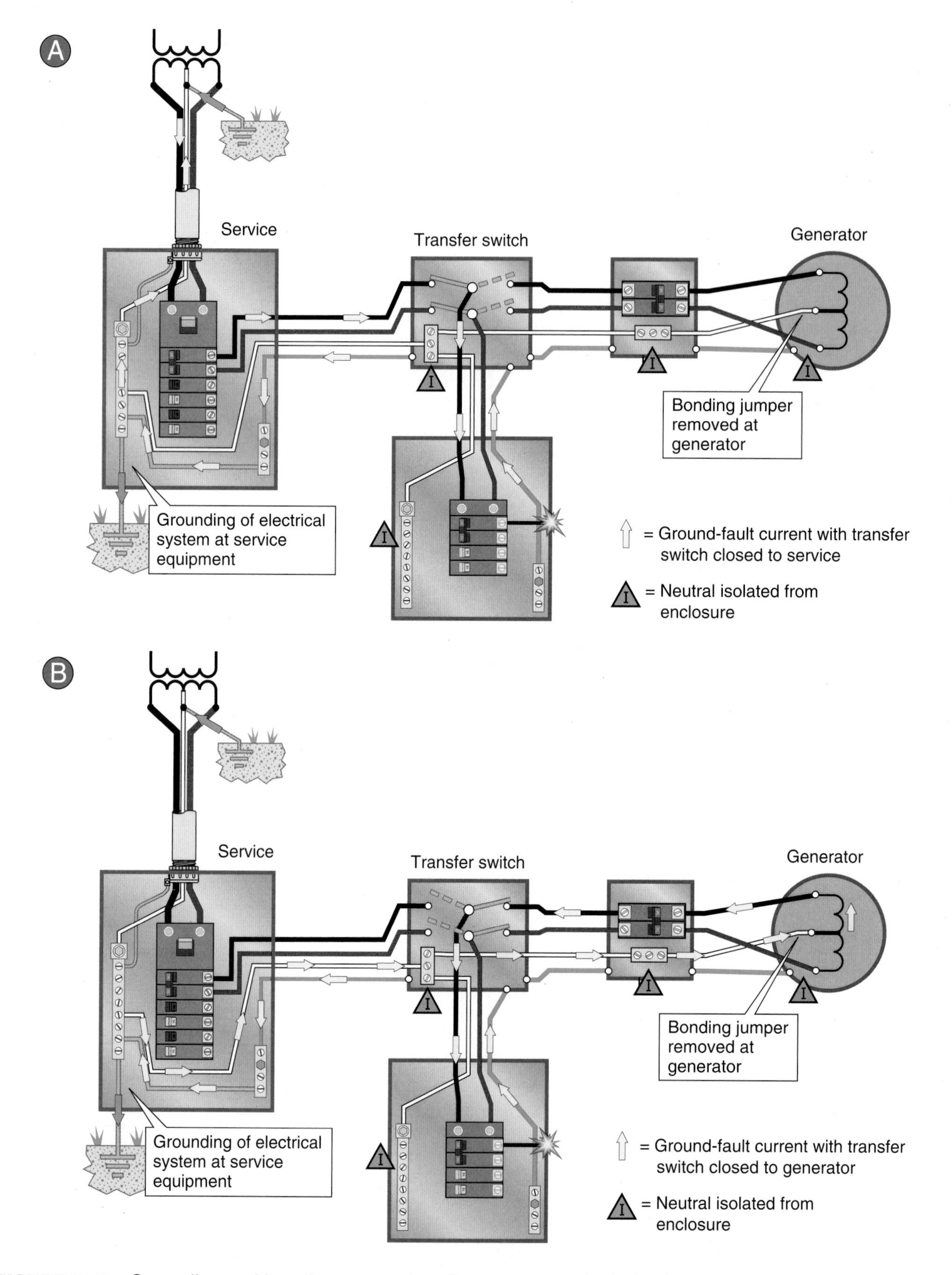

FIGURE 2-71 Grounding and bonding connections for nonseparately derived system with the transfer switch connected to the service and the permanently installed generator, *250.35(B)(2)*. (*© Cengage Learning 2012*)

shown in Figure 2-71, this bonding jumper is sized from *Table 260.66* on the basis of the size of the supply conductors. By reference to *250.102(C),* requirements are included for installation of supply conductors in a single raceway or in parallel raceways or cables. See the discussion in *250.30(A)(1)* of this book for sizing the supply-side bonding jumpers.

When the transfer switch is in the generator position, shown in Figure 2-71, the ground-fault return path (circuit) is over the equipment grounding conductor(s) to the service, over the main bonding jumper to the neutral, and returns to the generator source over the grounded system conductor.

The bonding jumper at the generator must be removed when the system is not separately derived. If the bonding jumper is not removed, the neutral will be connected to grounded equipment at two locations, and neutral current will flow over conductive paths such as metal conduit, cable trays, and metal piping.

Here's How: Let's assume a permanently installed generator has an 800-ampere overcurrent device that is installed remotely from the generator. Two 600-kcmil copper conductors are installed in parallel raceways from the generator to the distribution panelboard. The size of the bonding jumper is required to comply with *250.102(C).* This section provides that where bonding jumpers are installed in parallel raceways, they are to be sized according to *Table 250.66.* This table requires the bonding jumper in each raceway to be not smaller than 1/0 AWG. This ensures the bonding jumpers are adequate to carry fault current back to the source.

The bonding jumper on the load side of the overcurrent device is required to be not smaller than the equipment grounding conductor. As shown in Figure 2-71, this is determined from *Table 250.122,* based on the rating of the overcurrent device where the feeder originates.

250.36 HIGH-IMPEDANCE GROUNDED NEUTRAL SYSTEMS

The Requirements: High-impedance grounded neutral systems in which a grounding impedance, usually a resistor, limits the ground-fault current to a low value are permitted to be installed for 3-phase ac systems of 480 volts to 1000 volts if all the following conditions are met:

1. The conditions of maintenance and supervision ensure that only qualified persons service the installation.
2. Ground detectors are installed on the system.
3. Line-to-neutral loads are not served.

High-impedance grounded neutral systems are required to comply with the provisions of *250.36(A)* through *(G).*

Discussion: High-impedance grounded neutral systems are installed to increase system operational reliability and to reduce the downtime caused by outages from line-to-ground faults. See Figure 2-72 for the principles of operation. High-impedance grounded neutral systems have all of the advantages of an ungrounded system in that the first ground fault will not cause an overcurrent protective device to operate, but they do not have the disadvantages associated with ungrounded systems. These disadvantages include the possibility of transient overvoltages on ungrounded systems caused by sputtering or intermittent ground faults. Recall that the first ground fault on an ungrounded system simply grounds the system. Most of these first ground

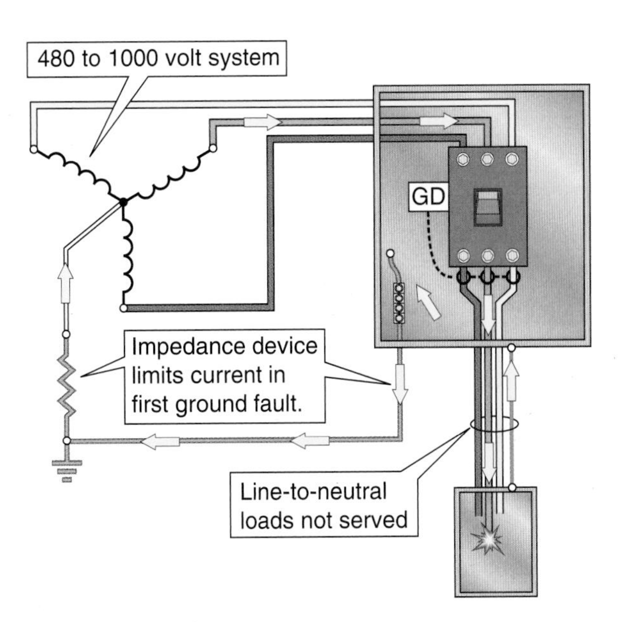

FIGURE 2-72 High-impedance grounded neutral systems, *250.36.* (© Cengage Learning 2012)

faults, however, are not bolted or solid faults but are sputtering or intermittent faults, because little current flows in the first ground fault. As a result, transient overvoltages of several times the system voltage can occur. These transient overvoltages can exceed the dielectric withstand rating of conductors. (See the discussion in *250.122* for additional information on the withstand rating of conductors.) The windings in motors and transformers often are vulnerable to transient overvoltages, which can lead to premature failure.

The three conditions in *250.36* must be complied with before a facility qualifies to have a high-impedance grounded neutral system. These conditions are the following:

1. The conditions of maintenance and supervision ensure that only qualified persons service the installation.

Discussion: This requirement appears in many locations in the *Code*. The management of the facility has the obligation to ensure that experienced people who are properly trained maintain the electrical system. See the definition of *qualified person* in *Article 100*.

2. Ground detectors are installed on the system.

Discussion: The *Code* is silent on the type or sophistication of ground detection system that is required. No guidance is provided on how the ground detection system is required to operate. Sophisticated ground detection equipment that is listed is available. Local and remote annunciation is available. Typically, for industrial occupancies, ground detection systems will sound an alarm or operate an indicating lamp, or both, to indicate a ground fault has occurred. The purpose of the alarm is to alert maintenance personnel so the first ground fault can be repaired before a second ground fault occurs.

3. Line-to-neutral loads are not served.

Discussion: Line-to-neutral loads are not permitted to be served, because the system is designed to limit the path for fault currents to a single path through the grounding resistor. These systems typically serve large industrial plants and serve motor loads or other phase-to-phase loads.

250.36(A) Grounding Impedance Location

The Requirements: The grounding impedance is required to be installed between the grounding electrode conductor and the system neutral point. Where a neutral point is not available, the grounding impedance must be installed between the grounding electrode conductor and the neutral point derived from a grounding transformer.

Discussion: As illustrated in Figure 2-69, the grounding impedance is installed between the neutral or grounded conductor and the grounding electrode conductor. As a result, the only path for ground-fault currents to take is through the grounding impedance. This grounding impedance limits the current, so an overcurrent device will not operate, and the equipment can continue to perform its function.

250.36(B) Grounded System Conductor

The Requirements: The grounded system conductor from the neutral point of the transformer or generator to its connection point to the grounding impedance is required to be fully insulated. The grounded system conductor must have an ampacity of not less than the maximum current rating of the grounding impedance and must not be smaller than 8 AWG copper or 6 AWG aluminum or copper-clad aluminum.

Discussion: The rating of the impedance device for the high-impedance grounded neutral system will usually limit the current to not more than 10 amperes. Therefore, the required minimum sizing of the grounded conductor is related not to its current-carrying ability but to providing a conductor that is large enough to withstand the physical or mechanical stress incurred in installation and maintenance of the system. The conductor is specifically required to be insulated. This rule modifies the provision in *230.41* that permits uninsulated service-entrance conductors under certain conditions. This insulation ensures a controlled return path for fault current through the impedance device.

250.36(C) System Grounding Connection

The Requirement: The system is not permitted to be connected to ground except through the grounding impedance.

Discussion: As previously mentioned, the connection of the grounded conductor to ground through the impedance device limits the fault current that can flow in the first ground fault (see Figure 2-73). This limitation helps ensure system and operational reliability.

The second ground fault, if it occurs on a different phase from the first one, will result in a phase-to-phase fault, and an overcurrent protective device will operate to take the faulted circuit off the line.

> **Informational Note:** The impedance is normally selected to limit the ground-fault current to a value slightly greater than or equal to the capacitive charging current of the system. This value of impedance will also limit transient overvoltages to safe values. For guidance, refer to criteria for limiting transient overvoltages in ANSI/IEEE 142-1991, Recommended Practice for Grounding of Industrial and Commercial Power Systems.

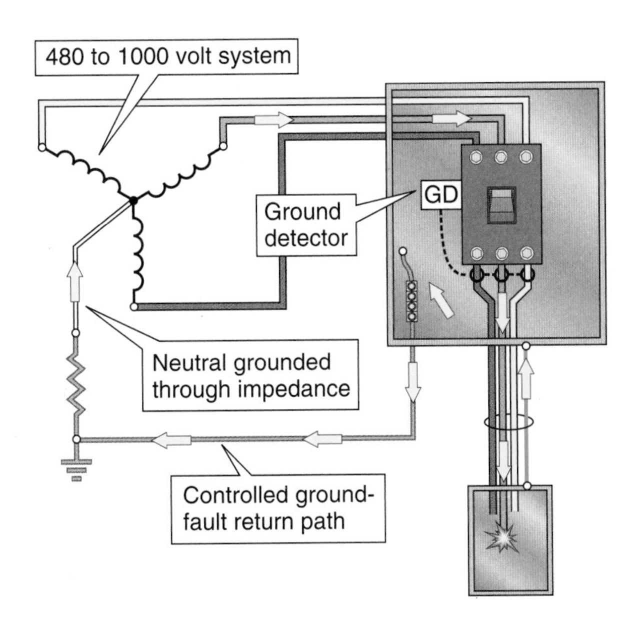

FIGURE 2-73 Controlled ground-fault return path in high-impedance grounded neutral systems, *250.36*. (*© Cengage Learning 2012*)

250.36(D) Neutral Point to Grounding Impedance Conductor Routing

The Requirement: The conductor connecting the neutral point of the transformer or generator to the grounding impedance is permitted to be installed in a separate raceway from the ungrounded conductors. It is not required to run this conductor with the phase conductors to the first system disconnecting means or overcurrent device.

Discussion: This requirement accommodates normal practice in industry to install the impedance device outside of the enclosure for the overcurrent protection. Because the impedance device often is a resistor, heat is produced as current flows through it. Locating the resistor in a ventilated enclosure outside the cabinet for the overcurrent protection is standard industry practice.

250.36(E) Equipment Bonding Jumper

The Requirement: The equipment bonding jumper (the connection between the equipment grounding conductors and the grounding impedance) is required to be an unspliced conductor run from the first system disconnecting means or overcurrent device to the grounded side of the grounding impedance.

Discussion: The equipment bonding jumper runs from the service cabinet to the load side of the grounding impedance (see Figure 2-74). Specific sizing requirements are in *250.36(G)*.

250.36(F) Grounding Electrode Conductor Location

The Requirement: The grounding electrode conductor is required to be attached at any point from the grounded side of the grounding impedance to the equipment grounding connection at the service equipment or first system disconnecting means.

Discussion: The grounding electrode conductor usually is connected either at the service equipment or at the grounding impedance. Either location is acceptable; the choice is a matter of design or engineering judgment.

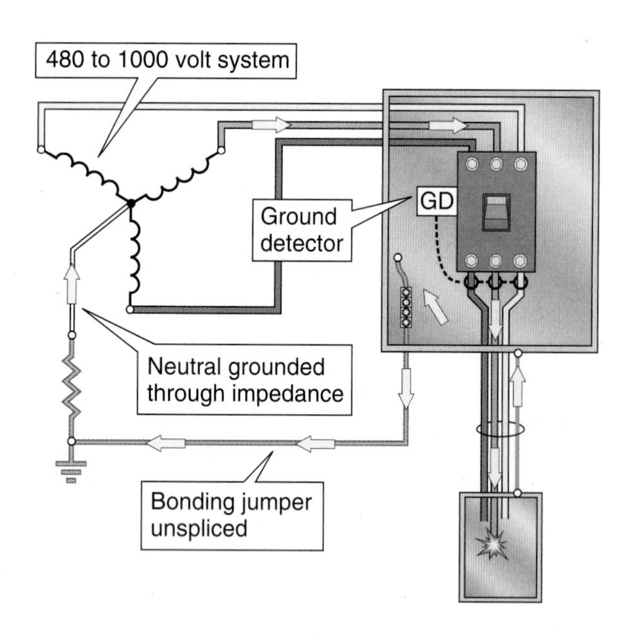

FIGURE 2-74 Unspliced bonding jumper for high-impedance grounded neutral systems, *250.36(E)*. (*© Cengage Learning 2012*)

250.36(G) Equipment Bonding Jumper Size

The Requirement: The equipment bonding jumper must be sized in accordance with (1) or (2).

1. Where the grounding electrode conductor connection is made at the grounding impedance, the equipment bonding jumper must be sized in accordance with *250.66*, based on the size of the service entrance conductors for a service or the derived phase conductors for a separately derived system.
2. Where the grounding electrode conductor is connected at the first system disconnecting means or overcurrent device, the equipment bonding jumper is required to be sized the same as the neutral conductor in *250.36(B)*.

Discussion: As illustrated in Figure 2-75, the point of connection of the grounding electrode conductor to the system determines the minimum size of the equipment bonding jumper. If the grounding electrode conductor is connected at the grounding impedance, the bonding jumper must be the same size as the grounding electrode conductor, as determined from *250.66*. If the grounding electrode conductor is connected at the first system disconnecting means or overcurrent device, the bonding jumper is sized the same as the neutral conductor.

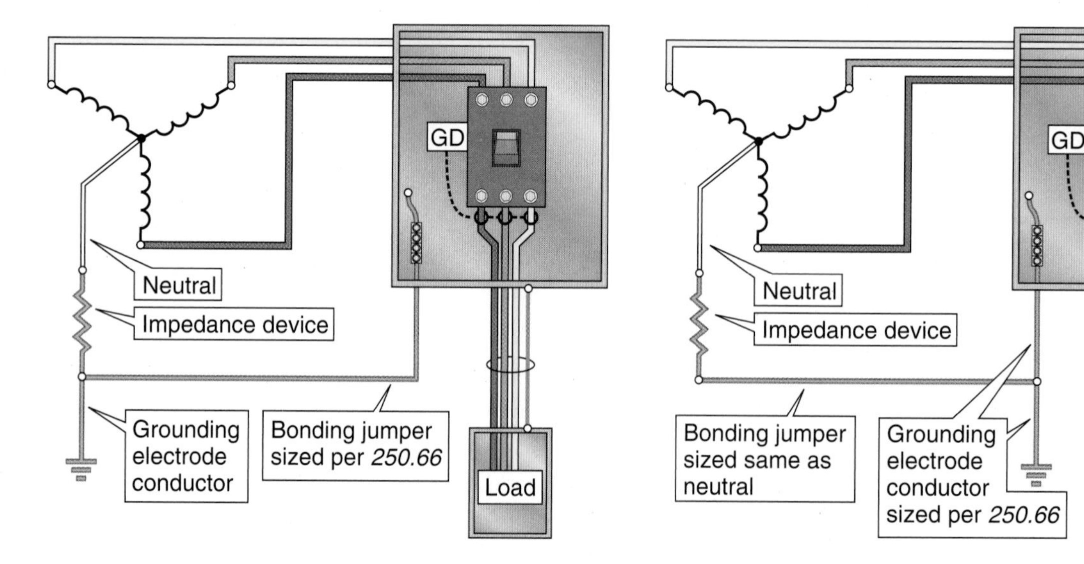

FIGURE 2-75 Size of equipment bonding jumper for high-impedance grounded neutral system, *250.36(G)*. (*© Cengage Learning 2012*)

REVIEW QUESTIONS

1. Which of the following electrical systems is *not* required to be grounded?
   a. 120-volt, 1-phase, 2-wire
   b. 120/240-volt, 1-phase, 3-wire
   c. 208Y/120-volt, 3-phase, 4-wire, wye-connected
   d. 240-volt, 3-phase, 3-wire, delta-connected
2. Which of the following electrical systems is *most likely* to be required to be grounded?
   a. 480Y/277-volt, 3-phase, 3-wire, wye-connected
   b. 480Y/277-volt, 3-phase, 4-wire, wye-connected
   c. 600Y/346-volt, 3-phase, 3-wire, wye-connected
   d. 480-volt, 3-phase, 3-wire, delta-connected
3. Which of the following electrical systems is likely to have a conductor with *a higher voltage to ground*?
   a. 120/240-volt, 1-phase, 3-wire
   b. 208Y/120-volt, 3-phase, 4-wire
   c. 120/240-volt, 3-phase, 4-wire
   d. 480Y/277-volt, 3-phase, 4-wire
4. Which of the following electrical systems is *not* likely to require a ground detector?
   a. 480Y/277-volt, 3-phase, 3-wire, wye-connected
   b. 480Y/277-volt, 3-phase, 4-wire, wye-connected
   c. 600Y/346-volt, 3-phase, 3-wire, wye-connected
   d. 480-volt, 3-phase, 3-wire, delta-connected
5. Which of the following statements about determining when a system is separately derived is most accurate?
   a. There is no bonding jumper from the secondary of the transformer to the enclosure.
   b. A bonding jumper is installed from the transformer secondary to the enclosure.
   c. A bonding jumper is not installed from the primary to the secondary of the transformer.
   d. A bonding jumper is not installed from the midpoint of the generator windings to the case.
6. Which of the following groups of systems is required to be grounded?
   a. Electrical systems used exclusively to supply industrial electric furnaces for melting, refining, tempering, and the like
   b. Separately derived systems used exclusively for rectifiers that supply only adjustable-speed industrial drives
   c. Separately derived systems supplied by transformers that have a primary voltage rating less than 1000 volts for control circuits
   d. High-impedance grounded neutral systems as specified in *250.36*

7. Which of the following categories of circuits or systems is required to be grounded?
   a. Circuits for electric cranes operating over combustible fibers in Class III locations
   b. Circuits in critical care areas of health care facilities
   c. Circuits for equipment within the electrolytic cell working zone in aluminum refineries
   d. Secondary circuits of 30-volt lighting systems
8. Which of the following locations is *not* permitted to be used to connect the grounding electrode at the service equipment?
   a. At the weatherhead where the service drop is connected to the service-entrance conductors
   b. In the meter socket enclosure supplied by an underground service lateral
   c. In a wireway from the metering equipment to the service equipment
   d. In a feeder panel if immediately adjacent to the service equipment
9. A grounding connection to the grounded (neutral) conductor is *not* permitted at or for which of the following locations?
   a. Separately derived systems
   b. Separate buildings supplied by a feeder
   c. A new branch circuit for an electric range
   d. Meters on the load side of the service disconnecting means but located near the service
10. Which of the following statements about a main bonding jumper is *not* true?
    a. It is permitted to consist of a wire, a busbar, or a screw.
    b. It typically is located in each section of a multisection switchboard.
    c. It provides the essential path for ground-fault current to return to the source.
    d. It may be bare or insulated.
11. If 240-volt delta connected systems are not grounded,
    a. the voltage to ground is considered 240 volts.
    b. the installation is not in compliance with the *NEC*.
    c. the system is required to be grounded.
    d. a grounding electrode for the equipment is not required.
12. Which of the following statements about ground detectors for systems that are permitted to operate ungrounded is *not* true?
    a. Ground detectors are permitted to consist of a system of lamps (light bulbs).
    b. Ground detectors are intended to indicate that one phase has become grounded.
    c. Ground detectors are available as listed equipment.
    d. Ground detectors must report the occurrence to a remote location.

13. Electrical systems that are permitted to be installed as ungrounded are usually not grounded so the first ground fault on the system will not result in operation of an overcurrent device.

    a. True

    b. False

14. Ungrounded electrical systems with conductors installed in grounded metal raceways and enclosures

    a. have no voltage to ground.

    b. have an approximate half system voltage to ground.

    c. present no shock hazard to personnel.

    d. cannot become grounded as there is no connection to the ground rod.

15. The owner of the premises can choose whether the following circuits are grounded or remain ungrounded: a crane that operates over combustible fibers, circuits in inhalation anesthetizing locations of hospitals, and 15-volt circuits for swimming pool lighting derived from a transformer.

    a. True

    b. False

16. Rather than making a grounding connection in each service disconnecting means for services that are double-ended, it is permitted to make a single grounding electrode connection at the tie point for the services.

    a. True

    b. False

17. The purpose behind permitting the grounding electrode connection to be made at the equipment grounding conductor terminal bar of the service rather than to the neutral is to provide a convenient connection point sometimes outside the service disconnecting means enclosure.

    a. True

    b. False

18. If the electrical system is grounded at the utility supply transformer, it is _______ that the grounded conductor be run to and be connected to the service disconnect.

    a. optional, based on the design of the electrical engineer,

    b. mandatory, based on *NEC* rules in *Article 250*,

19. The minimum size of the grounded service-entrance conductor is

    a. not smaller than the size of the grounding electrode.

    b. optional based on the design of the electrical engineer.

20. The minimum size of the grounded conductor for a 3-phase, 3-wire, corner-grounded, delta connected service is

    a. not smaller than the required grounding electrode conductor.

    b. not smaller than a neutral for a 3-phase, wye-connected system or the same ampacity.

c. not smaller than the ungrounded service conductors.

d. not specified as a grounded conductor is not required.

21. The minimum size of a main bonding jumper that is furnished by the manufacturer for listed service equipment

a. is permitted to be installed without calculation of proper size.

b. is only permitted to be installed if size is adequate according to *NEC* rules.

22. The system bonding jumper for a system that supplies more than one disconnecting means is sized

a. based on the full-load current rating of the transformer secondary windings.

b. based on the total circular mil area of the derived ungrounded conductors.

23. Because separately derived systems are not a service, it is generally permitted to ground the neutral at multiple locations on the load side of the separately derived system.

a. True

b. False

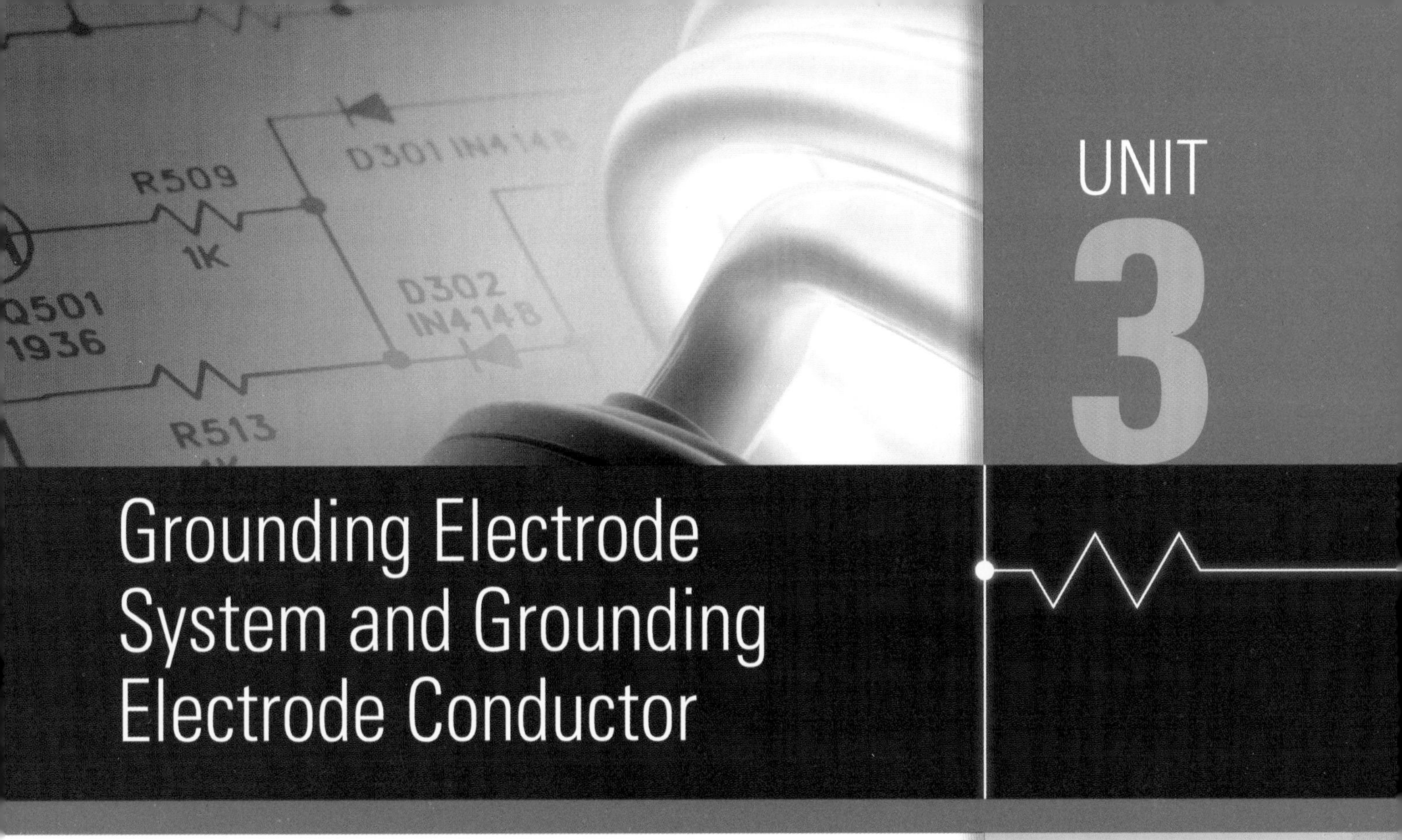

Grounding Electrode System and Grounding Electrode Conductor

OBJECTIVES

After studying this unit, the reader will understand

- the requirements for the grounding electrode system.
- grounding electrodes required to be used and those not permitted to be used.
- the installation requirements for the grounding electrode system.
- requirements for supplemental grounding electrodes.
- installation of auxiliary grounding electrodes.
- resistance requirements for rod pipe and plate electrodes.
- requirements for the use of a common grounding electrode.
- installation requirements for grounding electrode conductors.
- minimum size of grounding electrode conductors.
- methods for connecting grounding and bonding conductors to grounding electrodes.
- using building structural metal and water pipes as grounding electrode conductors.

250.50 GROUNDING ELECTRODE SYSTEM

The Requirements: All grounding electrodes as described in *250.52(A)(1)* through *(A)(7)* that are present at each building or structure served are required to be bonded together to form the grounding electrode system. If none of these grounding electrodes exist, one or more of the grounding electrodes specified in *250.52(A)(4)* through *(A)(8)* are required to be installed and used.

> ***Exception:*** *Concrete-encased electrodes of existing buildings or structures are not required to be part of the grounding electrode system if the steel reinforcing bars or rods are not accessible for use without disturbing the concrete.*

Discussion: Important concepts and requirements for properly installing the grounding electrode system are contained in this section. First, where two or more grounding electrodes are bonded together, a *grounding electrode system* is established (see Figure 3-1). The term *grounding electrode system* is not defined in the *NEC*. From a concept standpoint, it seems a system is created by two or more grounding electrodes that are bonded or connected together.

Another important concept is that the conductor that is used to connect grounding electrodes together is a *bonding conductor*, not a *grounding electrode conductor*. A grounding electrode conductor connects the grounding electrode system to the neutral or grounded conductor terminal bar in the service, at a separately derived system, or at the building disconnecting means for buildings or structures supplied by a feeder.

If more than one building or structure is on the premises and is supplied by a service or feeder, each building or structure is treated individually for creating a grounding electrode system. Each building or structure is treated as an island because all grounding electrodes that are at each building or structure must be bonded together to create a grounding electrode system.

All of the grounding electrodes described in *250.52(A)(1)* through *(A)(7)* that are *present* at

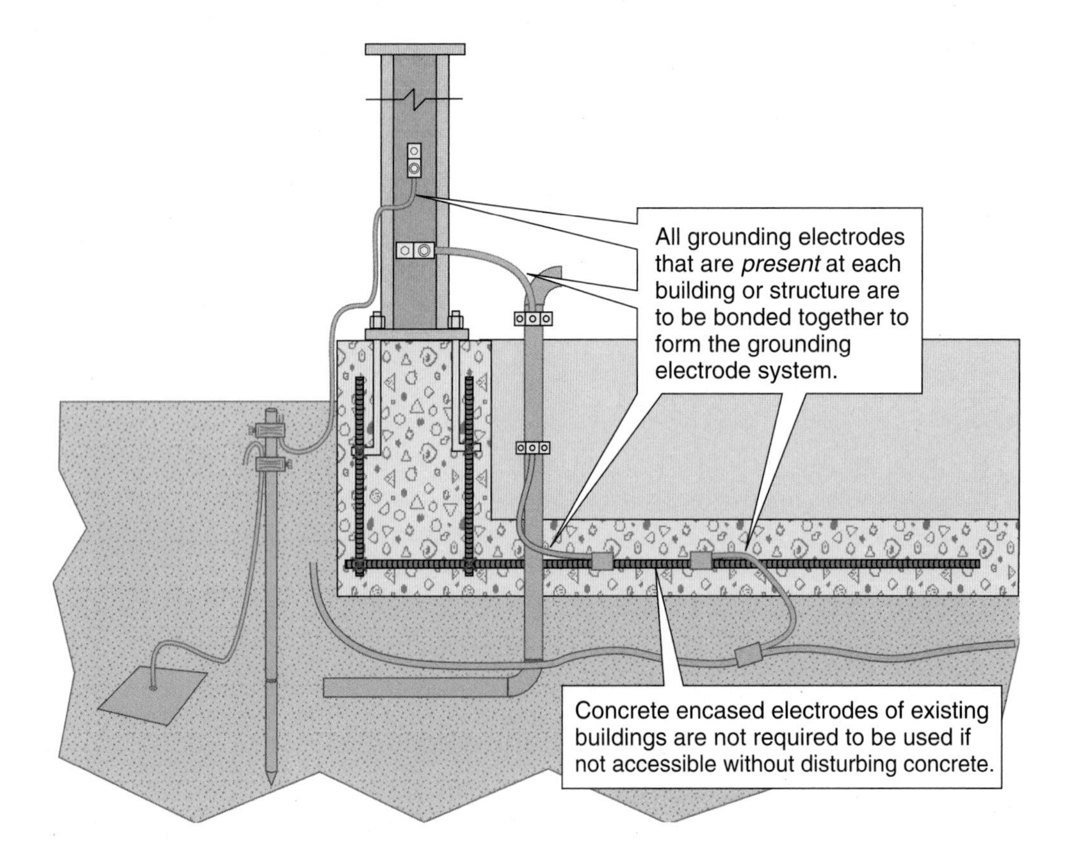

FIGURE 3-1 Grounding electrode system, *250.50*. (*© Cengage Learning 2012*)

each building or structure served are required to be bonded together to form a grounding electrode system. It is important to recognize that this section no longer uses the phrase "where available" to determine whether a grounding electrode is required to be connected. Thus, if the grounding electrode is *present* rather than *available,* it must be bonded to other electrodes to form the grounding electrode system.

It is common for electrical contractors to be on the job site during installation of underground piping systems for larger commercial and industrial facilities. Electricians thus have ready opportunity to connect to reinforcing steel before concrete is poured. Coordination between the building contractor and the electrical contractor or installer is essential at all job sites to provide an opportunity for bonding concrete-encased electrodes before the concrete is poured. For those installations that have an electrical inspection requirement, it is important to schedule an inspection and get approval before the concrete is poured over the reinforcing steel bars.

An exception from the general requirements is provided for concrete-encased electrodes of existing buildings or structures. Note that the exception does not apply to new buildings or to buildings under construction. The concrete-encased electrodes of existing buildings or structures are not required to be part of the grounding electrode system if the steel reinforcing bars or rods are not accessible for use without disturbing the concrete. Buildings are usually considered to be under construction until the authority having jurisdiction issues a certificate of occupancy. The building becomes *existing* after the certificate of occupancy is issued.

250.52 GROUNDING ELECTRODES

250.52(A) Electrodes Permitted for Grounding

Discussion: Although the title of this section is Electrodes *Permitted* for Grounding, the use of these electrodes becomes mandatory because of the language in *250.50* requiring all of these grounding electrodes to be bonded together to form the grounding electrode system. Some electrodes such as metal underground water pipes, metal frames of a building or structure, and concrete-encased electrodes usually are installed by trades other than electricians. Their use as a grounding electrode becomes a "gift" to the electrical system installer! Other electrodes such as ground rings, ground rods, pipe electrodes, and plate electrodes commonly are installed by electricians.

Some installation requirements are contained in the description of the grounding electrodes, such as the minimum length of the water pipe or location of the concrete-encased electrode encased in not less than 2 in. (50 mm) of concrete in the portion of the footing that is in direct contact with earth. Installation requirements that are not contained in the description of a grounding electrode are included in *250.53.*

250.52(A)(1) Metal Underground Water Pipe

The Requirements: A metal underground water pipe is required to be used as a grounding electrode where it is in direct contact with the earth for 10 ft (3.0 m) or more (including any metal well casing bonded to the pipe) and is electrically continuous (or made electrically continuous by bonding around insulating joints or insulating pipe) to the points of connection of the grounding electrode conductor and the bonding conductors.

Discussion: This section describes grounding electrodes that are required to be used if present at the building or structure being served. Underground metal water piping has been required to be used as a grounding electrode for many editions of the *NEC.* The rule applies whether the metal underground water pipe is a part of a public or community water distribution system or is from a well on the premises. (See Figure 3-2.)

There is no maximum or minimum size of water pipe to determine whether the metal water pipe is required to be used as a grounding electrode. The length of underground metal water pipe in contact with the earth is the qualifying condition. If the length of underground metal water piping in contact with the earth is less than 10 ft (3.0 m), it is not considered a grounding electrode for the purposes of this section.

The description of the underground metal water pipe does not include qualifiers regarding the type

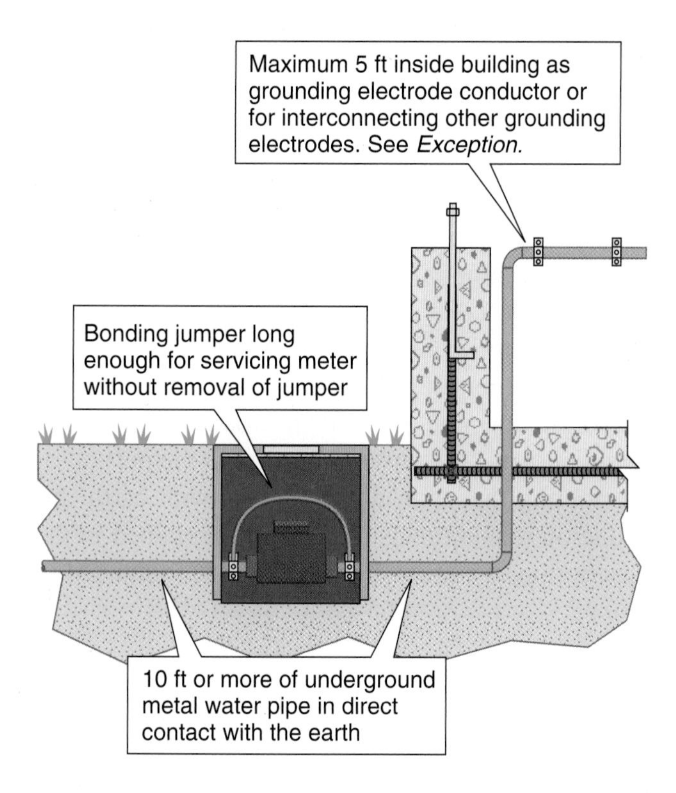

FIGURE 3-2 Metal underground water pipe grounding electrode, *250.52(A)(1)*. (*© Cengage Learning 2012*)

of underground metal water pipe such as galvanized steel, copper, or cast iron. If the piping is underground, metal, and used for water, regardless of the type of water and whether for domestic use or irrigation, drainage, or fire protection purposes, it must be used as a grounding electrode if it is in contact with the earth for 10 ft (3.0 m) or more.

Be aware that that at least two other NFPA standards prohibit underground metal piping for fire protection systems being used as a grounding electrode for electrical systems. The two NFPA standards are NFPA 13 *Standard for the Installation of Sprinkler Systems* and NFPA 24 *Standard for the Installation of Private Fire Service Mains and Their Appurtenances*. The section with the requirement is 10.6.8 in both Standards. It reads "In no case shall the underground piping be used as a grounding electrode for electrical purposes."

Some metal underground sprinkler or fire protection piping may be painted with an insulating asphaltic type material or be wrapped with a protective covering. If the piping is insulated from contact with earth, it obviously could not qualify as a grounding electrode as described in *NEC 250.52(A)(1)*. Verify with the local authority having jurisdiction what the enforcement practice is regarding connection to underground metal fire protection piping.

One last point on this issue: Bonding of interior metal water piping is required in *NEC 250.104(A)*. The bonding is required if the piping is considered to be "likely to become energized." A bonding connection to interior metal water piping will have the effect of treating underground metal piping as a grounding electrode unless an insulating or dielectric coupling is installed to separate the piping electrically.

250.52(A)(2) Metal Frame of the Building or Structure

The Requirements: The metal frame of the building or structure is required to be used as a grounding electrode where any of the following methods are used to make an earth connection:

1. At least one structural metal member is in direct contact with the earth for 10 ft (3.0 m) or more with or without concrete encasement
2. The hold-down bolts securing the structural steel column are connected to a concrete encased electrode that complies with *250.52(A)(3)* and is located in the support footing or foundation. The hold-down bolts shall be connected to the concrete-encased electrode by welding, exothermic welding, the usual steel tie wires, or other approved means.

Discussion: Methods of making an earth connection with structural metal members of the building are described in this *NEC* section and summarized in Figure 3-3. Two methods are recognized for creating an earth connection with the structural metal frame of a building or structure. If the installation of the structural metal (usually steel) by itself or the bonding to a concrete-encased grounding electrode does not comply with these rules, the structural metal member is not considered to be a grounding electrode. Even though structural members may not qualify as a grounding electrode, requirements are found in *250.104(C)* for bonding structural metal members of a building.

One major advantage of establishing structural metal members of the building as a grounding electrode is that the metal members can then be used to interconnect other grounding electrodes and also to

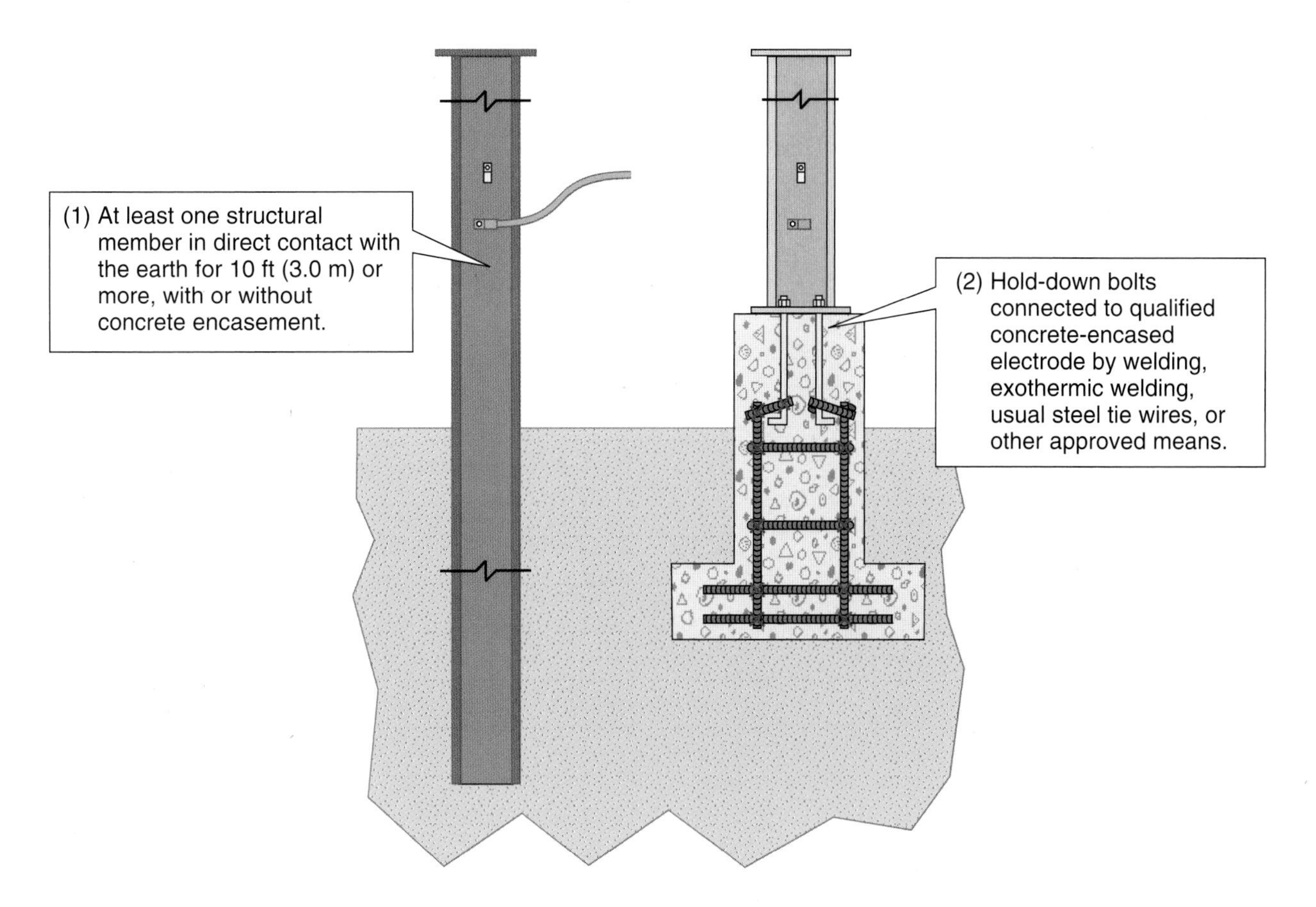

FIGURE 3-3 Metal frame of the building or structure grounding electrode, *250.52(A)(2)*. (*© Cengage Learning 2012*)

serve as a grounding electrode conductor for separately derived systems or for services that might be located at an upper floor level in the building. This is covered in *250.68(C)*.

Consider the following example: The service is located at one end of a large building where a grounding electrode conductor is connected to a structural metal member of the building. An underground metal water pipe grounding electrode supplies the opposite end of the building. A bonding connection can be made from the structural metal member grounding electrode conductor to the underground metal water pipe in the vicinity of the water pipe.

The structural metal grounding electrode can also be used for the grounding electrode for separately derived systems that may be installed at various locations in the building or structure. *Section 250.30(A)(4)* requires the structural metal member to serve as the grounding electrode for the separately derived system if it is in the vicinity of the separately derived system.

250.52(A)(3) Concrete-Encased Electrode

The Requirements: At least 20 ft (6.0 m) of either (1) or (2):

1. One or more bare or zinc-galvanized or other electrically conductive coated steel reinforcing bars or rods of not less than ½ in. (13 mm) in diameter, installed in one continuous 20-ft (6.0-m) length, or if in multiple pieces, connected together by the usual steel tie wires, exothermic welding, welding, or other effective means to create a 20-ft (6.0-m) or greater length; or
2. Bare copper conductor not smaller than 4 AWG.

Metallic components shall be encased by at least 50 mm (2 in.) of concrete and shall be located horizontally within that portion of a concrete foundation or footing that is in direct contact with the earth or within vertical foundations or structural components or members that are in direct contact with the earth.

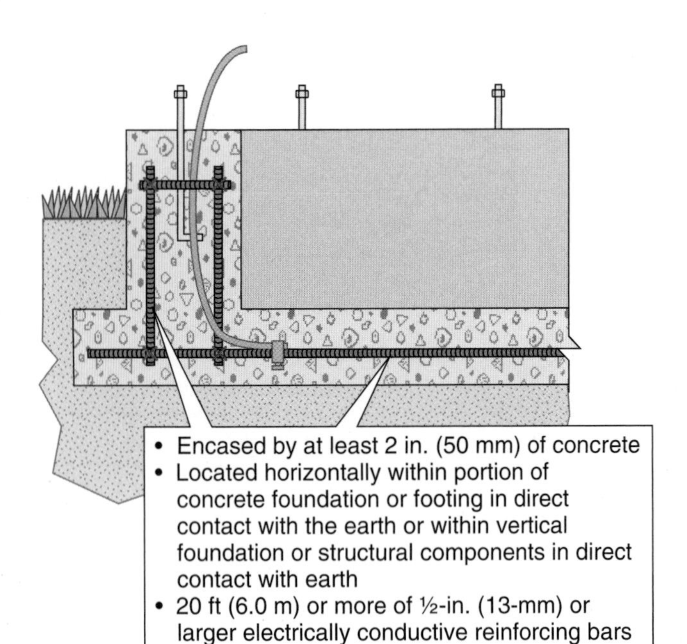

FIGURE 3-4 Concrete-encased grounding electrode, *250.52(A)(3)*. (*© Cengage Learning 2012*)

If multiple concrete-encased electrodes are present at a building or structure, it shall be permissible to bond only one into the grounding electrode system.

> ***Informational Note:*** Concrete installed with insulation, vapor barriers, films, or similar items separating the concrete from the earth is not considered to be in "direct contact" with the earth. See Figure 3-4.

Discussion: A concrete-encased electrode is one of the most effective for making a low-resistance connection to the earth. Another advantage of using a concrete-encased electrode is that the reinforcing steel is installed by other trades and usually is required by the building code, so there is little additional cost for use of the reinforcing steel as a concrete-encased electrode. In addition, a concrete-encased electrode is not required to be supplemented by an additional electrode if it is the only grounding electrode, such as is required for water pipe grounding electrodes.

The U.S. Army pioneered the use of concrete-encased electrodes in the early 1940s as the grounding electrode system for hundreds of ammunition bunkers and storage sheds in the Tucson and Flagstaff, Arizona, areas. Soil conditions were considered poor, and the climate was generally arid. A good earth connection was considered essential to discharge any static charge caused by wind and sandstorms or from lightning events from thunderstorms. Ground rods or large copper counterpoise systems were ruled out because of the effort to conserve essential materials during World War II. Initial tests of each building made in 1942 showed the concrete-encased electrode to have a resistance to earth of 5 ohms or less. The systems were tested again in 1960, and all readings were from 2 to 5 ohms. It was found that the foundations, even in the arid climate, maintained enough moisture to remain conductive.

Several other test installations of concrete-encased electrodes were made in the early 1960s in diverse locations. Their low-resistance connection to earth also was documented.

A concrete-encased electrode often is referred to as an "Ufer ground," after Herbert Ufer, who documented the effectiveness of such electrodes over several years. A research paper by Ufer titled "Investigation and Testing of Footing-Type Grounding Electrodes for Electrical Installations," presented at the IEEE Western Appliance Technical Conference on November 4, 1963, is reprinted in Appendix B of this book.

This section was changed during the processing of the 2008 *NEC* to recognize concrete-encased grounding electrodes that are installed vertically and within that portion of a concrete foundation or footing that is in direct contact with the earth. This section does not require that the vertical portion of a concrete foundation be in direct contact with the earth on both sides. Concrete foundation walls are often considered to be damp or wet as a result of their tendency to absorb and retain moisture. The hygroscopic nature of the concrete is one of the reasons it is so effective as a grounding electrode. The installation of concrete foundations and footings is under the jurisdiction and control of a building code and not the electrical code.

Another type of concrete-encased electrode that should be effective consists of vertical reinforcing steel bars that are connected by tie wire or welding to circular reinforcing steel to form a cage that is encased in concrete. The concrete must be in contact with the earth, so forms must be removed before this type of electrode would be considered effective.

Twenty feet or more of zinc galvanized or other electrically conductive coated steel reinforcing bars or rods not less than ½ in. (13 mm) in diameter must be encased by not less than 2 in. (50 mm) of concrete with the concrete in direct contact with the earth. This section does not require that the 20 ft (6.0 m) of steel reinforcing bars or rods be in one continuous length. Multiple lengths (this section places no minimum or maximum number of rods or bars) can be connected by tie wires, exothermic welding, or welding to create the 20 ft (6.0 m) length.

This section permits bonding only one of these electrodes into the grounding electrode system if multiple concrete-encased electrodes are present at a building or structure. Multiple concrete-encased electrodes can exist at buildings or structures if all of the reinforcing bars or rods are not connected together throughout the foundation or footings.

The *Informational Note* reminds us that objects like insulation or plastic vapor barriers isolate the concrete-encased electrode from contact with the earth so it no longer qualifies.

250.52(A)(4) Ground Ring

The Requirements: A ground ring is required to encircle the building or structure, be in direct contact with the earth, and consist of at least 20 ft (6.0 m) of bare copper conductor not smaller than 2 AWG. See Figure 3-5.

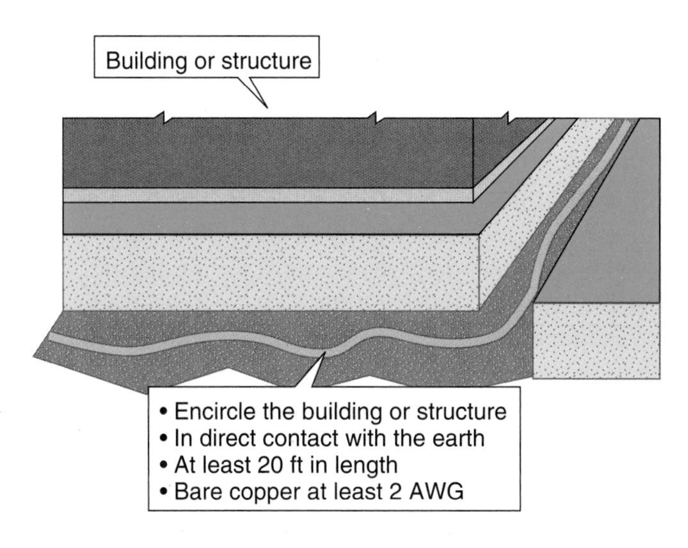

FIGURE 3-5 Ground ring grounding electrode, *250.52(A)(4)*. (*© Cengage Learning 2012*)

Discussion: Although 2 AWG is the minimum size of conductor permitted to be used as a ground ring, many consulting or design engineers will specify ground rings as large as 3/0 or 4/0 AWG or larger. The installation rules are at *250.53(F)*.

250.52(A)(5) Rod and Pipe Electrodes

The Requirements: Rod and pipe electrodes are required to be not less than 8 ft (2.44 m) in length and must consist of the following materials.

(a) Electrodes of pipe or conduit are not to be smaller than trade size ¾ (metric designator 21) and, where of iron or steel, must have the outer surface galvanized or otherwise metal-coated for corrosion protection.

(b) Rod-type grounding electrodes of stainless steel and copper or zinc-coated steel must be at least ⅝ in. (15.87 mm) in diameter unless listed.

See Figure 3-6.

Discussion: Ground rods, pipes, and conduit are required to be not less than 8 ft (2.5 m) in length

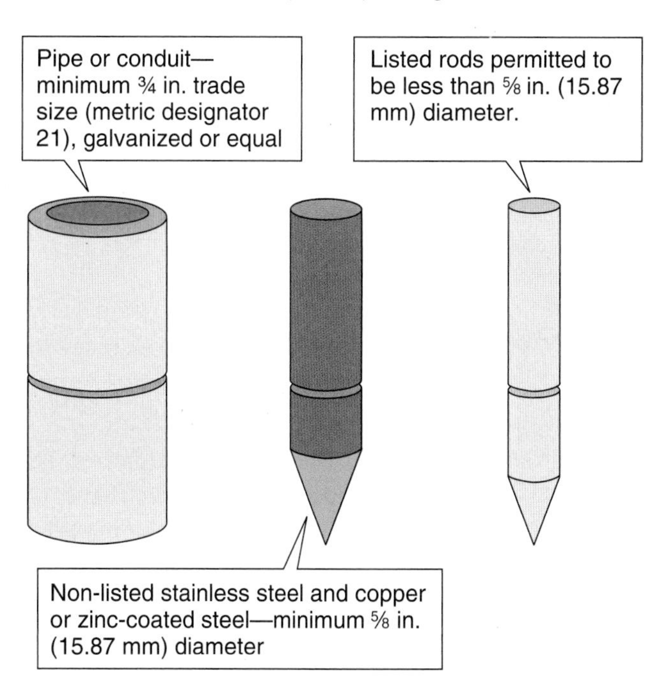

FIGURE 3-6 Rod and pipe grounding electrodes, *250.52(A)(5)*. (*© Cengage Learning 2012*)

regardless of how many are used. Ideally, at least a portion of the rod is installed below permanent moisture level. It is common for some installations, such as at data processing centers or telephone central offices, to use much longer ground rods or several rods that are threaded together and driven to a depth of 60 ft (18 m) or greater. The design engineer may specify a counterpoise system, which often consists of three sets of rods that are driven in a triangular configuration and bonded together at the top. With some of these installations, the specifications may require a resistance of 1 ohm or less. Other configurations such as a square, rectangle, or star are used as well.

As can be seen, *250.52(A)(5)(b)* will allow listed rods to have a diameter less than ⅝ in. (15.87 mm). No minimum diameter is given for listed rods. The UL product safety standard will dictate the minimum diameter. The diameter of listed rods may vary by the composition of the rod. Although some ground rods are permitted to be less than ⅝ in. (15.87 mm) in diameter, ¾- or 1-in rods are often installed if drivability is a consideration. Thinner ground rods may be very hard to drive in difficult soil conditions such as when large rocks or compact clay is present. Installation rules for ground rods are at *250.53(A), (B),* and *(G).*

For additional information, see the AEMC Installation and Testing Guide for ground rods, found in Appendix C in this book.

250.52(A)(6) Other Listed Electrodes

The Requirements: Other listed grounding electrodes are permitted.

Discussion: This section intends to recognize grounding electrodes other than those mentioned in other subsections of *250.52(A)*. Included is copper tubing grounding electrodes that are listed and have been available for several years. These are known as chemical rods because they typically have chemicals inside the approximately 2 in. (50.8 mm) copper tubing and weep or drain holes to permit the chemicals to leach into the soil to enhance the earth connection. These copper tubing grounding electrodes cannot be driven, so a hole must be excavated or bored to permit their installation. An access cover is usually provided to permit chemicals to be added as needed over time. A typical installation of the vertical and horizontal type of chemical rod is shown in Figure 3-7.

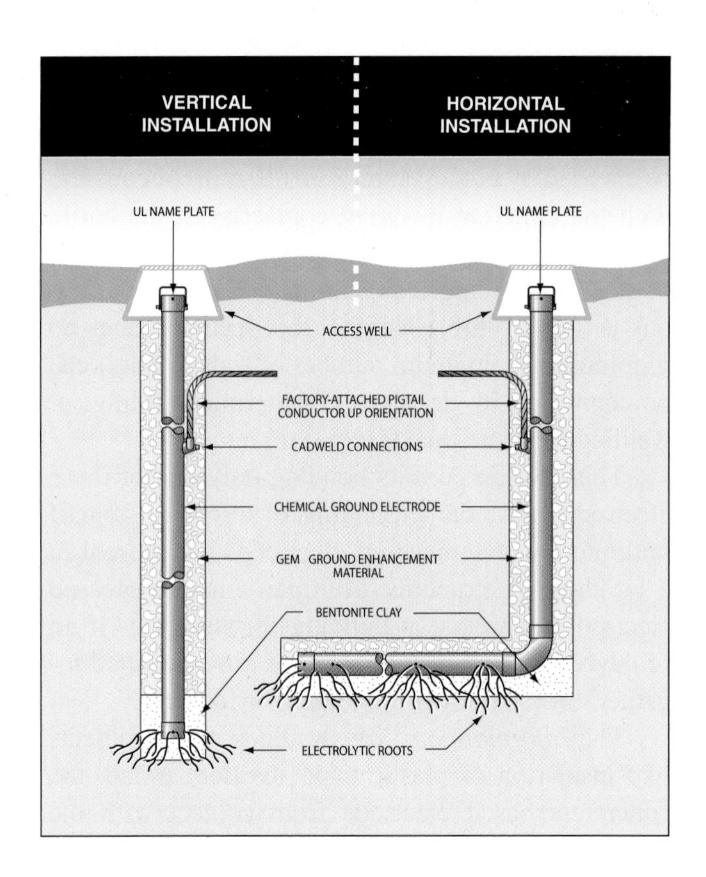

FIGURE 3-7 Other listed grounding electrodes, *250.52(A)(6)*. (*Courtesy of ERICO*)

It is important to carefully follow all manufacturers' installation instructions. This is critically important because no specific installation instructions are provided in the *NEC* for these listed grounding electrodes.

250.52(A)(7) Plate Electrodes

The Requirements: Each plate electrode is required to expose not less than 2 ft^2 (0.186 m^2) of surface to exterior soil. Electrodes of bare or conductively coated iron or steel plates must be at least ¼ in. (6.4 mm) in thickness. Solid, uncoated electrodes of nonferrous metal are to be at least 0.06 in. (1.5 mm) in thickness. See Figure 3-8.

Discussion: The common interpretation of the requirement to expose at least 2 ft^2 to the surface of the soil is that a plate electrode that measures 1 ft × 1 ft meets this rule, because both sides of the plate are considered to be in contact with the soil. Installation requirements are at *250.53(A), (B), (E),* and *(H).*

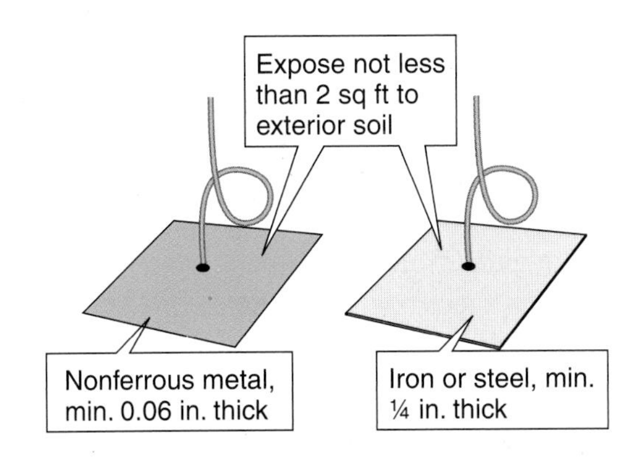

FIGURE 3-8 Plate-type grounding electrodes, *250.52(A)(7)*. (*© Cengage Learning 2012*)

FIGURE 3-9 Electrodes not permitted for grounding, *250.52(B)*. (*© Cengage Learning 2012*)

250.52(A)(8) Other Local Metal Underground Systems or Structures

The Requirements: Other local metal underground systems or structures such as piping systems, underground tanks, and underground metal well casings that are not bonded to a metal water pipe.

Discussion: No additional guidance is given to qualify the grounding electrodes described in this section. These grounding electrodes are required to be used for the grounding electrode system if the grounding electrodes described in *250.52(A)(1)* through *(A)(7)* are not present at the building or structure served. No distance from the building or structure served is given beyond which the underground objects are not required to be used as a grounding electrode.

250.52(B) Electrodes Not Permitted for Grounding

The Requirements: The following systems and materials are not permitted to be used as grounding electrodes:

1. Metal underground gas piping system
2. Aluminum

Discussion: These are the only electrodes not permitted to be used for the electrical system. Use of metal underground gas piping systems (see Figure 3-9) is not permitted because of the concern that electrical current might flow on the piping system, which could cause a spark if the pipe is cut or a flanged joint is taken apart.

Metal gas piping systems are required by *250.104(B)* to be bonded if installed in or attached to a building or structure. A dielectric coupling should be installed, preferably at the meter, to allow the aboveground portions of the metal gas piping systems to be bonded without affecting the underground portions. The dielectric coupling is an insulating fitting often installed by the supplier of the gas.

250.53 GROUNDING ELECTRODE SYSTEM INSTALLATION

250.53(A) Rod, Pipe, and Plate Electrodes

The Requirements: Rod, pipe, and plate electrodes are required to meet the requirements of *(A)(1)* through *(A)(3)*.

***(A)(1)* Below Permanent Moisture Level.** Where practicable, rod, pipe, and plate electrodes must be embedded below permanent moisture level. Rod, pipe, and plate electrodes must also be free from nonconductive coatings such as paint or enamel. See Figure 3-10.

Discussion: Notice that this *NEC* section does not use the term *where convenient* to describe the need to install the rod, pipe, or plate below permanent moisture level. The resistance of 8-ft (2.44-m) rods

FIGURE 3-10 Rod, pipe, and plate grounding electrode installation, *250.53(A)*. (*© Cengage Learning 2012*)

or pipe grounding electrodes will be high if they are installed in dry, rocky, or frozen soil. In areas where the frost line is several feet deep, it may be desirable to use other grounding electrodes, such as concrete-encased electrodes, that are installed below the frost line. Sectional ground rods are available in 10-ft (3.0-m) lengths that can be connected together as they are driven deeper than the minimum 8 ft (2.44 m) to achieve a low-resistance installation.

250.53(A)(2) **Supplemental Electrode Required.** A single rod, pipe, or plate electrode shall be supplemented by an additional electrode of a type specified in *250.52(A)(2)* through *(A)(8)*. The supplemental electrode shall be permitted to be bonded to one of the following:

1. The rod, pipe, or plate electrode
2. The grounding electrode conductor
3. The grounded service-entrance conductor
4. The nonflexible grounded service raceway
5. Any grounded service enclosure

Exception: If a single rod, pipe, or plate grounding electrode has a resistance to earth of 25 ohms or less, the supplemental electrode shall not be required.

Discussion: These requirements have been inserted in this section in the 2011 *NEC* and replace the language in previous *250.56*. *Section 250.56* was deleted in this revision. As you can see, the new rule turned the requirements of previous *250.56* on its head! Here's what I mean. The previous *250.56* required one grounding electrode of the rod, pipe, or plate type unless the test showed the resistance to ground to be more than 25 ohms. In that case, a supplemental grounding electrode was required. Now, a single grounding electrode of the rod, pipe, or plate type is required to be supplemented by a second grounding electrode unless the testing of the single grounding electrode shows the resistance of the earth connection to be 25 ohms or lower.

Figure 3-11 summarizes these requirements.

The grounding electrode system required by *250.50* is not required to meet any minimum level of resistance so far as the connection to earth is concerned. Perhaps it is assumed that the grounding electrode system, which often consists of several grounding electrodes that are bonded together, will provide a good connection to earth of reasonably low resistance.

For grounding electrodes consisting of rods, pipes, or plates, this section now requires two of the grounding electrodes to be installed and connected in parallel unless a resistance of not more than 25 ohms can be obtained with a single electrode. Any resistance at or below 25 ohms is acceptable.

Here's How: The resistance of the connection between the rod, pipe, or plate and the earth is tested with an earth resistance tester (see Figure 3-12). The resistance is not tested with a common voltmeter, ohmmeter, or ammeter or any combination of these individual meters. An earth resistance tester is specifically designed to measure the resistance of the connection between the grounding electrode and the earth. The most common instrument used is the 3-wire, fall-of-potential meter. Other types of equipment are available, including a snap-on earth resistance tester. Several manufacturers produce this type of equipment. For specific information on testing ground rods, see Appendix C in this book.

Two grounding electrodes of the rod, pipe, or plate are required to be installed unless the installer wants to take the time and expense to test a single grounding electrode and it has a resistance to ground of 25 ohms or less. The additional electrode is permitted to consist of any of the grounding electrodes

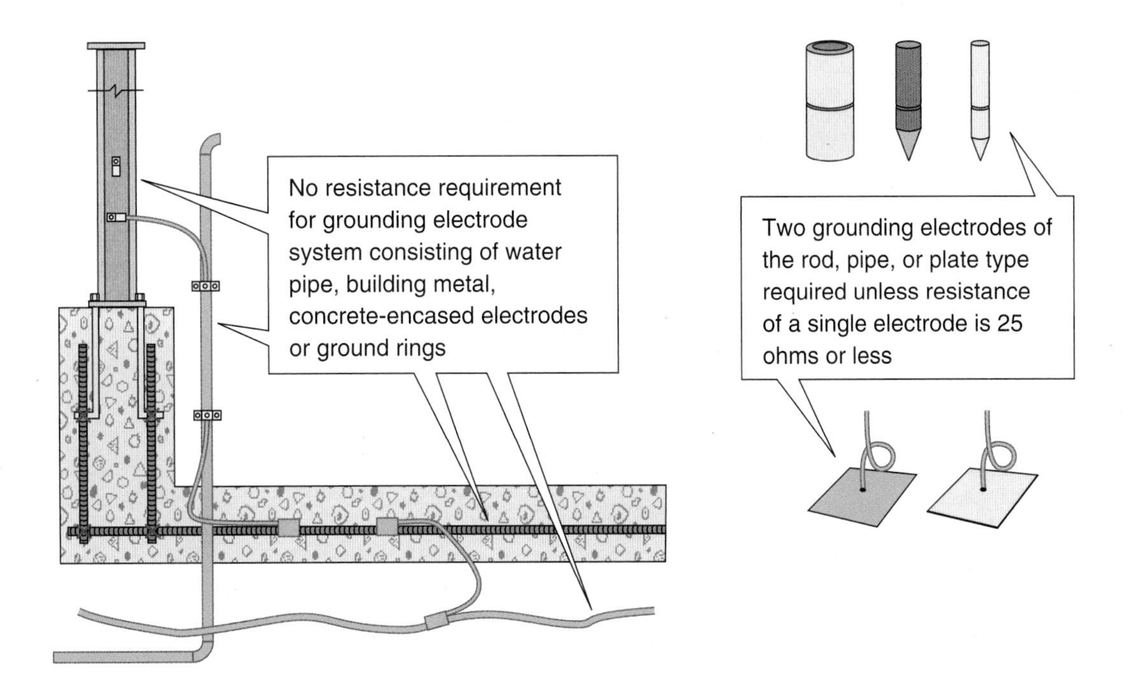

FIGURE 3-11 Resistance of rod, pipe, and plate electrodes, *250.53(A)(2)*. (© *Cengage Learning 2012*)

FIGURE 3-12 Earth resistance tester, *250.52(A)(2), Exception*. (*Courtesy of Chauvin Arnoux Inc./dba AEMC Instruments, Foxborough, MA*)

described in *250.52(A)(2)* through *(A)(8)*. This includes the following grounding electrodes:

- *(A)(2)* metal frame of the building or structure
- *(A)(3)* concrete-encased electrode
- *(A)(4)* ground ring
- *(A)(5)* rod or pipe electrodes
- *(A)(6)* other listed grounding electrodes
- *(A)(7)* plate electrodes
- *(A)(8)* other local metal underground systems or structures.

As can be seen, the water pipe-, structural metal-, and concrete-encased grounding electrodes are not permitted to augment or supplement the grounding electrode of the rod, pipe, or plate type. However, if a water pipe grounding electrode is required by *250.53(D)(2)* to be supplemented by another grounding electrode, a grounding electrode of the rod, pipe, or plate type is permitted to supplement the water pipe grounding electrode. If a grounding electrode of the rod, pipe, or plate type is used to supplement the water pipe grounding electrode, two of the rods, pipe, or plates are required to be installed unless it can be shown through testing that one of the electrodes has a resistance to earth of 25 ohms or less.

If multiple rod, pipe, or plate electrodes are installed to meet the requirements of this section, they must be installed at least 6 ft (1.8 m) apart. This requirement is to accommodate the sphere of influence of rods, pipes, and plates. The sphere of influence is illustrated in Figure 3-13. As can be seen, rings representing various shells of earth resistances radiate out from the rod, pipe, or plate. As the mass

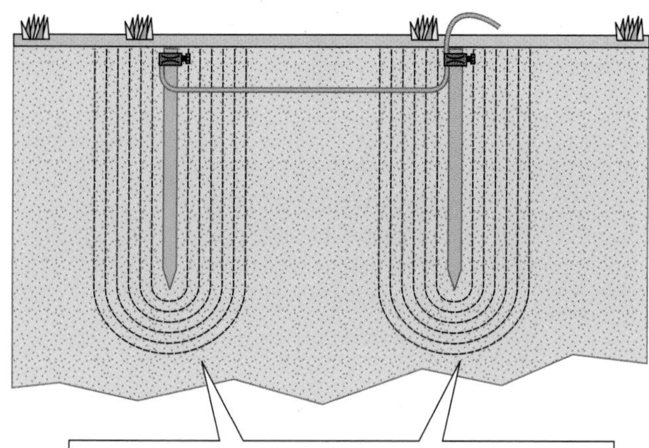

- Space grounding electrodes of the rod, pipe, or plate type not less than 6 ft (1.8 m) apart.
- Space ground rods so no overlapping sphere of influence
- Recommended to space rods not less than twice the length of the longest rod apart.

FIGURE 3-13 Ground rod sphere of influence, *250.53(A)(3)*. (*© Cengage Learning 2012*)

TABLE 3-1

Current Through Ground Rods at Various System Voltages

System Voltage	Resistance of Rod	Amperes
120	25	4.8
240	25	9.6
277	25	11.08
480	25	19.2
2400	25	96
4160	25	166.4
7200	25	288
12,470	25	508.8
100,000	25	4000

of the earth increases, the resistance decreases until any more reduction is negligible. This concept is discussed in detail in Appendix C. Note that it is not required to keep installing rods, pipes, or plates until the resistance to ground is 25 ohms. After the single supplemental rod, pipe, or plate is installed, no further grounding electrode is required, regardless of the resulting resistance.

Note that at typical voltages supplied for dwelling services, very little current will flow through a 25-ohm resistance. Not enough current will flow to trip a circuit breaker or cause a fuse to melt. The amount of current that will flow through the resistance is determined with a simple Ohm's law calculation. As can be seen in Table 3-1, very little current will flow in lower-voltage circuits. This supports the emphasis that equipment is never connected to grounding electrodes for the purpose of carrying fault current or clearing a fault through operation of an overcurrent protective device. At higher voltages such as on utility distribution systems, more current will flow, which will allow circuit-protective devices to operate. Much higher voltages, such as lightning events, will result in significant current flow.

Recall that for current to flow from an electrical system through a grounding electrode, a return path through the earth and from a grounding electrode to the source must be complete. If the source is from a utility transformer, the return path is often through one or more ground rods installed near the transformer. In a simple circuit, the impedance or resistance of the grounding electrode at the transformer is in series with the other circuit resistances, including that of the grounding electrode at the service. Most often, the utility primary conductor connects together all of the ground rods or grounding electrodes of their distribution system. As a result, the grounding electrodes are in parallel, resulting in a fairly low resistance. Therefore, the resistance of the grounding electrode system at the service represents most of the limiting effect of the earth return circuit.

> ***Informational Note:*** The paralleling efficiency of rods is increased by spacing them twice the length of the longest rod.

Discussion: Some manufacturers of ground rods specify that the rods be spaced apart by more than 6 ft (1.8 m), such as twice the length of the rod. This would result in 8-ft (2.5-m) rods being spaced at 16 ft (5 m) apart or more. Note also that the installation instructions supplied by the manufacturer of listed products must be followed. See *110.3(B)*.

250.53(B) Electrode Spacing

The Requirements: If more than one of the rod, pipe, or plate electrodes specified in *250.52(A)(5)* or *(A)(7)* are used, each electrode of one grounding system (including that used for strike termination

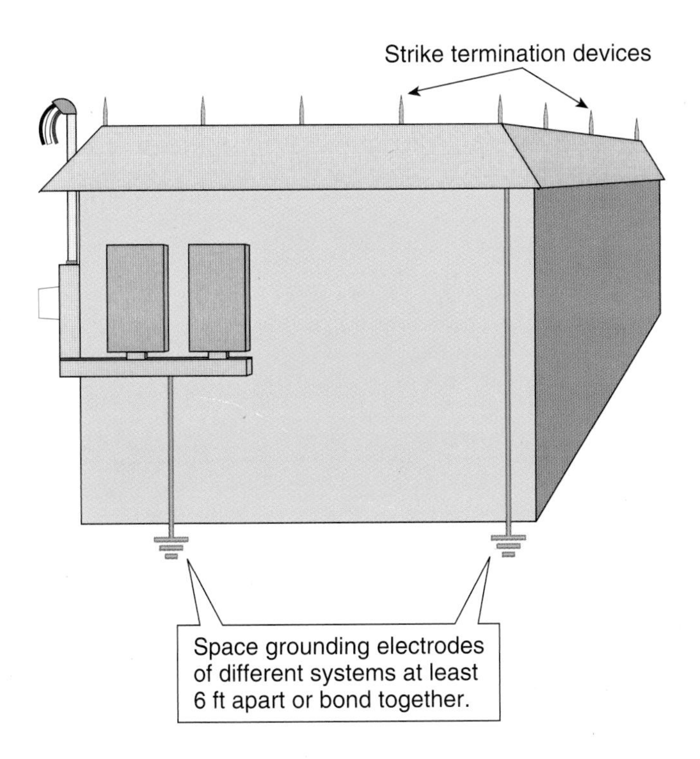

FIGURE 3-14 Grounding electrode spacing, *250.53(B)*. (© *Cengage Learning 2012*)

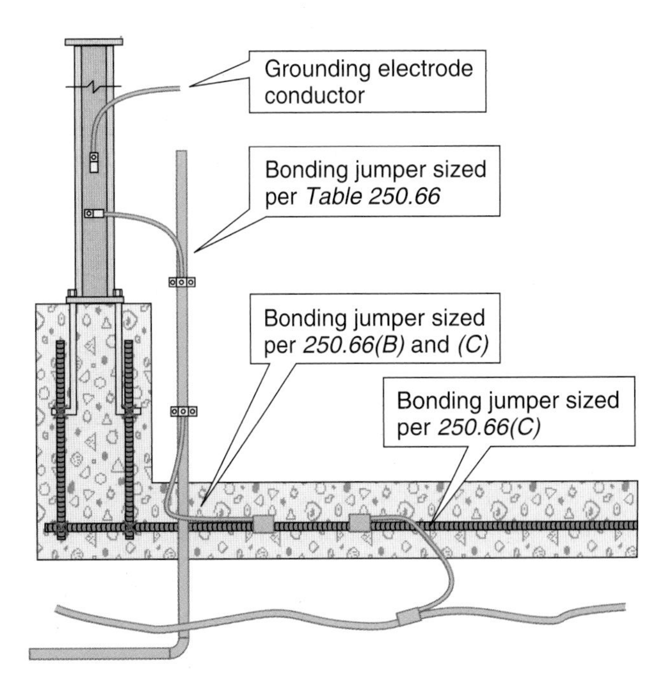

FIGURE 3-15 Installation of grounding electrode bonding jumpers, *250.53(C)*. (© *Cengage Learning 2012*)

devices) is required to be not less than 6 ft (1.83 m) from any other electrode of another grounding system. Two or more grounding electrodes that are bonded together are considered to be a single grounding electrode system.

Discussion: This section specifies that grounding electrodes of the rod, pipe, or plate type for one system are to be separated not less than 6 ft (1.83 m) from similar grounding electrodes of another system (see Figure 3-14). However, *250.106* requires that the ground terminals (grounding electrode) for the lightning protection system be bonded to the building or structure grounding electrode system. As a result, the grounding electrode for the building electrical service and the lightning protection system are bonded together, creating what is considered a single grounding electrode system.

250.53(C) Bonding Jumper

The Requirement: The bonding jumper(s) used to connect the grounding electrodes together to form the grounding electrode system is required to be installed in accordance with *250.64(A), (B),* and *(E),* must be sized in accordance with *250.66,* and connected in the manner specified in *250.70.* See Figure 3-15.

Discussion: This *NEC* section reinforces the earlier statement that the conductor used to connect the grounding electrodes together is a bonding conductor, not a grounding electrode conductor. Bonding conductors are not required to be installed in one continuous length, as is the general requirement for the grounding electrode conductor. The bonding conductor must be installed in compliance with the requirements for installing grounding electrode conductors in *250.64(A)* for aluminum conductors; in *(B),* which gives the requirements for protecting the grounding electrode conductor from physical damage; and in *(E),* which includes a requirement for bonding ferrous metal enclosures for grounding electrode conductors. Finally, the bonding jumper must be sized according to *250.66* for the type of grounding electrode being connected and *Table 250.66* for the size of the service-entrance conductor.

Note that the size of the bonding jumper depends on how the bonding jumper is connected. If the connection is in a daisy-chain fashion from a water pipe or building steel electrode to a ground ring, to a concrete electrode, and finally to a ground rod, the bonding jumper can be adjusted smaller in size, depending on the individual rules in *250.66.* If the order of the bonding jumper installation to the grounding electrodes is reversed from the ground rod to the water pipe or building steel, the largest bonding jumper required must be installed for the entire length. Also, note that the rules in this section coordinate with the rules in *250.64(F),* which permits one or more grounding electrode conductors to be run to the service either individually from each grounding electrode or collectively from two or more grounding electrodes. It becomes a design choice to bond two or more grounding electrodes or all of the electrodes together and run one, two, or more grounding electrode conductors to the service equipment or building disconnecting means.

Assume that an 800-ampere service is being installed with two 500-kcmil conductors in parallel. For purposes of using *Table 250.66,* add the 500-kcmil conductors together and consider them as a single 1000-kcmil conductor. Table 3-2 illustrates the minimum required size of the bonding conductor based on the assumed service-entrance conductors of 1000 kcmil.

Note: In the foregoing example, if the bonding conductors are installed in a daisy-chain manner but in a reverse direction from the rod to the water pipe, a 2/0 AWG would be required for all grounding electrodes.

TABLE 3-2

Sizing Grounding Electrode Bonding Jumpers

Grounding Electrode	Minimum-Size Bonding Jumper	*NEC* Section or Table
Water pipe to building steel	2/0 AWG copper	*Table 250.66*
Building steel to concrete-encased electrode	2 AWG copper*	*250.66(B) and (C)*
Concrete-encased electrode to ground ring	2 AWG copper†	*250.66(C)*
Ground ring to ground rod or ground plate	6 AWG copper	*250.66(A)*

*Would be 4 AWG if concrete-encased electrode were connected after the ground ring.
†Same size as for ground ring, up to 3/0 AWG per *250.66(C)*.

250.53(D) Metal Underground Water Pipe

To be used as a grounding electrode, metal underground water pipe must meet the requirements of *250.53(D)(1)* and *(D)(2).*

250.53(D)(1) Continuity

The Requirement: Continuity of the path to the buried portion of the pipe and the bonding connection to interior piping is not permitted to rely on water meters or filtering devices and similar equipment.

Discussion: Water meters are located in different locations according to the policies of the water utility or supplier. Some are located in a meter box outside the building, which may be located at the street or some other distance from the building. In some cases, the meter is located inside the building, usually within a few feet of the entrance into the building. Water filters usually are installed by the property owner and are located inside the building.

As illustrated in Figure 3-16, if the water meter is inside the building and 10 ft (3.0 m) or more of the metal water pipe is in direct contact with the earth, bonding around the meter or water filter is required if the meter or filter is between the 5-ft (1.52-m) maximum connection point inside the building and the end of the minimum 10-ft (3.0-m) length of the water pipe that is in contact with the earth.

250.53(D)(2) Supplemental Electrode Required

The Requirements: A metal underground water pipe is required to be supplemented by an additional electrode of a type specified in *250.52(A)(2)* through *(A)(8).* If the supplemental electrode is a rod, pipe, or plate type, it must comply with the rules in *250.52(A).* The supplemental electrode is required to be connected to one of the following:

1. the grounding electrode conductor,
2. the grounded service-entrance conductor,
3. a nonflexible grounded service raceway,
4. any grounded service enclosure
5. as provided by *250.32(B).*

Discussion: Any of the grounding electrodes specified in *250.52(A)* are permitted to supplement the underground water pipe electrode other than the water pipe

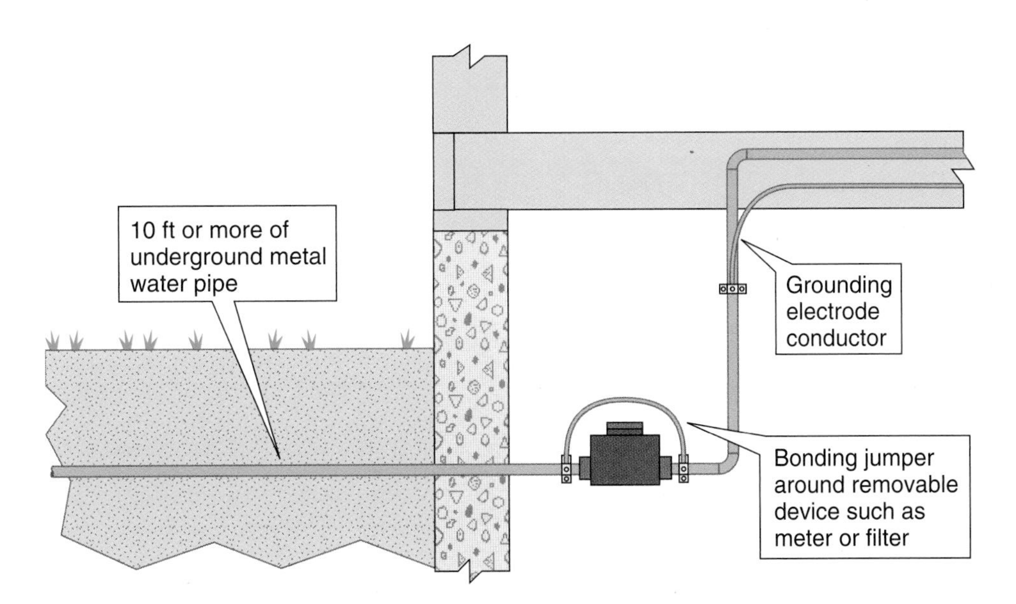

FIGURE 3-16 Metal underground water pipe grounding electrode: continuity, *250.53(D)(1)*. (*© Cengage Learning 2012*)

itself. The requirement for the supplemental grounding electrode is because repairs are often made to the metal water pipe with nonmetallic materials and the water pipe's ability to serve as a grounding electrode conductor to the underground water pipe grounding electrode would be lost. The supplemental electrode is permitted to be connected to the electrical system at several locations but is not permitted to be connected to the water pipe itself. (See Figure 3-17.)

Exception: *The supplemental electrode is permitted to be bonded to the interior metal water piping at any convenient point as covered in 250.68(C)(1), Exception.*

Discussion: For commercial, industrial, and institutional facilities with qualified maintenance staff, the supplemental grounding electrode is permitted to be connected further than 5 ft (1.52 m) from the

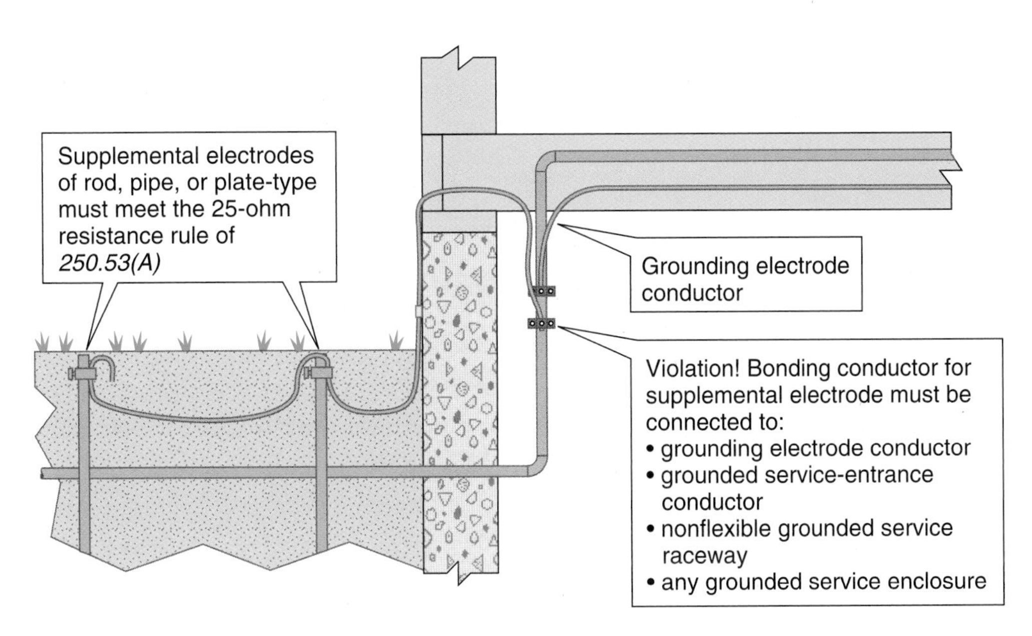

FIGURE 3-17 Supplemental grounding electrode for water pipe grounding electrode, *250.53(D)(2)*. (*© Cengage Learning 2012*)

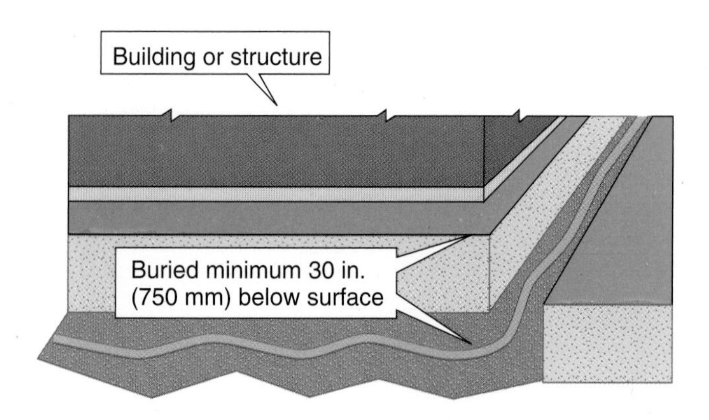

FIGURE 3-18 Installation of ground ring grounding electrode, *250.53(F)*. (*© Cengage Learning 2012*)

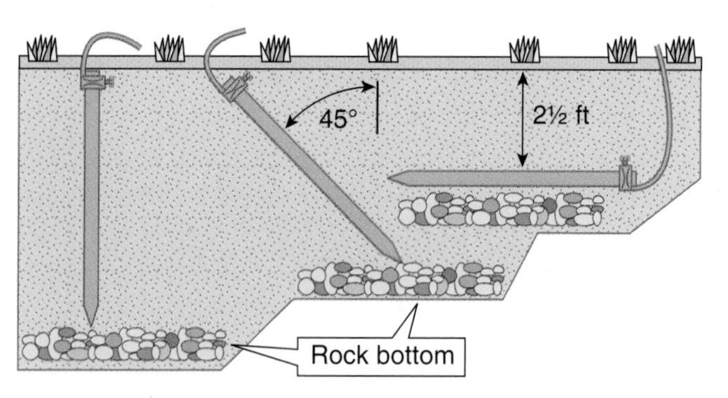

FIGURE 3-19 Installation of rod and pipe electrodes, *250.53(G)*. (*© Cengage Learning 2012*)

point of entrance. This presumes that the qualified staff will monitor the interior metal water piping to be certain that repairs or modifications are not made that will interrupt the path provided by the metal water pipe to the underground portion of the grounding electrode.

250.53(F) Ground Ring

The Requirement: The ground ring must be buried at a depth below the earth's surface of not less than 30 in. (750 mm). See Figure 3-18.

Discussion: The *NEC* wording does not specify any distance from the building or structure for the location of the ground ring. As a result, this decision is a design criterion. Ideally, the ground ring is buried at a depth that is below the frost line for the geographical area. The ground ring conductor will then be at a depth equal to the depth of the footing for the foundation and therefore at or below the level of permanent moisture. Keep in mind the description of a ground ring grounding electrode in *250.52(A)(4)* states the ground ring encircles the building or structure and is not permitted to be shorter than 20 ft for small structures such as a pole.

250.53(G) Rod and Pipe Electrodes

The Requirements: The rod and pipe must be installed such that at least 8 ft (2.44 m) of length is in contact with the soil. The electrode is to be driven to a depth of not less than 8 ft (2.44 m) except that, where rock bottom is encountered, the electrode is to be driven at an oblique angle not to exceed 45 degrees from the vertical. Where rock bottom is encountered at an angle up to 45 degrees, the electrode is permitted to be buried in a trench that is at least 30 in. (750 mm) deep. The upper end of the electrode is required to be flush with or below ground level unless the aboveground end and the grounding electrode conductor attachment are protected against physical damage, as specified in *250.10.*

Discussion: As can be seen in Figure 3-19, requirements are very specific for installation of rod and pipe electrodes. The goal is to place at least a portion of the rod or pipe below permanent moisture level to increase its effectiveness in making an earth connection.

Because the general requirement is that the rod or pipe be driven flush with or below the surface of the earth, the clamp is required to be suitable for direct earth burial. Specific requirements are located in *250.70* and provide that ground clamps must be listed for the materials of the grounding electrode and the grounding electrode conductor and, if to be used on pipe, rod, or other buried electrodes, also are required to be listed for direct soil burial or concrete encasement. Copper grounding electrode conductors are required to be used to connect those grounding electrodes, as the conductors terminate below or within 18 in. (450 mm) of the surface of the earth. Aluminum grounding electrode conductors are not suitable for connecting ground rods or pipes, because aluminum grounding electrode conductors

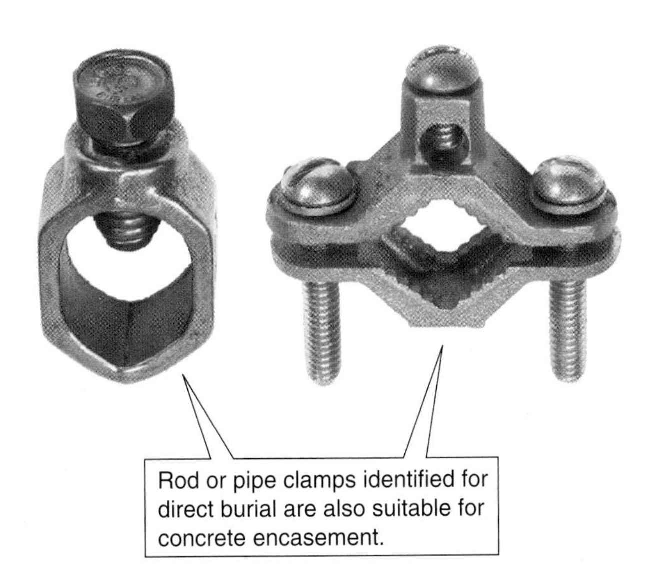

FIGURE 3-20 Ground rod clamps suitable for direct burial, *250.53(G)*. (*Courtesy of Thomas and Betts*)

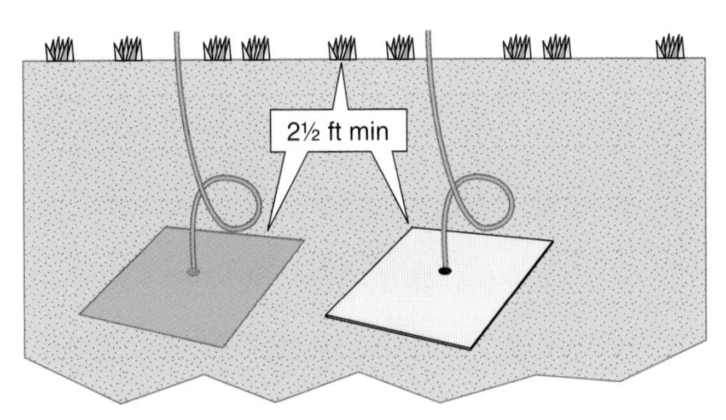

FIGURE 3-21 Installation of plate grounding electrodes, *250.53(H)*. (*© Cengage Learning 2012*)

are not permitted to be terminated within 18 in. (450 mm) of the earth. See *250.64(A)*.

Two types of ground clamp are shown in Figure 3-20. Additional information on ground clamps can be found in the Underwriters Laboratories *White Book* (*General Information Directory*) under the category "Grounding and Bonding Equipment (KDER)." It includes the following statement: "Ground clamps and other connectors suitable for use where buried in earth or embedded in concrete are marked for such use. This marking may be abbreviated 'DB' (for 'direct burial')." Some clamps include a tag containing this information; other clamps are simply marked "Direct Burial" or "DB." As indicated, clamps that are marked for direct burial also are suitable for concrete encasement without additional marking.

250.53(H) Plate Electrode

The Requirement: Plate electrodes are required to be installed not less than 30 in. (750 mm) below the surface of the earth. See Figure 3-21.

Discussion: *NEC 250.52(A)(7)* requires that the ground plate expose not less than 2 ft² of surface to the soil. Because the plate is directly buried, the connection of the grounding electrode conductor to the plate should be marked as being suitable for direct burial. The connector also should be suitable for connecting the material of the ground plate to the copper conductor. If the ground plate is listed by a qualified electrical testing laboratory, the connection means will have been examined as a part of the listing process.

Aluminum grounding electrode conductors are not suitable for connecting ground plates, because aluminum grounding electrode conductors are not permitted to be terminated within 18 in. (450 mm) of the earth. See *250.64(A)*.

250.54 AUXILIARY GROUNDING ELECTRODES

The Requirements: Auxiliary grounding electrodes are permitted to be connected to the equipment grounding conductors specified in *250.118* and are not required to comply with the electrode bonding requirements of *250.50* or *250.53(C)* or the resistance requirements of *250.56*. However, the earth is not permitted to be used as an effective ground-fault current path as specified in *250.4(A)(5)* and *250.4(B)(4)*.

Discussion: The name of this grounding electrode was changed from "supplementary" to "auxiliary" during the processing of the 2008 *NEC*. The reason for the change was to distinguish between the grounding electrode that is required to supplement the underground water pipe grounding electrode and this grounding electrode that is optional or *auxiliary*.

As the name indicates, auxiliary grounding electrodes are intended to be used to supplement the equipment grounding conductor installed with the branch circuit. Note that the auxiliary grounding

FIGURE 3-22 Auxiliary grounding electrodes, *250.54*. (*© Cengage Learning 2012*)

electrode is not permitted to supplement a feeder. This means that if a grounding electrode is used for connecting a feeder equipment grounding conductor to ground, it will be a required connection and all installation requirements must be complied with.

The auxiliary grounding electrode must not be used instead of the equipment grounding conductor (see Figure 3-22). To do so could create a serious shock hazard and would violate *250.4(A)(5)* and *250.4(B)(4)*. These sections require an effective ground-fault return path. Auxiliary grounding electrodes are bonded to the equipment grounding conductor near the equipment where the electrode is installed.

Auxiliary grounding electrodes often are specified by design engineers to be installed at parking lot lighting standards and function essentially to dissipate overvoltages, such as for lightning events.

Auxiliary grounding electrodes are not required to be bonded to the grounding electrode system, as provided in *250.50* or *250.53(C)*, and do not have to comply with the 25-ohm resistance requirements of *250.53(A)(2) Exception*.

Effect of Soil Moisture and Temperature on Grounding Electrodes

As mentioned briefly in the Introduction unit in this book, the moisture content and temperature of the soil can have a significant impact on the effectiveness of grounding electrode connections to the earth.

TABLE 3-3

Effect of Moisture Content on Earth Resistivity[1]

Moisture Content % By Weight	Resistivity, ohm-cm	
	Top Soil	Sandy Loam
0	1000×10^4	1000×10^6
2.5	250,000	150,000
5	165,000	43,000
10	53,000	18,500
15	17,000	10,500
20	12,000	6,300
30	6,400	4,200

[1]Military Handbook, *Grounding, Bonding, and Shielding for Electronic Equipments and Facilities,* Volume II of two volumes, Applications.

Additional information is provided in Appendix C in this book in the document titled, "Understanding Ground Resistance Testing."

For the effect of moisture content of the soil on earth resistivity, refer to Table 3-3.

Obviously, as the moisture content of the soil increases, the lower the resistance of the soil and the more effective the earth connection is likely to be. Take a few minutes and read through the article, "Investigation and Testing of Footing-Type Grounding Electrodes for Electrical Installations" included in Appendix B of this book. You will find that concrete-encased electrodes performed very well even in arid climates due to their ability to maintain moisture in the concrete footings.

Table 3-4 provides valuable information on the effect of soil temperature on earth resistivity. If your

TABLE 3-4

Effect of Temperature on Earth Resistivity[2]

Temperature °F	Temperature °C	Resistivity ohm-cm
68	20	7,200
50	10	9,900
32 (water)	0	13,800
32 (ice)	0	30,000
23	−5	79,000
5	−15	330,000

For sandy loam, 15.2% moisture.
[2]Military Handbook, *Grounding, Bonding, and Shielding for Electronic Equipments and Facilities,* Volume II of two volumes, Applications.

installation is in an area that experiences frozen soil, this information can be valuable in designing and installing a grounding electrode system to ensure an adequate earth connection.

Consider following the building code requirements for the depth of building foundations in the area of installation. Typically, building codes require the building or structure foundation to be below the local frost line. Grounding electrodes that are installed in nonfrozen soil will function better in making an earth connection.

As can be seen, ensuring a good earth connection with grounding electrodes can be a very complex problem. Many issues must be considered, including soil composition, moisture content, and temperature. Consultation with a geologist or other experts may be needed to design a grounding electrode system to meet specifications for the maximum resistance if required by the owner or design engineer.

250.58 COMMON GROUNDING ELECTRODE

The Requirements: Where an ac system is connected to a grounding electrode in or at a building or structure, the same electrode is required to be used to ground conductor enclosures and equipment in or on that building or structure. Where separate services, feeders, or branch circuits supply a building and are required to be connected to a grounding electrode(s), the same grounding electrode(s) must be used.

Two or more grounding electrodes that are effectively bonded together are considered to be a single grounding electrode system in this sense. See Figure 3-23.

Discussion: This section contains requirements clarifying that it is inappropriate to connect the system, such as the service, to one grounding electrode (system) and the equipment supplied from that system to another grounding electrode system. The importance of this rule cannot be overemphasized. The same grounding electrode must be used for both the system (the service or separately derived system) and the equipment. This rule prevents an earth return for fault current. As described earlier, an earth return almost always is a high-impedance and high-resistance path. This high-impedance path almost always results in an elevated voltage level for the equipment subject to a ground fault, which can cause a shock or electrocution hazard.

Note also that *250.6(D)* specifies that errors in data processing equipment do not justify grounding the data processing equipment, other than as required by *Article 250.* Simply connecting equipment such

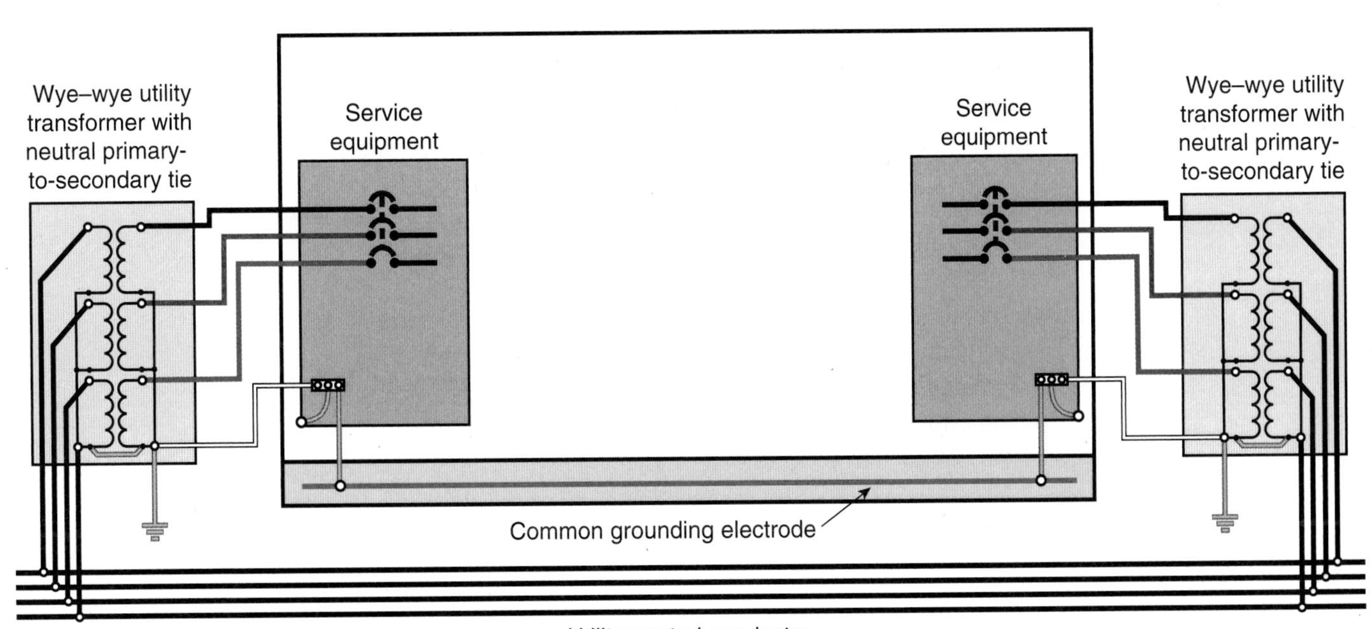

FIGURE 3-23 Common grounding electrode, *250.58.* (*© Cengage Learning 2012*)

as parking area lighting standards or any other equipment directly to a grounding electrode, rather than to an equipment grounding conductor run with the branch circuit, can create an extremely unsafe condition and should be avoided.

As shown in Figure 3-23, if two or more services or feeders supply a building, the common grounding electrode has the effect of bonding the neutral conductor of the two systems together. Under normal operation, this should not present a problem if the requirement of not connecting equipment to the neutral on the load side of the service is met. See *250.24(A)(5)*. It is not unusual to have electrical equipment in the building supplied by two or more services or feeders supplying the building or structure. Also, bonding conductors are connected to metal piping and structural systems from the two or more services or feeders. Without a doubt, some current from a ground fault will return to the source over the multiple common paths between the services or feeders. The lowest impedance path will carry the greatest fault current.

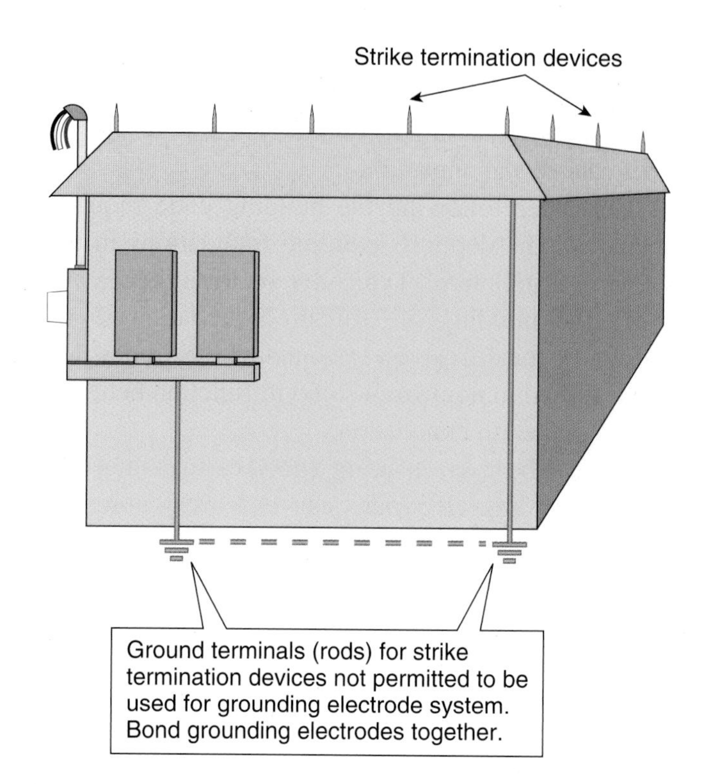

FIGURE 3-24 Use of air terminals, *250.60.* (*© Cengage Learning 2012*)

250.60 USE OF STRIKE TERMINATION DEVICES

The Requirements: Conductors and driven pipes, rods, or plate electrodes used for grounding strike termination devices are not permitted to be used in lieu of the grounding electrodes required by *250.50* for grounding wiring systems and equipment. This provision must not be interpreted to prohibit the required bonding together of grounding electrodes of different systems.

Discussion: The term *strike termination devices* refers to the terminals (sometimes referred to as *lightning rods*) used as a part of lightning protection systems (see Figure 3-24). Specific requirements for the design and installation of lightning protection systems are contained in NFPA 780 Lightning Protection Code.

This section simply requires that components of the lightning protection system not be used for the grounding electrode system for the electrical system. However, *250.106* requires that the ground terminals (grounding electrode) for the lightning protection system be bonded to the grounding electrode (system) for the electrical system. As indicated, when the grounding electrode system for the electrical system and the grounding electrode (earth terminal) for the lightning protection system are bonded together, the grounding electrodes become a grounding electrode system.

> ***Informational Note No. 1:*** See 250.106 for spacing from strike termination devices. See 800.100(D), 810.21(J), and 820.100(D) for bonding of electrodes.*

> ***Informational Note No. 2:*** Bonding together of all separate grounding electrodes will limit potential differences between them and their associated wiring systems.*

250.64 GROUNDING ELECTRODE CONDUCTOR INSTALLATION

The Requirements: Grounding electrode conductors at the service, at each building or structure supplied by a feeder(s) or branch circuit(s) or at a separately derived system are required to be installed as specified in *250.64(A)* through *(F)*.

250.64(A) Aluminum or Copper-Clad Aluminum Conductors

The Requirements: Bare aluminum or copper-clad aluminum grounding electrode conductors are not permitted to be used where in direct contact with masonry or the earth or where subject to corrosive conditions. Where used outside, aluminum or copper-clad aluminum grounding electrode conductors are not permitted to be terminated within 18 in. (450 mm) of the earth.

Discussion: The phrase *copper-clad aluminum* appears in this and a few other sections in the *NEC*. Copper-clad aluminum was used in limited amounts a few years ago but has fallen out of favor and is no longer in common distribution. Copper-clad aluminum was produced by applying a thin film of copper over an aluminum core. The concept was to produce a conductor that was lighter than copper but had the termination characteristics of copper.

As can be seen from this section, bare aluminum or copper-clad aluminum grounding electrode conductors are not permitted to be used if the conductor will be in direct contact with masonry or the earth or will be subject to corrosive conditions. The restrictions on bare aluminum and copper-clad aluminum conductors stem from concerns that the conductor would corrode and fail. Aluminum and copper-clad aluminum conductors that have an insulation suitable for the environment are acceptable for direct contact with masonry. Conductors with an insulation designation that includes a "W," such as THW, THWN, or XHHW, generally are suitable for direct contact with masonry. This should be verified by contacting the conductor manufacturer.

Other aluminum and copper-clad aluminum conductors have insulation that is suitable for direct earth burial. These include conductors identified as "UF" or "USE." If used outside, aluminum or copper-clad aluminum grounding conductors are not permitted to be terminated within 18 in. (450 mm) of the earth.

250.64(B) Securing and Protection Against Physical Damage

The Requirements: If exposed, a grounding electrode conductor or its enclosure must be securely fastened to the surface on which it is carried. Grounding electrode conductors are permitted to be installed on or through framing members. A 4 AWG or larger copper or aluminum grounding electrode conductor is required to be protected where exposed to physical damage. A 6 AWG grounding electrode conductor that is free from exposure to physical damage is permitted to be run along the surface of the building construction without metal covering or protection if it is securely fastened to the construction. Where not so installed, the conductor must be protected in rigid metal conduit (RMC), intermediate metal conduit (IMC), rigid polyvinyl chloride conduit (PVC), electrical metallic tubing (EMT), or cable armor.

Grounding electrode conductors smaller than 6 AWG must be protected by RMC, IMC, PVC, EMT, or cable armor.

Discussion: A number of significant safety requirements are included in this section for installing grounding electrode conductors. Note that the rules apply where the conductor is *exposed.* This term as defined in *Article 100* includes the space above suspended ceilings, so it is broader in application than where the conductor is simply visible. Note as well that the rules apply to conductors that are *exposed to physical damage* and *free from exposure to physical damage.* Although used in the *Code,* these terms are not defined. As a result, the installer and the electrical inspector must use their professional judgment to determine the methods necessary to protect conductors from physical damage. Previous editions of the *Code* have used the phrase *subject to severe physical damage* to indicate a degree of exposure. This term is no longer used. Grounding electrode conductors that are securely fastened to the structure usually are considered to be adequately protected. Additional means of protection should be applied, however, if grounding electrode conductors are installed on poles or if they are subject to contact by vehicles, machinery, lawn mowers, and so on, or if they will be in areas where digging may be conducted, such as in flower beds or around shrubbery.

Note the additional requirements for bonding ferrous metal enclosures for grounding electrode conductors in *250.64(E).* If the conductors are protected by PVC, bonding is not required, but Schedule 80 PVC conduit must be used (Common PVC conduit is Schedule 40).

250.64(C) Continuous

The Requirements: Except as provided in *250.30(A)(5)* and *(A)(6), 250.30(B)(1),* and *250.68(C)*, grounding electrode conductor(s) are required to be installed in one continuous length without a splice or joint. If necessary, splices or connections shall be made as permitted in *(1)* through *(4)*.

1. Splicing of the wire-type grounding electrode conductor is permitted only by irreversible compression-type connectors listed as grounding and bonding equipment or by the exothermic welding process.

Discussion: Irreversible compression-type connectors can be thought of as a system. The system consists of the connector (see Figure 3-25) and the compression tool used to apply the connector. The following practices should always be followed in making irreversible compression-type connections:

1. Use only the compression tool and compression-type connectors from the same manufacturer. Do not mix and match connectors and tools.
2. Select the connector for the type of conductor to be spliced, such as for copper or aluminum conductor. Some connectors are suitable for copper and aluminum; others, for only copper or aluminum.
3. Select the connector for the size of conductor to be spliced.
4. Use the appropriate compression tool for the connector to be applied.
5. Carefully follow all installation instructions from the manufacturer.

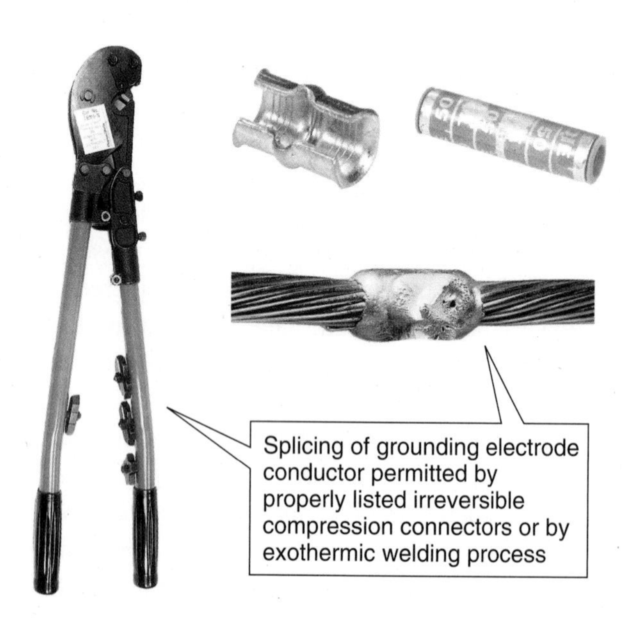

FIGURE 3-25 Splicing grounding electrode conductor, *250.64(C)(1)*. (*Courtesy of Thomas and Betts*)

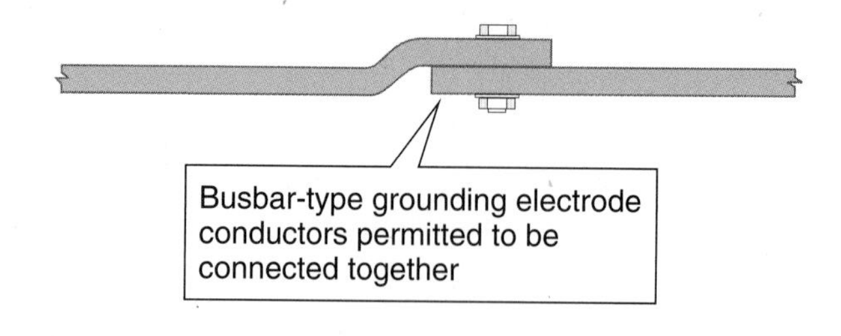

FIGURE 3-26 Splicing busbar grounding electrode conductors, *250.64(C)(2)*. (*© Cengage Learning 2012*)

More Requirements:

2. Sections of busbars are permitted to be connected together to form a grounding electrode conductor.

Discussion: This section permits connecting sections of busbars together to form a grounding electrode conductor (see Figure 3-26). Busbars typically are supplied in limited lengths, such as 10 ft (3.0 m). Connecting two or more lengths of busbar together may be necessary to reach the grounding electrode.

More Requirements:

3. Bolted, riveted, or welded connections of structural metal frames of buildings or structures.

Discussion: This section was added during the processing of the 2011 *NEC*. It coordinates with the permission to use the structural frame of a building or structure as a grounding electrode conductor in *250.68(C)(2)* (see Figure 3-27). The new section recognizes the usual methods of assembling these structural metal components of a building.

More Requirements:

4. Threaded, welded, brazed, soldered, or bolted-flange connections of metal water piping.

FIGURE 3-27 Bolted, riveted, or welded connections of structural metal frames of buildings permitted. (© *Cengage Learning 2012*)

FIGURE 3-28 Threaded, welded, brazed, soldered, or bolted-flange connections of metal water piping permitted. (© *Cengage Learning 2012*)

Discussion: This section was added during the processing of the 2011 *NEC*. It coordinates with the permission in *250.68(C)(1)* to use the interior metal water piping as a conductor to interconnect electrodes that are a part of the grounding electrode system. The new section recognizes the usual methods of assembling metal water piping systems. See Figure 3-28. The general rule is that the metal water piping system is limited to not more than 5 ft (1.52 m) for use as a conductor similar to a grounding electrode conductor. An exception allows an unlimited length of interior metal water piping to be used for these purposes in qualifying industrial, commercial, and institutional occupancies.

250.64(D) Service with Multiple Disconnecting Means Enclosures

The Requirements: If a service consists of more than a single enclosure as permitted in *230.71(A)*, grounding electrode connections are required to be made in accordance with *(D)(1)*, *(D)(2)*, or *(D)(3)*.

Discussion: This section has been expanded and reorganized to cover three different methods of connecting grounding electrode conductors or grounding electrode conductor taps to the service grounded (neutral) conductor. Now included are requirements for (1) common grounding electrode conductor and taps, (2) individual grounding electrode conductors, and (3) connections at a common location. A coordinating change was made to the title to become *Service with Multiple Disconnecting Means Enclosures.* It should be noted the installer can choose one of the three methods included in *(D)(1)*, *(D)(2)*, or *(D)(3)* for connecting the grounding electrode conductor if the service has more than one service disconnecting means enclosure.

Section 230.71(A) permits up to six service disconnecting means in separate enclosures to serve as the service disconnecting means. Several configurations can be installed. Although the *NEC* does not require metering equipment for measuring consumption of electric energy, provisions are made in the *Code* for this equipment. Many configurations of metering equipment can be installed, from individual meter socket enclosures, to current transformer enclosures, to metering at the primary voltage level.

Up to six sets of service-entrance conductors can originate individually from a service drop, underground service conductors, or from a service lateral or metering equipment. *Section 230.40, Exception No. 2,* coordinates with the rules in *230.71(A)* by allowing a set of service-entrance conductors to extend from a service drop, underground service conductors, or from a service lateral to the individual service disconnecting means enclosures.

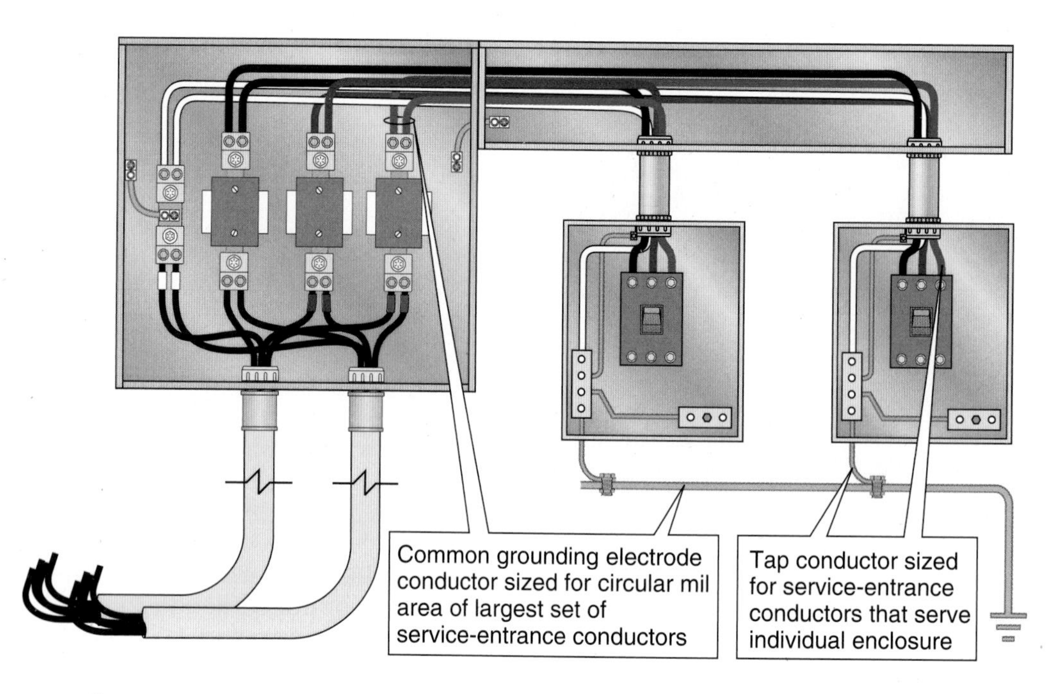

FIGURE 3-29 Common grounding electrode conductor and taps, *250.64(D)*. (*© Cengage Learning 2012*)

The additional methods of making grounding electrode conductor installations provided in this section, along with a number of methods for making connections to grounding electrodes permitted in *250.64(F)*, add many options for the installer. These additional options can also add confusion, because there are so many correct as well as incorrect methods of making these installations.

250.64(D)(1) Common Grounding Electrode Conductor and Taps

The Requirements: A common grounding electrode conductor and grounding electrode conductor taps shall be installed. The common grounding electrode conductor shall be sized in accordance with *250.66*, based on the sum of the circular mil area of the largest ungrounded service-entrance conductor(s).

If the service-entrance conductors connect directly to a service drop or lateral, the common grounding electrode conductor must be sized in accordance with *Table 250.66, Note 1*.

A grounding electrode tap conductor must extend to the inside of each service disconnecting means enclosure. The grounding electrode tap conductors must be sized in accordance with the grounding electrode conductors specified in *250.66* for the largest service-entrance conductor serving the individual enclosure. The tap conductors are required to be connected to the common grounding electrode conductor by one of the following methods in such a manner that the common grounding electrode conductor remains without a splice or joint:

1. Exothermic welding
2. Connectors listed as grounding and bonding equipment
3. Connections to an aluminum or copper busbar not less than ¼ in. × 2 in. (6 mm × 50 mm). The busbar is required to be securely fastened and be installed in an accessible location. Connections are required to be made by a listed connector or by the exothermic welding process. If aluminum busbars are used, the installation shall comply with *250.64(A)*. These rules are illustrated in Figures 3-29, 3-30, and 3-31.

Discussion: The term *common grounding electrode conductor* is not defined in the *Code* but refers to a grounding electrode conductor that is common to or supplies or serves more than one service

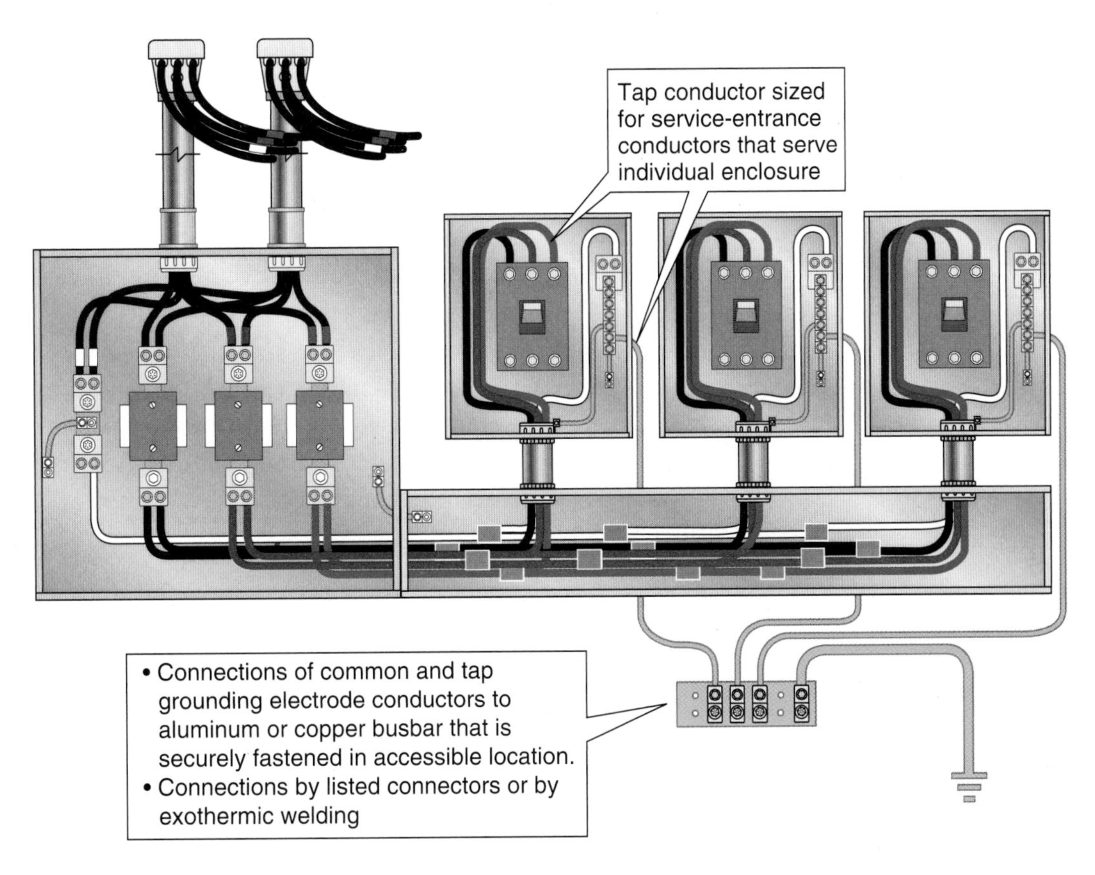

FIGURE 3-30 Sizing common grounding electrode conductor and taps, *250.64(D)*. (*© Cengage Learning 2012*)

disconnecting means enclosure. The term also is used in *250.30* for separately derived systems and applies to multiple separately derived systems that are connected to a common grounding electrode conductor.

Generally, the common grounding electrode conductor for service equipment is sized according to the sum of the circular mil areas of the largest ungrounded service-entrance conductors, in accordance with *NEC 250.66*. An important point in this discussion is if ground rod or ground plate grounding electrodes are the only grounding electrodes used, the grounding electrode conductor is not required to be larger than 6 AWG copper, according to *250.66(A)*. If a concrete-encased grounding electrode is the only grounding electrode used, the grounding electrode conductor is not required to be larger than 4 AWG copper, according to *250.66(B)*. Finally, if a ground ring grounding electrode is the only grounding electrode used, the grounding electrode conductor is not required to be larger than the size of the ground ring conductor, according to *250.66(C)*. See the discussion in *250.66(C)* in this book. These rules act as exceptions to the sizing rules in *Table 250.66* and apply to both tap and common grounding electrode conductors.

Here's How: Figure 3-29 represents an underground service with the service lateral terminating at the line side of current transformers used for electric utility metering. These current transformers are commonly used if the calculated load exceeds 200 amperes or as otherwise required by the electrical utility. The grounding electrode conductor tap conductors are selected from *260.66* on the basis of the size of the service-entrance conductors supplying the individual service disconnecting means. The minimum size of the common grounding electrode conductor is determined by adding the circular mil areas of the service-entrance conductors together, treating them as a single service-entrance conductor, then using *250.66* for determining the minimum size of grounding electrode conductor.

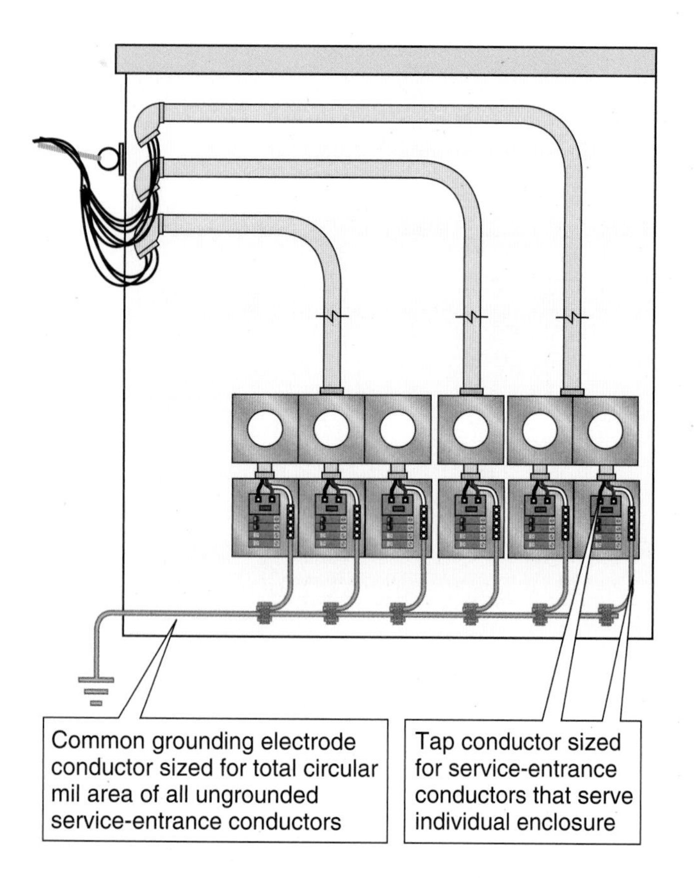

FIGURE 3-31 Grounding electrode conductor taps for 6 service disconnects, *250.64(D)*. (*Courtesy of Thomas and Betts*)

The following table illustrates the methods used for sizing the tap and common grounding electrode conductors shown in Figure 3-29.

Service disconnect	Service-Entrance Conductor Size	Service-Entrance Conductor Area	Grounding Electrode Conductor
No. 1, 200 Ampere	3/0 AWG copper	167,800 cm*	4 AWG (Tap)
No. 2, 400 Ampere	500 kcmil copper	500,000 cm	1/0 AWG (Tap)
	Total Area	667,800 cm	2/0 AWG (Common)

*Circular mil area determined from *NEC Table 8 of Chapter 9.*

Here's How: Figure 3-30 represents an overhead service supplying a current transformer-type metering cabinet with three service disconnecting means in separate enclosures. The service-entrance conductors connect to the service drop near the weatherheads and extend to the current transformer enclosure. A set of parallel service-entrance conductors extend from the load side of the current transformers and end in the wireway where connections are made to individual sets of service-entrance conductors that supply the individual service disconnecting means. Although three disconnecting means enclosures are shown in Figure 3-30, up to six separate service disconnecting means enclosures are permitted by *230.71(A)* to be grouped at one location. Assume the grounding electrode system requires full-size grounding electrode conductors.

The common grounding electrode conductor is sized according to the sum of the circular mil areas of the service-entrance conductors that supply the metering equipment and terminate in the wireway. Assume the service-entrance conductors in the wireway are two 500 kcmil copper that are connected in parallel. Add the two 500-kcmil conductors together to equal one 1000-kcmil conductor. Refer to *Table 250.66* and select a 2/0 AWG copper or 4/0 AWG aluminum grounding electrode conductor. This is the minimum size of the common grounding electrode conductor. Note that an identical procedure is used for service equipment supplied underground by a service lateral or underground service conductors.

The size of the tap grounding electrode conductors from each service disconnecting means to the common grounding electrode conductor is determined by reference to *250.66.* Simply find the size of the copper or aluminum service-entrance conductor in *Table 250.66.* Follow this size across the table to find the minimum size of the copper or aluminum grounding electrode conductor.

Here's How: Figure 3-31 represents a service installed with individual service-entrance conductors in conduits. This is common where a single service drop is attached to the building and from two to six service disconnecting means are individually supplied. The service for multiple-occupancy buildings is often installed this way.

The size of the common grounding electrode conductor is determined as follows: (1) add the circular mil areas of each set of service entrance conductors together, (2) consider this value to represent a single service-entrance conductor, and (3) refer to *Table 250.66* to determine the minimum size. In the following example, assume that the grounding electrode system consists of an underground metal water pipe and effectively grounded building steel that require full-size grounding electrode conductors.

Service Disconnect	Service-Entrance Conductor	Circular Mil Area
One	2 AWG copper	66,360 cm*
Two	3/0 AWG copper	167,800 cm*
Three	250 kcmil copper	250,000 cm
Four	500 kcmil copper	500,000 cm
Five	1/0 AWG copper	105,600 cm*
Six	2/0 AWG copper	133,100 cm*
	Total area	1,222,860 cm

*Circular mil area determined from *NEC Table 8 of Chapter 9.*

Refer to *Table 250.66* and find a minimum 3/0 AWG copper common grounding electrode conductor.

Discussion: The tap conductors are individually sized by reference to *250.66* for the service-entrance conductor serving the respective enclosures.

Here's How: The individual grounding electrode conductor taps are determined by reference to *Table 250.66,* with use of the size of the service-entrance conductor serving the respective disconnecting means. For this example, the grounding electrode system is assumed to consist of an underground metal water pipe and effectively grounded building steel.

Service Disconnect	Service-Entrance Conductor	GEC Tap Size
One	2 AWG copper	8
Two	3/0 AWG copper	4
Three	250 kcmil copper	2
Four	500 kcmil copper	1/0
Five	1/0 AWG copper	6
Six	2/0 AWG copper	4

Discussion: The tap conductors are required to be connected to the common grounding electrode conductor by exothermic welding, connectors that are *listed as grounding and bonding equipment,* or by the use of an aluminum or copper busbar and listed connections. In making the connections, the common grounding electrode conductor is required to remain without a splice or joint.

Exothermic welding can be used to make the connection. These connections are not required to be listed because they are applied in the field. The manufacturer's instructions must be followed to result in a proper connection. These instructions will include using the properly sized graphite mold and keeping the conductors dry and free from dirt or other contaminants.

Figure 3-29 shows the use of split-bolt type connectors for connecting the grounding electrode tap conductors to the common grounding electrode conductor. These split bolts must be sized correctly for the conductors to be connected and must be listed as grounding and bonding equipment. Some split bolts are so listed and some are not. Look for the marking on the split bolt or on the shipping container.

Figure 3-30 shows using a busbar and listed wire connectors such as lugs for making the connection of the grounding electrode tap conductors to the common grounding electrode conductor. The busbar must meet the conditions in *250.64(D)(1)(3)* as discussed earlier.

Figure 3-32 illustrates some of the pressure wire connectors permitted to be used for connecting the grounding electrode tap conductor to the common grounding electrode conductor. For pressure wire connectors such as lugs and split bolts, they are required to be "listed as grounding and bonding

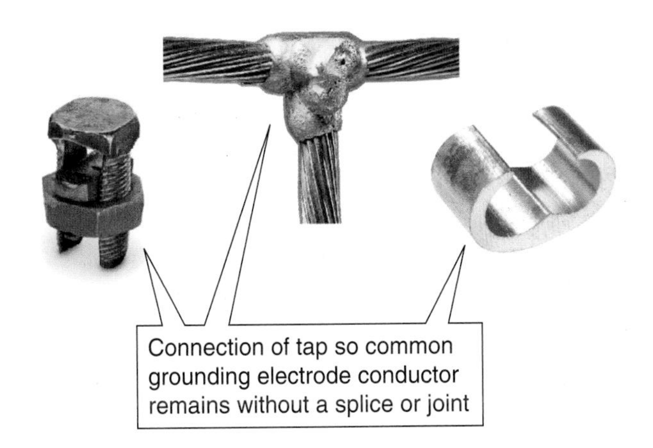

FIGURE 3-32 Connection of grounding electrode taps, *250.64(D).* (*Courtesy of Thomas and Betts*)

equipment." These connectors are tested and listed to UL 467. The UL *White Book* contains additional information on this equipment under Grounding and Bonding Equipment (KDER). The test includes a short-time current test in accordance with the UL standard. For example, a connector for 2 AWG wire must satisfactorily carry 3900 amperes for 6 seconds; whereas a wire connector for 3/0 AWG must carry 8030 amperes for 9 seconds. This test is intended to simulate carrying fault current. These pressure wire connectors include split bolts, ground clamps, grounding and bonding bushings and locknuts, and similar equipment. They are required to be identified as being suitable for connecting the grounding or bonding conductors such as copper or aluminum as well as for the size of conductors to be connected. Use of some type of tee or feed-through connector will allow proper connection. Also, a feed-through connector can be used to connect the common grounding electrode to the busbar where the taps are connected.

250.64(D)(2) Individual Grounding Electrode Conductors

More Requirements: A grounding electrode conductor is required to be connected between the grounded (neutral) conductor in each service equipment disconnecting means enclosure and the grounding electrode system. Each grounding electrode conductor is required to be sized in accordance with *250.66* based on the service-entrance conductor(s) supplying the individual service disconnecting means.

Discussion: This section represents the second of three options for connecting the system grounded (neutral) conductor to the grounding electrode or grounding electrode system. In this option shown in Figure 3-33, an individual grounding electrode conductor is installed from each service disconnecting means directly to the grounding electrode system. The grounding electrode conductor from each service disconnecting means is sized from *250.66* according to the size of service-entrance conductor(s) serving the service disconnecting means.

Another method for installing this "option 2" grounding electrode conductor is to install an individual grounding electrode conductor from each service disconnecting means to the copper or aluminum busbar, as provided in *250.64(D)(1)(3)*. In some respects, this method constitutes a common grounding electrode conductor from the grounding electrode(s) or system to the busbar and grounding electrode conductor taps to the individual service

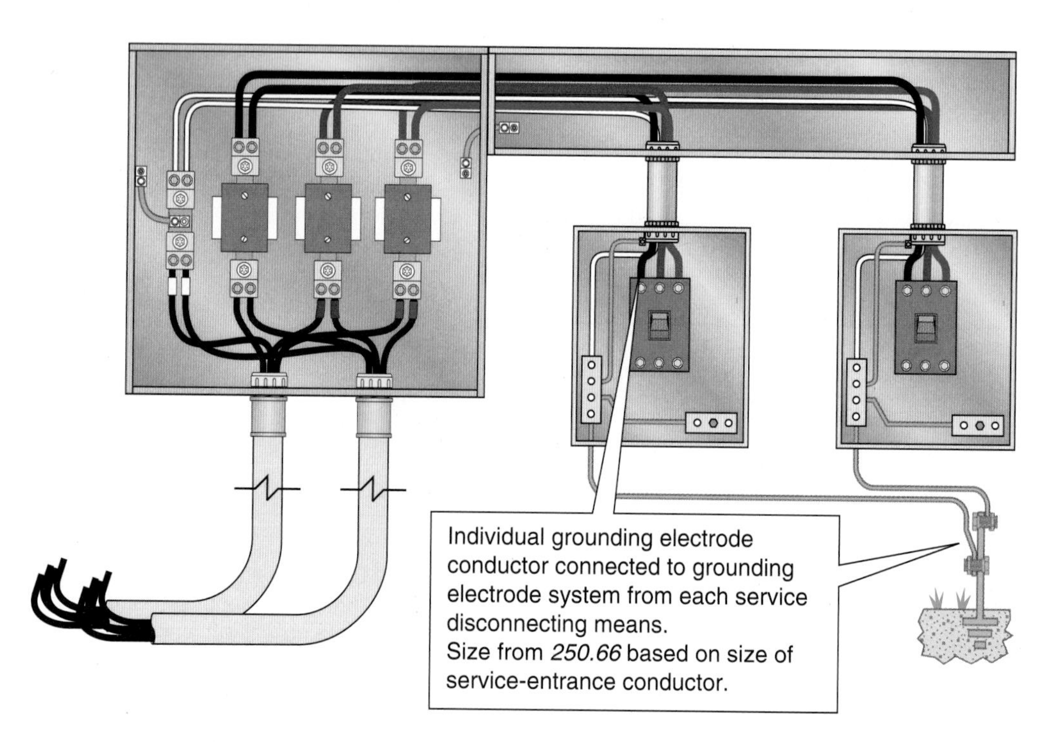

FIGURE 3-33 Individual grounding electrode conductors, *250.64(D)(2)*. (© Cengage Learning 2012)

disconnecting means enclosures. This method is covered in *250.64(D)(1)* above.

250.64(D)(3) Common Location

The Requirement: A grounding electrode conductor is required to be connected to the grounded service conductor(s) in a wireway or other accessible enclosure on the supply side of the service disconnecting means. The connection is required to be made with exothermic welding or a connector listed as grounding and bonding equipment. The grounding electrode conductor must be sized in accordance with *250.66* based on the service-entrance conductor(s) at the common location where the connection is made.

Discussion: This section represents the third of three options for connecting the system grounded (neutral) conductor to the grounding electrode or grounding electrode system. In this option shown in Figure 3-34, a single grounding electrode conductor is installed from an accessible location such as in a wireway directly to the grounding electrode system. This method is no doubt the least expensive to install from a cost and labor basis. However, not all installations have a wireway installed at the service, so other connections are recognized in *250.64(D)*.

The grounding electrode conductor from the wireway to the grounding electrode is sized from *250.66* on the basis of the size of service-entrance conductor(s) in the wireway.

250.64(E) Enclosures for Grounding Electrode Conductors

The Requirements: Ferrous metal enclosures for grounding electrode conductors are required to be electrically continuous from the point of attachment to cabinets or equipment to the grounding electrode and must be securely fastened to the ground clamp or fitting.

Nonferrous metal enclosures are not required to be electrically continuous. Ferrous metal enclosures that are not physically continuous from cabinets or equipment to the grounding electrode must be made electrically continuous by bonding each end of the raceway or enclosure to the grounding electrode conductor.

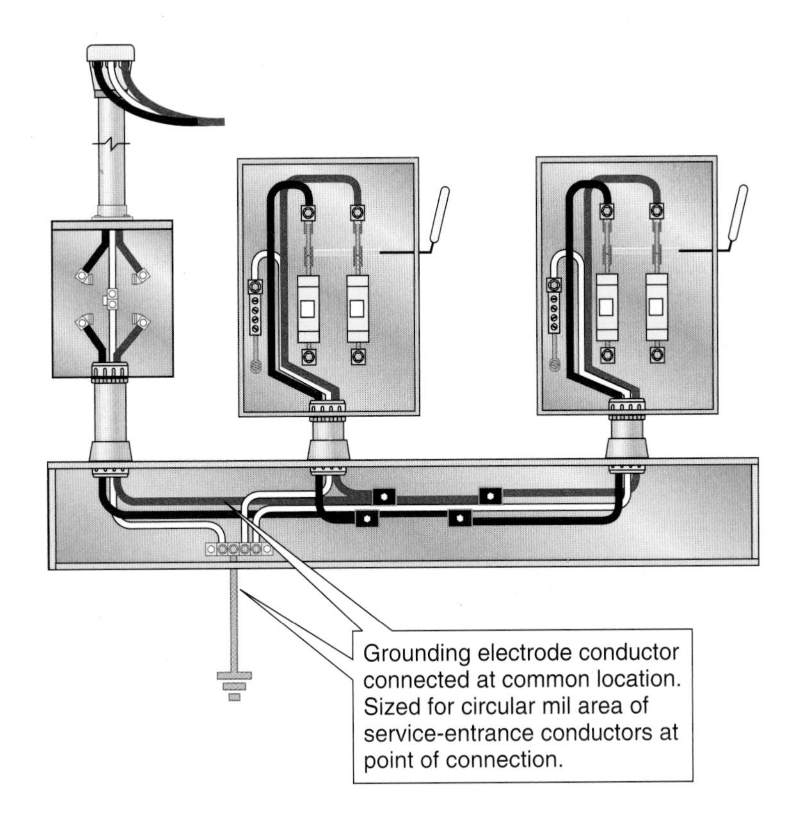

FIGURE 3-34 Connecting grounding electrode conductor at common location, *250.64(D)(3)*. (*© Cengage Learning 2012*)

Bonding methods in compliance with *250.92(B)* for installations at service equipment locations and with *250.92(B)(2)* through *(B)(4)* for other than service equipment locations shall apply at each end and to all intervening ferrous raceways, boxes, and enclosures between the cabinets or equipment and the grounding electrode. The bonding jumper for a grounding electrode conductor, raceway, or cable armor must be the same size as or larger than the enclosed grounding electrode conductor. If a raceway is used as protection for a grounding electrode conductor, the installation is required to comply with the requirements of the appropriate raceway article. See Figure 3-35.

Discussion: Changes have recently been made to this *NEC* section to clarify the bonding rules apply to only ferrous raceways, boxes, and enclosures between the service equipment and the grounding electrode. These principles for bonding should be applied not only to the service equipment but also to every location where the grounding electrode is protected in a ferrous raceway, including at a building or structure disconnecting means where supplied by a feeder as well as for separately derived systems.

What is a ferrous raceway? The word *ferrous* is defined as "containing or made of iron." As a result, "ferrous" materials are magnetic and can be checked with a magnet. The rules do not apply to nonferrous (nonmagnetic) raceways such as aluminum or rigid nonmetallic conduit. Note that Schedule 80 (extra-heavy-wall) PVC conduit is permitted to be used to protect conductors from physical damage and, of course, does not require the bonding provided for in this section. See the Underwriters Laboratories *General Information for Electric Equipment Directory* (*White Book*) or the *Electrical Construction Equipment Directory* (*Green Book*) in the product category DZYR for additional information. Installations must be made in accordance with *Article 352* and other applicable articles. Note that this PVC conduit has a reduced internal area because of the extra-heavy wall and thus is not permitted to contain the same number or size of conductors as Schedule 40 PVC.

Bonding of ferrous metal raceways that enclose grounding electrode conductors is necessary to create a parallel current path between the raceway and the conductor. The impedance of the circuit is approximately doubled when the grounding electrode conductor in a metal raceway is not bonded to the enclosure. The effect of induction during a high-current condition such as a lightning event has been described as creating a "choke coil." The grounding electrode conductor has been reported to burn off near where it exits the unbonded ferrous raceway as a result of this effect during high-current events.

The rules for bonding the ferrous metal raceway enclosing a grounding electrode can be summarized as follows:

1. The raceway must be electrically continuous from the point of attachment to cabinets or equipment to the grounding electrode.
2. The raceway must be securely fastened to the ground clamp or fitting at the grounding electrode.
3. Ferrous metal enclosures that are not physically continuous from cabinets or equipment to the grounding electrode must be made

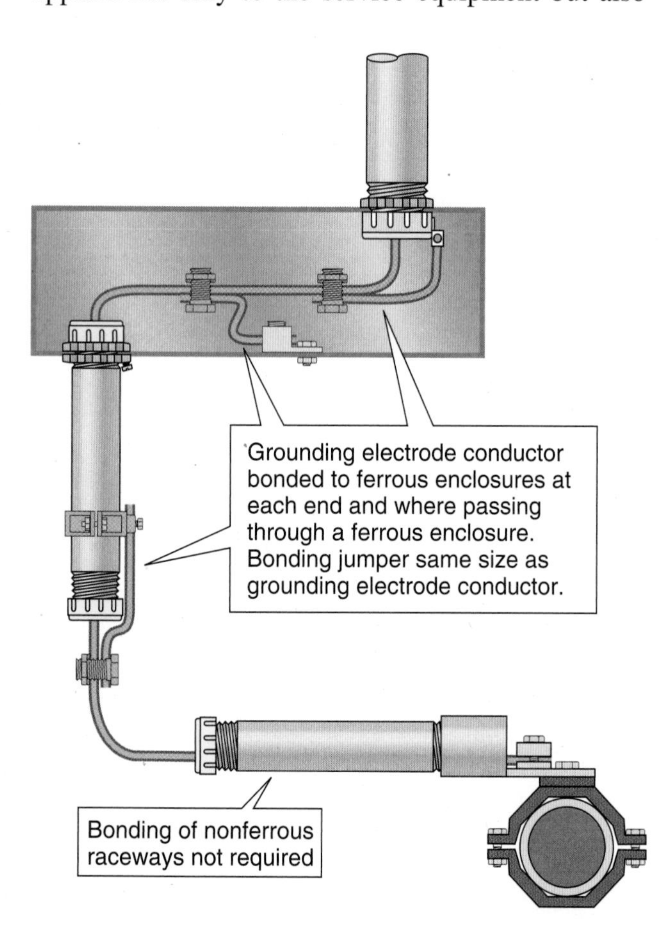

FIGURE 3-35 Bonding of ferrous enclosures for grounding electrode conductors, *250.64(E)*. (*© Cengage Learning 2012*)

electrically continuous by bonding each end of the raceway or enclosure to the grounding electrode conductor.

4. Bonding requirements apply at each end and to all intervening ferrous raceways, boxes, and enclosures between the service equipment and the grounding electrode.
5. The bonding jumper for a grounding electrode conductor, raceway, or cable armor must be of the same size as or larger than the required enclosed grounding electrode conductor.
6. Where a raceway is used as protection for a grounding electrode conductor, the installation must comply with the requirements of the appropriate *NEC* article regarding the raceway.

250.64(F) Installation to Electrode(s)

The Requirements: Grounding electrode conductor(s) and bonding jumpers interconnecting grounding electrodes are required to be installed as provided in *(F)(1)*, *(F)(2)*, or *(F)(3)*. The grounding electrode conductor must be sized for the largest grounding electrode conductor required among all the electrodes connected to it.

(F)(1) The grounding electrode conductor is permitted to be run to any convenient grounding electrode available in the grounding electrode system where the other electrode(s), if any, are connected by bonding jumpers in accordance with the rules in *250.53(C)*.

Discussion: The rules in this section are intended to support the provision of creating the grounding electrode system in *250.50* and the bonding of the grounding electrodes together in *250.53(C)*. See the discussion in *250.50* and Figure 3-1 for rules on bonding grounding electrodes together and the discussion in *250.53(C)* and Figure 3-15 for rules on installing and sizing bonding jumpers for grounding electrodes. As can be seen, a great deal of flexibility is allowed the installer of the grounding electrode system.

The grounding electrode conductor is permitted to connect to any of the grounding electrodes that make up the grounding electrode system so long as the bonding jumper between the grounding electrodes is sized for the largest grounding electrode conductor required. See the discussion in *250.53(C)* for additional information on sizing the bonding jumpers between the grounding electrodes.

More Requirements:

(F)(2) Grounding electrode conductor(s) are permitted to be run to one or more grounding electrode(s) individually.

Discussion: As shown in Figure 3-36, this section permits installing an individual grounding electrode conductor from the service, separately derived system, or building disconnecting means directly to each grounding electrode. The terminal bar in the equipment where the grounding electrode conductors originate serves to bond the grounding electrodes together.

The size of the grounding electrode conductor is determined from *250.66* according to the size of the service-entrance conductors, the size of the derived conductors for a separately derived system,

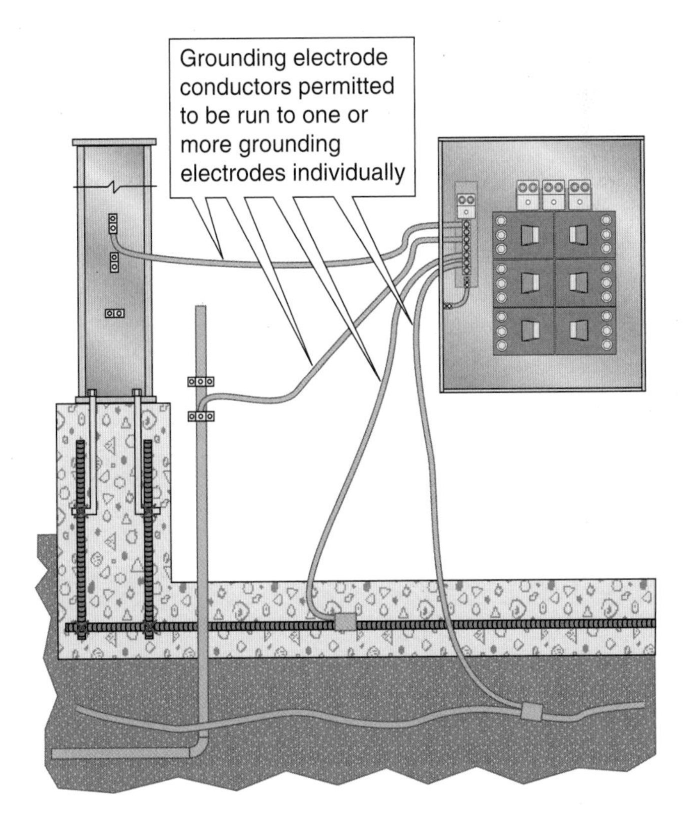

FIGURE 3-36 Individual grounding electrode conductor installation, *250.64(F)(2)*. (*© Cengage Learning 2012*)

or the size of the feeder conductors to a building or structure. See *250.24(A)* for services; *250.30(A)(5), (A)(6),* and *(B)(1)* for separately derived systems; and *250.32(E)* for buildings or structures supplied by a feeder. In addition, the maximum required size of the individual grounding electrode conductor is provided by the rules for ground rod and ground plate grounding electrodes in *250.66(A),* for concrete-encased grounding electrodes in *250.66(B),* and for ground rings in *250.66(C).* Larger grounding electrode conductors are permitted to be supplied if desired.

The following table illustrates the sizing of individual grounding electrode conductors for 200-ampere, 400-ampere, and 800-ampere supplies or services.

Grounding Electrode	Grounding Electrode Conductor for Service, Separately Derived System, or Feeder Supply		
	200 A 3/0 AWG	400 A[1] 500 kcmil	800 A[2] (2) 600 kcmil
Metal Water Pipe; Metal Frame of Building or Structure	4 AWG	1/0 AWG	3/0 AWG
Concrete-Encased Electrode	4 AWG	4 AWG	4 AWG
Ground Ring (2 AWG)	4 AWG	2 AWG	2 AWG
Rod, Pipe, Plate, Other Listed Electrodes	6 AWG	6 AWG	6 AWG

[1]500 kcmil permitted so long as load does not exceed 380 amperes [*240.4(B)*].
[2]Two 500-kcmil conductors in parallel permitted if load does not exceed 760 amperes [*240.4(B)*].

More Requirements:

(F)(3) Bonding jumper(s) from grounding electrode(s) and grounding electrode conductor(s) are permitted to be connected to an aluminum or copper busbar not less than ¼ in × 2 in. (6 mm × 50 mm). The busbar must be securely fastened and be installed in an accessible location. Connections are required to be made by a listed connector or by the exothermic welding process. The grounding electrode conductor is permitted to be connected to the busbar. If aluminum busbars are used, the installation is required to comply with *250.64(A).*

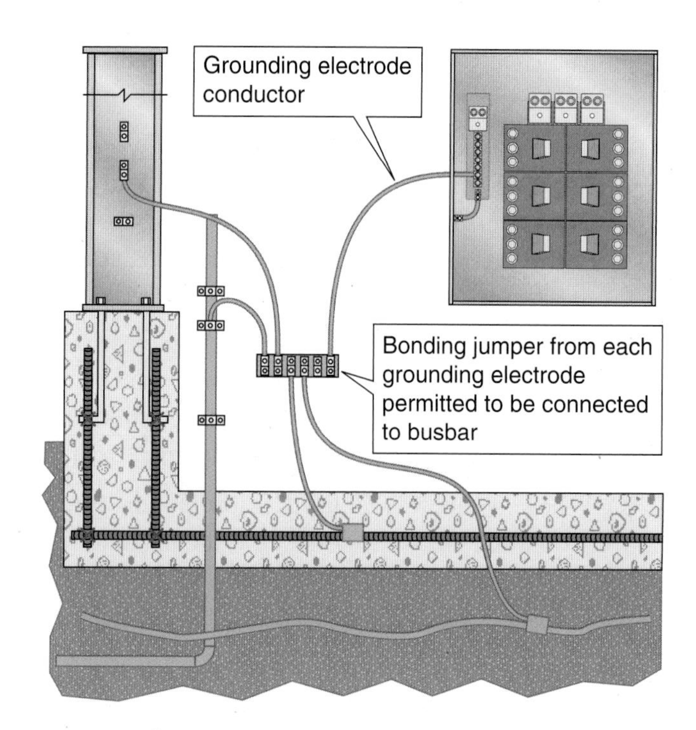

FIGURE 3-37 Using busbar for connecting bonding jumpers and grounding electrode conductors, *250.64(F)(3).* (*© Cengage Learning 2012*)

Discussion: Busbars are commonly installed at points where the grounding electrode conductor and bonding jumpers from grounding electrodes are connected together (see Figure 3-37). The busbar can be installed at any accessible location and is often installed near the service equipment switchboard or panelboard. In some cases, the busbar is installed to encircle the room in which the service equipment is installed. Connections to individual grounding electrodes are made at convenient locations and are permitted to be made with a listed connector or by means of the exothermic welding process. These connections are not required to be made with irreversible compression connectors, though these connectors are permitted to be used. See the discussion in *250.8* for additional information on exothermic welding. Though aluminum busbars are permitted as indicated, special precautions must be taken to protect the aluminum from adverse corrosive influences. See *250.64(A)* for these rules.

250.66 SIZE OF ALTERNATING-CURRENT GROUNDING ELECTRODE CONDUCTOR

The Requirements: The size of the grounding electrode conductor at the service, at each building or structure where supplied by a feeder(s) or branch circuit(s) or at a separately derived system of a grounded or ungrounded ac system must not be less than given in *Table 250.66,* except as permitted in *250.66(A)* through *(C).*

> ***Informational Note:*** See *250.24(C)* for size of ac system conductor brought to service equipment.

Discussion: Although this table is used for several purposes, the title indicates the primary purpose, which is sizing the grounding electrode conductor for ac systems. This table does not include a reference to the rating of an overcurrent protective device for determining the size of a grounding electrode conductor. *Table 250.122* uses the rating of the overcurrent protective device to select an equipment grounding conductor, whereas *Table 250.66* uses the size of a service-entrance conductor to select the size of a grounding electrode conductor. The size of separately derived conductors or feeder conductors are also used in this table.

To use this table, simply follow the column for a copper or aluminum service-entrance conductor, feeder, or branch circuit conductor down until the row includes the size of conductor, and follow that row across to find the minimum size of the copper or aluminum grounding electrode conductor.

Note 1 to the table requires that where a set of service-entrance conductors are installed to serve more than one service disconnecting means enclosure as permitted in *230.71(A)* and *230.40, Exception No. 2,* the circular mil area of the conductors be added together and that size be used in the table to select the grounding electrode conductor. This grounding electrode conductor could be connected in a wireway as shown in several figures in this book or could serve as the common grounding electrode conductor as shown in several other figures.

Note 2 to *Table 250.66* applies where there are no service-entrance conductors. Two examples of this are where service-lateral or underground service conductors terminate directly at service equipment and where busbar is used to supply the service equipment. Often, service-lateral

Table 250.66 Grounding Electrode Conductor for Alternating-Current Systems

Size of Largest Ungrounded Service-Entrance Conductor or Equivalent Area for Parallel Conductors[a] (AWG/kcmil)		Size of Grounding Electrode Conductor (AWG/kcmil)	
Copper	Aluminum or Copper-Clad Aluminum	Copper	Aluminum or Copper-Clad Aluminum[b]
2 or smaller	1/0 or smaller	8	6
1 or 1/0	2/0 or 3/0	6	4
2/0 or 3/0	4/0 or 250	4	2
Over 3/0 through 350	Over 250 through 500	2	1/0
Over 350 through 600	Over 500 through 900	1/0	3/0
Over 600 through 1100	Over 900 through 1750	2/0	4/0
Over 1100	Over 1750	3/0	250

Notes:
1. Where multiple sets of service-entrance conductors are used as permitted in *230.40*, Exception No.2, the equivalent size of the largest service-entrance conductor shall be determined by the largest sum of the areas of the corresponding conductors of each set.
2. Where there are no service-entrance conductors, the grounding electrode conductor size shall be determined by the equivalent size of the largest service-entrance conductor required for the load to be served.
[a]This table also applies to the derived conductors of separately derived ac systems.
[b]See installation restrictions in *250.64(A)*.

conductors (the underground conductors from the utility transformer to the service equipment) terminate in the meter base or current-transformer enclosure metering equipment, and service-entrance conductors are installed to the service disconnecting means.

For large services, busbar may be used from the transformer vault to service equipment. The busbar is not manufactured in standard circular mil sizes as conductors are but in dimensions such as ¼ × 2 in. (6.4 mm × 50 mm). For both of these methods, *Note 2* to *Table 250.66* requires that conductors be selected [from *Table 310.15(B)(16)*] that are adequate for the load to be served. These conductors are used in *Table 250.66* to select the minimum size of grounding electrode conductor.

In addition to sizing the grounding electrode conductor for ac systems, *Table 250.66* is used for sizing the following:

1. The grounded conductor brought to service equipment in *250.24(C)*
2. The main bonding jumper in *250.28*
3. The system bonding jumper in *250.28* and *250.30(A)(1)*
4. The supply-side bonding jumper for separately derived systems in *250.30(A)(2)*
5. The size of the grounded conductor for separately derived systems in *250.30(A)(3)*
6. The size of the bonding jumper on the supply side for nonseparately derived systems in *250.35(B)*
7. The grounding electrode bonding conductor in *250.53(C)*
8. The bonding jumper for ferrous raceways enclosing a grounding electrode conductor in *250.64(E)*
9. The supply-side bonding jumper on the supply side of an overcurrent device in *250.102(C)*
10. Bonding of metal water piping systems in *250.104(A)(1)*
11. Bonding of exposed structural metal in *250.104(C)*
12. Bonding on the secondary side of separately derived systems in *250.104(D)*

250.66(A) Connections to Rod, Pipe, or Plate Electrodes

The Requirement: Where the grounding electrode conductor is connected to rod, pipe, or plate electrodes as permitted in *250.52(A)(5)* or *250.52(A)(7)*, that portion of the conductor that connects directly to the grounding electrode is not required to be larger than 6 AWG copper wire or 4 AWG aluminum wire.

Discussion: When installing the grounding electrode conductor to a rod, pipe, or plate electrode, keep the restrictions in *250.64(A)* in mind. That section requires that terminations of aluminum grounding electrode conductors not be made within 18 in. (450 mm) of the earth. Thus, for all practical purposes, copper grounding electrode conductors are required for rod, pipe, or plate grounding electrodes.

As indicated in this section, 6 AWG copper is the largest grounding electrode conductor that is required to be installed where the conductor connects directly to the rod, pipe, or plate grounding electrode (see Figure 3-38). This rule applies irrespective of the size of the service and acts as an exception to the sizing rules in *Table 250.66.* The installation requirements of *250.64(B)* must be kept

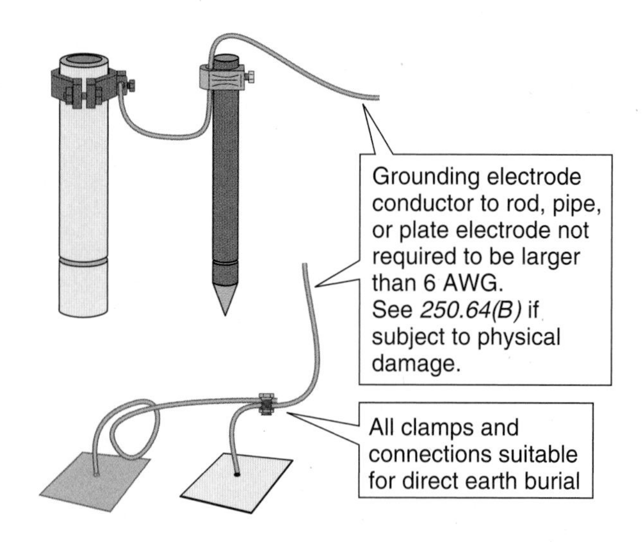

FIGURE 3-38 Connections to rod, pipe, or plate grounding electrodes, *250.66(A)* (*© Cengage Learning 2012*)

in mind. If the 6 or 4 AWG grounding electrode conductor will be subject to physical damage, it must be suitably protected. Because the phrases *exposed to physical damage* and *free from exposure to physical damage* are not defined terms in the *NEC*, the installer and the electrical inspector must exercise their professional judgment in the installation to prevent the grounding electrode conductor from being damaged.

Note that when a grounding electrode of the rod, pipe, or plate type is bonded to other grounding electrodes to form the grounding electrode system in accordance with *250.53(C)*, the bonding conductor is permitted to comply with the sizing rules of this section.

The rationale for not requiring a grounding electrode conductor larger than 6 AWG to a rod, pipe, or plate grounding electrode is that the 6 AWG grounding electrode conductor will carry all the current that can be dissipated into the earth by the grounding electrodes. Any current that flows through the grounding electrode conductor to the grounding electrode that is in excess of its allowable ampacity in *Table 310.15(B)(16)* will be of short duration and should not exceed the short-time withstand rating of the conductor.

250.66(B) Connections to Concrete-Encased Electrodes

The Requirements: Where the grounding electrode conductor is connected to a concrete-encased electrode as permitted in *250.52(A)(3)*, that portion of the conductor that is directly connected to the grounding electrode is not required to be larger than 4 AWG copper wire.

Discussion: The discussions in *250.66(A)* regarding the maximum size of the grounding electrode conductor apply to installations made to concrete-encased electrodes. Here, the maximum-size grounding electrode conductor required, irrespective of the size of the service, is a 4 AWG copper conductor. This section acts as an exception to the sizing rules in *Table 250.66*. Because the grounding electrode conductor will be encased in concrete, a copper conductor is required. See Figure 3-39.

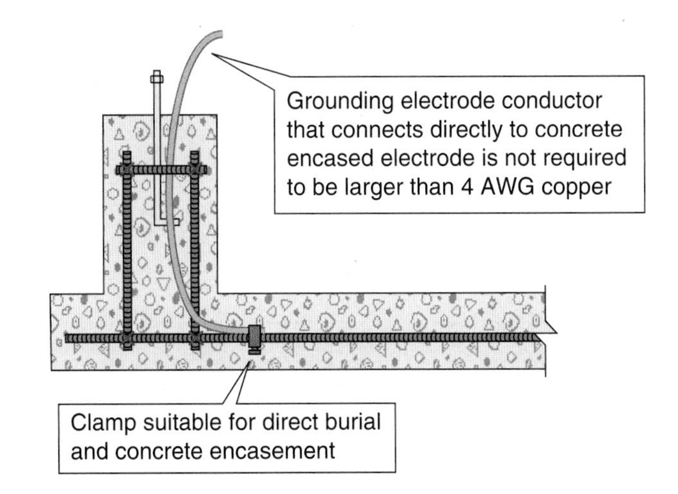

FIGURE 3-39 Grounding electrode conductor connections to concrete-encased electrodes, *250.66(B)*. (*© Cengage Learning 2012*)

250.66(C) Connections to Ground Rings

The Requirement: Where the grounding electrode conductor is connected to a ground ring as permitted in *250.52(A)(4)*, that portion of the conductor that connects directly to the grounding electrode is not required to be larger than the conductor used for the ground ring.

Discussion: The discussions in *250.66(A)* regarding the maximum size of the grounding electrode conductor apply to installations made to ground ring grounding electrodes as well (see Figure 3-40). The

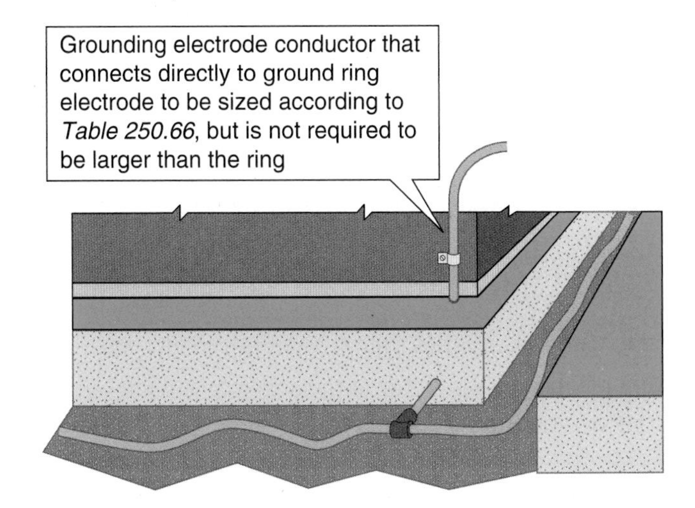

FIGURE 3-40 Grounding electrode conductor connections to ground ring electrodes, *250.66(C)*. (*© Cengage Learning 2012*)

TABLE 3-5

Sizing Grounding Electrode Conductor for Ground Rings

Service-Entrance Conductor (AWG/kcmil)	Ground Ring Conductor	Grounding Electrode Conductor
2 or smaller	2 or larger	8
2/0 or 3/0	2 or larger	4
Over 350–600	2	2
Over 350–600	4/0	1/0
Over 1100	2	2
Over 1100	4/0	3/0

rules for sizing the grounding electrode conductor for a ground ring are different from those for a rod, pipe, plate, or concrete-encased grounding electrode. Because the ground ring is not required to be larger than 2 AWG, the grounding electrode conductor is not required to be larger than 2 AWG even if a larger conductor is indicated in *Table 250.66*. If a ground ring conductor is installed that is larger than the size required for the grounding electrode conductor, the grounding electrode conductor from *Table 250.66* can be installed. Keep in mind that the grounding electrode conductor never has to be larger than the size of the ground ring or larger than 3/0 AWG. Refer to Table 3-5 for examples of the application of this section.

250.68 GROUNDING ELECTRODE CONDUCTOR AND BONDING JUMPER CONNECTION TO GROUNDING ELECTRODES

The Requirements: The connection of the grounding electrode conductor at the service, at each building or structure supplied by a feeder(s) or branch circuit(s) or at a separately derived system and associated bonding jumper(s), is required to be installed as specified in *250.68(A)* through *(C)*.

Discussion: A coordinating opening paragraph has been added to this section to ensure the connections to grounding electrodes are made properly, regardless of whether the connections are made at the service, at the building or structure disconnecting means, or at a separately derived system.

250.68(A) Accessibility

The Requirements: All mechanical devices (elements) used to terminate the connection of a grounding electrode conductor or bonding jumper to a grounding electrode are required to be accessible.

Exception No. 1: *An encased or buried connection to a concrete-encased, driven, or buried grounding electrode is not required to be accessible.*

Exception No. 2: *Exothermic or irreversible compression connections used at terminations, together with the mechanical means used to attach said terminations to fire-proofed structural metal whether or not the mechanical means is reversible, are not required to be accessible.*

Discussion: The general requirement is that the connection of the grounding electrode conductor or bonding jumper to the grounding electrode be accessible so the connection can be examined at some later time. The installation of an access point and cover may be required for this purpose, or the connection must be made at an accessible location. As can be seen in Figure 3-41, the connection of the grounding electrode conductor to a concrete-encased, driven, or buried grounding electrode is not required to be accessible. A new exception was added to the 2005 *NEC* to provide that the connection of the grounding electrode conductor to structural metal that is fireproofed is not required to be accessible where the method of connection is by an exothermic weld or irreversible compression connection. This type of connection will allow the integrity of the fireproofed structural metal to be maintained.

250.68(B) Effective Grounding Path

The Requirements: The connection of a grounding electrode conductor or bonding jumper to a grounding electrode is required to be made in a

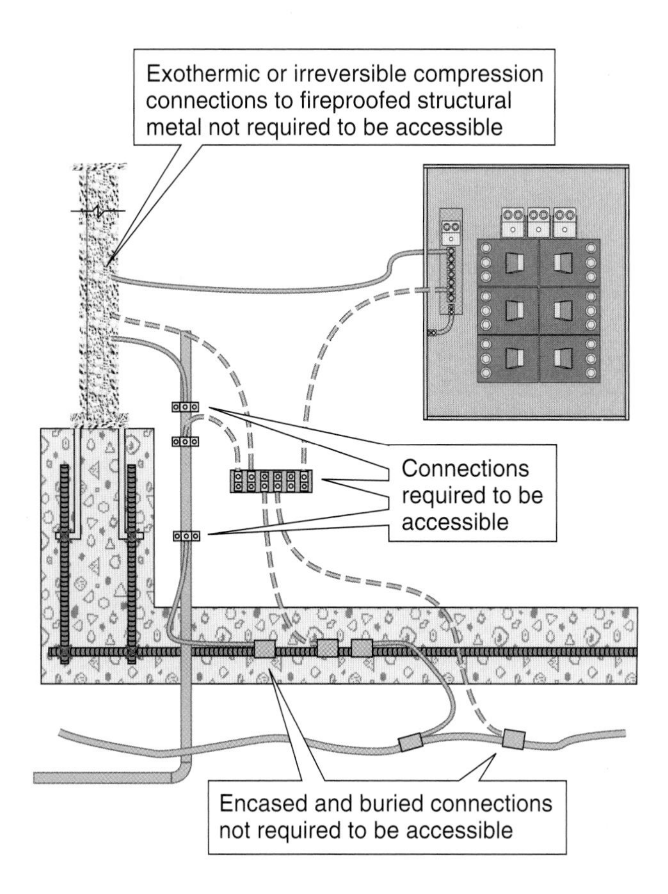

FIGURE 3-41 Grounding electrode conductor and bonding jumper connection to grounding electrodes, *250.68*. (*© Cengage Learning 2012*)

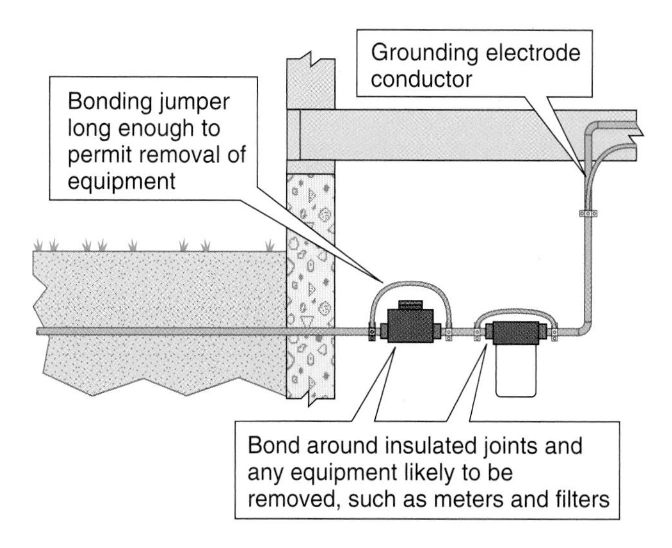

FIGURE 3-42 Effective grounding path, *250.68(B)*. (*© Cengage Learning 2012*)

manner that will ensure a permanent and effective grounding path. Where necessary to ensure the grounding path for a metal piping system used as a grounding electrode, bonding must be provided around insulated joints and around any equipment likely to be disconnected for repairs or replacement. Bonding conductors are required to be long enough to permit removal of such equipment while retaining the integrity of the grounding path.

Discussion: Bonding around equipment that is likely to be removed, such as water meters, water filters, and water treatment equipment, is necessary so the ground path is not interrupted when the equipment is removed for service or repair (see Figure 3-42).

The size of the bonding jumper is the same as required for the grounding electrode or bonding conductor.

250.68(C) Grounding Electrode Conductor and Bonding Jumper Connection Locations

The Requirements: Grounding electrode conductors and bonding jumpers shall be permitted to be connected at the following locations and be used to extend the connection to an electrode(s):

1. Interior metal water piping located not more than 1.52 m (5 ft) from the point of entrance to the building is permitted to be used as a conductor to interconnect electrodes that are part of the grounding electrode system.

Exception: *In industrial, commercial, and institutional buildings or structures if conditions of maintenance and supervision ensure that only qualified persons service the installation, interior metal water piping located more than 5 ft (1.52 m) from the point of entrance to the building is permitted as a bonding conductor to interconnect electrodes that are part of the grounding electrode system, or as a grounding electrode conductor, provided that the entire length, other than short sections passing perpendicularly through walls, floors, or ceilings, of the interior metal water pipe that is being used for the conductor is exposed.*

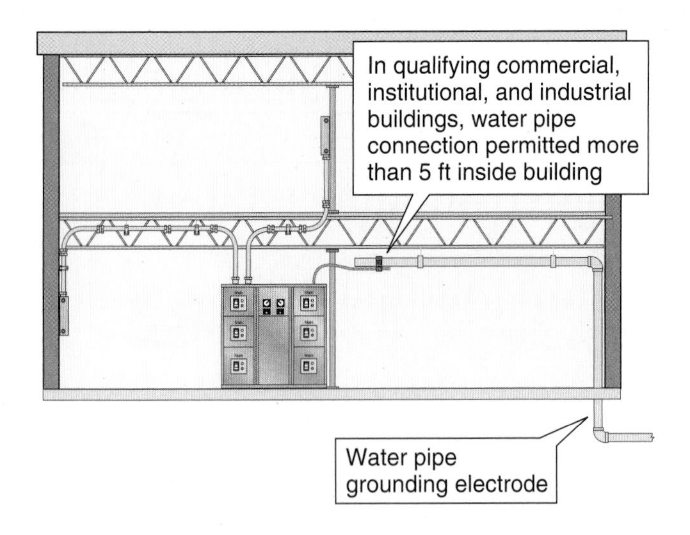

FIGURE 3-43 Water pipe grounding electrode exception, *250.68(C)(1), Exception.* (*© Cengage Learning 2012*)

Discussion: The 5-ft (1.52-m) restriction on the connection point inside the building is intended to reduce the likelihood that repairs are made to the metal water piping with a nonmetallic patch or coupling. The use of such nonconductive repair materials would interrupt the earth connection of the water piping electrode. See Figure 3-43.

The interior metal water pipe is permitted to be used as a conductor to interconnect other grounding electrodes in unlimited lengths (the 5-ft rule does not apply), provided that the following conditions are met:

1. It is an industrial, institutional, or commercial building or structure.
2. The management imposes conditions allowing only qualified persons to service the installation (*qualified person* is defined in *Article 100*).
3. The entire length of the interior metal water pipe is exposed, other than for short sections passing perpendicularly through walls, floors, or ceilings.

Note that the term *exposed* is defined in *Article 100* and does not mean "visible." For example, water pipe is considered exposed when located above lift-out panels for suspended ceilings. The concept included in the exception is that qualified persons will ensure that the path is continuous from the point where a connection is made to the metal water pipe to the underground portion of the grounding electrode.

2. The structural *metal* frame of a building that is directly connected to a grounding electrode as specified in *250.52(A)(2)* or *250.68(C)(2)(a), (b)* or *(c)* is permitted as a bonding conductor to interconnect the electrodes that are part of the grounding electrode system, or as a grounding electrode conductor.
   a. By connecting the structural metal frame to the reinforcing bars of a concrete-encased electrode as provided in *250.52(A)(3)* or ground ring as provided in *250.52(A)(4)*
   b. By bonding the structural metal frame to one or more of the grounding electrodes as defined in *250.52(A)(5)* or *(A)(7)* that comply with *250.53(A)(2)*
   c. By other approved means of establishing a connection to earth

Discussion: This part of the new section allows the structural metal frame of a building or structure to serve as a bonding means to connect conductors together or also to serve as a grounding electrode conductor. See Figure 3-44 for an example. To qualify as a conductor that can be used in this manner,

FIGURE 3-44 Metal frame of a building is permitted as a bonding means or a grounding electrode conductor if grounded properly. (*© Cengage Learning 2012*)

one of several options must be complied with, including these:

1. Verifying that the structural steel qualifies as a grounding electrode under the conditions in *250.52(A)(2)*.
2. Connecting the structural metal to a concrete encased grounding electrode or to a ground ring grounding electrode.
3. Bonding the structural metal to one or more ground rods or ground plates as provided in *250.52(A)(5)* or *(A)(7)*. Two ground rods or ground plates must be installed unless one provides a connection to earth of 25 ohms or lower.

250.70 METHODS OF GROUNDING AND BONDING CONDUCTOR CONNECTION TO ELECTRODES

The Requirements: The grounding or bonding conductor is required to be connected to the grounding electrode by exothermic welding, listed lugs, listed pressure connectors, listed clamps, or other listed means. Connections depending on solder are not permitted to be used.

Discussion: As can be seen, the only connection method not required to be listed is exothermic welding, because it is a field-applied process.

Figure 3-45 shows some methods for connecting the grounding or bonding conductor to electrodes.

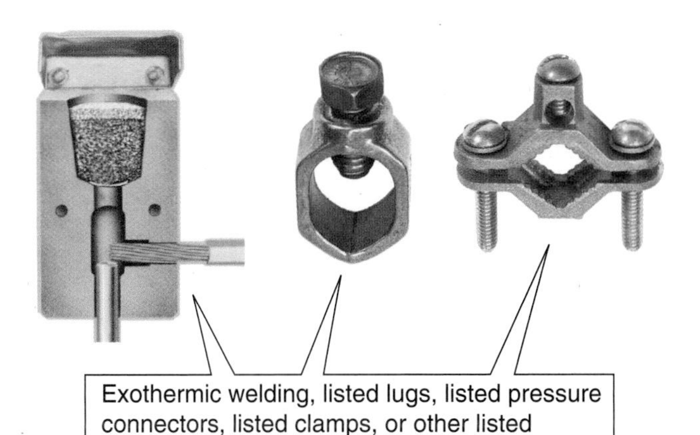

FIGURE 3-45 Methods of grounding and bonding conductor connection to electrodes, *250.70*. (*Courtesy of Thomas and Betts*)

More Requirements: Ground clamps must be listed for the materials of the grounding electrode and the grounding electrode conductor and, where used on pipe, rod, or other buried electrodes, shall also be listed for direct soil burial or concrete encasement.

Discussion: Clamps that are suitable for connecting only copper conductors are not required to be marked as such. Those that are suitable for connecting both copper and aluminum conductors are marked "AL/CU." Common materials for grounding electrodes include galvanized and copper water pipes, steel structural members, iron or steel concrete reinforcing rods, copper wires, and galvanized or copper-clad ground rods. The following information is from the Underwriters Laboratories *General Information Directory (White Book) in Grounding and Bonding Equipment* (KDER):

> *"Ground clamps and other connectors suitable for use where buried in earth or embedded in concrete are marked for such use. The marking may be abbreviated 'DB' (for 'direct burial')."*

Note that ground clamps and other connectors that are marked "DB" or "direct burial" also are suitable for being embedded in concrete without additional marking.

> *"Ground clamps are intended for use with ground rods and/or pipe electrodes in accordance with the* NEC *and are marked with the size of electrode and grounding electrode conductor with which the clamp is intended to be used. Clamps suitable for use on copper water tubing are marked 'Copper Water Tubing' or the equivalent, preceded or followed by the size of tubing. Ground rods, pipe electrodes, and water tubing trade sizes are stated in fractions such as ½, ⅝, and so forth.*

> *"Ground clamps intended for use with re-bar are marked with the size of re-bar with which the clamp is intended. Re-bar sizes may be specified in fractions, such as ½, ⅝, and so forth, or a number, such as 3, 4, 5, and so forth, where the number represents the numerator of a fraction when stated in eight-inch increments such as four = 4/8."**

*"Reprinted from the White Book with permission from Underwriters Laboratories Inc.® Copyright © 2010 Underwriters Laboratories Inc.®.

More Requirements: Not more than one conductor is permitted to be connected to the grounding electrode by a single clamp or fitting unless the clamp or fitting is listed for multiple conductors.

Discussion: This requirement supports the rule in *110.14(A)* that connectors and terminals are suitable for only one conductor unless otherwise marked.

More Requirements One of the following methods must be used for connecting grounding electrode conductors to grounding electrodes:

1. A pipe fitting, pipe plug, or other approved device screwed into a pipe or pipe fitting
2. A listed bolted clamp of cast bronze or brass, or plain or malleable iron
3. For indoor telecommunications purposes only, a listed sheet metal strap-type ground clamp having a rigid metal base that seats on the electrode and having a strap of such material and dimensions that it is not likely to stretch during or after installation
4. An equally substantial approved means

REVIEW QUESTIONS

1. Which of the following statements about creating a grounding electrode system is *not* true?
   a. All grounding electrodes that are available at the building must be bonded together.
   b. All grounding electrodes that are present at the building must be bonded together.
   c. If present, the following electrodes must be bonded together: underground metal water pipe and a ground ring.
   d. If present, the following electrodes must be bonded together: metal structural member electrode and a ground rod.
2. Which of the following is *not* a recognized grounding electrode in *250.52*?
   a. An underground metal water pipe in contact with the earth for 15 ft (4.5 m)
   b. A 2 AWG copper wire 20 ft (6.0 m) long buried in a ditch at least 2½ ft (750 mm) deep
   c. A steel structural frame connected to a concrete-encased electrode
   d. A ground plate buried at 30 in. (750 mm) below the surface
3. Which of the following statements about metal structural member grounding electrodes is *not* true?
   a. It is considered a grounding electrode if one or more of the steel members is in direct contact with the earth for 10 ft (3.0 m) or more.
   b. It is considered a grounding electrode if 10 ft (3.0 m) or more of one or more of the steel members is encased in concrete that is in direct contact with the earth.
   c. It is considered a grounding electrode if the metal frame is bonded to a concrete-encased grounding electrode.
   d. It is considered a grounding electrode if a steel column is driven into the ground 6 ft (1.8 m) to hardpan.

4. Which of the following statements about the installation of grounding electrodes is *not* true?
   a. A water pipe grounding electrode must be supplemented by at least one other grounding electrode.
   b. A concrete-encased grounding electrode must be at least 2½ ft (750 mm) below the earth's surface.
   c. A ground ring grounding electrode must encircle the building at a depth of 2½ ft (750 mm) or more.
   d. A pipe-type grounding electrode must be driven 8 ft (2.5 m) vertically where practicable.
5. The conductor that is used to connect grounding electrodes together to form the grounding electrode system is ____________________.
   a. a bonding conductor
   b. a grounding electrode conductor
   c. an equipment grounding conductor
   d. an equipment bonding jumper
6. Which of the following statements about grounding electrodes is true?
   a. An underground metal water pipe electrode is required to be supplemented by another grounding electrode.
   b. A concrete-encased electrode is required to be supplemented by another grounding electrode.
   c. An 8-ft (2.5-m) ground rod with a resistance of 24 ohms is required to be supplemented by another grounding electrode.
   d. Two ground plate grounding electrodes that are installed 6 ft (1.8 m) apart are required to be supplemented by another grounding electrode.
7. Which of the following statements about the installation of a ground rod is correct?
   a. Install vertically if practicable, then at not more than a 45-degree angle if not practicable to install vertically, then in a ditch at least 21½ ft (750 mm) deep if not practicable to install at not more than a 45-degree angle.
   b. In dry or arid regions, it is required to install rods that are longer than 8 ft to reach permanent moisture level.
   c. First measure the resistance of the first ground rod, and install additional rods until the resistance of the group is 25 ohms or less.
   d. It is permitted to install the rod vertically at not more than a 45-degree angle or in a ditch at least 2½ ft (750 mm) deep, at the installer's option.
8. Which of the following statements about the installation of an auxiliary grounding electrode is correct?
   a. It is permitted to be connected to equipment rather than an equipment grounding conductor when the manufacturer wants a "clean ground."
   b. It is permitted to be installed at parking lot lighting metal poles rather than connecting an equipment grounding conductor from the branch circuit.

c. It is permitted to supplement but not replace the equipment grounding conductor with the branch circuit.

d. The installer has the option of installing either the equipment grounding conductor or the local grounding electrode or both.

9. Which of the following statements about the installation of the grounding electrode conductor is true?

a. A grounding electrode conductor smaller than 6 AWG must be protected by installation in a conduit or cable armor.

b. A grounding electrode conductor smaller than 6 AWG is not permitted, as such sizes are too fragile.

c. 4 AWG is the minimum-size grounding electrode conductor to a ground ring grounding electrode.

d. 2 AWG is the minimum-size grounding electrode conductor to a concrete-encased grounding electrode.

10. Which of the following statements about splicing or connecting a grounding electrode conductor is *not* true?

a. It is permitted to splice or connect the grounding electrode conductor only by exothermic welding or irreversible compression connectors.

b. It is permitted to connect sections of copper busbars together to form the grounding electrode conductor.

c. It is permitted to connect grounding electrode conductors together by using a copper busbar not less than ¼ in. × 2 in. (6.4 mm × 50 mm).

d. Listed wire connectors or exothermic welding is permitted to be used to connect grounding electrode taps to the busbar.

11. Concrete-encased grounding electrodes that were not bonded to the grounding electrode system in existing buildings

a. are required to be bonded even if the concrete must be disturbed.

b. are not required to be bonded if the concrete must be disturbed to do so.

12. Because the title of *250.52* is "Electrodes Permitted for Grounding," the connection of these grounding electrodes is optional.

a. True

b. False

13. Which of the following underground water pipes is not recognized as a grounding electrode in *250.52(A)(1)*?

a. ¾-in. galvanized in contact with the earth for 22 ft

b. 1-in. copper in contact with the earth for 28 ft

c. 6-in. iron half-wrapped with corrosion-protection tape in contact with the earth for 40 ft

d. ¾-in. copper in contact with the earth for 12 ft

14. Plate grounding electrodes are required to be (circle the correct answer)

a. 2 ft square.

b. 1.414 ft square.

c. installed not less than 30 in. below earth surface.

d. installed at a level below the local frost line.

15. Chemical grounding electrodes are typically (circle the correct answer)

    a. constructed from 2-in. copper pipe, have weep holes, and are not suitable to be driven.

    b. constructed from galvanized water pipe, have weep holes, and are suitable to be driven.

16. Which of the following are permitted as a rod or pipe grounding electrode if 8-ft in length? (circle the correct answer)

    a. ¾-in. galvanized water pipe

    b. ⅝-in. galvanized ground rod

    c. ½-in. listed copper clad ground rod

    d. All of the above

17. If more than one grounding electrode of the pipe, rod, or plate type is installed, spacing not less than ___ apart is required.

    a. twice the length of the longest rod.

    b. 2½ ft

    c. 6 ft

    d. 8 ft

18. Which of the following statements is true about the resistance of grounding electrodes or grounding electrode systems?

    a. The 25-ohm resistance rule must be applied to electrodes of the water-pipe, ground-ring, and concrete-encased types.

    b. The 25-ohm resistance rule applies only to the rod, pipe, or plate-type grounding electrodes.

    c. Ground rods are not required to comply with the 25-ohm rule so long as they are driven 8 ft into the earth.

    d. The resistance of ground rods is typically measured with a volt-meter and an ohm-meter.

19. Which of the following statements is true about the use of grounding electrodes?

    a. A ground rod is required to serve as the grounding electrode for the service as well as for equipment supplied by the service.

    b. A ground rod is permitted to serve in place of the equipment grounding conductor so long as it has 8 ft in contact with the earth.

20. Copper, aluminum, and copper-clad aluminum grounding electrode conductors are permitted to be connected to an 8-ft ground rod.

    a. True

    b. False

UNIT

4

Enclosure, Raceway, and Service Cable Connections

OBJECTIVES

After studying this unit, the reader will understand

- grounding requirements for service raceways and enclosures.
- grounding requirements for other conductor enclosures and raceways.

250.80 SERVICE RACEWAYS AND ENCLOSURES

The Requirement: Metal enclosures and raceways for service conductors and equipment are required to be connected to the grounded system conductor if the electrical system is grounded or to the grounding electrode conductor for electrical systems that are not grounded.

Exception: *A metal elbow that is installed in an underground raceway and is isolated from possible contact by a minimum cover of 18 in. (450 mm) to any part of the elbow need not be connected to the grounded system conductor or grounding electrode conductor.*

Discussion: This *NEC* section generally requires that metal enclosures and raceways for service conductors be connected to the grounded system conductor if the system is grounded. If the system is ungrounded, the equipment is required to be connected to a grounding electrode. The term *grounded* means connected to earth or to some conducting body that extends the earth connection. Bonding requirements for services are located in *250.92*. The combination of grounding and bonding ensures that the equipment is maintained at earth potential and will serve as an effective fault return path. Figure 4-1 shows grounding and bonding of equipment.

The exception provides a practical solution to the requirement for grounding and bonding metal pulling elbows installed in runs of PVC conduit (see Figure 4-2). Metal pulling elbows often are installed in runs of PVC conduit so the ropes used for pulling conductors into the conduit do not cut through the inner radius of the conduit bend. It is often difficult or impossible to bond these underground pulling elbows. The 18-in. (450-mm) clearance required provides a reasonable isolation from contact in case the elbows become energized through a line-to-ground fault.

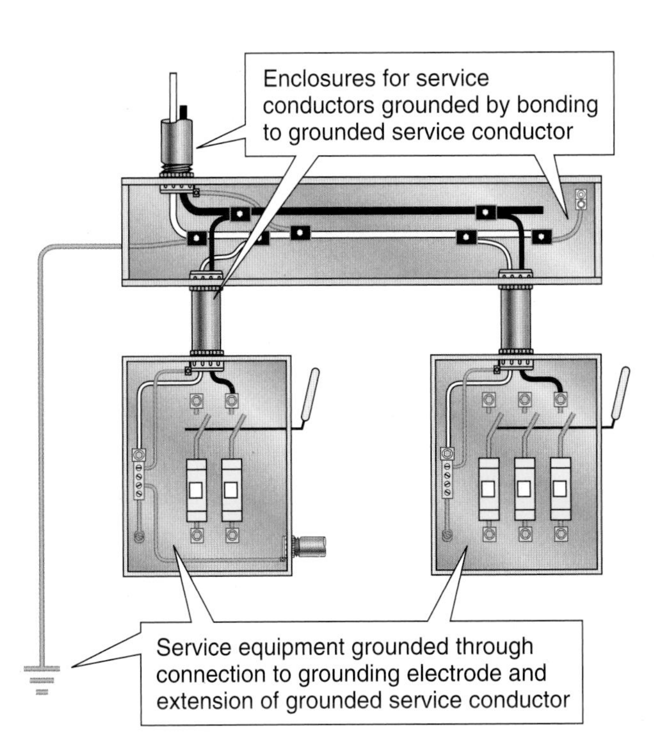

FIGURE 4-1 Grounding and bonding of equipment associated with the service, *250.80.* (*© Cengage Learning 2012*)

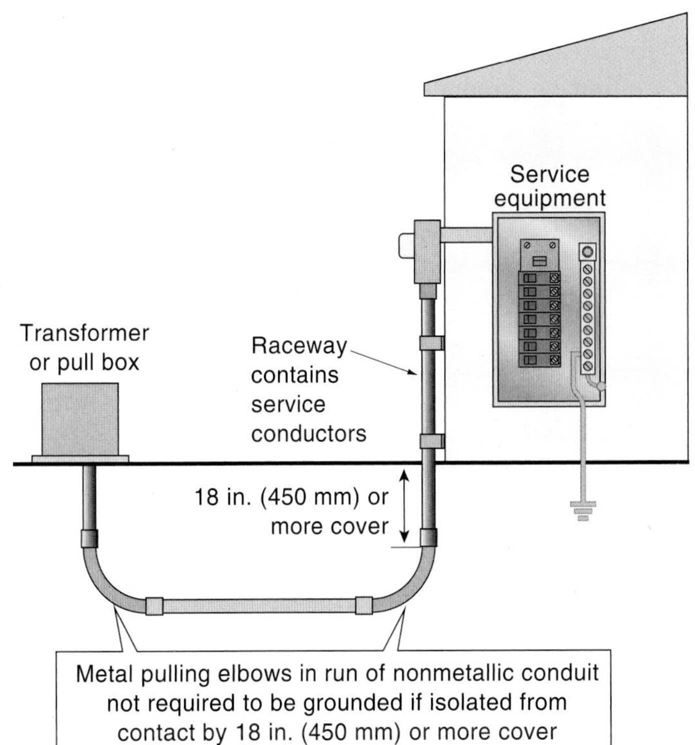

FIGURE 4-2 Grounding not required for isolated metal pulling elbows, *250.80 Exception.* (*© Cengage Learning 2012*)

250.86 OTHER CONDUCTOR ENCLOSURES AND RACEWAYS

The Requirement: Except as permitted by *250.112(I)*, metal enclosures and raceways for other than service conductors must be connected to the equipment grounding conductor.

Exception No. 1: *Metal enclosures and raceways for conductors added to existing installations of open wire, knob and tube wiring, and nonmetallic-sheathed cable are not required to be connected to the equipment grounding conductor where these enclosures or wiring methods comply with (1) through (4) as follows:*

1. *Do not provide an equipment ground;*
2. *Are in runs of less than 25 ft (7.6 m);*
3. *Are free from probable contact with ground, grounded metal, metal lath, or other conductive material; and*
4. *Are guarded against contact by persons.*

Discussion: This exception provides a way to use metal enclosures such as junction or outlet boxes for existing wiring systems that do not contain an equipment grounding conductor. As can be seen, safeguards are imposed to reduce the likelihood of a person's contacting a metal enclosure that is not grounded and has accidentally become energized.

Another Exception, No. 2: *Short sections of metal enclosures or raceways used to provide support or protection of cable assemblies from physical damage are not required to be grounded.*

Discussion: The *Code* is silent about what is meant by *short sections* of metal enclosures or raceways. As a result, this is a judgment call on the part of the AHJ. Some users and enforcers of the *Code* suggest that because metal raceways are normally supplied in 10-ft (3.0-m) lengths, a *short section* is one that is less than 10 ft (3.0 m) long.

Another Exception, No. 3: *A metal elbow is not required to be connected to the equipment grounding conductor if it is installed in a run of nonmetallic raceway and is isolated from possible contact by a minimum cover of 18 in. (450 mm) to any part of the elbow or it is encased in not less than 2 in. (50 mm) of concrete.*

Discussion: This exception is identical in application to the exception to *250.80* so far as the underground portion is concerned (see Figure 4-3). The exception excludes from bonding requirements any part of a pulling elbow that is encased in at least 2 in. (50 mm) of concrete. The rule regarding elbows encased in not less than 2 in. (50 mm) of concrete applies to installations made underground and above ground, as well as inside and outside of buildings.

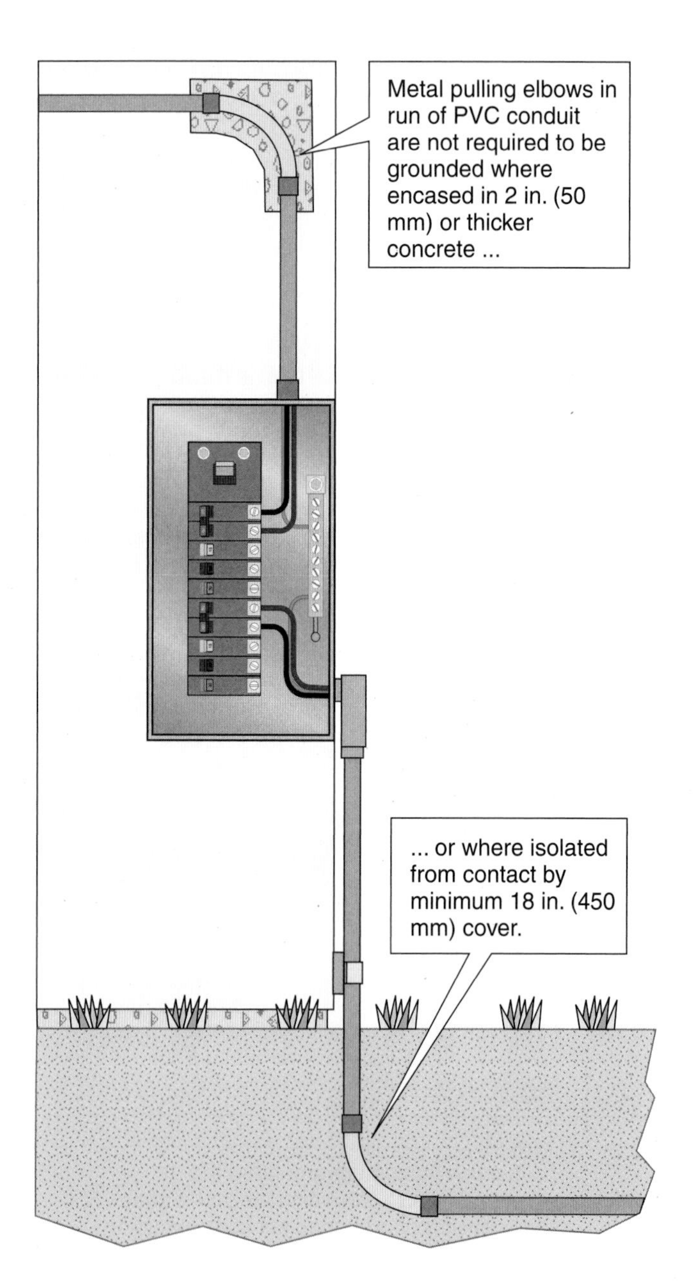

FIGURE 4-3 General rules for grounding and bonding, *250.86, Exception No. 3.* (*© Cengage Learning 2012*)

REVIEW QUESTIONS

1. Which of the following statements about metal raceways and enclosures for service conductors is true?
   a. To be considered grounded, the metal raceway must be connected directly to a grounding electrode.
   b. Metal raceways for service conductors are considered grounded if they are connected to a service enclosure that is grounded.
   c. The metal pulling elbows in a run of underground nonmetallic conduit are required to be grounded, irrespective of burial depth.
   d. The metal pulling elbows in a run of underground nonmetallic conduit are required to be grounded only if they are within 12 in. (304 mm) of the surface.
2. Which of the following statements is true about grounding metal raceways for conductors added to existing installations that do not contain an equipment grounding conductor?
   a. The exception does not apply if the circuit contains an equipment grounding conductor.
   b. The exception applies to runs more than 25 ft (7.6 m) long.
   c. The exception applies if the conduit contacts grounded metal surfaces.
   d. The exception applies if the conduit is subject to contact by persons.

UNIT 5

Bonding

OBJECTIVES

After studying this unit, the reader will understand

- general requirements for bonding.
- bonding requirements for services.
- bonding requirements for other enclosures.
- bonding in hazardous locations.
- material for, attachment of, and sizing equipment bonding jumpers.
- bonding of metal piping systems and exposed structural steel.
- bonding of lightning protection systems.

250.90 GENERAL

Bonding is required to be provided where necessary to ensure electrical continuity and the capacity to conduct safely any fault current likely to be imposed. Bonded system elements are shown in Figure 5-1.

Discussion: The concept of bonding and its critically important place in the safety system were discussed extensively in the Introduction to this book. In its simplest form, bonding completes a path for fault current to flow. Likewise, bonding extends the equipment grounding conductor path. Equipotential bonding to reduce differences in potential (voltage) is installed in special occupancies such as agricultural buildings or in special locations such as in patient care areas of health care facilities and for swimming pools. As we shall see, specific requirements are provided for bonding of service equipment and in hazardous locations (covered in Unit 8 of this book), but bonding is equally important for feeders and branch circuits.

250.92 SERVICES

250.92(A) Bonding of Services

The Requirements: The normally non-current-carrying metal parts of equipment indicated in *250.92(A)(1)* and *(A)(2)* are required to be bonded together. This includes the following:

1. The service raceways, cable trays, cablebus framework, auxiliary gutters, or service cable armor or sheath that enclose, contain, or support service conductors, except as permitted in *250.80.*
2. All enclosures containing service conductors, including meter fittings, boxes, or the like, interposed in the service raceway or armor.

FIGURE 5-1 Bonding of metal conduit, *250.90.* (*© Cengage Learning 2012*)

Discussion: The *NEC* imposes very specific requirements for bonding any and all non-current-carrying metal parts of equipment associated with the service. The reason for the extra emphasis is because conductors on the line or supply side of the service disconnecting means are not provided with overcurrent protection. As illustrated in Figure 5-2, the circuit is from the utility transformer to the service equipment, over the main bonding jumper, and returning to the transformer over the grounded service conductor. The electric utility does provide overcurrent protection on the line or supply side of their transformer, but such protection is not selected to protect conductors or equipment on the load or secondary side. Therefore, a high-capacity, low-impedance path must be provided on the secondary of the transformer to facilitate enough current to flow so the fuse on the supply side of the transformer will open.

Basically, any conductive equipment that supports or encases a service conductor is required to be bonded. Bonding joins the metallic parts into one low-impedance, high-capacity path for fault current to flow. See *250.102(C)* in this book for rules on sizing the bonding conductors.

250.92(B) Method of Bonding at the Service

The Requirements: Bonding jumpers meeting the requirements of this article are required to be used around impaired connections such as reducing washers or oversized, concentric, or eccentric knockouts. Standard locknuts or bushings are not permitted to be the only means for the bonding required by this section but are permitted to be installed to make a mechanical connection of the raceway(s).

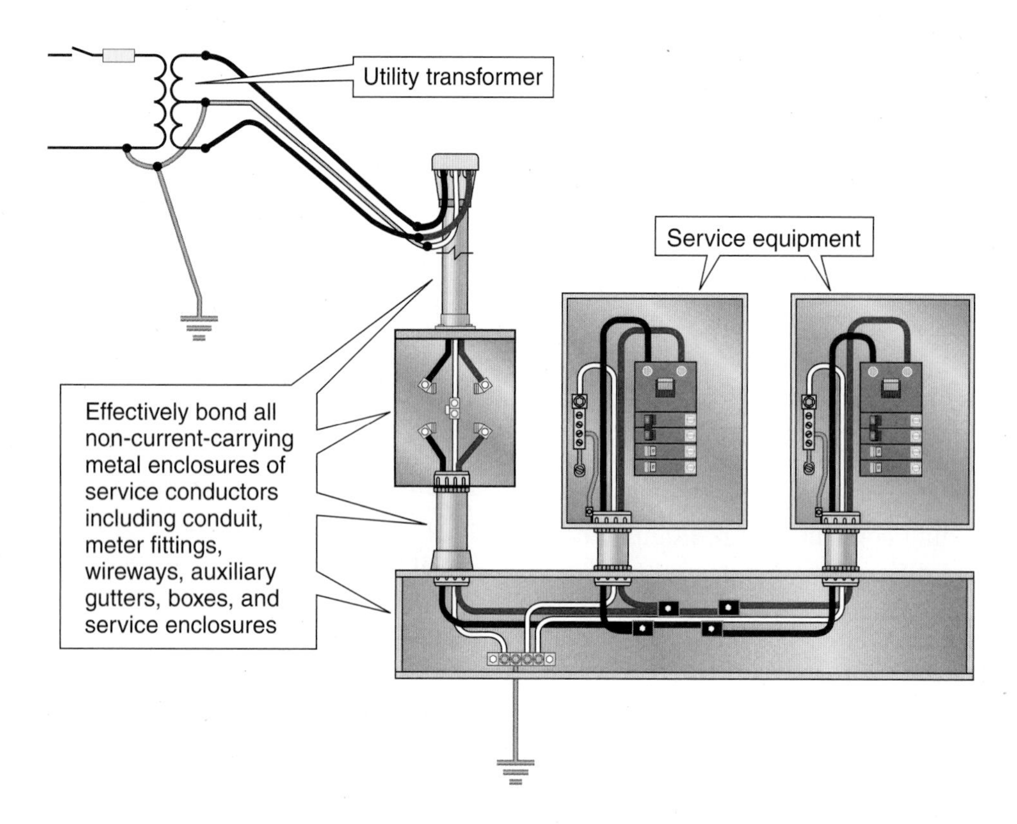

FIGURE 5-2 Bonding service equipment enclosures, *250.92(A)*. (*© Cengage Learning 2012*)

Electrical continuity at service equipment, service raceways, and service conductor enclosures must be ensured by one of the following methods:

250.92(B)(1) Bonding Equipment to the Grounded Service Conductor in a Manner Provided in *250.8*

Discussion: To illustrate conditions under which bonding is required for metal service enclosures, it should be noted that this section uses the term *impaired connections* and goes on to give four examples. These include reducing washers or oversized, concentric, or eccentric knockouts. See Figure 5-3 for an example of possible impaired connections. A bonding bushing and properly sized jumper must be used to bond metallic enclosures such as conduit or EMT to cabinets or other enclosures for service conductors.

As shown in Figure 5-4, the grounded service conductor (often a neutral) is permitted to be used to bond enclosures together on the line side of the service. This is one of the reasons why *250.24(C)* requires the neutral or grounded conductor to be sized not smaller than a grounding electrode conductor from *Table 250.66*. *Table 250.66* specifies the minimum-size conductor that is considered adequate for carrying fault current.

FIGURE 5-3 At the service equipment, when using metal conduit or EMT, bond around all *impaired connections* such as reducing washers, oversized, concentric or eccentric knockouts, or where paint has not been removed between the fitting and the enclosure. (*© Cengage Learning 2012*)

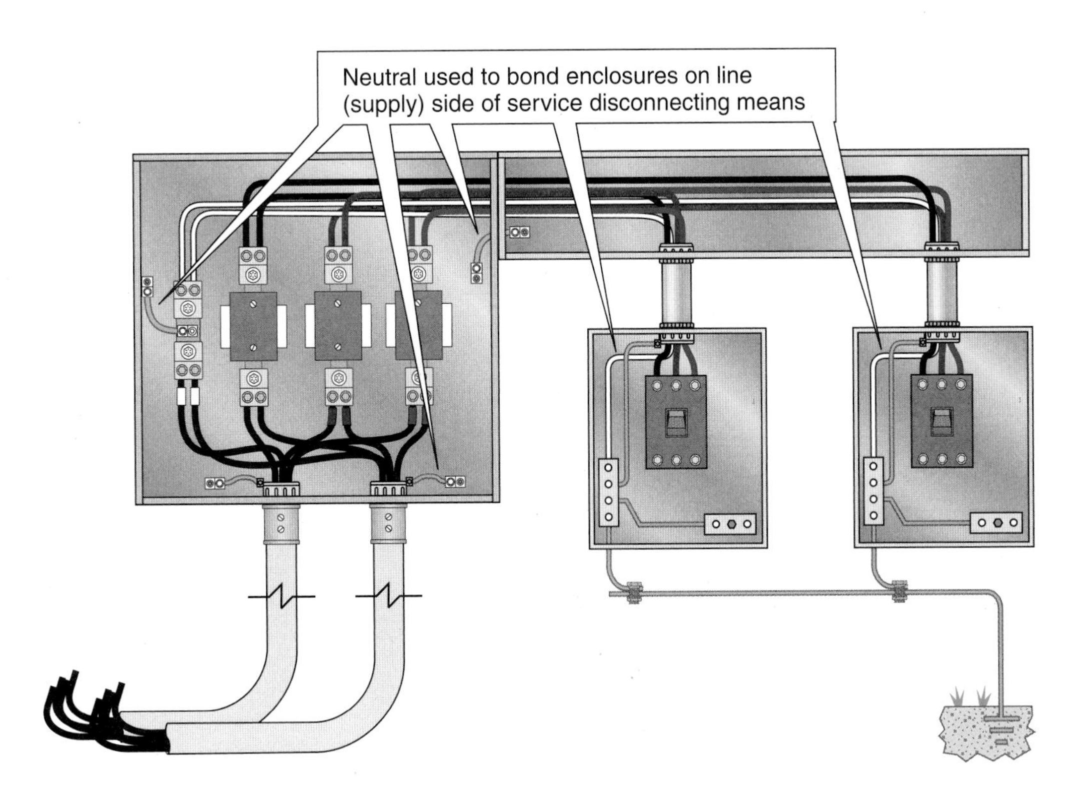

FIGURE 5-4 Use of neutral for bonding on supply side of service, *250.92(B)(1)*. (© Cengage Learning 2012)

The terminal for the neutral conductor in plug-in-type meter sockets is most often bonded to the enclosure at the factory. The main bonding jumper is installed in the service equipment. Then, the installation of the grounded (neutral) conductor bonds the enclosures together.

The current transformer cabinet at the left of Figure 5-4 is shown to be bonded to the service grounded (neutral) conductor. The conduits connected to the current transformer cabinet are bonded to the cabinet. A bonding jumper is shown connecting the wireway to the current transformer cabinet. A bonding jumper connects the supply conduits to the neutral in the respective service disconnecting means. Finally, the main bonding jumper connects the grounded service conductor to the service disconnecting means. Thus, all of the metal enclosures containing an ungrounded ("hot") service conductor are bonded together by use of the grounded service conductor.

Additional information on sizing the bonding jumper at each point is provided in *250.102(C)* in this book.

It should be noted that the nipples, conduits, and raceways are not required to be bonded at each end. The *Code* rule is that all metal enclosures containing or enclosing service conductors be bonded together, but it does not require that they be bonded at each end.

Section 250.8 provides a list of acceptable methods of connecting grounding and bonding equipment.

More Requirements: Bonding of remote meter equipment.

Discussion: Figures 5-5 and 5-6 show one acceptable way of complying with *NEC* requirements for bonding meter equipment installed remotely from service equipment. Other methods may comply as well. For underground- or overhead-supplied metering installations that are remote from the service equipment, such as on a meter pole or yard poles or current transformer cabinets, bonding the neutral conductor to the meter enclosure usually is the most practical method of bonding. This bonding connection provides the essential low-impedance path back to the source so fault current can be cleared by the overcurrent device on the line side of the utility transformer.

Some electric utilities also may require a connection to a grounding electrode at the metering

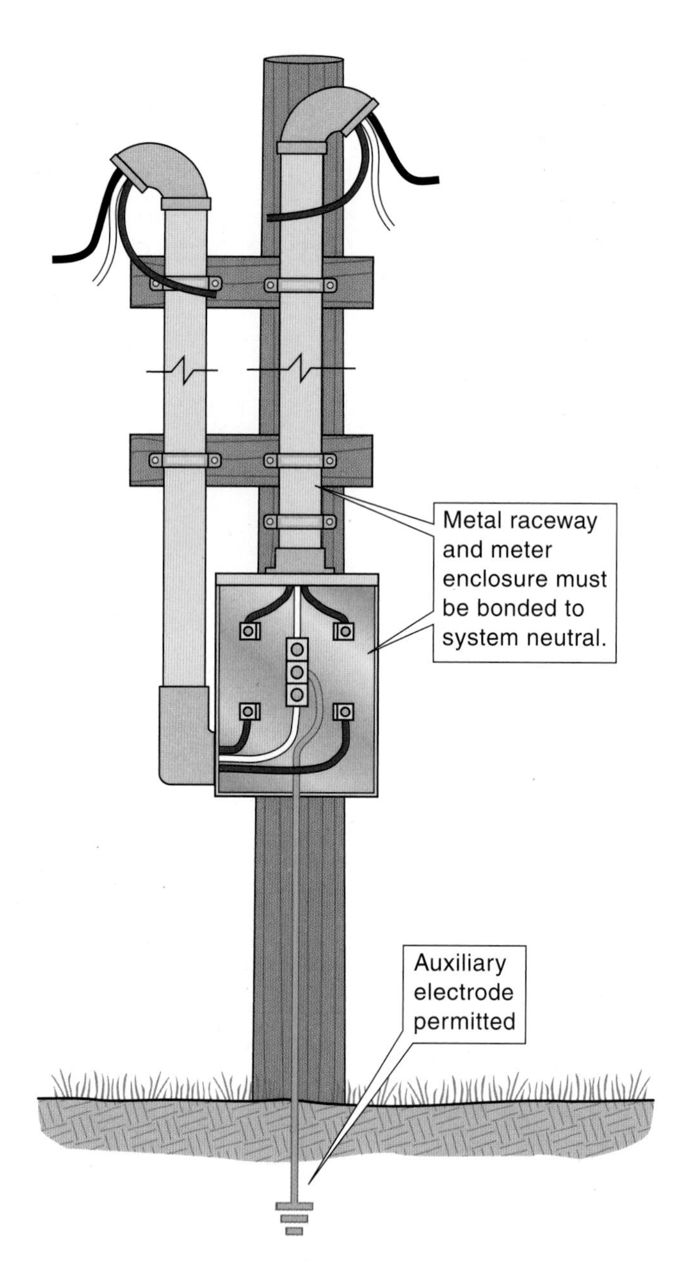

FIGURE 5-5 Bonding meter enclosures remote from and on the supply side of the service, *250.92(B)(1)*. (© Cengage Learning 2012)

equipment. Although this grounding electrode connection at the remote metering is not specifically required by *Article 250,* making such a connection on the line or supply side of the service disconnecting means should be acceptable.

A grounding electrode should not be relied on for bonding or grounding remote metering equipment without a connection to the neutral. Take a look at the methods for bonding equipment that contain service conductors. No mention is made of a connection to a grounding electrode to accomplish this purpose. Compare the installation you are making for compliance with *250.4(A)(3)* and *(5),* which requires that an effective ground-fault return path be provided. Similar requirements are found in *250.4(B)(2)* and *(4)* for ungrounded systems.

More Requirements: ***250.92(B)(2).*** Connections utilizing threaded couplings or threaded hubs on enclosures are permitted if made up wrenchtight.

Discussion: Threaded connections of conduits into threaded couplings and hubs are permitted without

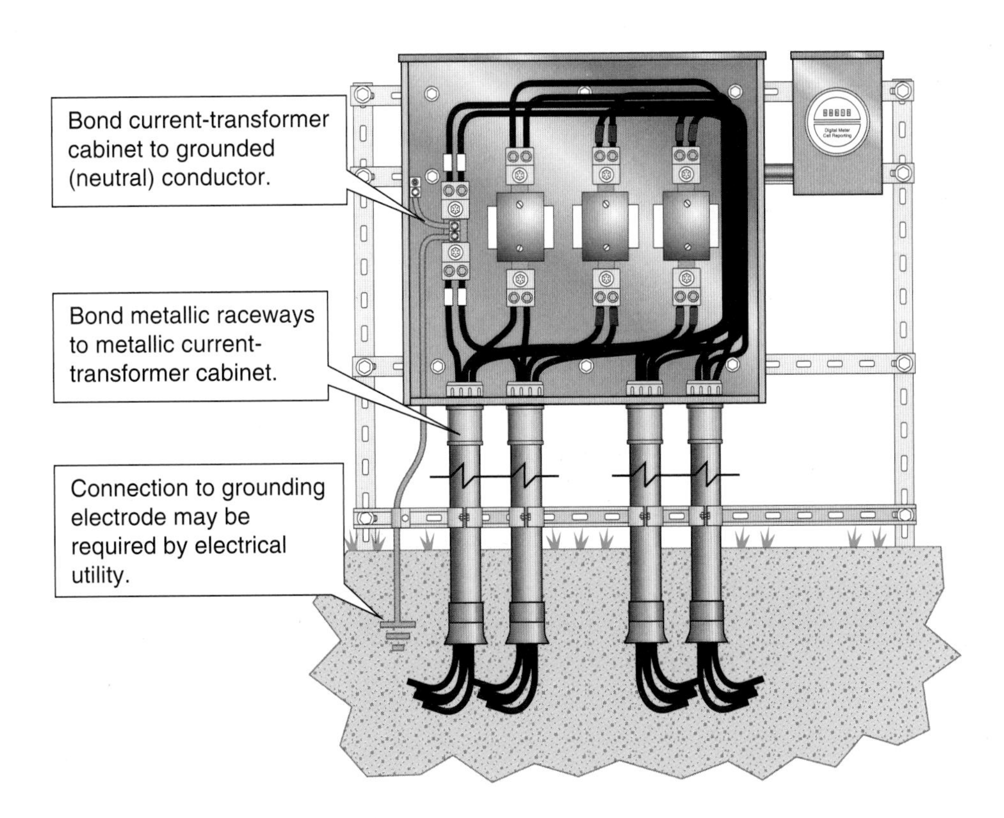

FIGURE 5-6 Bonding current transformer–type enclosures that are installed remote from and on the supply side of service equipment, *250.92(B)(1)*. (*© Cengage Learning 2012*)

a bonding jumper if made up wrenchtight. Making connections wrenchtight involves using properly sized pipe wrenches or similar tools and pulling the wrench with enough force so the connection is tight. The *Code* does not include a torque value for various sizes of conduit, but relies on good workmanship to ensure the connection is properly made. Tight connections are important because metal conduit and tubing will be required to carry a great deal of current if there is a line-to-ground fault on the line (supply) side of the service.

Conduit hubs are commonly furnished by manufacturers for equipment intended to be installed in a wet location. Electrical inspectors routinely accept conduit hubs for bonding conduit connections without additional bonding means where these hubs are supplied by manufacturers as an accessory for equipment such as meter bases and weatherproof service equipment. See Figure 5-7 for an illustration of threaded conduit couplings, bolt-on hubs, and weatherproof hubs used at service equipment.

FIGURE 5-7 Bonding connections utilizing threaded couplings or hubs, *250.92(B)(2)*. (*© Cengage Learning 2012*)

FIGURE 5-8 Weatherproof hubs of a type that may be suitable for bonding at a service. Follow manufacturer's installation instructions. *250.92(B)(1)*. (*Photos courtesy of Thomas & Betts*)

To be acceptable for bonding at service equipment, field-installed weatherproof hubs must be listed in compliance with UL 467, "Grounding and Bonding Equipment." Follow the manufacturer's installation instructions, which may include the installation of a bonding-type locknut or a bonding bushing and jumper. See Figure 5-8 for an example of weatherproof hubs that are designed to be installed by the electrician in field-punched holes in cabinets or similar enclosures.

More Requirements: *250.92(B)(3)*. Threadless couplings and connectors where made up tight for metal raceways and metal-clad cables.

Discussion: Threadless couplings and connectors are suitable for bonding metal conduit and electrical metallic tubing (EMT) if made up tight. Figure 5-9 shows threadless couplings on EMT. Threadless couplings are also available for use on rigid metal and intermediate metal conduit. Even though the *Code* uses the word *wrenchtight* in *250.92(B)(2)* and uses *tight* in *250.92(B)(3)*, the requirement for making a secure connection should be treated equally. One loose conduit or tubing fitting will result in a break in the ground-fault return path. A loose fitting can also result in arcing and sparking at connections or may create a shock hazard where energized equipment is at an elevated potential and another at earth potential.

FIGURE 5-9 Bonding connections using threadless couplings and connectors, *250.92(B)(3)*. (*© Cengage Learning 2012*)

More Requirements: *250.92(B)(4)* Other listed devices, such as bonding-type locknuts, bushings, or bushings with bonding jumpers.

Discussion: Figure 5-10 shows other fittings used for bonding metal conduits or EMT installed at service equipment. Bonding-type bushings are available in standard conduit sizes. Some bushings thread on conduit or fittings and other bushings are of the slip-on type. A set screw is provided to secure

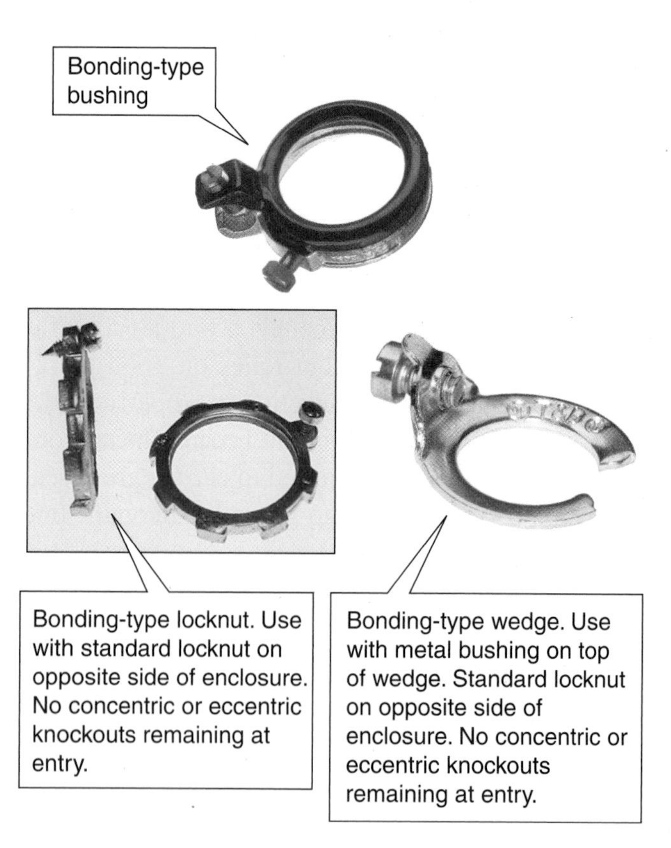

FIGURE 5-10 Bonding-type bushings, locknuts, and wedges, *250.92(B)(4)*. (*Courtesy of Thomas and Betts*)

the bushing to the conduit as well as to make a solid connection.

Bonding-type bushings are most commonly combination bonding/insulating bushings because they have an insulating throat designed to protect the conductors from sharp edges of conduit. The bushings are selected to match the size of conduit or fitting they are installed on. A properly sized bonding jumper must be installed from the bushing to the enclosure, to the neutral bus at the service equipment, or to the equipment grounding bus.

The size of the bonding jumper is determined from *NEC 250.102(C)*. Rules are provided for single bonding jumpers and for installations consisting of parallel runs of conduit or EMT.

In most cases, conduit-reducing washers impair the ground-fault return path, because they are usually installed without removing the paint on the enclosure where they are connected. Reducing washers are used as an example of an *impaired* path in *250.92(B)*. A bonding bushing and a bonding jumper are always required for installations that include reducing washers at the service. See *NEC 250.96* and Figure 5-11 for additional information on bonding around suspect conduit, EMT, and metal-armored cable connections.

Bonding-type locknuts are permitted only where there are no concentric or eccentric knockouts remaining where the connection is made. For example, if an installation is made with a 2-in. conduit to a panelboard where a combination set of knockouts of 1¼, 1½, 2, and 2½ in. exists, a bonding-type locknut is not acceptable because the 2½-in. knockout would remain. A bonding-type locknut is acceptable only where the size of the conduit matches the size of the knockout without extra eccentric or concentric knockouts remaining. Because a mechanical and electrical connection is required in *NEC 300.10,* a standard locknut is required on the opposite side of the enclosure.

Bonding-type wedges are available in conduit sizes. They are produced in two general styles: with one or two set screws. A bonding-type wedge with one set screw is shown in Figure 5-10. Those produced with one set screw are designed to be installed where the conduit size matches the knockout in the enclosure. The bonding wedge is intended to be installed in contact with the enclosure and with a metal bushing on top of it. The metal bushing must be of the thread-on type so the wedge can be tightened against the enclosure. A locknut should not be installed on top of the bonding wedge to secure it to the enclosure, as the teeth of the locknut can catch on the wedge and force one leg of the wedge out of the assembly.

A bonding-type wedge with two set screws may be installed for bonding purposes where concentric or eccentric rings remain at the connection. One of the set screws is used to secure the bonding wedge to the conduit. The other set screw is used to secure a lug (wire connector) to the wedge. A properly sized bonding jumper can then be connected to the bonding wedge. See the rules in *NEC 250.102(C)* for

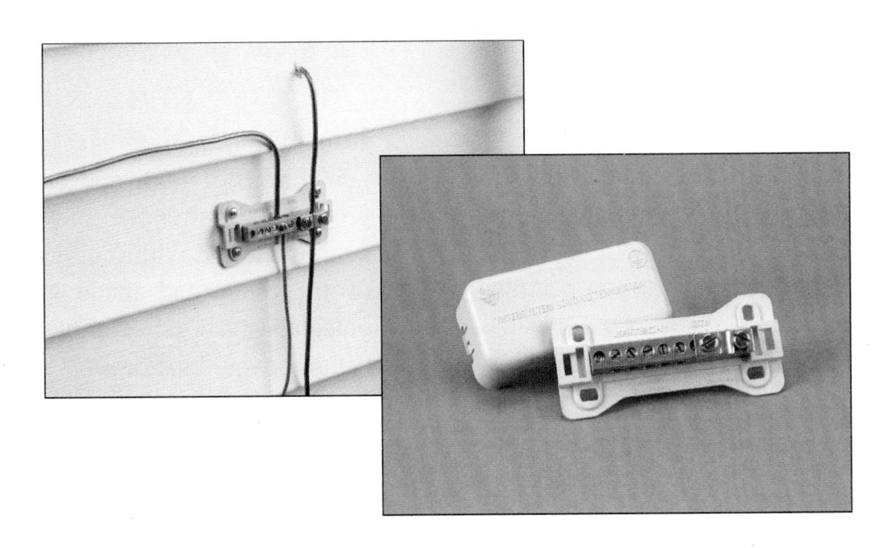

FIGURE 5-11 Intersystem bonding for other systems, *250.94.* (*Courtesy of Erico*)

sizing the bonding jumper. Once again, because a mechanical and electrical connection is required in *NEC 300.10,* a standard locknut is required on the opposite side of the enclosure.

250.94 BONDING FOR OTHER SYSTEMS

The Requirement: An intersystem bonding termination for connecting intersystem bonding conductors required for other systems shall be provided external to enclosures at the service equipment or metering equipment enclosure and at the disconnecting means for any additional buildings or structures. The intersystem bonding termination is required to comply with the following:

1. Be accessible for connection and inspection.
2. Consist of a set of terminals with the capacity for connection of not less than three intersystem bonding conductors.
3. Not interfere with opening the enclosure for a service, building or structure disconnecting means, or metering equipment.
4. At the service equipment, be securely mounted and electrically connected to an enclosure for the service equipment, to the meter enclosure, or to an exposed nonflexible metallic service raceway, or be mounted at one of these enclosures and be connected to the enclosure or to the grounding electrode conductor with a minimum 6 AWG copper conductor.
5. At the disconnecting means for a building or structure, be securely mounted and electrically connected to the metallic enclosure for the building or structure disconnecting means, or be mounted at the disconnecting means and be connected to the metallic enclosure or to the grounding electrode conductor with a minimum 6 AWG copper conductor.
6. The terminals are required to be listed as grounding and bonding equipment.

Discussion: The bonding means required by this *NEC* section is for bonding other systems such as for communications, radio and television equipment, and CATV systems to the electrical power and lighting system grounded conductor or equipment. (See Figure 5-11 for an example.) This section generally requires the installation of an accessible terminal bar at the service equipment or the building disconnecting means for additional buildings or structures. The conductor required for the bonding terminal bar is required to be not smaller than 6 AWG.

Bonding all the systems at one location ensures that all the grounded system conductors serving the same building or structure are at the same potential. This provision helps prevent serious equipment damage that can be caused by high-voltage flashover from local lightning events when equipment is supplied from two different systems. Equipment commonly supplied from two different systems includes televisions, stereo equipment, and facsimile equipment.

Exception: *In existing buildings or structures where any of the intersystem bonding and grounding conductors required by 770.100(B)(2), 800.100(B)(2), 810.21(F)(2), 820.100(B)(2), and 830. 100(B)(2) exist, installation of the intersystem bonding termination is not required. An accessible means external to enclosures for connecting intersystem bonding and grounding electrode conductors shall be permitted, provided at the service equipment and at the disconnecting means for any additional buildings or structures by at least one of the following means:*

1. *Exposed nonflexible metallic raceways*
2. *Exposed grounding electrode conductor*
3. *Approved means for the external connection of a copper or other corrosion-resistant bonding or grounding conductor to the grounded raceway or equipment.*

Informational Note No. 1: A 6 AWG copper conductor with one end bonded to the grounded nonflexible metallic raceway or equipment and with 6 in. (150 mm) or more of the other end made accessible on the outside wall is an example of the approved means covered in 250.94, Exception item (3).

Informational Note No. 2: See 770.100, 800.100, 810.21, 820.100 and 830.100 for bonding and grounding requirements for communications circuits, radio and television equipment, CATV circuits, and network-powered broadband communication systems.

Discussion: The three methods in the exception for providing the intersystem bonding terminations were the recognized methods in previous editions of the *NEC*. These methods now are permitted only for existing buildings or structures.

250.96 BONDING OTHER ENCLOSURES

250.96(A) General

The Requirements: Metal raceways, cable trays, cable armor, cable sheath, enclosures, frames, fittings, and other metal non-current-carrying parts that are to serve as equipment grounding conductors, with or without the use of supplementary equipment grounding conductors, are required to be bonded where necessary to ensure electrical continuity and the capacity to conduct safely any fault current likely to be imposed on them. Any nonconductive paint, enamel, or similar coating must be removed at threads, contact points, and contact surfaces or be connected by means of fittings designed so as to make such removal unnecessary. See Figure 5-12.

Discussion: The rules in *250.92* specifically require bonding for enclosures containing service conductors because the service conductors have only overload (not overcurrent) protection. This *NEC* section requires similar bonding if needed for the connections

FIGURE 5-12 Bonding other enclosures, *250.96(A)*. (*© Cengage Learning 2012*)

and equipment supplied by feeders or branch circuits to provide an effective ground-fault current path. As with bonding for services, every connection in the path that fault current would take must be regarded as suspect, and bonding should be considered. As previously discussed, only one loose conduit or electrical metallic tubing (EMT) connection can cause a break in the fault-current return path, leaving equipment at a dangerous voltage above ground.

Conduit or EMT connections to unpainted enclosures with locknuts that are made up tight should be satisfactory. If a painted enclosure is encountered, either the paint must be removed where the connection is made, or heavy, formed locknuts with sharp tabs to break the paint must be used. If the locknut does not break the paint, a bonding bushing or bonding locknut should be used to ensure a good connection that will carry any fault current likely to be imposed. For additional information on this subject, see the discussion at *250.12* and the UL report on the testing of conduit fittings in Appendix D.

Reducing Washers. Reducing washers are used when the knockout opening in an enclosure is larger than the conduit or cable fitting to be connected. Reducing washers are used in matching pairs with one inside the enclosure and one outside. They are available in many standard trade conduit opening sizes such as shown in Table 5-1 (note that metric equivalencies are not shown, for simplicity):

The *NEC* doesn't specifically cover the installation of reducing washers. Rather, at least so far as grounding and bonding issues are concerned, *Article 250* focuses on maintaining an effective ground-fault return path. So, after the installation of

TABLE 5-1

Reducing Washer Application Trade Size ½ in. to 2 in.

	½	¾	1	1¼	1½	2
½	•	•	•	•	•	•
¾	•		•	•	•	•
1	•	•		•	•	•
1¼	•	•	•		•	•
1½	•	•	•	•		•
2	•	•	•	•	•	

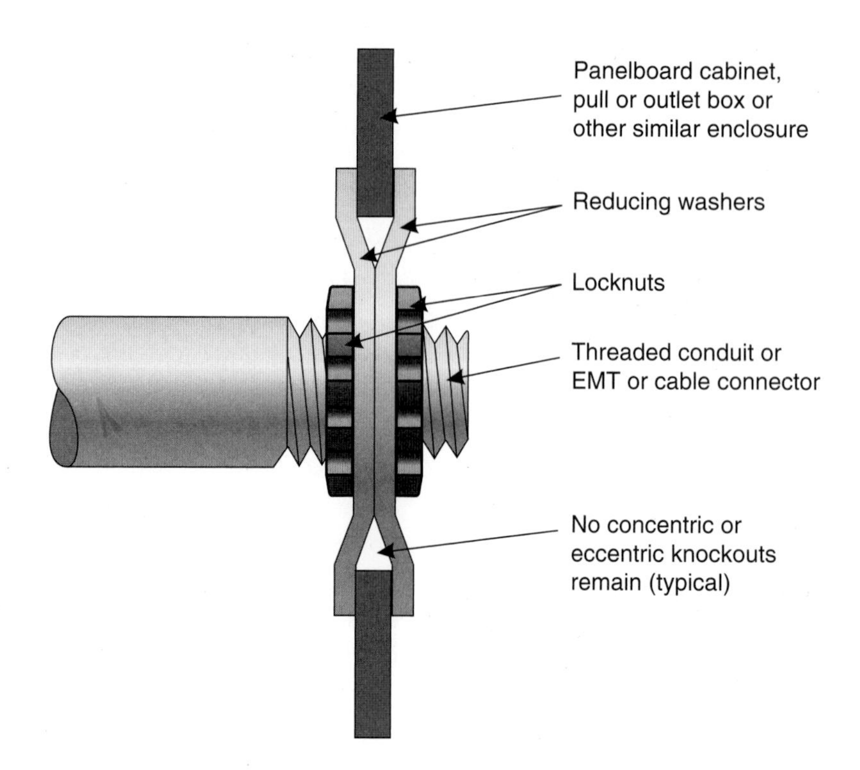

FIGURE 5-13 Installation requirements for reducing washers, *250.96(A)*. (*© Cengage Learning 2012*)

the reducing washers is made, the integrity and capability of the fault return path must not be reduced. See Figure 5-13 for a proper installation method for reducing washers.

Instructions on the application of listed reducing washers is included in the UL *White Book* under the category "Outlet Bushings and Fittings (QCRV)." It reads as follows: "Metal reducing washers are considered suitable for grounding for use in circuits over and under 250 V and where installed in accordance with ANSI/NFPA 70, *National Electrical Code*. Reducing washers are intended for use with metal enclosures having a minimum thickness of 0.053 in. for non-service conductors only. Reducing washers may be installed in enclosures provided with concentric or eccentric knockouts, only after all of the concentric and eccentric rings have been removed. However, those enclosures containing concentric and eccentric knockouts that have been Listed for bonding purposes may be used with reducing washers without all knockouts being removed."† This refers to outlet boxes that have especially designed combination trade size ½ and ¾ in. (metric designator 16 and 21) knockouts. See the discussion on bonding for over 250 volts and the photograph of the box construction features in Figure 5-14 in this book.

In discussions with UL representatives, they have clarified that the reference to installing the reducing washers in compliance with the *NEC* includes all

FIGURE 5-14 Outlet box listed with knockouts for 277 volts without bonding jumper, *250.97*. (*© Cengage Learning 2012*)

applicable rules, including neat and workmanlike installations (*110.12*), removing paint and all other compounds that would prevent a metal-to-metal connection (*250.12*), and providing an effective ground-fault return path [*250.4(A)(3)* and *(5)*].

As pointed out in the UL guide information, all remaining knockouts must be removed and a single set of reducing washers be installed. For example, a wireway has a set of concentric or eccentric knockouts that originally were for trade sizes ½ in. through 2 in. (metric designator 16 through 53). The knockouts through trade size 1½ (metric designator 41) have been removed. You want to connect a trade size 1 fitting to the wireway. Rather than installing a trade size 1½ × 1 (metric designator 41 × 27) set of reducing washers, it is required to remove the remaining knockouts and install a single set of 2 × 1 (metric designator 53 × 27) reducing washers.

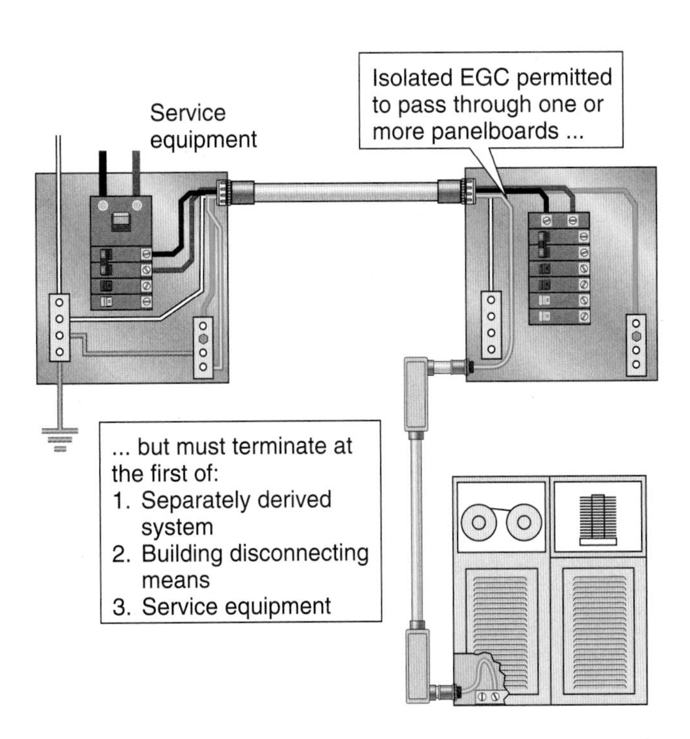

FIGURE 5-15 Isolated grounding circuits (EGC = equipment grounding conductor), *250.96(B)*. (*© Cengage Learning 2012*)

250.96(B) Isolated Grounding Circuits

The Requirement: If installed for the reduction of electrical noise (electromagnetic interference) on the grounding circuit, an equipment enclosure supplied by a branch circuit is permitted to be isolated from a raceway containing circuits supplying only that equipment by one or more listed nonmetallic raceway fittings located at the point of attachment of the raceway to the equipment enclosure. The installation of the metal raceway must comply with provisions of this article and must be supplemented by an internal insulated equipment grounding conductor installed in accordance with *250.146(D)* to ground the equipment enclosure.

> **Informational Note:** Use of an isolated equipment grounding conductor does not relieve the requirement for grounding the raceway system.

Discussion: This requirement functions like an exception to the general rule that the metal raceway that supplies the branch circuit to equipment must be connected at both ends to provide the equipment grounding required and provide the fault-current return path back to the source. As can be seen, the purpose of the special rules is to provide for isolation of equipment from the effects of electrical noise (electromagnetic interference) on the grounding circuit. (See Figure 5-15.)

Important rules in the section include the following:

1. The equipment is supplied by a branch circuit and not by a feeder or service.
2. The isolation applies to a raceway containing circuits supplying only that individual equipment.
3. The isolation is permitted to be accomplished by one or more listed nonmetallic raceway fittings located at the point of attachment of the raceway to the equipment enclosure.
4. The installation of the metal raceway is required to comply with provisions of *Article 250*. This means the metal raceway is required to be connected to a grounded enclosure even though the insulated equipment grounding conductor is routed through it.
5. The raceway must be supplemented by an internal insulated equipment grounding conductor installed in accordance with *250.146(D)* to ground the data-processing or other equipment enclosure.

The reference to *250.146(D)* permits the isolated equipment grounding conductor to bypass one or more panelboards where the branch circuit originates on its way back toward the service or source of a separately derived system where the branch circuit originates. Note also that the isolated equipment grounding conductor is not permitted to pass the building or structure disconnecting means on its way back to the service that may be in another building or structure.

See also the discussion at *250.6(D)* for limitations on alterations to prevent objectionable currents.

250.97 BONDING FOR OVER 250 VOLTS

The Requirement: For circuits of over 250 volts to ground, the electrical continuity of metal raceways and cables with metal sheaths that contain any conductor other than service conductors is required to be ensured by one or more of the methods specified for services in *250.92(B)*, except for *(1)*.

Discussion: The system to which this rule commonly applies is 480Y/277-volt, 3-phase, 4-wire. (An example is shown in Figure 2-5.) The voltage to ground for this system is 277 volts. Other system voltages that do not have to comply with the rules of this section are 208Y/120 volt, 3-phase, 4-wire; 120/240-volt, 1-phase, 3-wire; and 240/480-volt, 1-phase, 3-wire.

"Normal" bonding methods are considered adequate for most circuits. Special assured bonding is required for some systems and circuits, including the service *(250.92)*, over-250-volts-to-ground circuits *(250.97)*, and hazardous locations *(250.100)*. For these over-250-volts-to-ground circuits, the purpose for additional bonding is to eliminate the shock hazard created by the higher-voltage circuit. Equipment can be energized by a fault that is downstream from an open or a broken connector or coupling. Such a condition can result in a hazardous shock exposure to a person who may contact equipment energized at different potentials, which could be as high as 277 volts.

Exception: *If oversized, concentric, or eccentric knockouts are not encountered, or where a box or enclosure with concentric or eccentric knockouts is listed to provide a reliable electrical bonding connection, the following methods are permitted:*

1. *Threadless couplings and connectors for cables with metal sheaths*
2. *Two locknuts, on rigid metal conduit or intermediate metal conduit, one inside and one outside of boxes and cabinets*
3. *Fittings with shoulders that seat firmly against the box or cabinet, such as electrical metallic tubing connectors, flexible metal conduit connectors, and cable connectors, with one locknut on the inside of boxes and cabinets*
4. *Listed fittings*

Discussion: This exception has become the rule for these installations. Few installations of these over-250-volt branch circuits are made with service-type bonding, such as bonding bushings with bonding jumpers or bonding-type locknuts. Most installations include cable connectors or EMT connectors with a shoulder on the outside and a locknut or a snap-in connector that does not require a locknut. See Figure 5-16 for a metal box with several wiring methods connected without a bonding jumper being required.

Outlet or junction boxes are commonly available that have an eccentric-style knockout that is listed

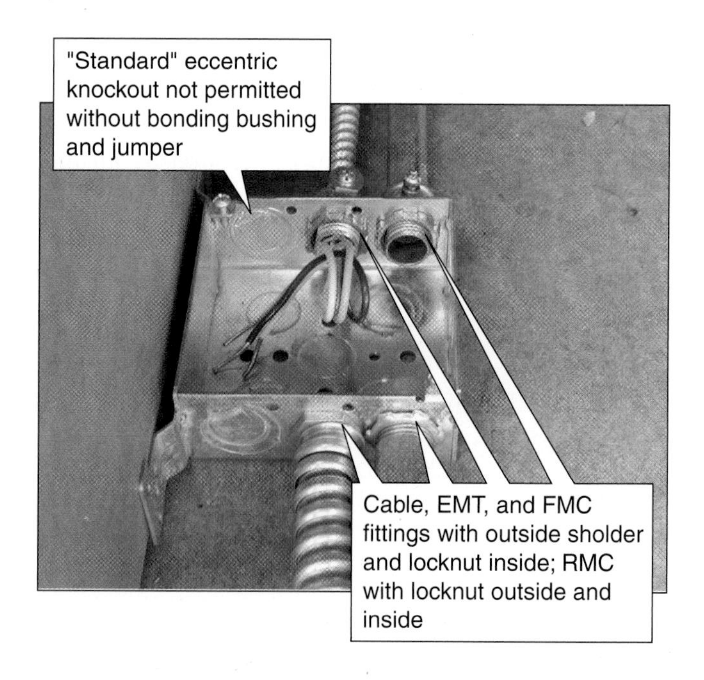

FIGURE 5-16 Bonding for over 250 volts, *250.97*. (*© Cengage Learning 2012*)

for direct connection of over-250-volt branch circuits without a bonding jumper (see Figure 5-14). (Concentric knockouts have a common center point for all knockouts. This results in one edge of the knockouts being common. Eccentric knockouts have the center point offset for each knockout.) These eccentric knockouts for the listed boxes are usually of the combination type, including both trade size ½ (metric designator 16) and trade size ¾ (metric designator 21) knockouts. The trade size ½ (metric designator 16) knockout is fairly easy to remove with normal electrical hand tools. The trade size ¾ (metric designator 21) knockout is attached to the box in a more secure manner such as by spot welding, which makes it more difficult to remove and helps ensure the connection of the trade size ½ (metric designator 16) cable or EMT fitting. This configuration may not be individually identified on the box, because the product safety standard allows the identification to be on the shipping carton. The boxes are fairly easy to identify after one becomes familiar with the construction features.

250.98 BONDING LOOSELY JOINTED METAL RACEWAYS

The Requirement: Expansion fittings and telescoping sections of metal raceways are required to be made electrically continuous by equipment bonding jumpers or other means. See Figure 5-17.

Discussion: Listed expansion fittings are available for use on rigid metal conduit (Type RMC) and intermediate metal conduit (Type IMC). Some of these fittings include an internal bonding jumper to ensure an effective ground-fault current path is maintained through the fitting. Verify with the manufacturer the bonding capability of any expansion fitting. The *Guidelines for Installing Steel Conduit/Tubing* (©2001, Conduit Section of the Steel Tube Institute of North America) contains the following information in 4.7.8:" Additional bonding considerations: Expansion fittings and telescoping sections of metal raceways shall be listed for grounding, and shall be made electrically continuous by the use of equipment bonding jumpers." Thus, even though the expansion fitting is listed for grounding, the Guideline requires a bonding jumper to ensure the low-impedance path. Verify with the manufacturer of the expansion fitting to be installed whether an external bonding jumper is required or whether the fitting provides positive bonding by design features.

Other methods of providing for expansion or for movement such as at seismic joints of buildings or structures include flexible metal conduit (Type FMC) or liquidtight flexible metal conduit (Type LFMC) or nonmetallic conduit (see Figure 5-18). A bonding jumper is required across the Type FMC conduit as provided in *250.118(5)* and across the Type LFMC conduit as required in *250.118(6)*. A bonding jumper is always required across liquidtight flexible

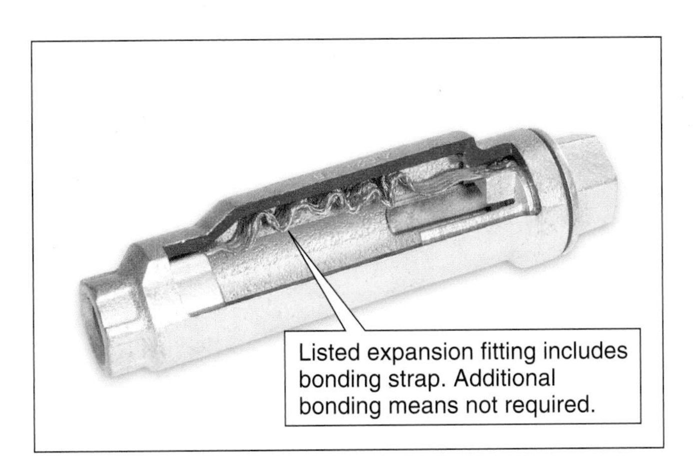

FIGURE 5-17 Bonding loosely jointed metal raceways, *250.98*. (*Courtesy of Thomas and Betts*)

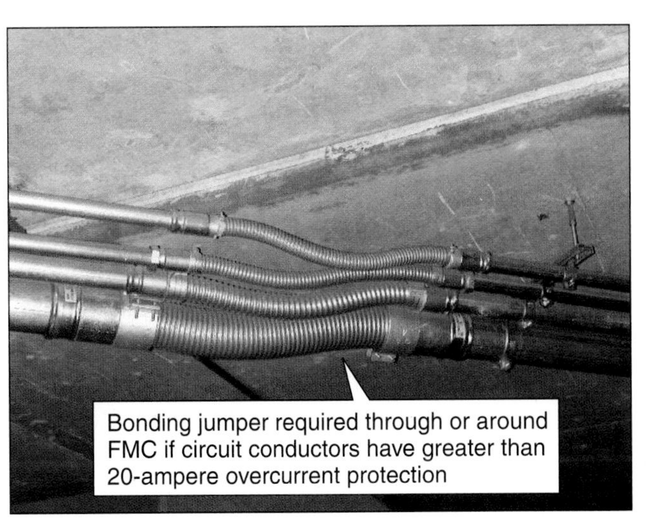

FIGURE 5-18 Bonding at seismic or expansion joints, *250.98*. (*© Cengage Learning 2012*)

nonmetallic conduit (Type LFNC). The bonding jumper is limited in length to 6 ft if it is installed outside the flexible conduit. See *250.102(E)*. The minimum size of the bonding jumper is determined from *250.102(C)* if the expansion fitting is on the supply side of the service and from *250.102(D)* if the fitting is on the load side of the service.

250.100 BONDING IN HAZARDOUS (CLASSIFIED) LOCATIONS

The Requirement: Regardless of the voltage of the electrical system, the electrical continuity of non-current-carrying metal parts of equipment, raceways, and other enclosures in any hazardous (classified) location as defined in *500.5* is required to be ensured by any of the methods specified in *250.92(B)(2)* through *(4)* that are approved for the wiring method used. One or more of these bonding methods must be used whether or not supplementary equipment grounding conductors of the wire type are installed.

Specific requirements for bonding and providing the effective ground-fault return path are contained in the following *Code* locations:

501.30 Class I locations (flammable vapors)
502.30 Class II locations (combustible dust)
503.30 Class III locations (fibers and flyings)

See Unit 8 of this book for a discussion on and illustrations of these requirements.

250.102 BONDING CONDUCTORS AND JUMPERS

250.102(A) Material

The Requirement: Equipment bonding jumpers are required to be of copper or other corrosion-resistant material. A bonding jumper is required to consist of a wire, bus, screw, or similar suitable conductor. See Figure 5-19.

Discussion: Although not specifically mentioned here, the installation rules for aluminum conductors in *250.64(A)* should be followed. Restrictions are placed on the installation of aluminum conductors to maintain the integrity of the conductor and connection.

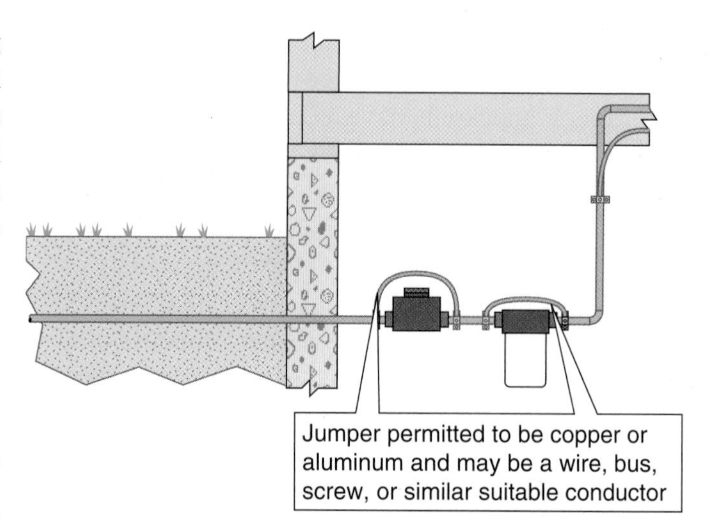

FIGURE 5-19 Equipment bonding jumpers: material, *250.102(A)*. (*© Cengage Learning 2012*)

250.102(B) Attachment

The Requirements: Bonding jumpers are required to be attached in the manner specified by the applicable provisions of *250.8* for circuits and equipment and by *250.70* for grounding electrodes.

Discussion: Permitted methods include terminal bars, exothermic welding, listed pressure connectors listed as grounding and bonding equipment, machine screw-type fasteners that engage not less then two threads or that are secured by a nut, thread-forming machine screws that engage not less than two threads in the enclosure, connections that are part of a listed assembly, and other listed means (see Figure 5-20).

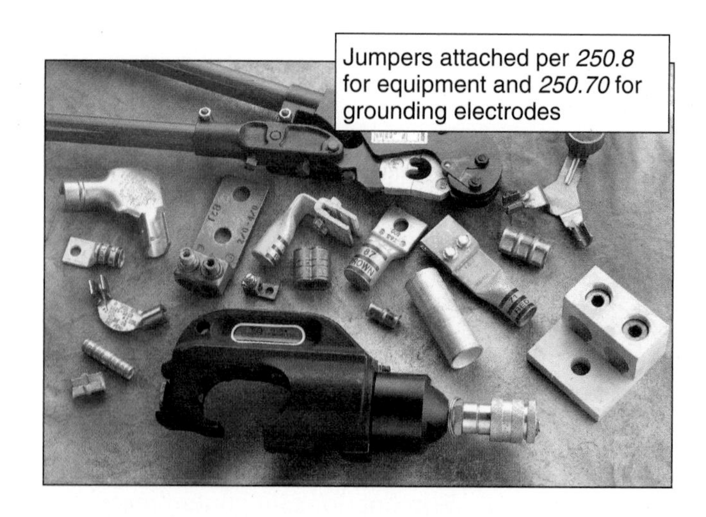

FIGURE 5-20 Equipment bonding jumpers: attachment, *250.102(B)*. (*Courtesy of Thomas and Betts*)

Connection devices or fittings that depend solely on solder are not permitted to be used. Sheet metal screws, drywall screws, and so forth are not permitted to be used to connect grounding conductors or connection devices such as lugs or equipment grounding terminal bars to enclosures.

250.102(C) Size—Supply-Side Bonding Jumper

(1) Size for Supply Conductors in a Single Raceway or Cable

The Requirements: The supply-side bonding jumper is not permitted to be smaller than the sizes shown in *Table 250.66* for grounding electrode conductors. If the ungrounded supply conductors are larger than 1100-kcmil copper or 1750-kcmil aluminum, the bonding jumper must have an area not less than 12½% of the area of the largest set of ungrounded supply conductors.

Discussion: It is important to recall that for bonding jumpers on the line or supply side of an overcurrent device, be it service equipment or for a separately derived system, the size of the bonding jumper is based on the size of the supply conductor, not on the rating of the overcurrent device (see Figure 5-21). For service-entrance conductor sizes identified in *Table 250.66,* simply refer to the table. Although the primary purpose of the table is for selecting the size of a grounding electrode conductor, this table also is used to select the minimum size of bonding conductors on the line or supply side of the service disconnecting means and for conductors on the supply side of a separately derived system.

For example, if the service-entrance or derived conductors are 500 kcmil, by reference to *Table 250.66,*

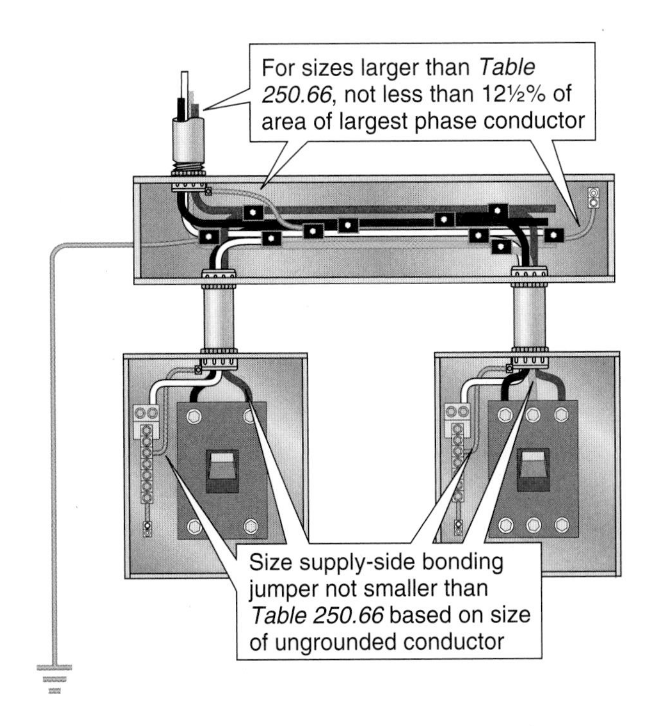

FIGURE 5-21 Supply-side bonding jumper on supply side of service, *250.102(C)*.
(*© Cengage Learning 2012*)

we find the minimum size of the supply-side bonding jumper is 1/0 AWG.

For conductors or sets of conductors larger than those shown in the table, the 12½% rule is invoked. Refer to *Table 8* of *NEC Chapter 9* for the circular mil area of conductors sized 1/0, 2/0, 3/0, and 4/0 AWG. Table 5-2 in this book identifies some common sizes of services in which sets of parallel conductors might be used, the total circular mil area of the conductors, and the minimum size of the bonding jumper. These sizing rules also apply to derived conductors for a separately derived system if it is

TABLE 5-2

Size of Supply-Side Bonding Jumpers for 800 through 2000 Ampere Services

Service Size	Service-Entrance Conductors	Total Circular Mil Area	Minimum Area of Bonding Jumper	Next Standard Size*	Parallel Conduits Are Used†
800 Amp	2 sets, 500 kcmil	1000 kcmil		2/0 *(Table 250.66)*	(2) 1/0
1000 Amp	3 sets, 400 kcmil	1200 kcmil	150,000 cm	3/0	(3) 1/0
1200 Amp	4 sets, 350 kcmil	1400 kcmil	175,000 cm	4/0	(4) 2
1600 Amp	5 sets, 350 kcmil	1750 kcmil	218,750 cm	250 kcmil	(5) 2
2000 Amp	6 sets, 400 kcmil	2400 kcmil	300,000 cm	300 kcmil	(6) 1/0

* All conductors are in one raceway such as a wireway, or a single bonding jumper is used for multiple conduits.
† Conductors are in individual raceways with an individual bonding jumper to the raceway.

determined that bonding is required because of uncertain or suspect connections of metallic raceways.

250.102(C)(2) Size for Parallel Conductor Installations

More Requirements: If the ungrounded supply conductors are paralleled in two or more raceways or cables, and an individual supply-side bonding jumper is used for bonding these raceways or cables, the size of the supply-side bonding jumper for each raceway or cable is required to be selected from *Table 250.66* based on the size of the ungrounded supply conductors in each raceway or cable. A single supply-side bonding jumper installed for bonding two or more raceways or cables shall be sized in accordance with *(C)(1)*.

Discussion: This requirement is illustrated in the foregoing table and in Figure 5-22. As can be seen, the bonding jumper for parallel conductor installations is based on the size of the conductors in the individual conduit or cable. This always results in a bonding jumper that is smaller and much more manageable to install than where a single bonding jumper is installed in a daisy-chain fashion for multiple conduits or cables.

If a single supply-side bonding jumper is to be installed to bond more than one raceway or cable, the rule requires that the size of ungrounded supply conductors be installed according to the rules for single supply-side bonding jumpers in *250.102(C)(1)*.

250.102(D) Size—Equipment Bonding Jumper on Load Side of an Overcurrent Device

The Requirement: The equipment bonding jumper on the load side of an overcurrent device is required to be sized, as a minimum, in accordance with *250.122*.

Discussion: Due to the changes in this section for the 2011 *NEC*, these rules apply to *250.122* rather than only *Table 250.122*. Section *250.122* provides that the equipment grounding conductor or bonding

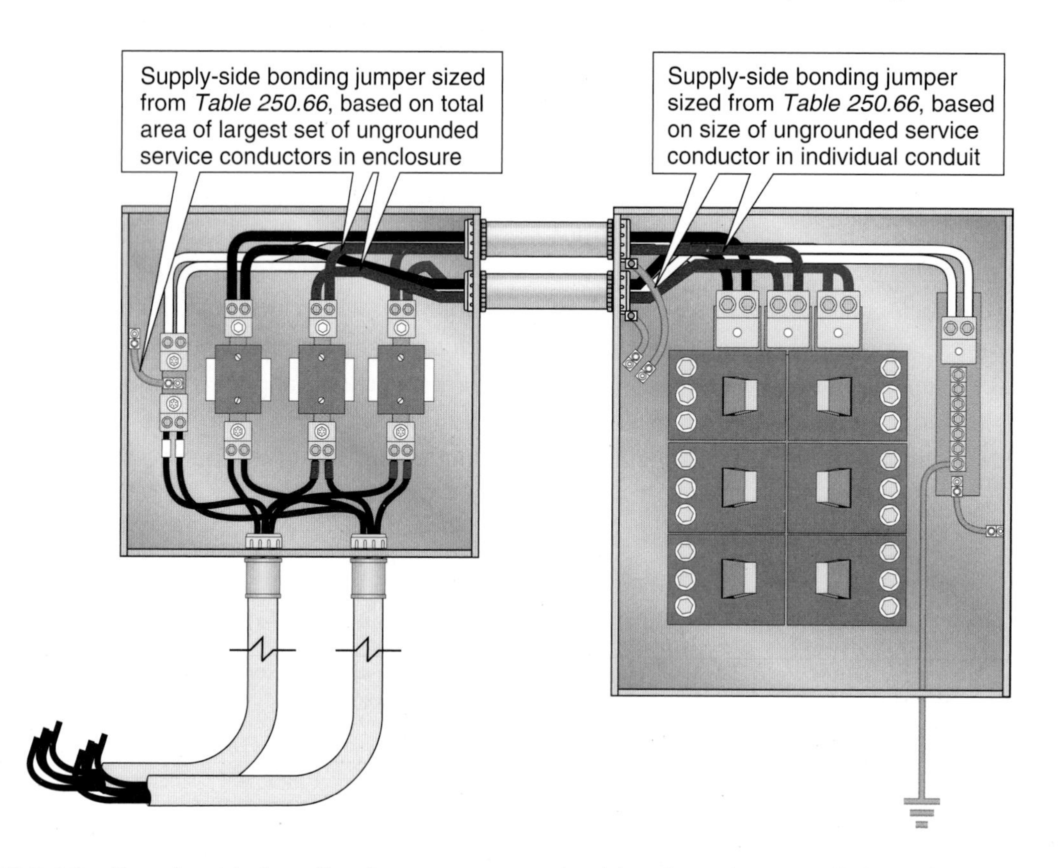

FIGURE 5-22 Supply-side bonding jumper on supply side of service: parallel conduits or cables, *250.102(C)*. (© Cengage Learning 2012)

jumper is not required to be larger than the circuit conductors. As a result, that requirement or provision was removed from this section.

Table 250.122 is used for sizing bonding jumpers on the load side of the service. This table is based on the rating of the overcurrent device on the supply side of the feeder or branch circuit. It then assigns a minimum-size copper or aluminum equipment grounding conductor for the feeder or circuit. The conductor is referred to as the equipment bonding jumper for application of the rule in *250.102(D)*.

As indicated, *250.122(A)* provides that the equipment bonding jumper is not required to be larger than the ungrounded ("hot") circuit conductors. This rule often applies when smaller motors are supplied and the branch-circuit, short-circuit, and ground-fault device is oversized to allow the motor to start and run.

Here's How: For example, a 2-horsepower (hp), 230-volt, 1-phase motor is installed. *Table 430.248* indicates that the full-load current to be used for selection of the branch-circuit conductors and overcurrent protective device is 12 amperes. The motor branch-circuit conductor must be sized not less than 125% of 12 amperes, or 15 amperes. [See *430.22(A)*.] A 14 AWG conductor is selected from *Table 310.15(B)(16)*. A circuit breaker used as the branch-circuit, short-circuit, and ground-fault device is permitted to be sized at 250% of the motor full-load current from Table *430.248*, or

$$12 \times 2.5, \text{ or } 30, \text{ amperes}$$

[*430.52(C)(1)* and *Table 430.52*]. Thus, in this example, a 14 AWG equipment bonding jumper is permitted for a circuit that would otherwise require a 10 AWG bonding jumper, according to *Table 250.122*.

More Requirements: A single common continuous equipment bonding jumper is permitted to connect two or more raceways or cables if the bonding jumper is sized in accordance with *250.122* for the largest overcurrent device supplying circuits in the raceways or cables.

Discussion: As shown in Figure 5-23, a single equipment bonding jumper is permitted to be installed to bond several raceways. If installed to comply with this section, the bonding jumper is sized from *250.122*, including *Table 250.122*, on the basis of the rating of the overcurrent device on the supply side of the conductors contained in the raceway or cable. Because the largest overcurrent device is 350 amperes, *Table 250.122* requires a 3 AWG copper or a 1 AWG aluminum equipment bonding jumper.

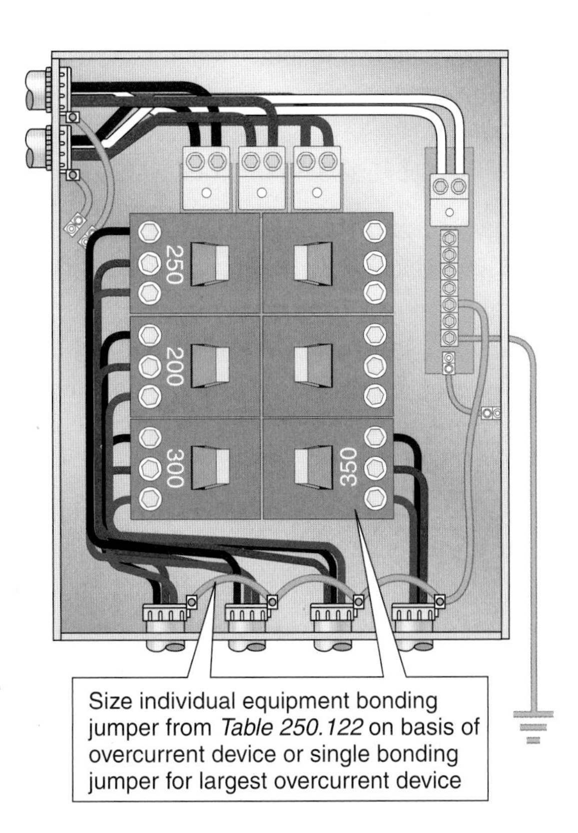

FIGURE 5-23 Sizing of equipment bonding jumper on load side of service, *250.102(D)*. (*© Cengage Learning 2012*)

The installer must consider the size of lug on the bonding bushing if this bonding method is employed. For example, if a 250-ampere and 60-ampere feeder or branch circuit are installed, *Table 250.122* requires a minimum 4 AWG bonding jumper for the conduit or cable containing the conductors protected by the 250-ampere overcurrent device and permits a bonding jumper not smaller that 10 AWG for the conduit or cable containing the conductors protected by the 60-ampere overcurrent device. It would be unusual for the bonding bushing for the smaller conduit to have a lug large enough to accept the 4 AWG bonding jumper. This may require the installation of individual bonding jumpers sized for the rating of the overcurrent device.

250.102(E) Installation

The Requirements: The equipment bonding jumper is permitted to be installed inside or outside of a raceway or enclosure.

1. **Inside a Raceway or Enclosure.** If installed inside of a raceway, the equipment bonding jumpers or conductors are required to comply with *250.119* and *250.148.*
2. **Outside a Raceway or Enclosure.** If installed on the outside, the length of the equipment bonding jumper is generally not permitted to exceed 6 ft (1.8 m) in length and must be routed with the raceway or enclosure.

Exception: *An equipment bonding jumper or supply-side bonding jumper longer than 6 ft (1.8 m) is permitted at outside pole locations for the purpose of bonding or grounding isolated sections of metal raceways or elbows installed in exposed risers of metal conduit or other metal raceway and for bonding grounding electrodes and is not required to be routed with the raceway or enclosure.*

Discussion: There is no limit to the length allowed for equipment bonding jumpers to be installed inside a raceway or enclosure. Nothing in the *Code* indicates when these conductors should be called equipment grounding conductors rather than equipment bonding jumpers. Typically, equipment bonding jumpers are shorter than equipment grounding conductors.

It is not uncommon for electrical designers to require an equipment grounding conductor to be installed in every raceway even though the raceway is metal and recognized in the *NEC* as an equipment grounding conductor. This is done to ensure an effective ground-fault return path.

The maximum length for a bonding jumper to be installed on the outside of an enclosure or raceway generally is limited to 6 ft (1.8 m). This length limitation relates to keeping the impedance of the ground-fault return path as low as practicable. See Figure 5-24 for the installation of equipment bonding conductors external to enclosures.

Installation of Equipment Bonding Jumper

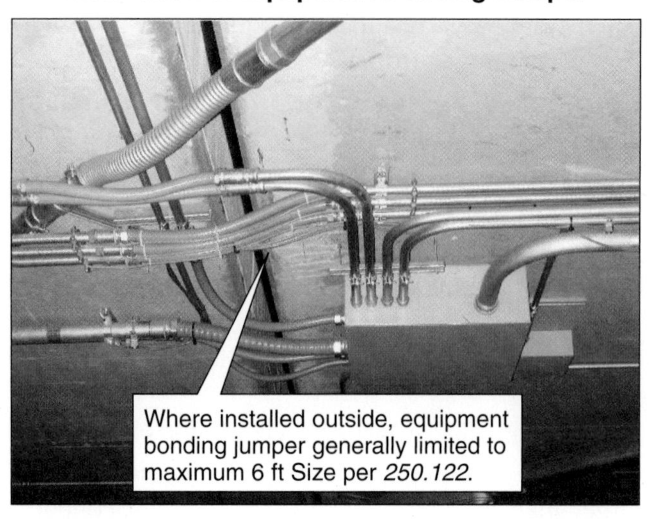

FIGURE 5-24 Installation of equipment bonding jumper, *250.102(E)*. (*© Cengage Learning 2012*)

The exception permits an equipment bonding jumper or supply-side bonding jumper of unlimited lengths if necessary for bonding raceways or elbows installed in exposed risers of metal conduit or other metal raceways on poles in outside locations. This allows the exposed metal raceway to be bonded to the system grounded (often a neutral) conductor. Equipment such as isolated enclosures or metal conduit should never be bonded solely by connection to a grounding electrode. This is simply such a high-resistance path that little current would flow, which would leave the equipment at a dangerous voltage level above ground.

The exception also permits unlimited lengths for the purpose of bonding grounding electrodes together, and these are not required to be routed with a raceway or enclosure.

250.104 BONDING OF PIPING SYSTEMS AND EXPOSED STRUCTURAL STEEL

250.104(A) Metal Water Piping

The Requirements: The metal water piping system is required to be bonded as provided in *(A)(1), (A)(2),* or *(A)(3)* of this section. The bonding jumper(s)

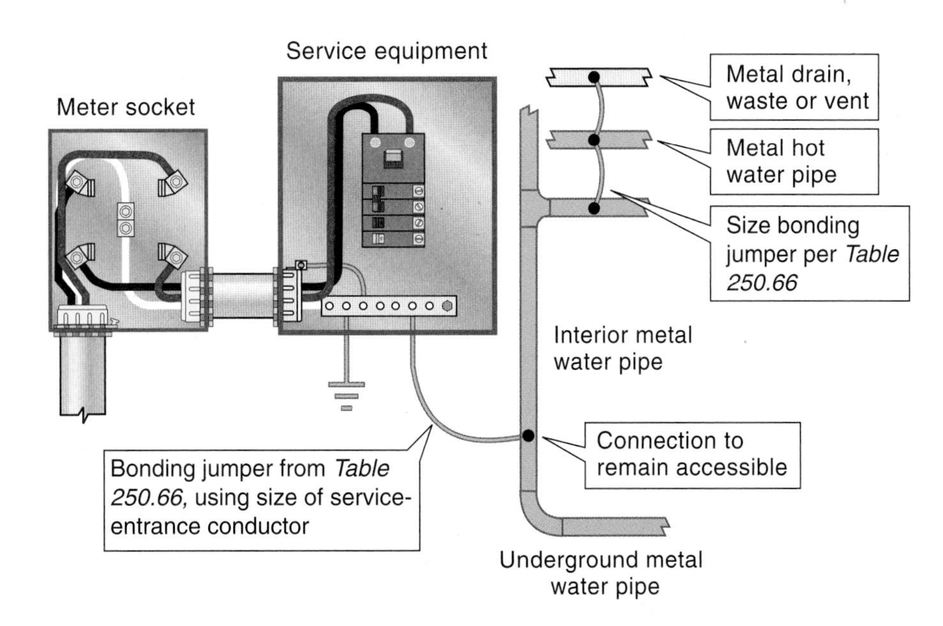

FIGURE 5-25 Bonding of metal water piping, *250.104(A)(1)*. (© Cengage Learning 2012)

must be installed in accordance with *250.64(A)*, *(B)*, and *(E)*. The points of attachment of the bonding jumper(s) must be accessible (see Figure 5-25).

250.104(A)(1) General

The Requirements: Metal water piping system(s) installed in or attached to a building or structure is (are) required to be bonded to the service equipment enclosure, the grounded conductor at the service, the grounding electrode conductor if of sufficient size, or to the one or more grounding electrodes used. The bonding jumper(s) is (are) to be sized in accordance with *Table 250.66,* except as permitted in *250.104(A)(2)* and *(A)(3)*.

Discussion: As indicated, this section requires bonding of metal water piping systems and applies if the metal water piping system is not being used as the grounding electrode as required by *250.50.* Connecting a grounding electrode conductor to a metal water piping system so it will serve as the grounding electrode also satisfies the rule for bonding the metal water piping system. Unless modified by the rules in *250.104(A)(2)* and *(A)(3),* the size of the bonding jumper is the same size as the required grounding electrode conductor. *Table 250.66* is used to select the bonding jumper in accordance with the size of the service conductor at the building or structure service equipment. The procedures previously described are used for sizing the bonding conductor. Note, however, that the bonding jumper to the metal water piping system is not required to be larger than 3/0 AWG.

A frequently asked question is whether metallic stub-outs for valves and supply lines to plumbing fixtures are required to be bonded when the piping system in the building or structure is nonmetallic. The answer is no, because the nonmetallic piping *system* in the building or structure is not conductive.

250.104(A)(2) Buildings of Multiple Occupancy

The Requirement: In buildings of multiple occupancy where the metal water piping system(s) installed in or attached to a building or structure for the individual occupancies is (are) metallically isolated from all other occupancies by use of nonmetallic water piping, the metal water piping system(s) for each occupancy is (are) permitted to be bonded to the equipment grounding terminal of the panelboard or switchboard enclosure (other than service equipment) supplying that occupancy. The bonding jumper is required to be sized in accordance with *Table 250.122,* based on the rating of the overcurrent protective device for the circuit supplying the occupancy.

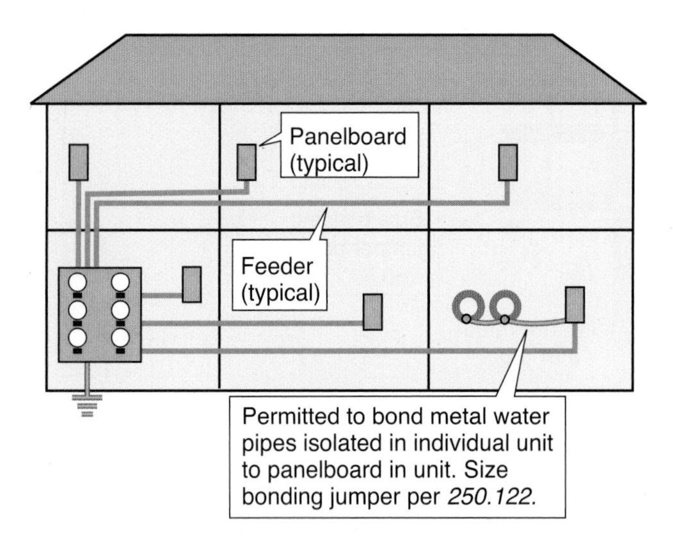

FIGURE 5-26 Bonding of metal water piping systems in buildings of multiple occupancy, *250.104(A)(2)*. (*© Cengage Learning 2012*)

Discussion: This subsection functions as an exception to the general rule for installing and sizing the bonding jumper to the metal water piping system (see Figure 5-26). As indicated, this rule applies to multiple-occupancy buildings in which each unit is isolated from the others by nonmetallic piping, but the water piping in each unit is metal and thus is required to be bonded.

When these conditions apply, the bonding jumper to the water piping system is connected at the equipment grounding terminal bar in the panelboard that supplies the individual unit, rather than to the service equipment. The size of the bonding jumper is determined from *Table 250.122* in accordance with the rating of the feeder that supplies the individual dwelling unit.

Here's How: For example, the dwelling units are supplied by a 100-ampere feeder from the service equipment. *Table 250.122* permits an 8 AWG copper or 6 AWG aluminum bonding jumper to be installed from the equipment grounding terminal bar in the panelboard to the metallic water piping system.

Discussion: The provisions in this *NEC* section do not apply if a multiple-occupancy building is supplied by a metal water piping system and the metal piping is electrically continuous to and between the individual units. If this is the situation, a single bonding connection from the service equipment to the metal water piping system is required. As previously mentioned, if the metal water piping system is used as a grounding electrode, the connection of the grounding electrode conductor to the metal water piping system serves as the bonding means required in *250.104(A)*.

250.104(A)(3) Multiple Buildings or Structures Supplied by a Feeder(s) or Branch Circuit(s)

The Requirements: The metal water piping system(s) installed in or attached to a building or structure is (are) required to be bonded to the building or structure disconnecting means enclosure if located at the building or structure, to the equipment grounding conductor run with the supply conductors, or to the one or more grounding electrodes used. The bonding jumper(s) must be sized in accordance with *250.66,* based on the size of the feeder or branch circuit conductors that supply the building. The bonding jumper is not required to be larger than the largest ungrounded feeder or branch circuit conductor supplying the building.

Discussion: Special rules are provided for bonding metal water piping systems in a building or structure supplied by a feeder or branch circuit. Here, the size of the supply conductors is used in *Table 250.66* to determine the minimum size of the bonding jumper required to be connected to the metal water piping system. The bonding jumper is permitted to be connected to the building or structure disconnecting means, to the equipment grounding conductor run with the feeder or branch circuit, or to the grounding electrode system at the building or structure. See Figure 5-27.

Here's How: Assume the overcurrent protection of the feeder supplying the building is 1200 amperes. Four sets of 350-kcmil copper conductors in parallel are installed to the building disconnecting means. Add the circular mil area of the conductors together to equal one 1400-kcmil conductor. By reference to *Table 250.66,* it is found that a 3/0 copper or 250-kcmil aluminum bonding jumper is required. In this example, the bonding jumper is the same size as if it were selected from *Table 250.122* as an equipment grounding conductor.

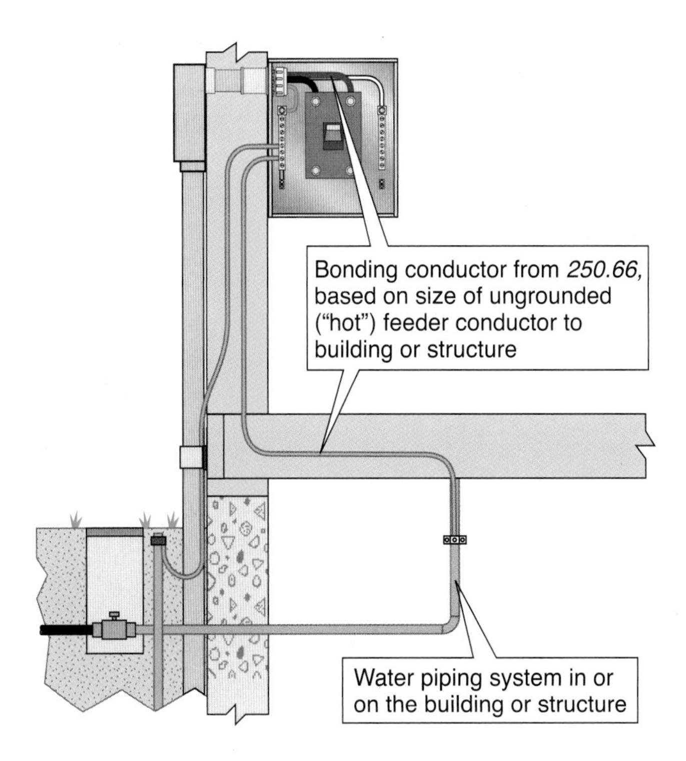

FIGURE 5-27 Bonding of metal water piping in multiple-occupancy buildings or structures supplied by a feeder or branch circuit, *250.104(A)(3)*. (© *Cengage Learning 2012*)

250.104(B) Other Metal Piping

The Requirements: If installed in or attached to a building or structure, metal piping system(s), including gas piping, that is likely to become energized is required to be bonded to the service equipment enclosure, the grounded conductor at the service, the grounding electrode conductor if of sufficient size, or to the one or more grounding electrodes used. The bonding jumper(s) is (are) required to be sized in accordance with *250.122,* using the rating of the circuit that is likely to energize the piping system(s). The equipment grounding conductor for the circuit that is likely to energize the piping is permitted to serve as the bonding means. The attachment point of the bonding jumper(s) is required to be accessible.

Discussion: This *NEC* section uses language that requires some interpretation by the installer and by the AHJ. The phrase *likely to become energized* requires exercise of professional judgment to determine when the metal piping system is required to be bonded. Often, AHJs simply require these metal piping systems to be bonded, because they are conductive and may be energized accidentally, even though there is no electrical equipment in the vicinity of the piping.

Code Panels are required to comply with the *NEC Style Manual* in the use of terms and are required to avoid the use of terms that are vague and unenforceable. *Likely* is identified as one of those terms. Yet, the Style Manual explains that the term *likely to become energized* means "failure of insulation on." Because metal piping systems and metal structural building frame members are usually not insulated, the metal water piping system and structural building frame members should be bonded for safety.

As indicated, the equipment grounding conductor for the circuit that supplies the equipment may be used as the conductor to bond the metallic piping. Care should be taken to protect smaller bonding conductors from physical damage if they are installed external to equipment.

> **Informational Note No. 1:** This note states that bonding all piping and metal air ducts within the premises will provide additional safety. However, Article 250 does not require that other piping or metal air duct systems be bonded. See Figure 5-28.

> **Informational Note No. 2:** Additional information for gas piping systems can be found in Section 7.13 of the National Fuel Gas Code, NFPA 54-2009.

FIGURE 5-28 Bonding other metal piping, *250.104(B)*. (© *Cengage Learning 2012*)

This reference is to the rule in the *National Fuel Gas Code,* where you will find a requirement that corrugated stainless steel tubing (CSST) is required to be bonded near the point of entry into the building with a conductor not smaller than 6 AWG copper.

250.104(C) Structural Metal

The Requirements: Exposed structural metal that is interconnected to form a metal building frame and is not intentionally grounded and is likely to become energized is required to be bonded. The bonding connection can be to the service equipment enclosure, the grounded conductor at the service, the grounding electrode conductor where of sufficient size, or the one or more grounding electrodes used. The bonding jumper(s) must be sized in accordance with *Table 250.66* and installed in accordance with *250.64(A), (B),* and *(E).* The point of attachment of the bonding jumper(s) is required to be accessible.

Discussion: This *NEC* section also uses the phrase *likely to become energized,* so the installer and the AHJ must decide whether the exposed structural metal is required to be bonded. Because the structural metal is conductive and could become a shock hazard if accidentally energized, the metal is generally required to be bonded if the structure is supplied by an electrical system. See Figure 5-29 and the discussion on "likely to become energized" in *250.104(B).*

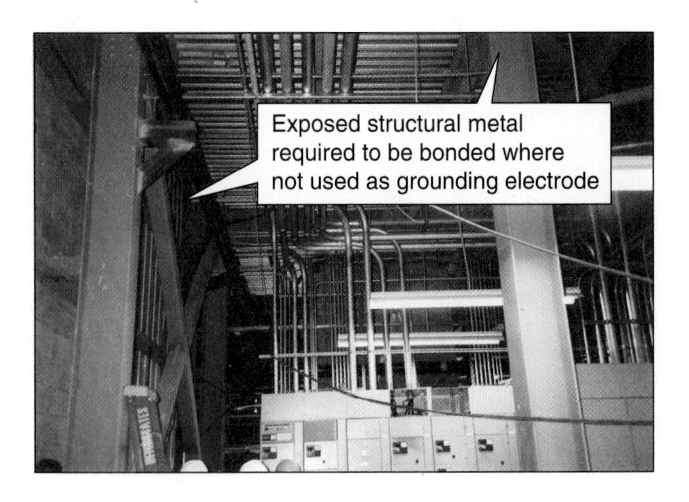

FIGURE 5-29 Bonding of exposed structural metal, *250.104(C).* (*© Cengage Learning 2012*)

The bonding jumper is sized from *Table 250.66* just as if the exposed structural steel is being used as a grounding electrode for the grounding electrode system. Likewise, the point of attachment of the bonding jumper to the structural metal is required to be accessible.

250.104(D) Separately Derived Systems

The Requirements: Metal water piping systems and structural metal members that are interconnected to form a building frame are required to be bonded to separately derived systems in accordance with *(D)(1)* through *(D)(3).* See Figure 5-30.

250.104(D)(1) Metal Water Piping System(s)

The Requirements: The grounded conductor of each separately derived system is generally required to be bonded to the nearest available point of the metal water piping system(s) in the area served by each separately derived system. This connection is required to be made at the same point on the separately derived system where the grounding electrode conductor is connected. Each bonding jumper must be sized in accordance with *Table 250.66,* based on the largest ungrounded conductor of the separately derived system.

Exception No. 1: *A separate bonding jumper to the metal water piping system is not required where the metal water piping system is used as the grounding electrode for the separately derived system and the water piping system is in the area served by the separately derived system.*

Exception No. 2: *A separate water piping bonding jumper is not required where the metal frame of a building or structure is used as the grounding electrode for a separately derived system and is bonded to the metal water piping in the area served by the separately derived system.*

Discussion: These requirements ensure that the metal water piping is bonded to the separately derived system in the area served if the metal water piping and metal frame of a building or structure are not used as a grounding electrode for the separately derived system. As indicated in the exceptions, a bonding jumper is not required to be run to the metal

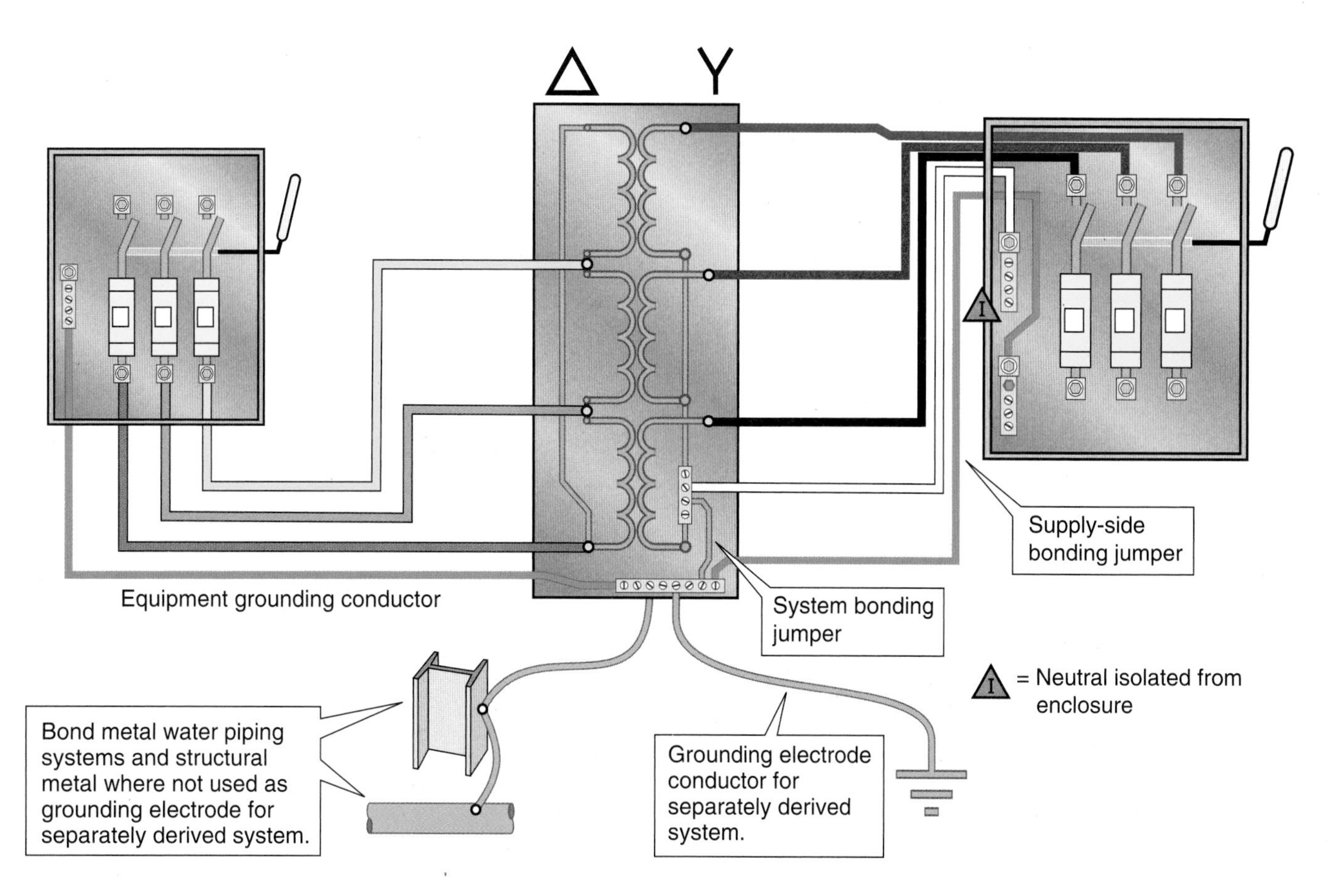

FIGURE 5-30 Bonding metal water piping systems and structural metal, *250.104(D)*. (*© Cengage Learning 2012*)

water piping system or to the structural metal system if it is used as a grounding electrode for the separately derived system, as this would be an unnecessary duplication.

Here's How: Assume metal water pipes and structural metal are in the area served from a separately derived system. The metal water pipes and structural metal do not qualify as a grounding electrode, so they are not connected to the separately derived system for that purpose. *Sections 250.104(C)(1)* and *(2)* require the metal water pipes and structural metal to be bonded with a bonding conductor sized from *Table 250.66.* If the derived conductors from the separately derived system are 250 kcmil, *Table 250.66* requires a 2 AWG copper or 1/0 AWG aluminum bonding conductor to be connected to the metal water pipes and structural metal. A single bonding jumper can be connected to either the water pipe or metal structural member and another jumper installed from the first metal object to the other one.

250.104(D)(2) Structural Metal

The Requirement: Where exposed structural metal that is interconnected to form the building frame exists in the area served by the separately derived system, it is required to be bonded to the grounded conductor of each separately derived system. This connection must be made at the same point on the separately derived system where the grounding electrode conductor is connected. Each bonding jumper is required to be sized in accordance with *Table 250.66,* based on the largest ungrounded conductor of the separately derived system.

Exception No. 1: *A separate bonding jumper to the building structural metal is not required where the metal frame of a building or structure is used as the grounding electrode for the separately derived system.*

Exception No. 2: *A separate bonding jumper to the building structural metal is not required where* *the water piping of a building or structure is used as the grounding electrode for a separately derived system and is bonded to the building structural metal in the area served by the separately derived system.*

Discussion: This is a parallel requirement to that for metal water piping systems and simply restates the requirements in reverse order from that for rules that apply to bonding of metal water piping systems. These requirements ensure that the building structural metal is bonded to the separately derived system in the area served if the building structural metal is not used as a grounding electrode for the separately derived system. As indicated in the exceptions, a bonding jumper is not required to be run to the building structural metal or to the metal water piping system if it is used as a grounding electrode for the separately derived system.

250.104(D)(3) Common Grounding Electrode Conductor

The Requirement: If a common grounding electrode conductor is installed for multiple separately derived systems as permitted by *250.30(A)(6),* and exposed structural metal that is interconnected to form the building frame or interior metal piping exists in the area served by the separately derived system, the metal piping and the structural metal member are required to be bonded to the common grounding electrode conductor.

Exception: *A separate bonding jumper from each derived system to metal water piping and to structural metal members is not required where the metal water piping and the structural metal members in the area served by the separately derived system are bonded to the common grounding electrode conductor.*

Discussion: This requirement incorporates the concept of bonding metal water pipes and metal structural members to the *common grounding electrode* for separately derived systems (see Figure 5-31). Common grounding electrode conductors are permitted to be installed when there are multiple separately derived systems in the building or structure and the water pipe and exposed structural metal in the vicinity of the separately derived system are not

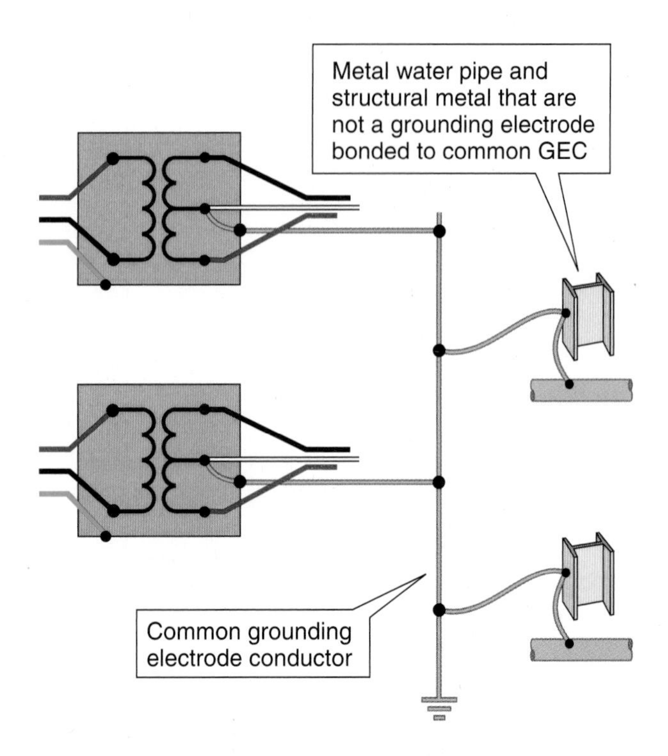

FIGURE 5-31 Bonding metal water piping and structural metal to common grounding electrode conductor, *250.104(D)(3).* (*© Cengage Learning 2012*)

considered a grounding electrode for the separately derived system. The common grounding electrode and taps outlined in *250.30(A)(6)* are permitted to be installed as an option to running multiple grounding electrode conductors from each separately derived system to the grounding electrode system for the building or structure.

250.106 LIGHTNING PROTECTION SYSTEMS

The Requirements: The lightning protection system ground terminals (grounding electrodes) are required to be bonded to the building or structure grounding electrode system.

Discussion: As discussed in *250.60* in this book, the standard for lightning protection systems uses the term *ground terminals* to equate to the commonly used term *grounding electrode* in the *NEC*. This *NEC* section simply requires the ground terminal and grounding electrode system to be bonded together. Once bonded together, they become a component of

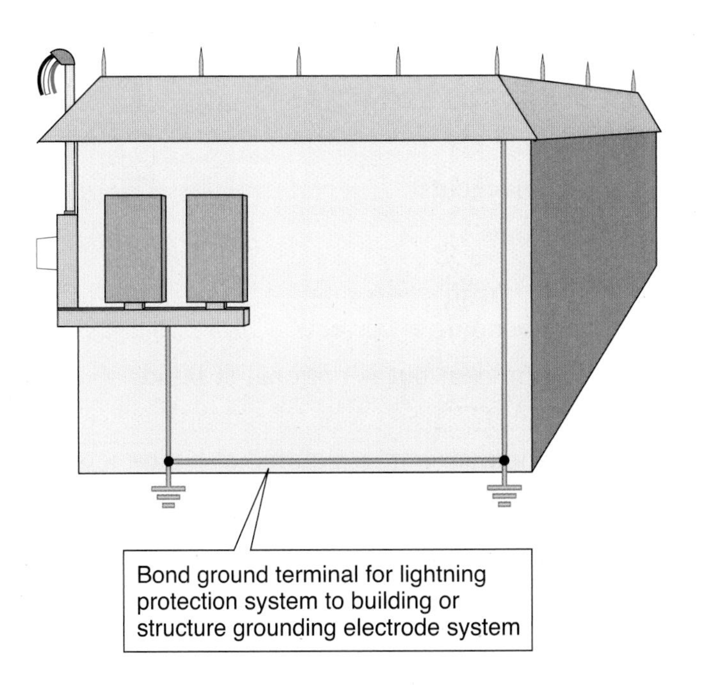

FIGURE 5-32 Bonding lightning protection systems, *250.106*. (*© Cengage Learning 2012*)

the grounding electrode system required by 250.50. See Figure 5-32.

Informational Note No. 1: See 250.60 for use of strike termination devices. For further information, see NFPA 780-2011, Standard for the Installation of Lightning Protection Systems, which contains detailed information on grounding, bonding, and sideflash distance from lightning protection systems.

Informational Note No. 2: Metal raceways, enclosures, frames, and other non-current-carrying metal parts of electric equipment installed on a building equipped with a lightning protection system may require bonding or spacing from the lightning protection conductors in accordance with *NFPA 780-2011, Standard for the Installation of Lightning Protection Systems.*

REVIEW QUESTIONS

1. Which of the following statements best describes bonding of electrical equipment?
   a. It completes the path for fault current to flow.
   b. It connects enclosures together and thus reduces any shock hazard.
   c. It extends the equipment grounding connection throughout the installation.
   d. All of these statements are essential descriptions of bonding.
2. Which of the following enclosures for service conductors are required to be bonded?
   a. Service raceways, cable trays, and auxiliary gutters
   b. Meter fittings, boxes, or similar enclosures
   c. Any metallic raceway or armor enclosing a grounding electrode conductor
   d. All of the above
3. Which of the following statements best describes the reason the *Code* specifically requires bonding of conductive enclosures for service conductors?
   a. Equipment on the line or supply side of the service disconnecting means is not provided with overcurrent protection.
   b. An effort is being made to keep all equipment at the same potential to reduce shock hazard.
   c. Such bonding will allow overcurrent protection provided by the utility company on the secondary of its transformer to clear.
   d. Such bonding will facilitate the operation of the main circuit breaker in the service equipment.

4. Which of the following methods is *not* permitted for bonding metallic enclosures for service conductors?
   a. Bonding-type bushings with a bonding jumper
   b. Bonding-type locknuts if there are no concentric knockouts
   c. Standard locknuts if there are no concentric knockouts
   d. Bonding wedges where there are no concentric knockouts
5. Which of the following statements about intersystem bonding is correct?
   a. Intersystem bonding is required at the service equipment but is optional at buildings supplied by a feeder.
   b. Intersystem bonding is optional at the service equipment but is required at buildings supplied by a feeder.
   c. Intersystem bonding is required at the service equipment and at buildings supplied by a feeder as well.
   d. Intersystem bonding is required at the main distribution for mobile home parks but is not required at the disconnecting means at the mobile home.
6. Which of the following is *not* recognized for use for intersystem bonding for existing installations?
   a. Exposed flexible metallic conduit
   b. Exposed grounding electrode conductor
   c. Connection to grounded service equipment
   d. A 6 AWG copper conductor tail
7. Metal raceways, cable trays, cable armor, cable sheath, enclosures, frames, fittings, and other metal non-current-carrying parts that are to serve as grounding conductors are *not* required to be bonded if a supplemental equipment grounding conductor is installed inside for grounding equipment.
   a. True
   b. False
8. Which of the following statements about isolated equipment grounding conductors is *not* true?
   a. The equipment is permitted to be supplied by either a branch circuit or a feeder.
   b. The isolation is permitted to be accomplished by one or more listed nonmetallic raceway fittings.
   c. The electrical isolation is permitted from a raceway containing conductors supplying only an individual equipment.
   d. The raceway must be supplemented by an internal insulated equipment grounding conductor.
9. Bonding of enclosures and raceways is generally required for circuits from which of the following systems?
   a. 120/240-volt, 1-phase, 3-wire
   b. 208Y/120-volt, 3-phase, 4-wire
   c. 120/240-volt, 3-phase, 4-wire
   d. 480Y/277-volt, 3-phase, 4-wire

10. Which of the following is *not* permitted for the bonding required by *250.97?*
    a. Threadless couplings and connectors for cables with metal sheaths
    b. Two locknuts, on rigid metal conduit or intermediate metal conduit, one inside and one outside of boxes and cabinets
    c. Fittings with shoulders that seat firmly against the box or cabinet, with one locknut on the inside of boxes and cabinets
    d. One locknut outside the enclosure and one metal bushing inside the enclosure for rigid metal conduit or intermediate metal conduit
11. Which of the following statements is correct about sizing the supply-side bonding jumper on the line or supply side of the service equipment?
    a. It is sized from *Table 250.66* in accordance with the size or rating of the overcurrent protective device in the service equipment.
    b. It is sized from *Table 250.122* in accordance with the size rating of the overcurrent protective device in the service equipment.
    c. It is sized from *Table 250.66* in accordance with the size of the service-entrance conductors on the supply side of the service.
    d. It is sized from *Table 250.122* in accordance with the size of the service-entrance conductors on the supply side of the service.
12. Which of the following statements is correct about sizing the bonding jumper on the load side of the service equipment?
    a. It is sized from *Table 250.66* in accordance with the rating of the overcurrent protective device in the service equipment.
    b. It is sized from *Table 250.122* in accordance with the rating of the overcurrent protective device in the service equipment.
    c. It is sized from *Table 250.66* in accordance with the size of the service-entrance conductors on the supply side of the service.
    d. It is sized from *Table 250.122* in accordance with the size of the service-entrance conductors on the supply side of the service.
13. Which of the following statements is correct about sizing the bonding jumper to water piping in the building where the service equipment is located?
    a. It is sized from *Table 250.66* in accordance with the rating of the overcurrent protective device in the service equipment.
    b. It is sized from *Table 250.122* in accordance with the rating of the overcurrent protective device in the service equipment.
    c. It is sized from *Table 250.66* in accordance with the size of the service-entrance conductors on the supply side of the service.
    d. It is sized from *Table 250.122* in accordance with the size of the service-entrance conductors on the supply side of the service.
14. Which of the following statements is correct about sizing the bonding jumper to water piping in the building supplied by a feeder from the service equipment?
    a. It is sized from *Table 250.66* in accordance with the rating of the overcurrent protective device for the feeder.

b. It is sized from *Table 250.122* in accordance with the rating of the overcurrent protective device for the feeder.

c. It is sized from *Table 250.66* in accordance with the size of the feeder conductors to the building.

d. It is sized from *Table 250.122* in accordance with the size of the feeder conductors to the building.

15. Which of the following statements is correct about sizing the bonding jumper to gas piping in the building where the service equipment is located?

    a. It is sized from *Table 250.66* in accordance with the rating of the overcurrent protective device for the circuit that might energize the piping.

    b. It is sized from *Table 250.122* in accordance with the rating of the overcurrent protective device for the circuit that might energize the piping.

    c. It is sized from *Table 250.66* in accordance with the size of the circuit conductors that might energize the piping.

    d. It is sized from *Table 250.122* in accordance with the size of the circuit conductors that might energize the piping.

16. A 1¼-in. bonding-type locknut is permitted to be used for bonding a metal conduit nipple used at a service with which of the following combination knockouts?

    a. ¾, 1, 1¼, 1½, 2 in.

    b. ¾, 1, 1¼, 1½ in.

    c. ¾, 1, 1¼ in.

    d. All of the above

17. The intersystem bonding means is required to provide not less than ____ terminals.

    a. 2

    b. 3

    c. 4

    d. 6

18. Equipment bonding jumpers are permitted to consist of

    a. wire.

    b. bus.

    c. screw.

    d. any of the above

19. Which of the following is the minimum-size bonding jumper to metal water piping systems in a multi-occupancy building in which the water piping of each unit is isolated from the others? The 400-A building service is supplied by 500-kcmil service-entrance conductors and each occupancy is supplied by a 100-ampere feeder.

    a. 3/0 AWG copper

    b. 1/0 AWG copper

    c. 3 AWG copper

    d. 8 AWG copper

20. The minimum size of the bonding jumper from the building disconnecting means to the metal water pipe if the building is supplied by a 600-ampere feeder consisting of two 350-kcmil copper conductors in parallel is

    a. 3/0 AWG copper.

    b. 2/0 AWG copper.

    c. 1/0 AWG copper.

    d. 1 AWG copper

21. Structural metal in the area served by a separately derived system, but is not used as a grounding electrode for the separately derived system, is required to be bonded with a conductor sized in accordance with *Table 250.66.*

    a. True

    b. False

22. The grounding electrode (earth terminal) for a lightning protection system is required to be completely separate from the grounding electrode system for the electrical service and the electrodes are not permitted to be bonded together.

    a. True

    b. False

UNIT 6

Equipment Grounding and Equipment Grounding Conductors

OBJECTIVES

After studying this unit, the reader will understand

- equipment fastened in place or connected by permanent wiring methods that is required to be connected to an equipment grounding conductor.
- types of acceptable equipment grounding conductors.
- identification requirements for equipment grounding conductors.
- installation requirements for equipment grounding conductors.
- the minimum size of equipment grounding conductors.
- rules for continuity of equipment grounding conductors.
- means of identification of wiring device terminals.

250.110 EQUIPMENT FASTENED IN PLACE (FIXED) OR CONNECTED BY PERMANENT WIRING METHODS

The Requirements: Exposed normally non-current-carrying metal parts of fixed equipment supplied by or enclosing conductors or components that are likely to become energized are required to be connected to an equipment grounding conductor under any of the following conditions:

1. Where within 8 ft (2.5 m) vertically or 5 ft (1.5 m) horizontally of ground or grounded metal objects and subject to contact by persons
2. Where located in a wet or damp location and not isolated
3. Where in electrical contact with metal
4. Where in a hazardous (classified) location as covered by *Articles 500* through *517*
5. Where supplied by a wiring method that provides an equipment ground, except as permitted by *250.86, Exception No. 2,* for short sections of metal enclosures
6. Where equipment operates with any terminal at over 150 volts to ground

Exception No. 1: *If exempted by special permission, metal frames of electrically heated appliances that have the frames permanently and effectively insulated from ground shall not be required to be grounded.*

Exception No. 2: *Distribution apparatus, such as transformer and capacitor cases, mounted on wooden poles at a height exceeding 8 ft (2.5 m) above ground or grade level are not required to be grounded.*

Exception No. 3: *Listed equipment is not required to be connected to an equipment grounding conductor if protected by a system of double insulation or its equivalent. Where such a system is employed, the equipment is required to be distinctively marked.*

Discussion: It is obvious that the exposed *non-current-carrying metal parts* are capable of carrying current if accidentally energized; therefore, they are required to be grounded or connected to an equipment grounding conductor so as to provide the effective ground-fault current path required in *250.4(A)* and *250.4(B)*. The phrase "likely to become energized" is a subjective one that requires interpretation by the installer and the AHJ. See the discussion of this phrase at *250.104(B)* in this book. Most often, this section is interpreted to require that equipment be connected to an equipment grounding conductor if any of the conditions in the section apply for equipment located in a building or structure that is supplied by electricity.

250.112 SPECIFIC EQUIPMENT FASTENED IN PLACE (FIXED) OR CONNECTED BY PERMANENT WIRING METHODS

The Requirements: Except as permitted in *250.112(F)* and *(I)*, exposed, normally non-current-carrying metal parts of equipment described in *250.112(A)* through *(K)*, and normally non-current-carrying metal parts of equipment and enclosures described in *250.112(L)* and *(M)*, are required to be connected to an equipment grounding conductor, regardless of voltage.

(A) **Switchboard Frames and Structures.** Switchboard frames and structures supporting switching equipment, except frames of 2-wire dc switchboards where effectively insulated from ground.

(B) **Pipe Organs.** Generator and motor frames in an electrically operated pipe organ, unless effectively insulated from ground and the motor driving it.

(C) **Motor Frames.** Motor frames, as provided by *430.242.*

Discussion: Specific requirements for grounding motors and related equipment can be found at *Article 430, Part XIII*. Although the general requirements for equipment grounding in *Article 250* apply to motors and associated equipment, specific requirements are in *Article 430*.

More Requirements:

(D) **Enclosures for Motor Controllers.** Enclosures for motor controllers unless attached to ungrounded portable equipment.

(E) **Elevators and Cranes.** Electric equipment for elevators and cranes.

(F) **Garages, Theaters, and Motion Picture Studios.** Electric equipment in commercial garages, theaters, and motion picture studios, except pendant lampholders supplied by circuits not over 150 volts to ground.

(G) **Electric Signs.** Electric signs, outline lighting, and associated equipment as provided in *600.7.*

Discussion: Specific requirements can be found at *600.7* for grounding of signs and related equipment. These rules amend or supplement the general requirements in *Article 250.*

More Requirements:

(H) **Motion Picture Projection Equipment.** Motion picture projection equipment.

(I) **Power-Limited Remote-Control, Signaling, and Fire Alarm Circuits.** Equipment supplied by Class 1 circuits is required to be grounded unless operating at less than 50 volts. Equipment supplied by Class 1 power-limited circuits, Class 2, and Class 3 remote-control and signaling circuits, and by fire alarm circuits, is required to be grounded if the system is required to be grounded by *Part II* or *Part VIII* of this article.

(J) **Luminaires.** Luminaires are required to be grounded as provided in *Part V* of *Article 410.*

Discussion: *Part V* of *Article 410* contains the specific requirements for grounding luminaires or connecting luminaires to an equipment grounding conductor. *Section 410.40* contains the general rules that luminaires and lighting equipment are required to be grounded as required in *Article 250* and *Part V* of *Article 410.* This amounts to a circular reference, as *Article 250* points to *Part V* of *Article 410* for grounding of luminaires in *250.112(J).*

Section 410.42 generally requires luminaires with exposed conductive parts be connected to an equipment grounding conductor or be insulated from the equipment grounding conductor and other conducting surfaces or be inaccessible to unqualified personnel.

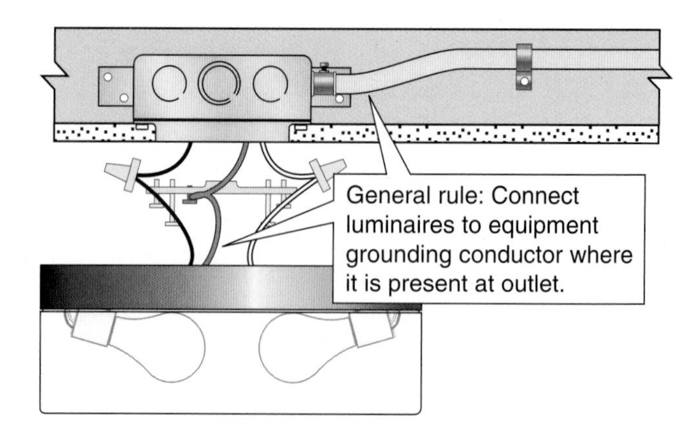

FIGURE 6-1 Grounding luminaires, *250.112(J).* (*© Cengage Learning 2012*)

Section 410.44 contains requirements for the methods of grounding and states luminaires and equipment are to be mechanically connected to an equipment grounding conductor as specified in *250.118* and sized in accordance with *250.122* (if of the wire type) (see Figure 6-1). *Section 250.118* contains the list of conductors or wiring methods permitted to be used as an equipment grounding conductor. This section also imposes limitations on the use of some wiring methods such as flexible metal conduit and liquidtight flexible metal conduit.

Section 410.44, Exception No. 1, provides that luminaires made of insulating materials that are directly wired or attached to outlets supplied by a wiring method that does not provide a ready means for grounding are not required to be connected to an equipment grounding conductor. These rules impose severe restrictions on the design or style of luminaires that are permitted to be installed without connection to an equipment grounding conductor.

Section 410.44, Exception No. 2, provides that replacement luminaires are permitted to be grounded by an equipment grounding conductor installed in accordance with *250.130(C)* (see Figure 6-2). That section permits an equipment grounding conductor to be installed separately from the branch-circuit conductors under specific conditions. See the discussion on *250.130(C)* for these provisions.

Exception No. 3 to 410.44 allows replacement luminaires that are GFCI protected to be installed without connection to an equipment grounding conductor if an equipment grounding conductor does not exist in the outlet where the fixture is connected

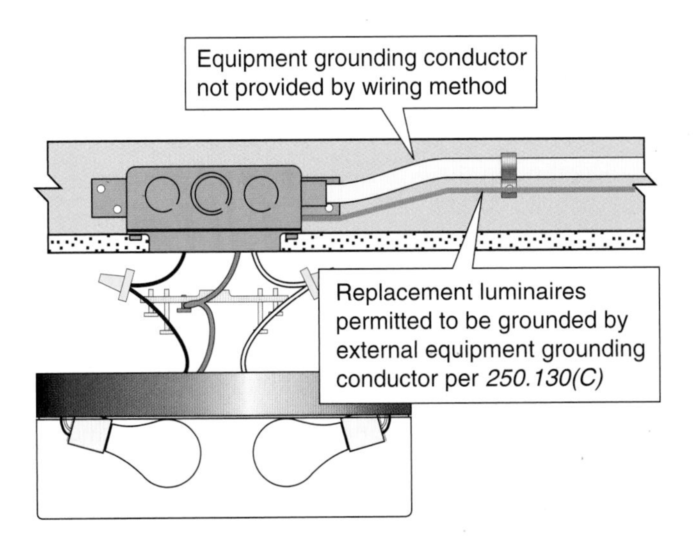

FIGURE 6-2 Replacement luminaires grounded with equipment grounding conductor external to wiring method, *410.44, Exception No. 2.* (*© Cengage Learning 2012*)

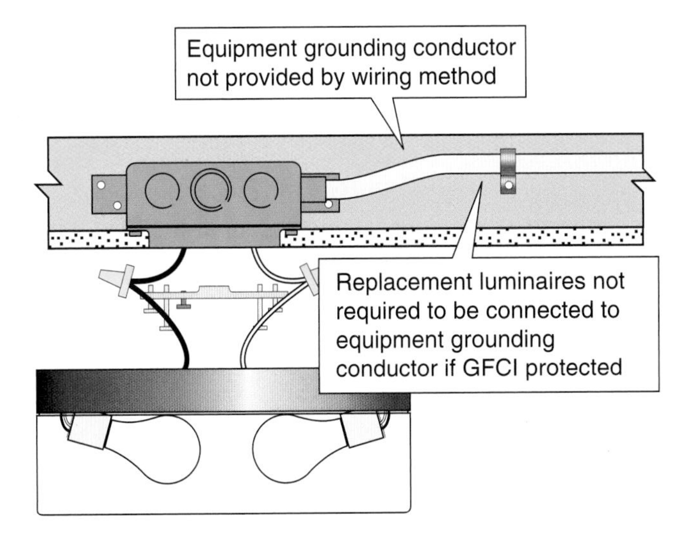

FIGURE 6-3 Replacement luminaires protected by GFCI device, *410.44, Exception No. 3.* (*© Cengage Learning 2012*)

(see Figure 6-3). However, the requirements of *110.3(B)* must be factored into the application of this exception. *Section 110.3(B)* requires that listed equipment be installed and used in accordance with any instructions included in the listing. Manufacturers of listed luminaires often provide installation instructions with their product that must be followed. These instructions often include a requirement that the luminaire be grounded or connected to an equipment grounding conductor. To ignore these requirements and protect the luminaire with a GFCI device would be a violation of the rules in *110.3(B).*

More Requirements:

(K) **Skid-Mounted Equipment.** Permanently mounted electrical equipment and skids are required to be connected to an equipment bonding jumper sized as required by *250.122.*

(L) **Motor-Operated Water Pumps.** Motor-operated water pumps, including the submersible type.

(M) **Metal Well Casings.** Where a submersible pump is used in a metal well casing, the well casing is required to be connected to the pump circuit equipment grounding conductor.

Discussion: As shown in Figure 6-4, motor-operated water pumps of the type typically mounted at the top of a well casing or a submersible type are required to be connected to a properly sized equipment grounding conductor. See the discussion on sizing equipment grounding conductors for motor circuits at *250.102(D)* and *250.122(D)* in this book.

In addition, a metal well casing for submersible water pumps is required to be connected to the equipment grounding conductor for the submersible pump.

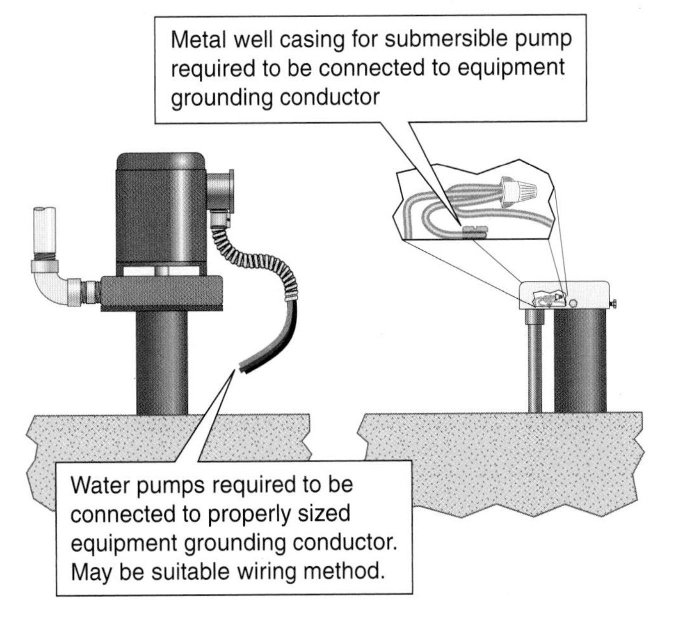

FIGURE 6-4 Pumps and metal well casings, *250.112(L)* and *(M).* (*© Cengage Learning 2012*)

The connectors should be suitable for a wet location if a twist-on wire connector is used for connections inside the well casing, due to the damp environment.

250.118 TYPES OF EQUIPMENT GROUNDING CONDUCTORS

The Requirements: The equipment grounding conductor run with or enclosing the circuit conductors is required to be one or more or a combination of the following:

1. A copper, aluminum, or copper-clad aluminum conductor. This conductor is permitted to be solid or stranded; insulated, covered, or bare; and in the form of a wire or a busbar of any shape.

Discussion: This section provides the general requirements for wire and busbar equipment grounding conductors. The minimum size of equipment grounding conductors is given in *Table 250.122* and applies generally. Some specific requirements are contained in other articles such as in *517.13* for patient care areas of health care facilities and in *Article 680* for swimming pools, hot tubs, and spas. Aluminum equipment grounding conductors have restrictions on their installation due to concern for corrosion of the conductor. See *250.120* for these requirements.

Copper-clad aluminum was produced for a few years in the 1970s in branch-circuit sizes. It consists of an aluminum conductor core with a copper outer sleeve. It was intended to be an alternative to copper that was less expensive and lighter weight, and at the same time be suitable for connections to wiring devices that are intended for copper conductors.

Even though this section permits the equipment grounding conductor to be bare or covered, some *NEC* sections require the equipment grounding conductor to be insulated. The following *NEC* sections require an insulated equipment grounding conductor for the purpose indicated:

- *250.96(B),* for isolated grounding circuits
- *250.146(D),* for isolated ground receptacles
- *517.13(B),* in patient care areas of health care facilities
- *518.4(A),* if the wiring method used in fire-rated construction for assembly occupancies does not itself qualify as an equipment grounding conductor in *250.118*
- *520.5(A),* in Type AC cable used as the wiring method in theaters and similar occupancies
- *530.11,* in Type AC cable used as wiring method in motion picture and television studios and similar locations
- *547.5(F),* direct-buried equipment grounding conductors for agricultural buildings
- *550.33(A),* the equipment grounding conductor in a feeder for a mobile or manufactured home
- *553.8(C),* in the feeder to floating buildings
- *555.15(B),* branch-circuit wiring for marinas and boatyards
- *555.15(E),* feeder for marinas and boatyards
- *605.8(A),* in flexible cord to freestanding office partitions
- *Article 680,* Swimming Pools, Fountains, and Similar Installations, including several sections with requirements for an insulated equipment grounding conductor
- *682.31(A), (B),* and *(C),* for feeders and branch circuits for electrical equipment associated with Natural and Artificially Made Bodies of Water.

More Requirements:

2. Rigid metal conduit

Note: See the *UL Guide Information* on FHIT systems for equipment grounding conductors installed in a raceway that are part of an electrical circuit protective system or a fire rated cable listed to maintain circuit integrity.

3. Intermediate metal conduit

Note: See the *UL Guide Information* on FHIT systems for equipment grounding conductors installed in a raceway that are part of an electrical circuit protective system or a fire rated cable listed to maintain circuit integrity.

4. Electrical metallic tubing

Note: See the *UL Guide Information* on FHIT systems for equipment grounding conductors installed in a raceway that are part of an electrical circuit protective system or a fire-rated cable listed to maintain circuit integrity.

Discussion: Rigid metal conduit (RMC), intermediate metal conduit (IMC), and electrical metallic tubing (EMT) are recognized as equipment grounding conductors in this section. Though these wiring methods are not specifically listed to serve as an equipment grounding conductor, they have a long history of use for that purpose. *NEC 342.2* contains the following statement in the definition of IMC: "A steel threadable raceway of circular cross section designed for the physical protection and routing of conductors and cables and for use as an equipment grounding conductor when installed with its integral or associated coupling and appropriate fittings."* A similar statement is included for RMC in *344.2* and for EMT in *358.2.*

As has previously been emphasized in this book, it is most important to use good workmanship practices in the installation of metal conduit and tubing that are to be used as an equipment grounding conductor. The metal conduit and tubing become a current-carrying conductor when they serve as the return path for ground faults. At times, this ground-fault current can be extremely large. Loose or broken couplings or fittings can either interrupt the fault return path or emit arcs and sparks as current is flowing over the metal raceway.

An equipment grounding conductor of the wire type should be installed where an uncertainty exists about the integrity of the fault path provided by the metal conduit or EMT. Several sections in the *Code* require the installation of an insulated equipment grounding conductor in feeders or branch circuits. See the partial list in the discussion on *250.118(1)* in this book. This requirement applies even though a metal conduit or EMT is installed.

An Informational Note has been added following *250.120(A)* that affects rigid metal conduit (RMC), intermediate metal conduit (IMC), and electrical metallic tubing (EMT). Even though steel conduit and tubing are recognized as equipment grounding conductors in *250.118,* there are times that a supplemental (wire) equipment grounding conductor is specified by the designer or installer for various reasons.

In the listing for electrical circuit protective systems (identified in the *UL Fire Resistance Directory* as "FHIT") that are used as fire protection measures, UL has determined that if the user or designer chooses to install an enclosed equipment grounding conductor, it must be sized in accordance with *250.122,* and the insulation material type must be the same as that of the fire-rated circuit integrity (CI) cable. Use of other types of insulation on the enclosed equipment grounding conductors or bare conductors can constitute a safety hazard and violates the fire-rating listing of the system. Compatibility of materials in a fire-rated system is a requirement of the UL standard. Some materials can provide a carbon residue that is conductive or can generate conductive gases that can cause premature failure in a fire condition.

For RMC, IMC, and EMT, no maximum length of conduit or tubing is given in the *NEC* beyond which the conduit or tubing is no longer considered as providing an *effective ground-fault return path* as defined in *250.2.* However, it is recognized that all conductors, including steel conduit and tubing, offer resistance to the flow of electricity, so they have a practical limitation on their effective length.

To determine the maximum permissible length of conduit or tubing to serve as an equipment grounding conductor, the Steel Tubing Institute commissioned a research project conducted by Georgia State University. For this project, a computer model was developed and data were verified by actual field tests. The software, called GEMI (for "Grounding and Electromagnetic Interference"), is available for download at http://www.steelconduit.org. Individual settings can be made in the software for various factors such as the temperature of the conduit and the conductor, the fault-clearing current, and the arc voltage.

For temperature settings, the conductor usually will operate somewhere between the ambient temperature and the conductor insulation temperature or temperature rating of the termination, whichever is lower. The termination temperature usually is 75°C.

The setting for the fault-clearing current is determined by the type of overcurrent device. For typical molded-case circuit breakers, the instantaneous trip value is usually from 5 to 10 times the trip setting of the breaker. For a 225-ampere breaker, the clearing current would be from 1125 to 2250 amperes. Circuit breakers operate according to the inverse-time principle—that is, the greater the current that flows through the breaker, the faster the breaker

*Reprinted with permission from NFPA 70-2011.

will operate. If uncertain about the trip values of the breaker, consult the time–current curve of the breaker manufacturer. For time-delay, current-limiting fuses, the typical instantaneous clearing time is from 2 to 3 times the circuit rating. Again, the greater the current that flows in the circuit, the faster the fuse will operate to clear the fault.

For the arc voltage, the default setting in GEMI is 40 volts. This is the voltage that is used to maintain the arc. To maintain a steady arc, fixed carbon blocks were used in the test circuit. Additional work is being done to simulate real-life scenarios, in which the energized conductor and the grounded surface may be separated only by the thickness of the insulation. Eustace Soares, in his classic work *Grounding Electrical Distribution Systems for Safety,* suggested that the arc voltage should be 50 volts.

Table 6-1 provides the maximum length of RMC, IMC, and EMT used as an equipment grounding conductor for various circuit or feeder ampacities using

- an overcurrent device that opens at 5 times its rating (indicated as 5×) to simulate circuit breakers.
- an overcurrent device that opens at 2 times its rating (indicated as 2×) to represent cartridge fuses.
- 122°F (50°C) temperature for the conduit [operating hotter than 86°F (30°C) ambient].
- 167°F (75°C) temperature for the conductor (a temperature of 77° to 86°F (25°C to 30°C) for an unloaded conductor).
- 40-volt drop at the arc.

Note: It is important to consult the time–current curve for the specific overcurrent protective device to determine its opening or clearing characteristics. All manufacturers of circuit breakers and fuses have time–current curves available for their products. Some may be available for viewing or downloading from the Internet. The available fault current is an important factor in determining the time it will take for an overcurrent device to open. Overcurrent protective devices are inverse-time devices—that is, the more current that flows in the circuit, the faster it will operate. It is desirable for enough fault current to flow in the faulted circuit so the overcurrent device will open at its instantaneous point, rather than during its time-delay mode.

Another Point: Not all faults are arcing faults. Many faults are bolted—that is, they are solidly connected to the other conductor or to the grounded object, rather than through an arc. If the fault is solidly connected, all of the energy is available to flow through the circuit, which will result in more rapid opening of the overcurrent device if an effective ground-fault return path is provided.

TABLE 6-1

Maximum Effective Length of Metal Conduit and Tubing

Circuit: 120 V:	Size of Conduit*	EMT		IMC		RMC	
		OC 5×	OC 2×	OC 5×	OC 2×	OC 5×	OC 2×
30 A	½	251	590	282	579	272	555
60 A	¾	250	574	288	564	280	534
100 A	1	250	555	286	537	274	507
125 A	1¼	341	608	356	572	331	543
150 A	1½	291	602	312	563	292	538
200 A	2	294	605	315	561	293	538
250 A	2	266	560	299	525	280	498
300 A	2½	297	584	287	520	274	499
350 A	2½	268	540	268	480	253	461
400 A	3	276	549	269	491	257	471

Typical systems are 120/240-V, 1-phase, 3-wire and 208Y/120-V, 3-phase, 4-wire.
*Trade sizes of conduit in inches, sized for 3 wires.

Discussion: The *NEC* does not contain any rules on the minimum size of conduit or EMT that may be used as an equipment grounding conductor for various circuits. It is thought that if the conduit or EMT is sized for the number of conductors according to *Table 1* of *NEC Chapter 9,* there is adequate metal in the conduit or tubing to serve as the equipment ground-fault return path.

More Requirements:

5. Listed flexible metal conduit (FMC) meeting all the following conditions:
   a. The conduit is terminated in listed fittings.
   b. The circuit conductors contained in the conduit are protected by overcurrent devices rated at 20 amperes or less.
   c. The combined length of flexible metal conduit and flexible metallic tubing and liquidtight flexible metal conduit in the same ground-fault current path does not exceed 6 ft (1.8 m).
   d. If used to connect equipment where flexibility is necessary to minimize the transmission of vibration from equipment or to provide flexibility for equipment that requires movement after installation, an equipment grounding conductor is required to be installed.*

Discussion: As indicated in this section, the use of listed FMC as an equipment grounding conductor is limited to accommodate its ability to carry fault current (see Figure 6-5). There is no maximum length for the use of Type FMC conduit if a properly sized equipment grounding conductor is installed inside the flex and is used to ground equipment. As indicated in *(5)(d)* of this rule, an equipment grounding conductor is required to be installed where FMC is used to connect equipment if flexibility "is necessary to minimize the transmission of vibration from equipment or to provide flexibility for equipment that requires movement"* after installation. The installer and the AHJ must decide on the meaning of the phrase "flexibility is necessary after installation." Connections to equipment that is stationary will not necessarily require an equipment grounding conductor. Connections to equipment that will move from time to time or articulate in use will require an equipment grounding conductor.

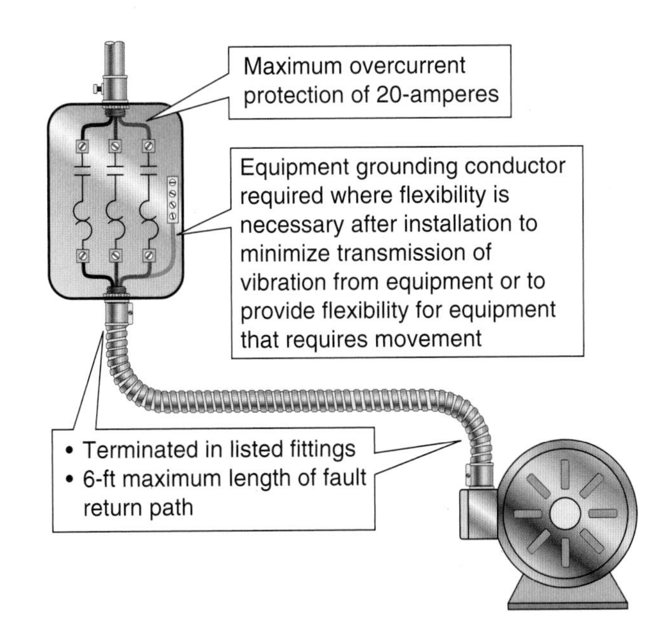

FIGURE 6-5 FMC as equipment grounding conductor, *250.118(5)*. (*© Cengage Learning 2012*)

If flexible metal conduit is installed in patient care areas of health care facilities, the rules in *NEC 517.13* must be complied with. This section requires two independent equipment grounding conductors and fault return paths. *NEC 250.118(5)* recognizes FMC as an equipment grounding conductor up to 6 ft (1.8 m) long with overcurrent protection of contained conductors not exceeding 20 amperes. As a result, 6 ft (1.8 m) becomes the maximum length of FMC in patient care areas even with an insulated equipment grounding conductor inside the FMC. The reason is that if the FMC is longer than 6 ft (1.8 m), it is not recognized as an equipment grounding conductor. The patient care area wiring would then be supplied by only one of the two equipment grounding conductors required.

More Requirements:

6. Listed liquidtight flexible metal conduit (LFMC) meeting all the following conditions is recognized as an equipment grounding conductor:
   a. The conduit is terminated in listed fittings.
   b. For trade sizes ⅜ through ½ (metric designators 12 through 16), the circuit conductors contained in the conduit are protected by overcurrent devices rated at 20 amperes or less.

*Reprinted with permission from NFPA 70-2011.

c. For trade sizes ¾ through 1¼ (metric designators 21 through 35), the circuit conductors contained in the conduit are protected by overcurrent devices rated not more than 60 amperes and there is no flexible metal conduit, flexible metallic tubing, or liquidtight flexible metal conduit in trade sizes ⅜ through ½ (metric designators 12 through 16) in the grounding path.

d. The combined length of flexible metal conduit and flexible metallic tubing and liquidtight flexible metal conduit in the same ground return path does not exceed 6 ft (1.8 m).

e. If used to connect equipment where flexibility is necessary to minimize the transmission of vibration from equipment or to provide flexibility for equipment that requires movement after installation, an equipment grounding conductor is required to be installed.*

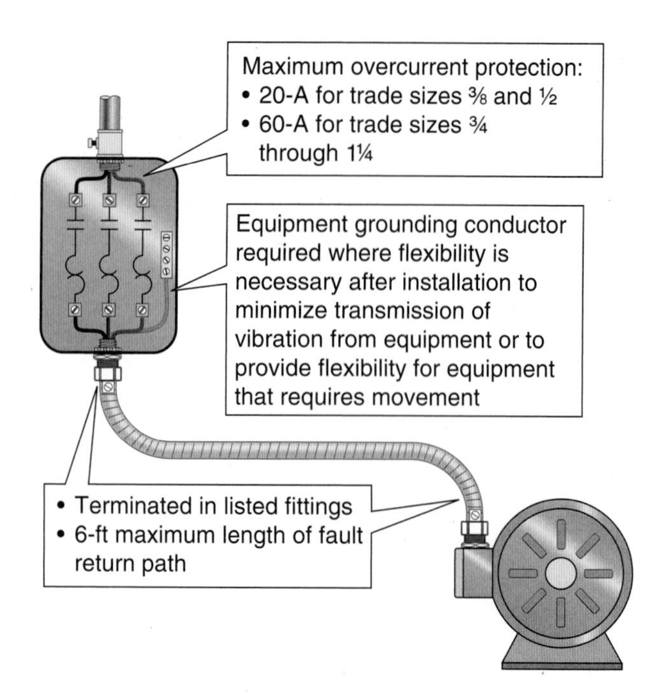

FIGURE 6-6 LFMC as equipment grounding conductor, *250.118(6)*. (*© Cengage Learning 2012*)

Discussion: Listed liquidtight flexible metal conduit (LFMC) is identified as an equipment grounding conductor in *250.118(6)*. However, requirements are included to recognize its limited capability of carrying fault current. The permitted use of this wiring method as an equipment grounding conductor applies only to the product that is listed in compliance with UL-360, which is the product safety standard. Additional installation requirements or limitations are contained in the *Underwriters Laboratories General Information Directory* in category DXHR.

Unlisted LFMC should not be used as an equipment grounding conductor. An equipment grounding conductor of the wire type that is sized in accordance with *Table 250.122* should be installed through the LFMC with the circuit conductors to provide a ground-fault return path and to ensure proper equipment grounding. Figure 6-6 summarizes the use of Type LFMC as an equipment grounding conductor. The rules on using LFMC as an equipment grounding conductor are summarized as follows:

- The fittings must be listed. Listed fittings have been tested to ensure their ability to carry fault current as provided in the product standard.
- For trade sizes through ½ in. (metric designators 12 through 16), the circuit conductors contained in the conduit must be protected by overcurrent devices rated at 20 amperes or less.
- For trade sizes ¾ through 1¼ (metric designators 21 through 35), the circuit conductors contained in the conduit must be protected by overcurrent devices rated not more than 60 amperes. Flexible metal conduit, flexible metallic tubing, or liquidtight flexible metal conduit in trade sizes through ½ (metric designators 12 through 16) are not permitted in this grounding path.
- The combined length of flexible metal conduit and flexible metallic tubing and liquidtight flexible metal conduit in the same ground return path does not exceed 6 ft (1.8 m).
- An equipment grounding conductor is required to be installed if the LFMC is used to connect equipment where flexibility is necessary to minimize the transmission of vibration from equipment or to provide flexibility for equipment that requires movement after installation.

Because the *Code* is silent on the meaning of the phrase "flexibility is necessary after installation," the electrical inspector may be called on to decide what

*Reprinted with permission from NFPA 70-2011.

the phrase means. Connections to equipment that is stationary would not necessarily require an equipment grounding conductor. Connections to equipment that would move from time to time or articulate in use would require an equipment grounding conductor.

No maximum length of LFMC is given in this *Code* rule if a properly sized equipment grounding conductor is installed inside.

More Requirements:

7. Flexible metallic tubing where the tubing is terminated in listed fittings and meeting the following conditions:
   a. The circuit conductors contained in the tubing are protected by overcurrent devices rated at 20 amperes or less.
   b. The combined length of flexible metal conduit and flexible metallic tubing and liquidtight flexible metal conduit in the same ground return path does not exceed 6 ft (1.8 m).

More Requirements:

8. Armor of Type AC cable as provided in *320.108.*

Discussion: Type AC cable was originally known as BX cable. It is a factory assembly of two or more individually insulated conductors with a paper wrap. A bare 16 AWG aluminum bonding conductor is installed between the insulated, paper-wrapped conductors and the metal armor. An overall spiral interlocked armor is placed over the circuit and bonding conductors. Type AC cable with and without an insulated equipment grounding conductor is shown in Figure 6-7. The combination of the metal cable armor and the aluminum bonding conductor in listed cables qualifies as an equipment grounding conductor. Cables that have an additional copper equipment grounding conductor not smaller than 12 AWG are permitted as the wiring method in certain patient care areas in accordance with *517.13* and *517.30(C)(3)*, as well as in other occupancies.

More Requirements:

9. The copper sheath of mineral-insulated, metal-sheathed cable.

Discussion: Mineral-insulated, metal-sheathed cable (Type MI) is covered in *Article 332*. It is defined in *332.2* as "A factory assembly of one or more conductors insulated with a highly compressed refractory mineral insulation (magnesium oxide) and enclosed in a liquidtight and gastight continuous copper or alloy steel sheath." Type MI cable is constructed with one or more conductors surrounded by insulation within an overall metallic sheath. Typical construction of Type MI cable is shown in Figure 6-8. Cable rated 600 V is labeled in sizes 16 AWG to 500-kcmil single conductor, 16 to 4 AWG two and three conductor, 16 to 6 AWG four conductor, and 16 to 10 AWG seven-conductor constructions. Cable rated 300 V is labeled in two, three, four, and seven conductor, sizes 18 to 16 AWG, for use on signaling circuits.

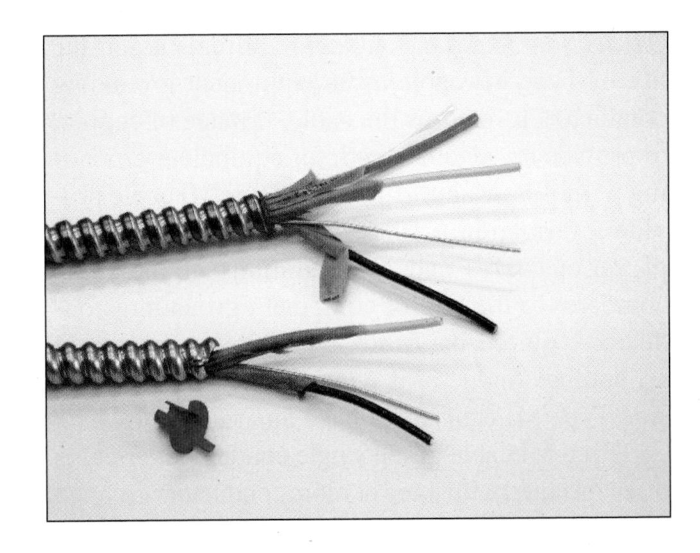

FIGURE 6-7 Type AC cable as an equipment grounding conductor, *250.118(8).* (*© Cengage Learning 2012*)

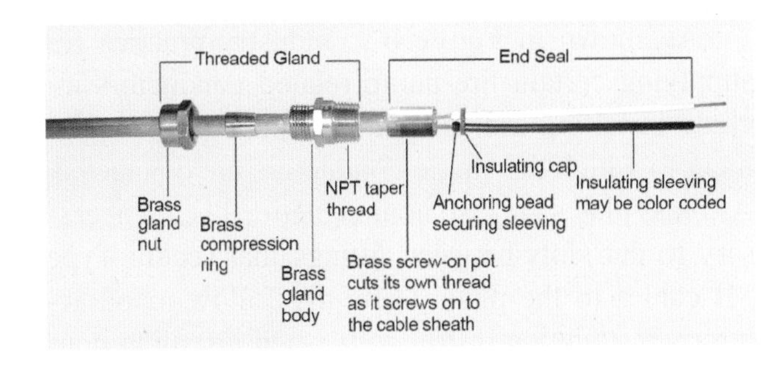

FIGURE 6-8 Type MI cable as equipment grounding conductor, *250.118(9).* (*Photo courtesy of Tyco Thermal Controls*)

Section 332.108 contains requirements on the use of Type MI cable as an equipment grounding conductor. It requires the cable, if made of copper, to provide an adequate path for equipment grounding purposes. Manufacturers of Type MI cable provide information to show how the copper sheath of various sizes and configurations of the cable compares to the size of equipment grounding conductor required by *Table 250.122.* An equipment grounding conductor is required to be provided inside Type MI cables that have an armor of steel.

Type MI cable of the single conductor variety is often installed with three or more conductors grouped together to supply fire pumps or other equipment. These conductors terminate at distribution or control equipment in individual connectors with termination kits. A typical installation is shown in Figure 6-9. Special action must be taken to reduce heating surrounding metal by induction if these terminations are at a magnetic enclosure. *Section 300.20* addresses this issue and requires one of the following steps to be taken to reduce the heating effect:

1. cutting slots in the metal between the individual holes through which the individual conductors pass or
2. passing all the conductors in the circuit through an insulating wall sufficiently large for all of the conductors of the circuit.

An Informational Note offers the following information: "Because aluminum is not a magnetic metal, there will be no heating due to hysteresis; however, induced currents will be present. They will not be of sufficient magnitude to require grouping of conductors or special treatment in passing conductors through aluminum wall sections."

As shown in Figure 6-9, industry practice for supplying certain fire-pump related conductors include installing a brass plate in the wall of equipment to which the single-conductors are connected. As brass plates are also nonmagnetic, it is not necessary to cut slots between fittings that secure Type MI cables to the plate. Carefully follow manufacturer's installation instructions when installing nonmagnetic plates for termination of single-conductor Type MI cables.

*Reprinted with permission from NFPA 70-2011.

FIGURE 6-9 Terminating single conductor Type MI cable, *250.118(9).*
(*Photo courtesy of Tyco Thermal Controls*)

More Requirements:

10. Type MC cable that provides an effective ground-fault current path in accordance with one or more of the following:
    a. It contains an insulated or uninsulated equipment grounding conductor in compliance with *250.118(1).*
    b. The combined metallic sheath and uninsulated equipment grounding/bonding conductor of interlocked metal tape–type MC cable that is listed and identified as an equipment grounding conductor.
    c. The metallic sheath or the combined metallic sheath and equipment grounding conductors of the smooth or corrugated tube type MC cable that is listed and identified as an equipment grounding conductor.*

Discussion: Type MC cable is manufactured in three broad types with variation of construction in some types:

- Interlocked metal tape
- Smooth tube
- Corrugated tube

See Figure 6-10 for a photo of interlocked armor and corrugated tube Type MC cables.

Listed Type MC cable is manufactured to comply with the product safety standard UL 1569. The

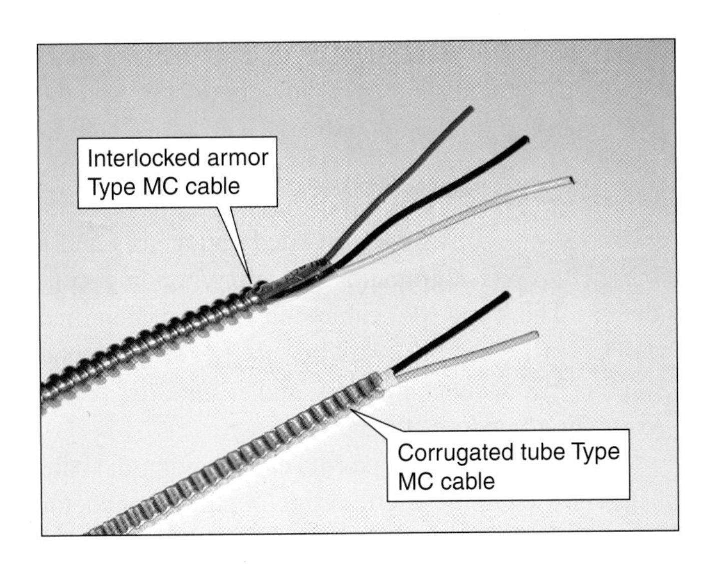

FIGURE 6-10 Interlocked armor and corrugated tube Type MC cables, *250.118(10)*. (*© Cengage Learning 2012*)

following information is from the UL *General Information Directory* in product category (PJAZ) Metal-Clad Cable:

"This category covers Type MC metal-clad cable. It is rated for use up to 2000 V, and Listed in sizes 18 AWG through 2000 kcmil for copper, 12 AWG through 2000 kcmil for aluminum, or copper-clad aluminum, and employs thermoset or thermoplastic-insulated conductors. It is intended for installation in accordance with Article 330 of ANSI/NFPA 70, *National Electrical Code (NEC)*.

"The cable consists of one or more insulated circuit conductors, a grounding path (grounding conductor, metal sheath, or combination thereof) as described below, one or more optional optical fiber members, and an overall metal sheath. The metal sheath is an interlocked metal tape, a corrugated metal tube, or a smooth metal tube. The metal sheath of single conductor cable is nonferrous. A nonmetallic jacket may be provided under and/or over the metal sheath. Cable with metal armor, rated 2400 to 35,000 V is covered under Medium-voltage Power Cable (PITY) and is marked 'Type MV' or 'MC.'

"Cable with interlocked armor that has been determined to be suitable for use as a grounding means has interlocked aluminum armor in direct contact with a single, full-sized, bare aluminum grounding/bonding conductor.

"This cable is marked to indicate that the armor/grounding conductor combination is suitable for ground. The equipment grounding conductor required within all other cable with interlocked armor may be insulated or bare, may be sectioned, and is located in the cable core but not in contact with the armor. Any additional grounding conductors of either design have green insulation. One insulated grounding conductor may be unmarked, one other may have only a yellow stripe and the balance have surface markings that indicate they are additional equipment grounding conductors or isolated grounding conductors.

"The sheath of the smooth or corrugated tube Type MC cable or a combination of the sheath and a supplemental bare or unstriped green insulated conductor is suitable for use as the ground path required for equipment grounding. The supplemental grounding conductor may be sectioned. When sectioned, all sections are identical. Each additional green insulated grounding conductor has either a yellow stripe or a surface marking or both to indicate that it is an additional equipment or isolated grounding conductor. Additional grounding conductors, however marked, are not smaller than the required grounding conductor.

"Cable with an interlocked armor that is intended as a ground path is marked 'armor is grounding path component,' and is provided with installation instructions."*

As can be seen, Type MC cable is produced in several constructions, from 2-wire with ground for simple branch circuits, to 35,000 volts. For our discussion, the Type MC cable with spiral interlocked armor will be referred to as "traditional" Type MC. Generally, the armor of traditional Type MC cable with interlocked metal-tape sheath is not identified

*Reprinted from the White Book with permission from Underwriters Laboratories Inc.®, Copyright © 2010 Underwriters Laboratories Inc.®.

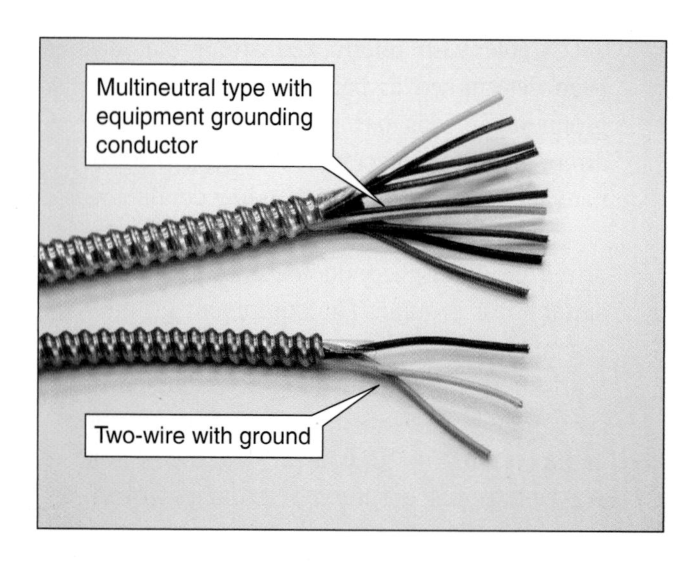

FIGURE 6-11 "Traditional" interlocked armor Type MC cable contains equipment grounding conductor, *250.118(10)*. (*© Cengage Learning 2012*)

as an equipment grounding conductor (See Figure 6-11). Typical cable construction includes an insulated copper equipment grounding conductor that is sized to comply with *Table 250.122*.

More than one variation of Type MC cable is available that has a bare 10 AWG aluminum equipment grounding/bonding conductor installed under and in constant contact with the aluminum armor (see Figure 6-12). The combination of the armor and the equipment grounding/bonding conductor is recognized as an effective equipment ground-fault return path. For traditional Type MC cable, a plastic wrap surrounds all the circuit conductors and the equipment grounding conductor. The cables identified as "A" and "B" in Figure 6-12 show a construction of Type MC cable that has a plastic material extruded on each of the insulated conductors and a bare 10 AWG aluminum grounding/bonding conductor. The Type MC cable construction identified as "C" in Figure 6-12 has the plastic wrap around all the insulated conductors and a bare 10 AWG grounding/bonding conductor.

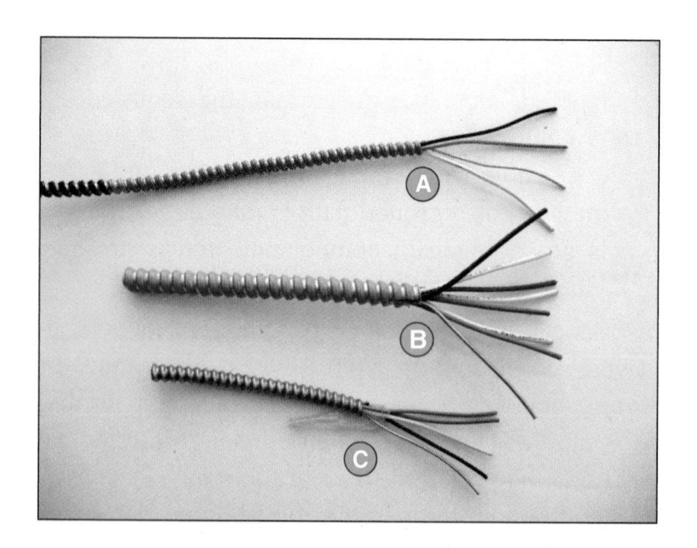

FIGURE 6-12 Type MC cable with aluminum bonding/grounding conductor, *250.118(10)*. (*© Cengage Learning 2012*)

Fittings used for Type MC cable having the bare 10 AWG equipment grounding/bonding conductor must be identified and listed for grounding, as the armor or sheath assembly function as an equipment grounding conductor path. Type MC cable of this construction is suitable for wiring in patient care areas of health care facilities if it has an additional insulated equipment grounding conductor. See *517.13* for additional information.

The *UL Guide Information* on the selection and marking of connectors for Type MC cable is contained in the UL *White Book* under the category "Metal-Clad Cable Connectors, Type MC (PJOX)." It reads in part: "**Connector Selection**—Connectors are intended to be selected in accordance with the size and type of cable for which they are designated. Bronze connectors are intended for use only with cable employing corrugated copper tube. Aluminum connectors are intended for use only with cable employing corrugated aluminum, interlocking aluminum, or smooth aluminum tube, unless marked otherwise on the carton (see PRODUCT MARKINGS below).

"Use in Concrete—Fittings made of aluminum are not considered suitable for use in concrete or cinder fill unless protected with asphalt paint or the equivalent. Fittings suitable for use in concrete are identified by a marking on the carton.

"Grounding—Metal-clad cable connectors for use with metal-clad interlocking armor ground cable, corrugated aluminum or copper tube, or smooth aluminum tube, are considered suitable for grounding for use in circuits over and under 250 V and where installed in accordance with ANSI/NFPA 70, *National Electrical Code*.

"Dry and Wet Locations—Nonmetallic parts, such as glands or seals, are suitable for use at a temperature of 90°C in dry and wet locations. The fittings are suitable for use in dry or wet locations unless marked otherwise (see PRODUCT MARKINGS below).

"Use with Armored Cable—Metal-clad cable connectors also suitable for use with armored cable, Type AC, are so marked on the device or carton. Listed armored cable, Type AC, is covered under Armored Cable Connectors, Type AC (AWSX).

Product Markings

"Metal-clad cable fittings or the smallest unit shipping cartons are marked with

1. "the range of cable diameters and the type of cable sheath (corrugated, interlocking or smooth),
2. "the material of the sheath (aluminum, copper, or steel) for which they have been investigated,
3. "'Concrete-tight' if suitable for use in poured concrete, and
4. "'For Type AC Cable' (or equivalent wording) if suitable for that use.

"See the following table for additional carton markings. Metal-clad cable fittings suitable for use only in dry locations are marked 'Dry Locations' on the device and smallest unit carton.

Type of Metal-Clad Cable	Abbreviation
Metal-clad interlocking armor cable	MCI
Metal-clad interlocking armor ground cable	MCI-A
Metal-clad continuous smooth sheath armor cable"†	MCS

Carefully follow the manufacturer's installation instructions for termination of the aluminum equipment grounding/bonding conductor. These instructions typically include a recommendation that the aluminum conductor be bent back about 120° and cut off before the cable connector is installed on the cable armor. Figure 6-13 shows the acceptable way of cutting the aluminum grounding/bonding conductor and connecting the cable to outlet boxes with listed connectors.

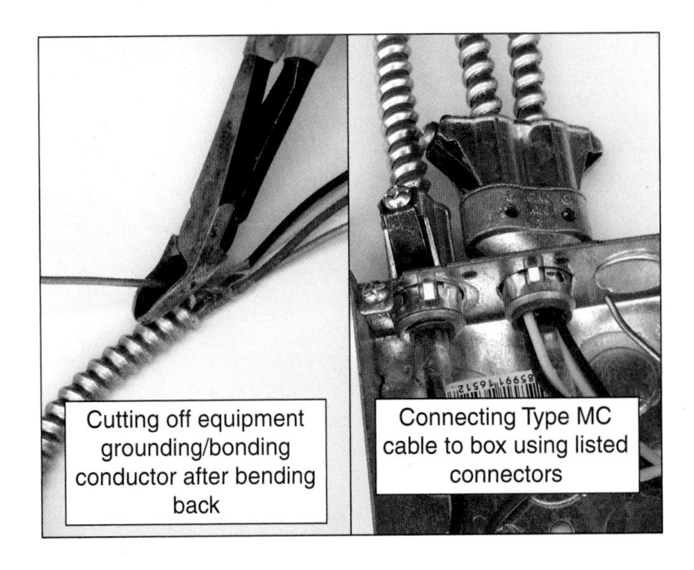

FIGURE 6-13 Terminating Type MC cable having an aluminum equipment grounding/bonding conductor. (*© Cengage Learning 2012*)

Wire connectors that are suitable for aluminum must be used if the aluminum grounding/bonding conductor is connected in the box, cabinet, or other enclosure. Copper and aluminum conductors are not permitted to be in direct contact in a connector unless the connector is identified for that function.

The sheath of the smooth or corrugated tube Type MC cable or a combination of the sheath and a supplemental bare or unstriped green insulated conductor is suitable for use as the required equipment grounding conductor (see Figure 6-14). The supplemental grounding conductor may be sectioned. (A sectioned equipment grounding conductor is one of the required size for the typical overcurrent device located on the supply side of the cable that is divided into two or more equal sizes. This is often done to facilitate the construction of feeder-size Type MC cables because it is easier to form the cable in a circular manner.) When sectioned, all sections are identical. Each additional green insulated grounding conductor has either a yellow stripe or a surface marking or both to indicate that it is an additional equipment or isolated grounding conductor. Additional grounding conductors, however marked, are not smaller than the required grounding conductor.

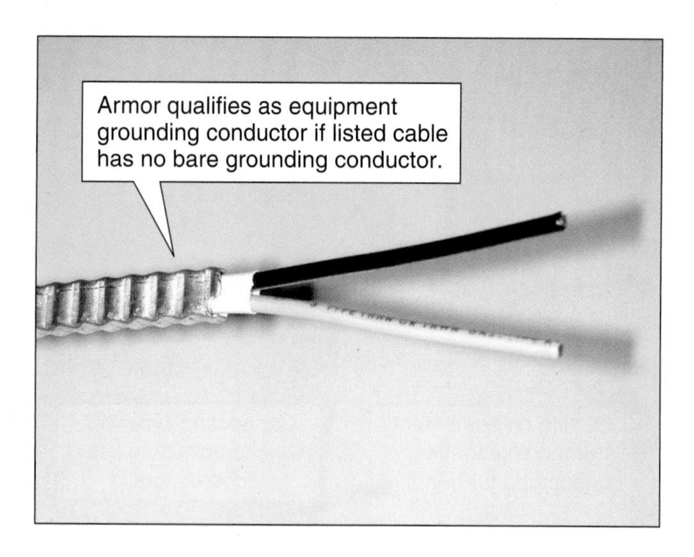

FIGURE 6-14 Corrugated sheath Type MC cable as equipment grounding conductor, *250.118(10)*. (*© Cengage Learning 2012*)

More Requirements:

11. Cable trays as permitted in *392.3(C)* and *392.7.*
12. Cablebus framework as permitted in *370.3.*

More Requirements:

13. Other listed electrically continuous metal raceways and listed auxiliary gutters.

Discussion: As shown in Figure 6-15, metal wireways and auxiliary gutters are examples of other listed electrically continuous metal raceways and listed auxiliary gutters that are suitable for use as an equipment grounding conductor. The *UL Guide Information* at ZOYX contains the following statement: "This category covers metallic and nonmetallic wireways, auxiliary gutters, and associated fittings for installation in accordance with *Articles 366, 376, 378* and *645* of ANSI/NFPA 70, *National Electrical Code (NEC)*. Metallic wireways installed in accordance with the product markings and manufacturer's instructions are suitable for use as equipment grounding conductors, and are Listed for grounding."

FIGURE 6-15 Listed wireways used as an equipment grounding conductor, *250.118(13)*. (*© Cengage Learning 2012*)

One important issue that must be considered when it is intended to use metal wireways or auxiliary gutters as an equipment grounding conductor is the fact that these enclosures are typically painted. The paint on these enclosures is an insulator. Thus, the assembly of sections, elbows, and closure fittings must be connected together in such a manner that good electrical continuity is maintained. This may require scraping the paint from connection points. If metallic conduit or tubing is connected to the metal wireway or auxiliary gutter, once again a reliable mechanical and electrical connection must be made. See *250.96(A)* and *300.10.* This may require scraping paint from connection points or installing a bonding locknut or bushing to ensure a reliable connection that will carry fault current.

If conductors are installed in a nonmetallic raceway from a metal wireway or auxiliary gutter, an equipment grounding conductor must be installed in the raceway with the feeder or branch-circuit conductors. A lug or terminal bar for connecting these equipment grounding conductors can be mounted in good electrical contact with the wireway.

More Requirements:

14. Surface metal raceways listed for grounding.

Discussion: The Underwriters Laboratories Guide information for surface metal raceways has the following statement in product category RJBT, "Surface metal raceways are considered suitable for grounding for use in circuits over and under 250 V and where installed in accordance with the *NEC*." These raceways are covered in *Article 386. Section 386.60* has the following requirement: "*386.60* Grounding. Surface metal raceway enclosures providing a transition from other wiring methods are

FIGURE 6-16 Surface metal raceways listed for grounding, 250.118(14). (*Photo Courtesy of Wiremold*)

required to have a means for connecting an equipment grounding conductor." Transition fittings that comply with this requirement are commonly available for extending the wiring method from surface as well as concealed boxes and from other wiring methods. Many types of fittings are used when wiring with the surface metal raceway method. A properly sized equipment grounding conductor must be installed inside the surface metal raceway if the ground-fault return path is suspect for any reason. See Figure 6-16 for a photo of a surface metal raceway installation.

250.119 IDENTIFICATION OF EQUIPMENT GROUNDING CONDUCTORS

The Requirements: Unless required elsewhere in the *NEC*, equipment grounding conductors are permitted to be bare, covered, or insulated. Individually covered or insulated equipment grounding conductors are required to have a continuous outer finish that is either green or green with one or more yellow stripes except as permitted in this section. Conductors with insulation or individual covering that is green, green with one or more yellow stripes, or otherwise identified as permitted by this section are not to be used for ungrounded ("hot") or grounded (often neutral) circuit conductors.

Discussion: The identification of equipment grounding conductors is an essential safety issue. The requirements in the opening paragraph of this section apply to equipment grounding conductors in sizes through 6 AWG. *Section 250.119(A)* applies to conductors larger than 6 AWG. Recent changes to this section limit or restrict the use of conductors with insulation or individual covering that is green, green with one or more yellow stripes, or otherwise identified as an equipment grounding conductor. These conductors so identified are not permitted to be used as ungrounded ("hot") or grounded (often neutral) conductors. See Figure 6-17.

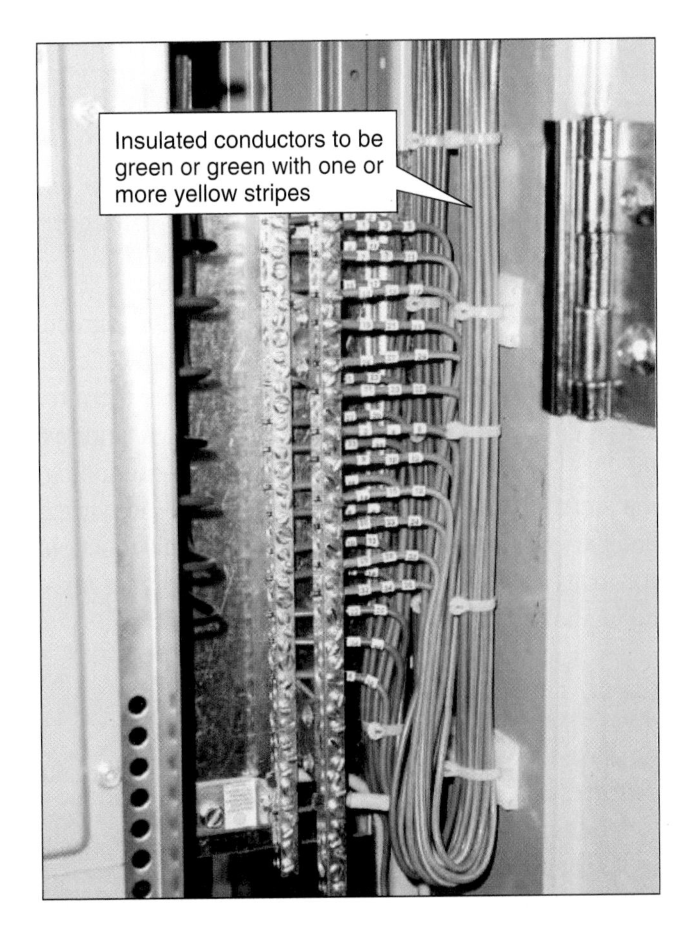

FIGURE 6-17 Identification of equipment grounding conductors 6 AWG and smaller, 250.119. (© *Cengage Learning 2012*)

FIGURE 6-18 Green insulation in low-voltage circuits, *250.119, Exception.* (*© Cengage Learning 2012*)

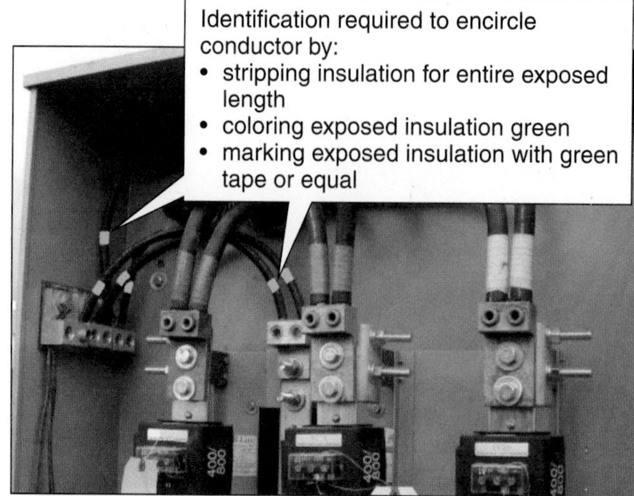

FIGURE 6-19 Identification of equipment grounding conductors larger than 6 AWG, *250.119(A).* (*© Cengage Learning 2012*)

An Exception: *Power-limited Class 2 or Class 3 cables, power-limited fire alarm cables, or communications cables containing only circuits operating at less than 50 volts where connected to equipment not required to be grounded in accordance with 250.112(I) shall be permitted to use a conductor with green insulation or green with one or more yellow stripes for other than equipment grounding purposes.**

Discussion: Many low-voltage systems use a green insulated conductor for low voltage power applications. See Figure 6-18. For example, air conditioner thermostat circuit wiring uses a green insulated conductor as the "hot" or power conductor for the air-conditioning fan supply. This color coding in the thermostat wiring and the associated cable has been used for this function for many years without any problems involving mistaking this conductor as a grounding conductor.

250.119(A) Conductors Larger Than 6 AWG

The Requirements: Equipment grounding conductors larger than 6 AWG are required to comply with *(A)(1)* and *(A)(2)* (see Figure 6-19).

1. An insulated or covered conductor (larger than 6 AWG) is permitted to be permanently identified as an equipment grounding conductor at the time of installation at each end and at every point where the conductor is accessible.

Exception: *Conductors larger than 6 AWG are not be required to be marked in conduit bodies that contain no splices or unused hubs.**

2. The identification is required to encircle the conductor and must be accomplished by one of the following:
   a. Stripping the insulation or covering from the entire exposed length
   b. Coloring the insulation or covering green at the termination
   c. Marking the insulation or covering with green tape or green adhesive labels at the termination.

250.119(B) Multiconductor Cable

The Requirements: When the conditions of maintenance and supervision ensure that only qualified persons service the installation, one or more insulated conductors in a multiconductor cable, at the time of installation, is (are) permitted to be permanently identified as equipment grounding conductors at each end and at every point where the conductors are accessible by one of the following means:

1. Stripping the insulation from the entire exposed length

*Reprinted with permission from NFPA 70-2011.

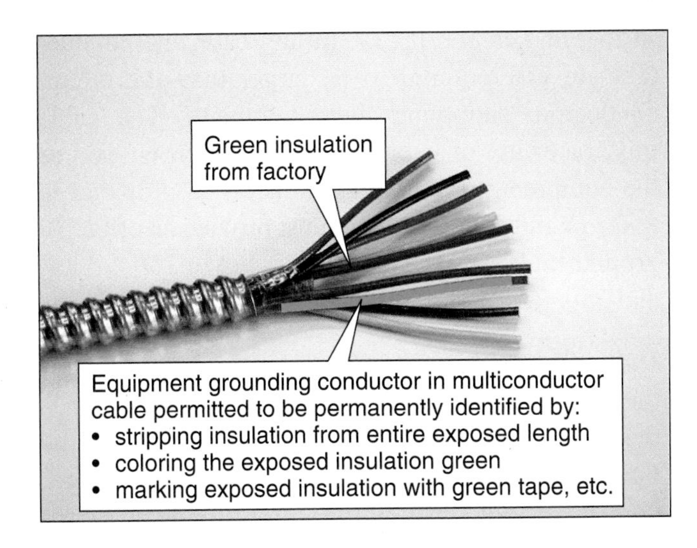

FIGURE 6-20 Identification of equipment grounding conductor in multiconductor cable, *250.119(B)*. (*© Cengage Learning 2012*)

FIGURE 6-21 Installation of raceway, cable trays, cable armor, cablebus, or cable sheaths, *250.120(A)*. (*© Cengage Learning 2012*)

2. Coloring the exposed insulation green
3. Marking the exposed insulation with green tape or green adhesive labels

See Figure 6-20.

250.120 EQUIPMENT GROUNDING CONDUCTOR INSTALLATION

The Requirements: An equipment grounding conductor is required to be installed in accordance with *250.120(A), (B),* and *(C).*

250.120(A) Raceway, Cable Trays, Cable Armor, Cablebus, or Cable Sheaths

The Requirements: If the equipment grounding conductor consists of a raceway, cable tray, cable armor, cablebus framework, or cable sheath, or if it is a wire within a raceway or cable, it is required to be installed in accordance with the applicable provisions in the Code, using fittings for joints and terminations approved for use with the type of raceway or cable used. All connections, joints, and fittings must be made tight using suitable tools. (See Figure 6-21.)

Discussion: Specific requirements are in place for installation of equipment grounding conductors as the conductors become current carrying during ground-fault conditions. Loose connections either can result in metallic equipment being isolated and becoming a shock hazard or can create arcing and sparking at connections that are loose. Neither condition is acceptable. Be certain that positive metal-to-metal connections are made at every connection point to ensure an effective ground-fault return path bas been created.

250.120(B) Aluminum and Copper-Clad Aluminum Conductors

The Requirements: Equipment grounding conductors of bare or insulated aluminum or copper-clad aluminum are permitted. Bare conductors are not permitted to come in direct contact with masonry or the earth or where subject to corrosive conditions. Aluminum or copper-clad aluminum conductors are not permitted to be terminated within 18 in. (450 mm) of the earth.

Discussion: These restrictions for aluminum and copper-clad aluminum conductors are intended to provide an additional level of protection to ensure a satisfactory performance of these conductors. As can be seen, these restrictions are similar to those in *250.64(A)* for grounding electrode conductors.

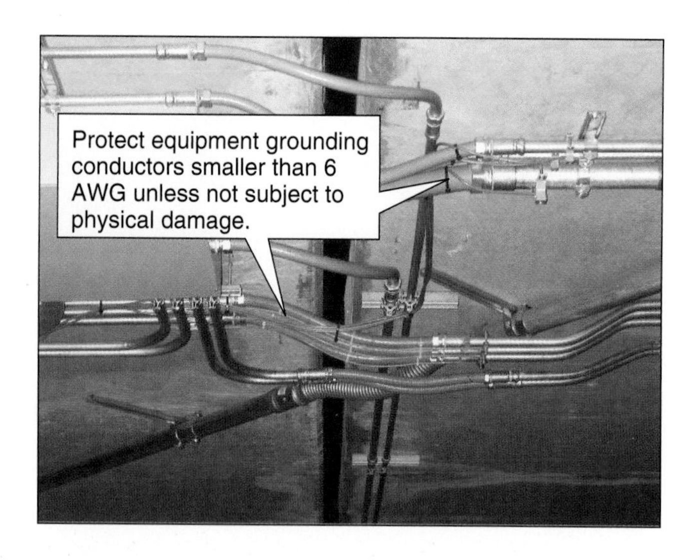

FIGURE 6-22 Protection of equipment grounding conductors smaller than 6 AWG, *250.120(C)*. (*© Cengage Learning 2012*)

250.120(C) Equipment Grounding Conductors Smaller Than 6 AWG

The Requirements: Equipment grounding conductors smaller than 6 AWG are required to be protected from physical damage by a raceway or cable armor except if run in hollow spaces of walls or partitions, if not subject to physical damage, or if protected from physical damage.

Discussion: Smaller equipment grounding conductors are more fragile and subject to damage or breakage during and after initial installation. Additional methods of protection are justified (see Figure 6-22). Common applications are for bonding metal piping systems such as for water and gas piping and for bonding together the equipment grounding terminal bars in patient care areas of health care facilities, as provided in *517.14*.

250.122 SIZE OF EQUIPMENT GROUNDING CONDUCTORS

250.122(A) General

The Requirements: Copper, aluminum, or copper-clad aluminum equipment grounding conductors of the wire type are not permitted to be smaller than shown in *Table 250.122*. In any case, the conductors are not required to be larger than the circuit conductors supplying the equipment. If a cable tray, raceway, or cable armor or sheath is used as the equipment grounding conductor as provided in *250.118* and *250.134(A)*, it must provide an effective ground-fault return path as provided in *250.4(A)(5)* and *250.4(B)(4)*.

Discussion: The title of *Table 250.122* indicates the size of the equipment grounding conductor is the "minimum" size. This concept is emphasized in the table note, indicating a larger conductor might be needed to comply with the requirement to provide an effective fault-current return path. It is an obvious failure of design if the equipment grounding conductor is so small that while carrying fault

Table 250.122 Minimum Size Equipment Grounding Conductors for Grounding Raceway and Equipment

Rating or Setting of Automatic Overcurrent Device in Circuit Ahead of Equipment, Conduit, etc., Not Exceeding (Amperes)	Size (AWG or kcmil)	
	Copper	Aluminum or Copper-Clad Aluminum*
15	14	12
20	12	10
60	10	8
100	8	6
200	6	4
300	4	2
400	3	1
500	2	1/0
600	1	2/0
800	1/0	3/0
1000	2/0	4/0
1200	3/0	250
1600	4/0	350
2000	250	400
2500	350	600
3000	400	600
4000	500	750
5000	700	1200
6000	800	1200

Note: Where necessary to comply with 250.4(A)(5) or (B)(4), the equipment grounding conductor shall be sized larger than given in this table.

*See installation restrictions in 250.120.

Reprinted with permission from NFPA 70-2011

current it overheats and damages adjacent insulated conductors. Other undesirable effects of the equipment grounding conductor's carrying too much current can be annealing the copper, melting, and damaging the terminal. When the terminal is damaged, the equipment grounding conductor can become loose after the conductor and terminal cools.

Table 250.122 can be used as is for many installations in which the available fault current and operating characteristics of the overcurrent device prevent the equipment grounding conductor from overheating to the point of damage. The equipment grounding conductors are not sized the same as for the circuit conductors (except for small sizes); rather, they are sized according to their ability to carry a lot of fault current for a short time until the overcurrent device opens the faulted circuit. This can be thought of as the short-time rating of the conductor. [The longtime rating of the conductor is the allowable ampacity in *Table 310.15(B)(16)*].

The short-circuit withstand rating of the equipment grounding conductor can be verified using Publication P-32-382, *Short-Circuit Characteristics of Cable,* from the Insulated Cable Engineers Association. (See Figure 6-23.) If the equipment grounding conductor selected from *Table 250.122* is beyond the thermal damage curve in the table in Publication P-32-382 two options are available: (1) increase the size of the conductor, or (2) replace the overcurrent device with one that operates faster, thereby limiting the available fault current. Manufacturers of fuses and circuit breakers will furnish a time–current curve for the overcurrent device on request.

Figure 6-23 is an insulated conductor thermal damage chart from the Insulated Cable Engineers Association (ICEA) Publication 32-382. The table is calculated for a maximum conductor operating temperature of 75°C and a maximum short-circuit temperature of 150°C. Although the chart shows clearing times from 100 cycles to 1 cycle, other times can be calculated by using the formula shown on the chart. A similar formula is available from the Insulated Cable Engineers Association (ICEA) for aluminum conductors. The chart shows short-circuit current in thousands of amperes on the left and conductor sizes across the bottom. To use the chart, simply determine the short-circuit current available at the beginning of the circuit and follow that line across until it intersects with a diagonal line indicating the opening time of the overcurrent device. The vertical line at that intersection indicates the minimum-size conductor that should be used for that application.

The short-circuit current at any point in the electrical system can be determined by using one of several software programs available, including one furnished online at http://www.bussmann.com. Calculations can be performed beginning at the utility supply transformer and proceeding through the service conductors to the service equipment, through one or more feeders, and finally through branch circuits. The operating characteristics of the overcurrent device, including the opening time, can be determined from the manufacturer.

The conductor withstand rating chart indicates a 10 AWG copper conductor is suitable for not more than 4300 amperes if the overcurrent device will open the faulted circuit in one cycle or 0.0167 seconds. The 10 AWG conductor should not be used where the available fault current exceeds approximately 3000 amperes if the overcurrent device will not open in less than 2 cycles or 0.0333 seconds.

If 40,000 amperes of short-circuit current is available and the overcurrent device will clear the faulted circuit in 1 cycle, a 1/0 AWG conductor is required. If the overcurrent device will not clear the fault in less than 2 cycles, a 3/0 equipment grounding conductor is required for the 40,000 amperes of short-circuit current.

From a logical standpoint, there is no reason to increase the size of the equipment grounding conductor to exceed the size of the supply or ungrounded conductors. This concept is included in *250.122(A)* where it states, "... but in no case shall they be required to be larger than the circuit conductors supplying the equipment."

250.122(B) Increased in Size

The Requirement: If ungrounded conductors are increased in size, equipment grounding conductors, where installed, are also required to be increased in

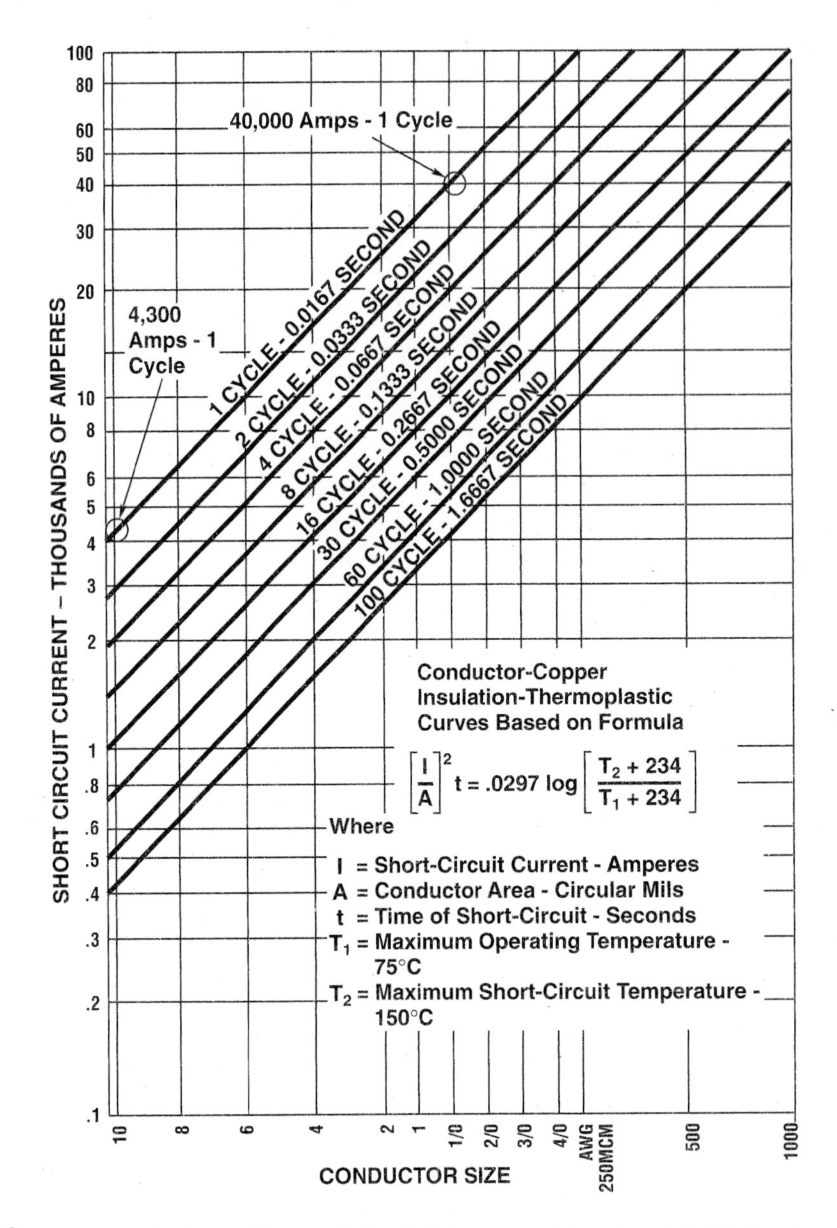

FIGURE 6-23 Copper conductor withstand chart. (*Courtesy of Insulated Cable Engineers Association Inc.*)

size proportionately according to circular mil area of the ungrounded conductors.

Discussion: Although the requirements of this rule are quite broad, the rule is aimed specifically at installations with any of the following conditions:

1. The ungrounded ("hot") conductors are increased in size to adjust for voltage drop.
2. The ungrounded ("hot") conductors are increased in size to adjust for high ambient temperatures.
3. The ungrounded ("hot") conductors are increased in size to adjust for more than three current-carrying conductors in a raceway or cable.

The heading of *Table 310.15(B)(16)* gives the conditions of use for which the table's allowable ampacities are calculated. The conditions of use are as follows:

1. There are not more than three current-carrying conductors in a raceway, cable, or earth (directly buried), and

2. The ambient temperature where the conductors are located is not more than 86°F (30°C).

An *adjustment* (for an excessive number of current-carrying conductors) and *correction* (for conductors installed where the ambient temperature is higher than those in the table) must be made for conductors installed outside of these conditions of use. The term *derating* is often used to describe *correction* and *adjustment.*

If the ungrounded conductors are increased in size greater than that required in *Table 310.15(B)(16)* for any reason, a corresponding increase must be made in the size of the equipment grounding conductor in the same proportion and based on the circular mil area of the conductors. (See Figure 6-24.)

Here's How: A 200-ampere feeder is being installed. Because of concern for voltage drop and high ambient temperature, the size of the ungrounded conductor is to be increased from 3/0 AWG to 300 kcmil. What size must the equipment grounding conductor be?

Here is the formula for increasing the size of the equipment grounding conductor, or EGC:

Circular mil area of increased phase conductor/ Required phase conductor = Ratio of increase

Area of required EGC × ratio of increased phase conductor = required EGC

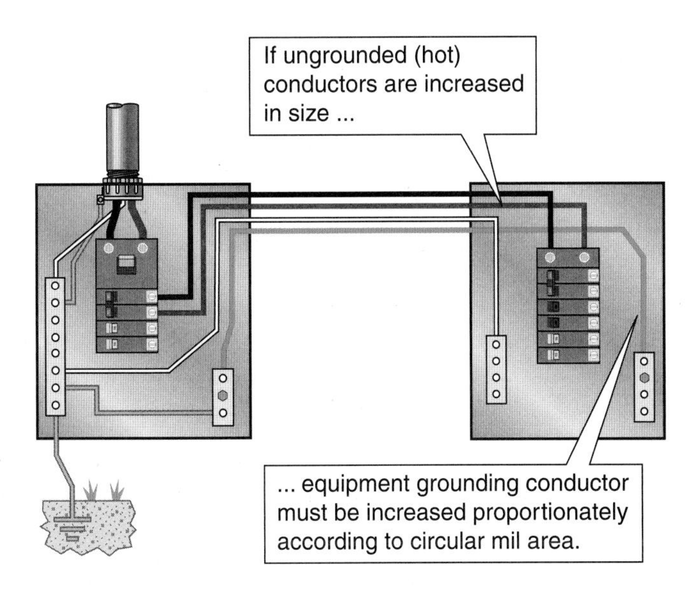

FIGURE 6-24 Equipment grounding conductor increased in size, *250.122(B)*. (*© Cengage Learning 2012*)

Example:

300,000 ÷ 167,800 = 1.788 (ratio of increase)

(167,800 cm is area of 3/0 conductor from *Table 8* of *Chapter 9*.)

26,240 × 1.788 = 46,917 cm (minimum area of EGC)

(26,240 cm is the area of a 6 AWG conductor from *Table 8* of *Chapter 9* and is the EGC required in *Table 250.122*.)

The next standard size larger than 46,917 cm = 3 AWG from *Table 8* of *NEC Chapter 9*.

250.122(C) Multiple Circuits

The Requirement: If a single equipment grounding conductor is run with multiple circuits in the same raceway, cable, or cable tray, it is required to be sized for the largest overcurrent device protecting conductors in the raceway, cable, or cable tray. Equipment grounding conductors installed in cable trays are required to meet the minimum requirements of *392.3(B)(1)(c)*.

Discussion: This section permits several branch circuits to be installed in a common raceway or cable and a single equipment grounding conductor to be installed. The equipment grounding conductor is permitted to be installed if it is sized according to *Table 250.122* for the largest overcurrent device for the branch circuit installed in the common raceway or cable. See Figure 6-25.

Requirements related to equipment grounding conductors installed in cable trays were added to this section in recent editions of the *NEC*. A reference is included to *392.3(B)(1)(c)*, where the requirement is that equipment grounding conductors installed as single conductors in cable trays are not permitted to be smaller than 4 AWG. Additional requirements for using the metallic cable tray as an equipment grounding conductor can be found in *392.7*. *Table 392.7(B)* gives the minimum cross-sectional area of metal required if the cable tray system is to be used as an equipment grounding conductor, on the basis of the overcurrent device protecting conductors supported by the cable tray.

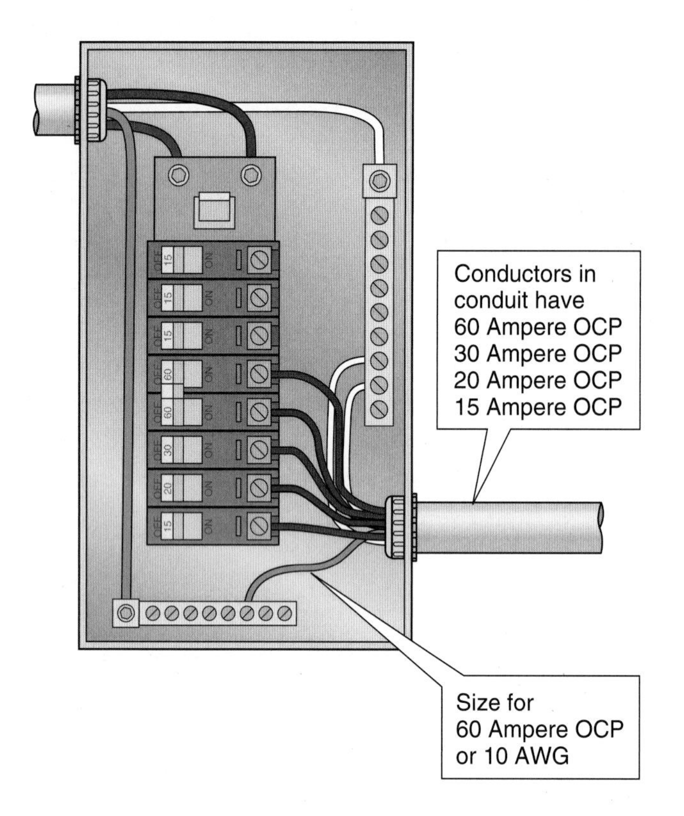

FIGURE 6-25 Size of equipment grounding conductor for multiple circuits, *250.122(C).* (*© Cengage Learning 2012*)

250.122(D) Motor Circuits

The Requirement: Equipment grounding conductors are required to be installed in compliance with *250.122(D)(1)* or *(D)(2).*

1. **General.** The equipment grounding conductor is required to be not smaller than determined by *250.122(A),* based on the rating of the branch circuit short circuit and ground-fault protection.

Discussion: *Section 250.122(A)* provides that the equipment grounding conductor of the wire type is required to be sized not smaller than given in *Table 250.122* but is not required to be larger than the circuit conductors.

The equipment grounding conductor for motors that do not have an instantaneous trip circuit breaker or motor circuit protector as their overcurrent device is sized in accordance with the rating of the branch-circuit short-circuit and ground-fault overcurrent device. Recall that the overcurrent device for motor branch circuits is allowed to be increased above the rating of the branch-circuit conductor to allow the motor to start and run. It is common to size the branch-circuit conductors for the motor at 125% of the full-load current from the applicable table in *Article 430* in accordance with *430.22(A).* According to *430.52(C)(1)* and *Table 430.52,* a standard inverse-time circuit breaker can be 250% of the motor full-load rating. A dual-element fuse is permitted to be sized at 175% of the motor full-load rating. If the value derived from this calculation does not result in a standard rating for an overcurrent device, the next higher overcurrent device rating is permitted to be selected as provided in *430.52(C)(1), Exception No. 1.* See *240.6(A)* for the standard ampere ratings of fuses and fixed-trip circuit breakers.

Here's How: For example, a 5-hp, 208-volt, 3-phase motor is installed. *Table 430.250* indicates that the full-load current to be used for selection of the branch-circuit conductors and overcurrent device is 16.7 amperes. The motor branch-circuit conductor must be not less than 125% of 16.7 amperes or 20.9 amperes. (See *430.22.*) If all the terminals in the circuit are rated for 75°C, a 12 AWG conductor can be selected from that column of *Table 310.15(B)(16),* as it has an allowable ampacity of 25 amperes. If it cannot be verified that all the terminals for connection of the branch-circuit conductors are rated 75°C, select a 10 AWG copper conductor. A circuit breaker used as the branch-circuit short-circuit and ground-fault device is permitted to be sized at 250% of the motor full-load current from *Table 430.250,* or

$$16.7 \times 2.5 = 41.8 \text{ amperes}$$

[See *430.52(C)(1)* and *Table 430.52.*] The next standard rating of an overcurrent device in *240.6(A)* is 45 amperes. From *Table 250.122,* a 10 AWG copper equipment grounding conductor is required. According to the opening paragraph of *250.122,* however, the equipment grounding conductor is not required to be larger than the circuit conductors supplying the equipment, so a 12 AWG equipment grounding conductor is selected so long as the temperature rating of the conductor and terminals are compatible.

More Requirements:

2. **Instantaneous-Trip Circuit Breaker or Motor Circuit Protector.** If the overcurrent device consists of an instantaneous trip circuit

breaker or a motor short-circuit protector, the equipment grounding conductor must be sized not smaller than given by *250.122(A)*, using the maximum permitted rating of a dual element time-delay fuse selected for branch-circuit short-circuit and ground-fault protection in accordance with *430.52(C)(1), Exception No. 1.*

Discussion: Instantaneous trip circuit breakers and motor circuit protectors are required to be installed as a part of a listed assembly. As can be seen in the exception to *430.52(C)(3)*, the rating of the instantaneous trip circuit breaker is in some cases permitted to be as much as 1700% of the full-load motor current. For these installations, *250.122(D)(2)* requires the size of the equipment grounding conductor to be based on the rating of a dual element fuse selected according to *430.52(C)(1), Exception No. 1*, rather than on the rating of the instantaneous trip circuit breaker.

Here's How: For example, a 50-hp, 460-volt, 3-phase motor is installed with an instantaneous trip circuit breaker or motor circuit protector as the branch-circuit short-circuit and ground-fault protection. *Table 430.250* indicates the full-load current to be used for selection of the branch-circuit conductors and overcurrent device is 65 amperes. The motor branch-circuit conductor must be not less than 125% of 65 amperes or 81 amperes. (See *430.22*.) Assuming all terminations are rated 75°C, a 4 AWG copper conductor is selected from *Table 310.15(B)(16)* because it has an allowable ampacity of 85 amperes.

Regardless of the rating of an instantaneous circuit breaker or a motor circuit protector, the sizing of an equipment grounding conductor must not be smaller than given in *250.122(A)* with use of the maximum permitted rating of a dual-element time-delay fuse selected for branch-circuit short-circuit and ground-fault protection in accordance with *430.52(C)(1), Exception No. 1.* If a dual-element fuse were to be used as the branch-circuit short-circuit and ground-fault device, it is permitted to be sized at 175% of the motor full-load current from *Table 430.250*, or

$$65 \times 1.75 = 113.8 \text{ amperes}$$

[See *430.52(C)(1)* and *Table 430.52*]. *Section 430.52(C)(1), Exception No. 1*, provides that if the calculation of the current from *Table 430.52* does not result in a "standard" overcurrent device rating, the next larger standard size is permitted. The next standard rating of an overcurrent device in *240.6(A)* greater than 114 amperes is 125 amperes. From *Table 250.122*, a 6 AWG equipment grounding conductor is required.

250.122(F) Conductors in Parallel

The Requirement: If conductors are run in parallel in multiple raceways or cables as permitted in *310.10(H)*, the equipment grounding conductors, where used, are required to be run in parallel in each raceway or cable. Each parallel equipment grounding conductor is required to be sized on the basis of the ampere rating of the overcurrent device protecting the circuit conductors in the raceway, cable, or cable tray in accordance with *Table 250.122*.

Discussion: Note, this rule applies whether the installation of the equipment grounding conductors is voluntary or mandatory. The application of the rule usually is for the installation of feeders. Where the conduit or tubing is recognized as an equipment grounding conductor in *250.118*, the installation of an additional equipment grounding conductor of the wire type is not required. Some local inspection regulations require a supplemental equipment grounding conductor to be installed in all metal raceways, especially those that are installed in underground or corrosive environments. In other cases, the electrical design engineer specifies the installation of a supplemental equipment grounding conductor in all raceways, even those of metal. Obviously, an equipment grounding conductor is required in nonmetallic raceways and in almost all metallic cables.

If the feeder conductors are in parallel in multiple raceways [two or more conductors connected together at both ends in accordance with *310.10(H)*] the equipment grounding conductors, where installed, are also required to be installed in parallel. See Figure 6-26.

Note that this rule requires a full-size equipment grounding conductor to be installed in each raceway or cable in accordance with the sizing requirements of *Table 250.122*. The equipment grounding conductor is based on the rating of the overcurrent

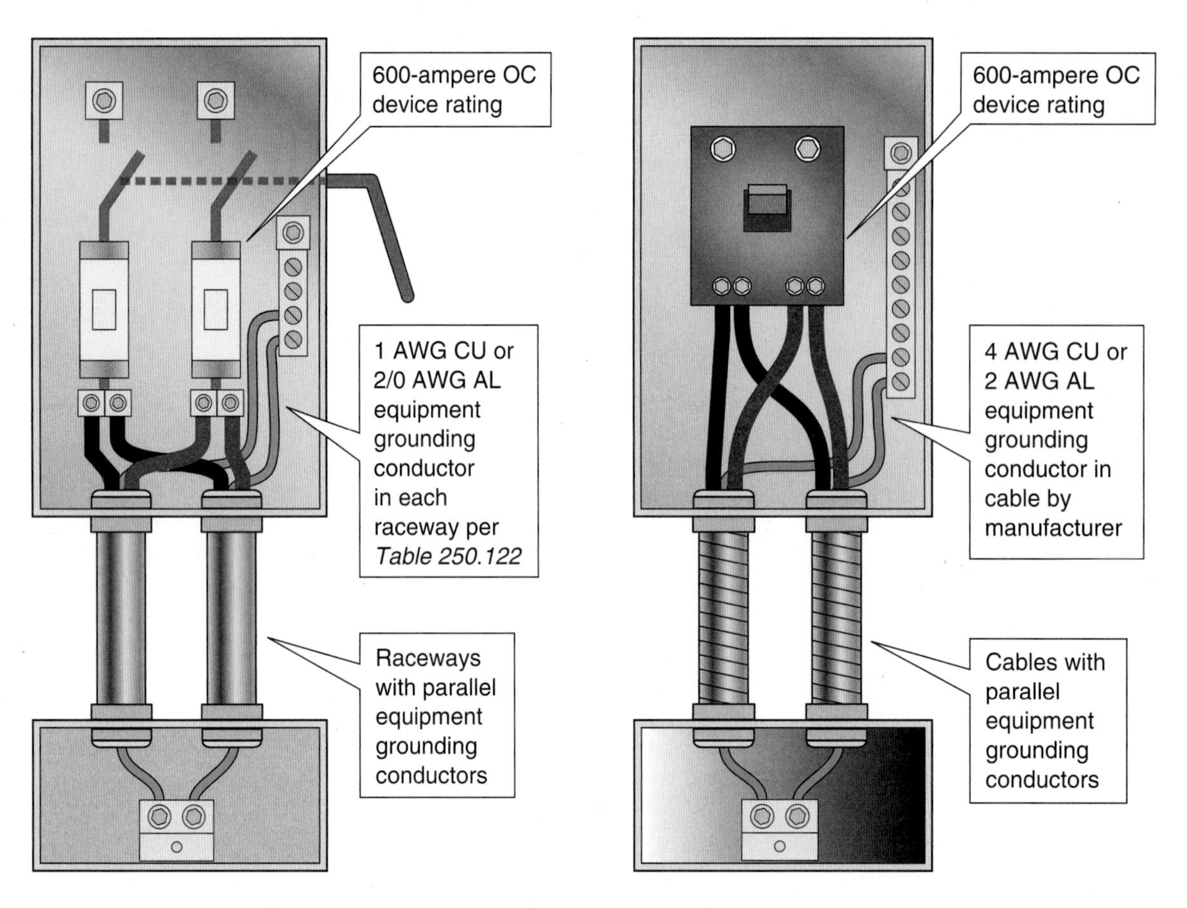

FIGURE 6-26 Equipment grounding conductors in parallel, *250.122(F)*. (*© Cengage Learning 2012*)

device on the supply side of the conductors. Space must be provided in the conduit or tubing for all circuit conductors as well as the equipment grounding conductor.

Although the requirements of this section may pose little difficulty for installations in raceways, some additional problems may arise for installation of listed Type MC cables. Listed cables are manufactured with equipment grounding conductors that are sized according to *Table 250.122* or, in some cases, one size larger. For example, a cable intended for use with a 200-ampere overcurrent device will have a 4 AWG copper or 2 AWG aluminum equipment grounding conductor. (See Table 6.1 of UL 1569, the product safety standard for Type MC cables.) The circuit conductors usually will be 3/0 AWG copper or 250-kcmil aluminum. The problem (or challenge!) arises when two or more of these cables are installed in parallel. *Table 250.122* requires a 3 AWG copper or 1 AWG aluminum equipment grounding conductor in each cable when the overcurrent device on the supply side of the conductors is 400 amperes. See Table 6–2 for an explanation of the typical cable construction and *NEC* requirements. Obviously, many other combinations of conductors can be installed in parallel.

Some manufacturers of listed cable produce cables with oversized equipment grounding conductors to accommodate the rules in this section.

TABLE 6-2

Equipment Grounding Conductors for Cables

Overcurrent Protection	Number of 200-A Cables in Parallel	Minimum Size of Equipment Grounding Conductor in Each Cable	Typical Size of Equipment Grounding Conductor in Listed Cable
200 A	1	6	4
400 A	2	3	4
600 A	3	1	4
800 A	4	1/0	4

The manufacturer should be consulted to determine product availability for a specific application. In addition, some manufacturers indicate a willingness to manufacture specific sizes of listed cables with a custom size of equipment grounding conductor. However, a minimum length of cable for the custom order may apply, and a significant lead time must usually be allowed.

250.122(G) Feeder Taps

The Requirement: Equipment grounding conductors installed with feeder taps are not permitted to be smaller than shown in *Table 250.122,* based on the rating of the overcurrent device ahead of the feeder, but are not be required to be larger than the tap conductors.

Discussion: This requirement was added to the 2005 *NEC*. It provides direction for the installation of equipment grounding conductors for feeder taps installed according to *240.21*. (See Figure 6-27.)

Two of the most common feeder taps are the 10-ft (3.0-m) and 25-ft (7.5-m) taps. For the 10-ft (3.0-m) feeder tap, the upstream overcurrent device is not permitted to be larger than 10 times the ampacity of the smaller conductor. If tap conductors are installed with the maximum permitted overcurrent protection, the equipment grounding conductor will be of the same size as for the feeder conductors.

Here's How: For example, a 400-ampere overcurrent device protects the feeder. A set of tap conductors sized at 8 AWG are connected to the larger feeder. *Table 250.122* requires a 3 AWG copper equipment grounding conductor, based on the 400-ampere overcurrent device. The provisions of *250.122(G)* permit an 8 AWG equipment grounding conductor for the tapped circuit as this is the same size as the feeder conductors.

For the 25-ft (7.5-m) tap rule, the ampacity of the smaller conductors must be not less than one-third the rating of the upstream overcurrent protective device.

Here's How: For example, a 600-ampere overcurrent device protects the feeder. The tap conductors must have an ampacity of 200 amperes or more (one-third of 600 amperes). Size 3/0 AWG copper conductors are selected for the phase conductors. *Table 250.122* requires a 1 AWG copper equipment grounding conductor for the 600-ampere overcurrent device. The provisions of *250.122(G)* require the 1 AWG equipment grounding conductor for the tapped circuit, based on the rating of the 600-ampere overcurrent device. A 200-ampere overcurrent device would result in a 6 AWG equipment grounding conductor.

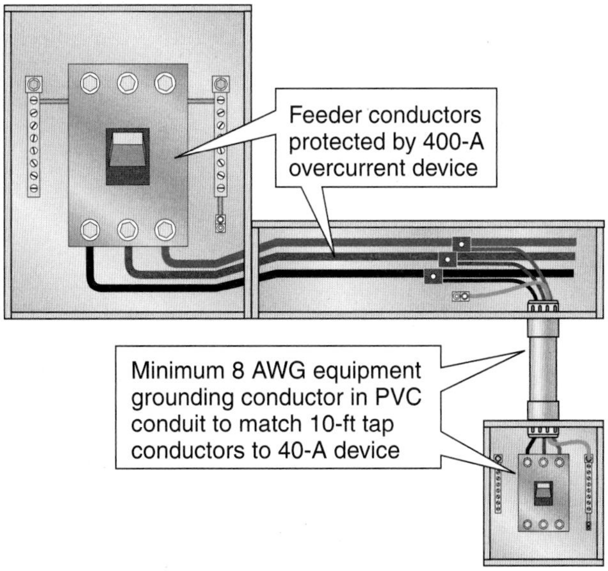

FIGURE 6-27 Equipment grounding conductor for feeder taps, *250.122(G)*. (*© Cengage Learning 2012*)

250.124 EQUIPMENT GROUNDING CONDUCTOR CONTINUITY

250.124(A) Separable Connections

The Requirements: Separable connections such as those provided in drawout equipment or attachment plugs and mating connectors and receptacles must provide for first-make, last-break of the equipment grounding conductor. First-make, last-break is not required where interlocked equipment, plugs, receptacles, and connectors prevent energization without grounding continuity.

Discussion: The design of attachment plugs demonstrates the requirements of this section. The

FIGURE 6-28 Equipment grounding conductor continuity, *250.124.* (*© Cengage Learning 2012*)

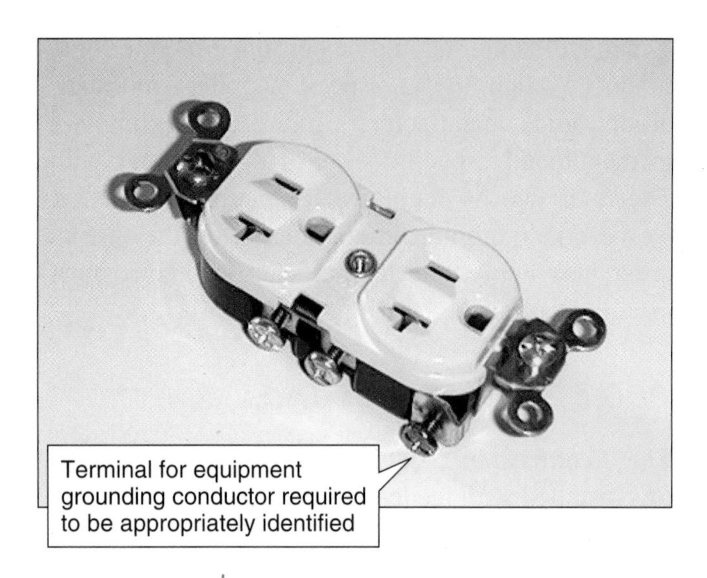

FIGURE 6-29 Identification of wiring device terminals, *250.126.* (*© Cengage Learning 2012*)

grounding pin or terminal is longer for standard plugs and receptacles, so the grounding pin makes contact first when inserted and maintains contact longer when being withdrawn (see Figure 6-28). For twist-lock attachment caps, the design of the grounding blade may be wider than the other poles. The receptacle is designed so contact is made with the grounding pole when it is inserted into the receptacle before the attachment cap is twisted.

250.126 IDENTIFICATION OF WIRING DEVICE TERMINALS

The Requirement: The terminal for the connection of the equipment grounding conductor is required to be identified by one of the following:

1. A green, not readily removable terminal screw with a hexagonal head.
2. A green, hexagonal, not readily removable terminal nut.
3. A green pressure wire connector. If the terminal for the grounding conductor is not visible, the conductor entrance hole is required to be marked with the word "green" or "ground," the letters "G" or "GR," a grounding symbol shown or otherwise identified by a distinctive green color. If the terminal for the equipment grounding conductor is readily removable, the area adjacent to the terminal must be similarly marked.

See Figure 6-29.

Information Note: This is one example of a symbol used to identify the termination point for an equipment grounding conductor. See Figure 6-30.

Discussion: This requirement applies to terminals for wiring devices such as switches and receptacles but does not apply to terminal bars installed in panelboards. The most common connection for wiring devices is the captive hexagonal green grounding screw. Note that the manufacturer of the wiring device has established tightening torque values for these connections. Other wiring devices such as timers and dimmers may be supplied with a grounding wire pigtail, rather than a grounding screw.

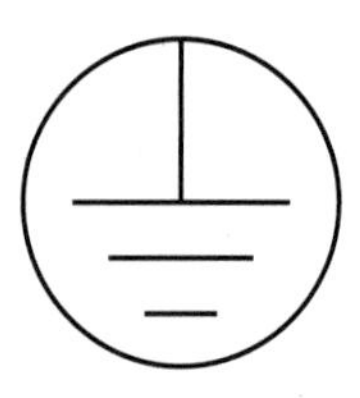

FIGURE 6-30 Example of symbol used to identify termination point for equipment grounding conductors, *250.126 FPN.* (*Reprinted with permission from NFPA 70-2011.*)

REVIEW QUESTIONS

1. Which of the following is *not* a true statement regarding the requirement for grounding of equipment?
   a. When within 8 ft (2.5 m) vertically or 5 ft (1.5 m) horizontally of ground or grounded metal objects and subject to contact by persons
   b. When in a hazardous (classified) location as covered by *Articles 500* through *517*
   c. When supplied by a metal-clad, metal-sheathed, metal-raceway, or other wiring method that provides an equipment ground
   d. When equipment operates with any terminal at over 50 volts to ground
2. Specific requirements for grounding luminaires are located in which of the following *NEC* sections?
   a. *250.112(J)*
   b. *250.122*
   c. *410.42*
   d. *600.7*
3. When a submersible pump is used in a metal well casing, the well casing is required to be bonded to
   a. the pump circuit equipment grounding conductor.
   b. the pump circuit equipment grounding conductor only if the water pipe is nonmetallic.
   c. the grounding electrode conductor.
   d. the metal water pipe from the pump to the building.
4. Which of the following statements about the use of listed flexible metal conduit as an equipment grounding conductor is *not* true?
   a. The conduit must be terminated in listed fittings listed for grounding.
   b. The overcurrent device for contained conductors must not exceed 30 amperes.
   c. The total length in the ground-fault return path must not exceed 6 ft.
   d. If flexibility is required after installation, an equipment grounding conductor is required.
5. Which of the following statements about the use of listed Type LFMC as an equipment grounding conductor is *not* true?
   a. The conduit must be terminated in listed fittings listed for grounding.
   b. For trade sizes ⅜ through ½, the overcurrent protection of contained conductors must not exceed 20 amperes.
   c. For trade sizes ¾ through 1¼, the overcurrent protection of contained conductors must not exceed 60 amperes.
   d. An equipment grounding conductor is not needed if flexibility is required after installation for listed LFNC.

6. What is the maximum permitted length in the *NEC* for 1¼-in. Type IMC–containing conductors protected by a 125-ampere current-limiting fuse?
   a. 608 ft
   b. 572 ft
   c. 356 ft
   d. No maximum
7. Individually covered or insulated conductors that are either green or green with one or more yellow stripes are
   a. required to be used as an equipment grounding conductor.
   b. permitted to be used as an ungrounded ("hot") conductor if phase-taped black.
   c. permitted to be used as a grounded (neutral) conductor if phase-taped white.
   d. required to be used as a grounding electrode conductor.
8. Which of the following statements about equipment grounding conductors larger than 6 AWG is *not* true?
   a. They are required to have green insulation.
   b. They are permitted to be bare.
   c. They are permitted to be phase-taped green.
   d. Any identification used must encircle the conductor.
9. Which of the following statements about aluminum equipment grounding conductors is true?
   a. Aluminum equipment grounding conductors are not permitted to be direct-buried.
   b. Bare aluminum equipment grounding conductors are permitted to come in direct contact with masonry.
   c. Aluminum equipment grounding conductors are not permitted to be terminated within 18 in. of the earth.
   d. Aluminum equipment grounding conductors smaller than 6 AWG are not permitted.
10. When ungrounded ("hot") conductors are increased in size, equipment grounding conductors are
    a. required to be sized with use of *Table 250.122* in accordance with the rating of the overcurrent device.
    b. required to be increased the same number of sizes as for the phase conductors.
    c. required to be increased in size in proportion to the phase conductors.
    d. required to be of the same size as for the phase conductors.
11. Which of the following phrases is *not* true regarding the primary purpose of increasing the size of equipment grounding conductors?
    a. When the phase conductors are increased in size to adjust for voltage drop
    b. When the phase conductors are increased in size to adjust for high ambient temperatures
    c. When the phase conductors are increased in size to adjust for more than three current-carrying conductors in a raceway or cable
    d. When the phase conductors are increased in size for high harmonic currents

12. Which of the following statements about sizing the equipment grounding conductors is true?
    a. They are sized from *Table 250.122* in accordance with the rating of the overcurrent device.
    b. They are of the same size as for the branch-circuit or feeder conductors.
    c. They are sized from *Table 250.66* in accordance with the rating of the overcurrent device.
    d. They are sized in accordance with the rating of the running overload protection for continuous duty motors.
13. Which of the following statements about sizing equipment grounding conductors for parallel circuits is true?
    a. When all parallel conductors are installed in a common raceway, an equipment grounding conductor must be installed for each set of parallel conductors.
    b. When parallel conductors are installed in separate cables, a full-size equipment grounding conductor must generally be installed in each cable.
    c. When parallel conductors are installed in separate raceways, the equipment grounding conductor is sized according to the ampacity of the conductors in the raceway.
    d. The equipment grounding conductor must be sized not smaller than the load rating of the equipment at the termination of the circuit.
14. Which of the following statements about sizing equipment grounding conductors is true when feeder conductors are tapped in accordance with the rules in *240.21*?
    a. Equipment grounding conductors are sized in accordance with the rating of the overcurrent device at the termination of the 10-ft tap approval.
    b. Equipment grounding conductors are sized in accordance with the rating of the overcurrent device at the termination of the 25-ft tap rule.
    c. Equipment grounding conductors are sized in accordance with the rating of the overcurrent device on the line side of the tapped conductors.
    d. Equipment grounding conductors are required to be sized the same as for the feeder tap conductors.
15. The frames of motors are required to be connected to an equipment grounding conductor in accordance with *250.110*.
    a. True
    b. False
16. Electric signs are required to be grounded in accordance with the rules in 600.7 rather than according to Article 250.
    a. True
    b. False
17. List the three broad types of Type MC cable.
    a. ______________________
    b. ______________________
    c. ______________________
18. The UL Product Standard to which listed Type MC cable is manufactured is ________.

19. Equipment grounding conductors may have to be sized larger than given in *Table 250.122* for which of the following reasons?

    (1) The conductor is not capable of withstanding the available short-circuit current.

    (2) The circuit conductors are increased in size.

    a. Reason (1) only
    b. Reason (2) only
    c. Both reasons (1) and (2)
    d. Neither reasons (1) nor (2)

20. For general motor circuits, the size of the equipment grounding conductors is required to be based on

    a. *Table 250.122,* based on the rating of the branch-circuit, short-circuit, and ground-fault protection.
    b. *Table 250.122,* based on the size of the motor branch-circuit conductors.
    c. *Table 250.122*, based on the rating of the motor running overload protection.
    d. *Table 250.66*, based on the size of the motor branch-circuit conductors.

21. The minimum size of an equipment grounding conductor for feeder taps is determined

    a. based on the rating of the overcurrent protection at the load end of the feeder tap.
    b. based on the ampacity of the tap conductors from *Table 310.16.*
    c. based on the ampacity of the feeder to which the tap conductors are connected.
    d. based on the rating of the overcurrent protection on the supply side of the feeder conductor.

UNIT 7

Methods of Equipment Grounding

OBJECTIVES

After studying this unit, the reader will understand

- equipment grounding conductor connections.
- methods permitted for nongrounding receptacle replacement and branch-circuit extensions.
- rules for equipment fastened in place or connected by permanent wiring methods.
- grounding frames of ranges, dryers, and associated equipment.
- use of grounded circuit conductor for grounding equipment.
- connecting receptacle grounding terminals.
- bonding of general-use snap switches.
- continuity and attachment of equipment grounding conductors to boxes.

250.130 EQUIPMENT GROUNDING CONDUCTOR CONNECTIONS

The Requirements: Equipment grounding conductor connections at the source of separately derived systems are required to be made in accordance with *250.30(A)(1)*. Equipment grounding conductor connections at service equipment must be made as indicated in *250.130(A)* or *(B)*. For replacement of nongrounding-type receptacles with grounding-type receptacles and for branch-circuit extensions only in existing installations that do not have an equipment grounding conductor in the branch circuit, connections as indicated in *250.130(C)* are permitted.

Discussion: As discussed earlier for grounded systems, the main bonding jumper for the service and the system bonding jumper for separately derived systems provides the critical link between the equipment grounding conductor and the grounded system conductor. This is true whether the equipment grounding conductor is of the wire type or is metallic conduit such as RMC or IMC or metal tubing (EMT). For ungrounded systems, the equipment grounding conductor, regardless of whether it is of the wire type or the metallic raceway type, is connected to the service or building disconnecting means enclosure. The grounding electrode(s) are connected at that same point.

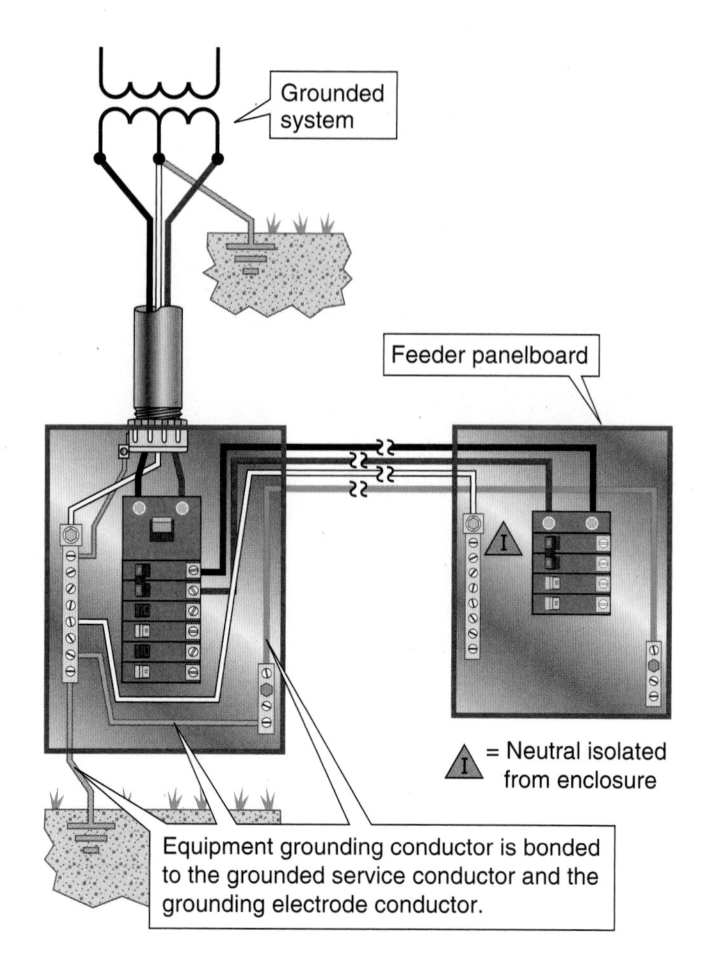

FIGURE 7-1 Equipment grounding conductor connections for grounded systems, *250.130(A)*. (*© Cengage Learning 2012*)

250.130(A) for Grounded Systems

The Requirement: The connection shall be made by bonding the equipment grounding conductor to the grounded service conductor and the grounding electrode conductor.

Discussion: The main bonding jumper or system bonding jumper provides the link to connect the equipment grounding conductor to the grounded service conductor. The grounded system conductor provides the low-impedance path to return ground-fault current back to the source. Although the connection at the service includes a connection to the grounding electrode conductor, very little current will flow through this path because it is a much higher-impedance path than through the grounded conductor. (See Figure 7-1.)

250.130(B) for Ungrounded Systems

The Requirement: The connection is required to be made by bonding the equipment grounding conductor to the grounding electrode conductor.

Discussion: As discussed at *250.4(B)*, the enclosure and equipment grounding conductors are not connected to the electrical supply system if the system is ungrounded. The equipment grounding conductors and the grounding electrode conductor(s) are connected to the service or building or structure disconnect enclosure for the ungrounded electrical supply system. This connection attempts to keep the equipment for the ungrounded system at earth potential, which reduces the shock hazard to personnel. (See Figure 7-2.)

FIGURE 7-2 Equipment grounding conductor connections for ungrounded systems, *250.130(B)*. (*© Cengage Learning 2012*)

250.130(C) Nongrounding Receptacle Replacement or Branch-Circuit Extensions

The Requirement: The equipment grounding conductor of a grounding-type receptacle or a branch-circuit extension is permitted to be connected to any of the following:

1. Any accessible point on the grounding electrode system as described in *250.50*
2. Any accessible point on the grounding electrode conductor
3. The equipment grounding terminal bar within the enclosure where the branch circuit for the receptacle or branch circuit originates
4. For grounded systems, the grounded service conductor within the service equipment enclosure
5. For ungrounded systems, the grounding terminal bar within the service equipment enclosure

Discussion: The installation of receptacles, both grounding-type and nongrounding-type, is covered in *Article 406*. Basically, grounding-type receptacles are required to be installed if an equipment grounding conductor is available at the outlet where the receptacle is located in compliance with *406.4(A)*.

Special rules are provided for replacement of receptacles in *406.4(D)*.

1. If a grounding means exists in the receptacle enclosure or an equipment grounding conductor is installed as provided in *250.130(C)*, grounding-type receptacles are required to be installed and be connected to the equipment grounding conductor in accordance with *406.4(C)* or *250.130(C)*.
2. If a grounding means does not exist in the receptacle enclosure, and nongrounding-type receptacles are being replaced, three options are available:
   a. Another nongrounding-type receptacle is permitted to be installed. (See rule (3)(b).)
   b. A nongrounding-type receptacle is permitted to be replaced with a GFCI-type receptacle, even though an equipment grounding conductor is not available in the box. These receptacles are required to be marked "No Equipment Ground," to inform the user that despite the equipment grounding terminal present on the receptacle, an equipment grounding connection is not provided. (Some authorities refer to this feature as a "phantom ground.") It is not permitted to install an equipment grounding conductor from the GFCI-type receptacle to any outlet supplied from the GFCI device, because no equipment grounding conductor is available in the outlet.
   c. A nongrounding-type receptacle is permitted to be replaced with a grounding type receptacle if it is supplied through a GFCI device. If this is done, the grounding type receptacle is required to be marked "GFCI Protected" and "No Equipment Ground." This marking serves to inform the user that even though the receptacle has an equipment grounding terminal, an equipment grounding connection is not provided, and that GFCI protection is provided upstream.

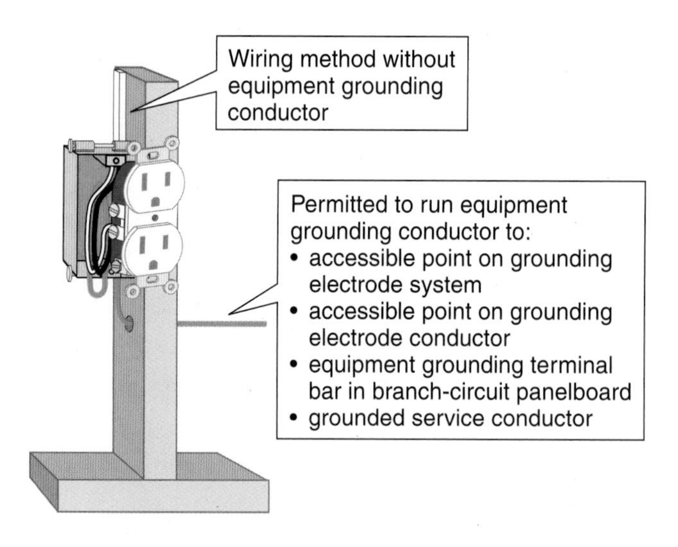

FIGURE 7-3 Nongrounding receptacle replacement, *250.130(C)*. (*© Cengage Learning 2012*)

3. If the receptacle is replaced at a location where a ground-fault circuit interrupter is required, GFCI protection is required by the installation of either a circuit breaker to protect the branch circuit or a GFCI receptacle.

See Figures 7-3 and 7-4.

CAUTION: Many types of equipment are required to be grounded by *Article 250.* For the requirements in *Article 250,* see the conditions under which equipment is required to be connected to an equipment grounding conductor in *250.110,* the list of equipment required to be connected to an equipment grounding conductor if fastened in place or connected by a permanent wiring method in *250.112,* and the equipment required to be connected to an equipment grounding conductor where connected by cord and plug in *250.114.*

In addition, manufacturers' installation instructions often specifically require equipment to be grounded or connected to an equipment grounding conductor, such as microwave ovens and refrigerators. Compliance with these installation instructions is required by *110.3(B).*

Other sections in the *Code* may have specific requirements for providing a grounding-type receptacle or grounding-type connection. For example, *517.13(B)* requires grounding-type receptacles be installed in patient care areas. These receptacles are required to be connected to not less

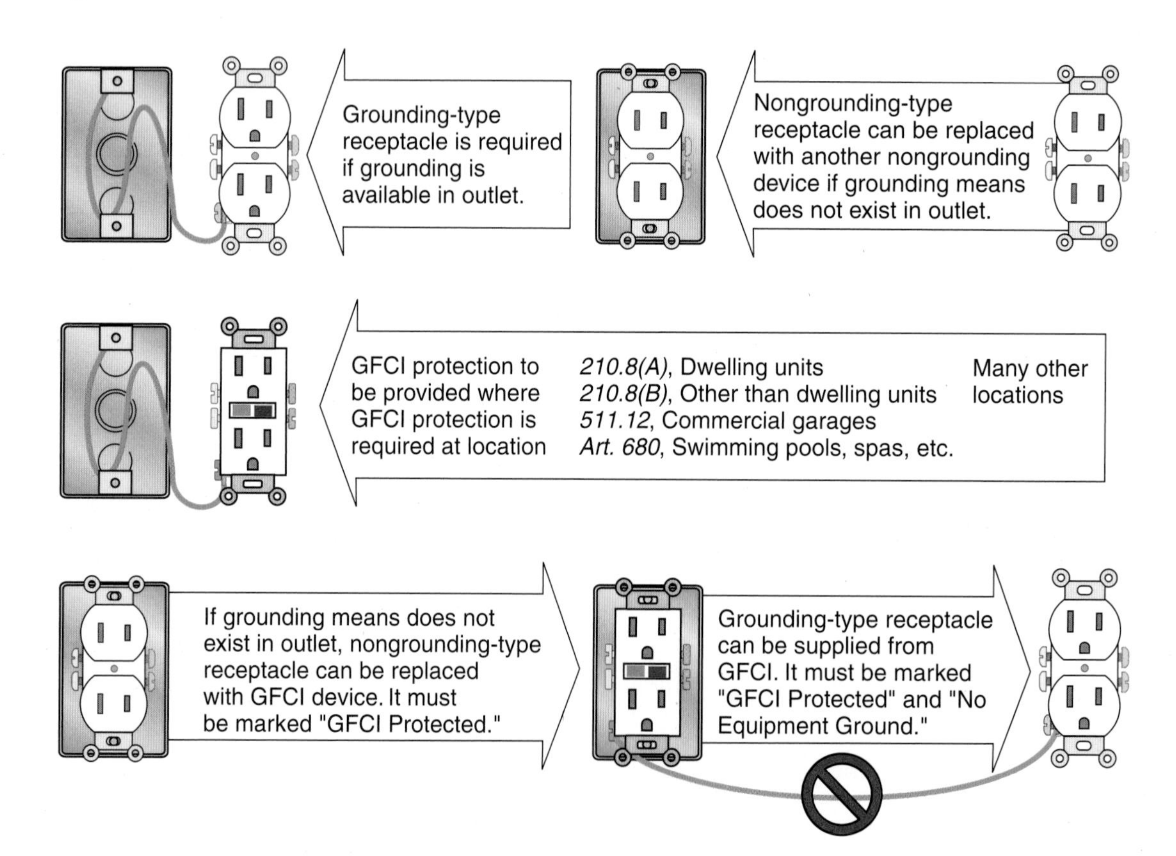

FIGURE 7-4 Replacement of nongrounding receptacles and GFCI protection, *250.130(C)* and *406.3(D)*. (*© Cengage Learning 2012*)

than two equipment grounding paths, the wiring method and an insulated equipment grounding conductor. A metal outlet or device box is required to be grounded by *314.4*. It would obviously be a violation of *Code* rules to install a metal box and then attempt to comply with the requirements for connecting the equipment to an equipment grounding conductor by locating a GFCI device on the supply side of the box or receptacles in a patient care area.

It is fairly obvious that a GFCI device is not permitted to substitute for the specific equipment grounding requirements of the equipment manufacturer or if equipment grounding is specifically required in other sections of the *Code*.

Discussion: Branch-circuit extensions are what the name implies—an extension of an existing branch circuit (see Figure 7-5). Even though the existing branch circuit may not have an equipment grounding conductor as a part of the wiring method, the extension of the branch circuit is considered new work, and the branch circuit must provide an equipment grounding conductor. Fortunately, this section of the *Code* provides a method to include an equipment grounding conductor for the new work.

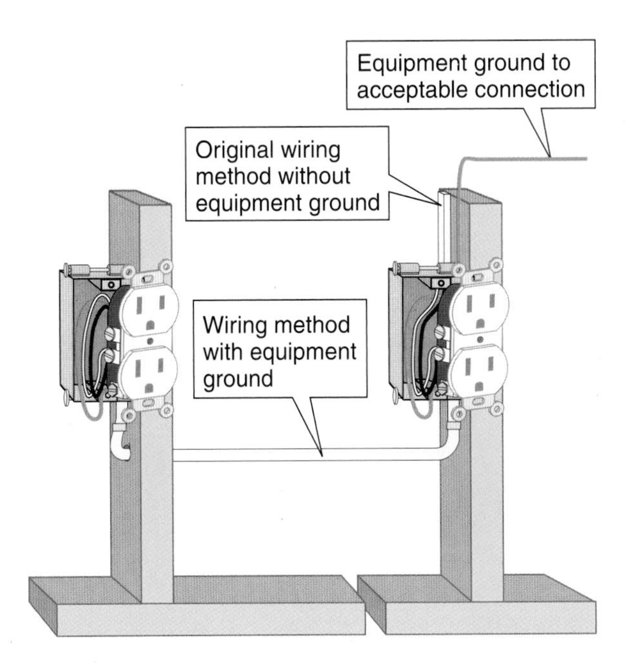

FIGURE 7-5 Branch-circuit extension, *250.130(C)*. (*© Cengage Learning 2012*)

Here's How: A junction or device box is installed at the point on the existing branch circuit where the extension is to originate. An equipment grounding conductor is installed back (on the system) to the point at which the branch circuit originates or to one of the other locations permitted in *250.130(C)*. The equipment grounding conductor does not necessarily have to be installed in a cable or a conduit so long as it is protected from physical damage. See *250.120* for installation requirements for equipment grounding conductors, including *Part (C)* for equipment grounding conductors smaller than 6 AWG. Installing an equipment grounding conductor remote from the wiring method is less than ideal, because the ground-fault return path is not run with the circuit conductors. However, this provision is a practical application of the *Code* rules to accommodate an existing condition.

250.134 EQUIPMENT FASTENED IN PLACE OR CONNECTED BY PERMANENT WIRING METHODS (FIXED)—GROUNDING

The Requirement: Unless grounded by connection to the grounded circuit conductor as permitted by *250.32, 250.140,* and *250.142,* non-current-carrying metal parts of equipment, raceways, and other enclosures, if grounded, must be connected to an equipment grounding conductor by one of the following methods.

Discussion: *NEC 250.134* does not simply require that all non-current-carrying metal parts of equipment be connected to an equipment grounding conductor. It recognizes

1. the permission in *250.32(B,) Exception,* to connect equipment at existing buildings or structures supplied by a feeder or branch circuit to the grounded-circuit conductor under specific conditions, and
2. the permission to connect the frame of existing ranges or dryers to the grounded-circuit conductor, and finally,
3. to use the grounded-circuit conductor for grounding (and bonding) equipment associated with the service and, to a limited extent, equipment that is downstream from the service equipment.

250.134(A) Equipment Grounding Conductor Types

The Requirement: Equipment that is required to be grounded (connected to an equipment grounding conductor) is permitted to be connected to any of the equipment grounding conductors permitted or recognized by *250.118.*

250.134(B) With Circuit Conductors

The Requirement: Equipment that is required to be grounded is permitted to be grounded by connecting it to an equipment grounding conductor contained within the same raceway, cable, or otherwise run with the circuit conductors.

Exception: *As provided in 250.130(C), the equipment grounding conductor is permitted to be run separately from the circuit conductors.*

Discussion: The *NEC* recognizes here the importance of all of the circuit conductors being run together, to provide a low-impedance path for fault current. See the extensive discussion at *250.4(A)(5)* in this book as well as the information on the subject in Appendix A.

The exception allows the equipment grounding conductor to be run separately from the circuit conductors for replacing nongrounding-type receptacles with grounding type and for branch circuit extensions. See the discussion of these issues at *250.130(C)* in this book.

250.136 EQUIPMENT CONSIDERED GROUNDED

The Requirements: Under the conditions specified in *250.136(A)* and *(B),* the non-current-carrying metal parts of the equipment are considered to be grounded.

250.136(A) Equipment Secured to Grounded Metal Supports

The Requirements: Electrical equipment secured to and in electrical contact with a metal rack or structure provided for its support and connected to an equipment grounding conductor by one of the means indicated in *250.134.* The structural metal frame of a building is not permitted to be used as the required equipment grounding conductor for ac equipment. See Figure 7-6.

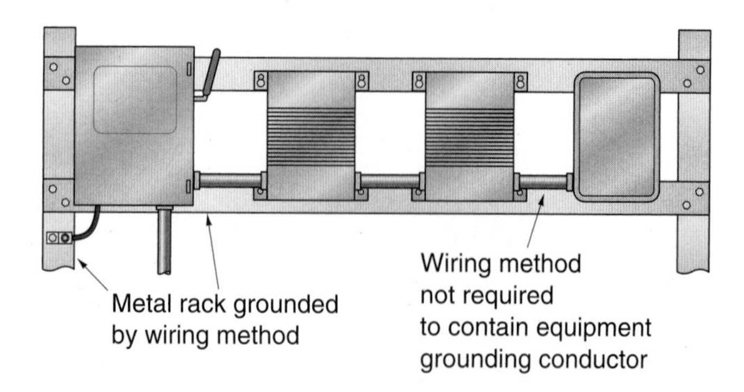

FIGURE 7-6 Use of metal rack to ground equipment, *250.136(A).* (*© Cengage Learning 2012*)

Discussion: The permitted method of grounding equipment by securing it to a metal rack that is properly grounded by connecting it to an equipment grounding conductor supports the similar rule in *250.112(K)* for grounding of skid-mounted equipment. However, note the requirement that the building not be used as an equipment grounding conductor. This rule supports the requirement in *250.134* for installing an equipment grounding conductor, as well as the rule in *300.3(B)* for keeping all of the circuit conductors together, including the ungrounded phase conductors, the grounded conductor (if used), and the equipment grounding conductor. These rules work together to accomplish the effective fault-current return path required in *250.4(A)(5).*

250.140 FRAMES OF RANGES AND CLOTHES DRYERS

The Requirement: Frames of electric ranges, wall-mounted ovens, counter-mounted cooking units, clothes dryers, and outlet or junction boxes that are part of the circuit for these appliances are required to be connected to an equipment grounding conductor

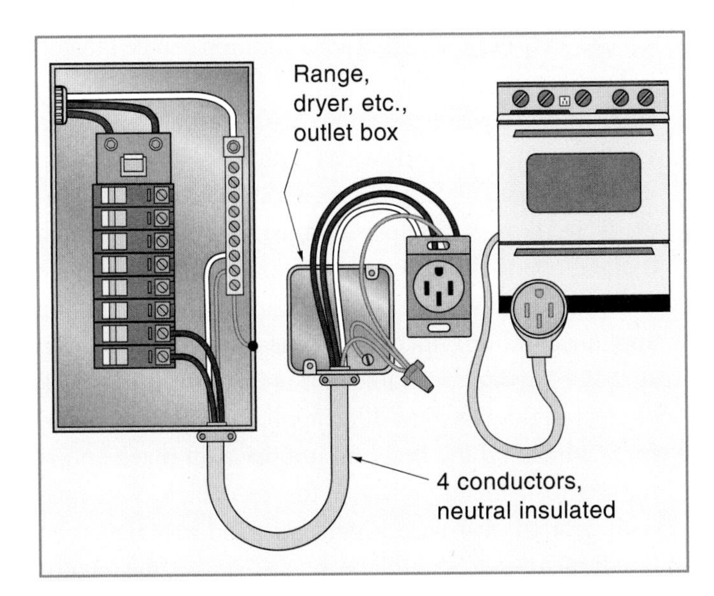

FIGURE 7-7 Range and dryer equipment grounding for branch circuits installed upon or after the adoption of the 1996 *NEC, 250.140.* (© *Cengage Learning 2012*)

in the manner specified by *250.134* or *250.138.* See Figure 7-7.

Discussion: The general rule is these appliances and the outlet or junction boxes that are in the supply circuit be connected to an equipment grounding conductor that is run with the branch-circuit conductor. Thus, for plug-in type appliances, the receptacles and pigtails must be 3-pole with ground. These often are referred to as 4-wire receptacles and pigtails. The outlet boxes and frames of the appliances are grounded by the equipment grounding conductor. The neutral conductor, if supplied, is not permitted to be connected to the equipment on the supply side of the receptacle or junction box or to the appliance itself.

In addition, the neutral conductor is required to be an insulated conductor.

A Bit of History: For many years, beginning in the early 1940s, it was permitted to connect the frames of electric ranges, wall-mounted ovens, counter-mounted cooking units, clothes dryers, and outlet or junction boxes that are part of the circuit for these appliances to the branch-circuit neutral. A 3-wire branch circuit was run without an equipment grounding conductor. Likewise, 3-wire receptacles and appliance pigtails were installed. The historical reason given was an effort to conserve the copper material needed for the extra equipment grounding conductor on behalf of the war effort during World War II.

This practice continued until the processing of the 1996 *NEC,* when the present requirement for an additional equipment grounding conductor was reinstated. One of the proposed reasons for eliminating the practice of connecting the frame of the appliances to the neutral was "The war has been over for more than 48 years!"

CAUTION: Some appliances are not intended for connection with a receptacle and pigtail. These appliances typically are furnished with an "appliance whip" by the manufacturer. This "whip" usually consists of supply conductors in a flexible conduit or metal-armored cable and is intended to be connected to the branch circuit for the appliance at a junction box.

CAUTION: Ranges and dryers that are intended to be supplied through a receptacle and a pigtail are shipped from the manufacturing plant with the bonding jumper from the neutral to the frame in place. This is to ensure that the frame of the range or dryer will be grounded regardless of whether it is supplied by a 3-wire or a 4-wire pigtail. This bonding jumper should be disconnected when connecting the 4-wire pigtail, to prevent the neutral from being grounded to the frame.

Exception: *For existing branch-circuit installations only where an equipment grounding conductor is not present in the outlet or junction box, the frames of electric ranges, wall-mounted ovens, counter-mounted cooking units, clothes dryers, and outlet or junction boxes that are part of the circuit for these appliances are permitted to be connected to the grounded-circuit conductor if all the following conditions are met.*

1. *The supply circuit is 120/240-volt, single-phase, 3-wire; or 208Y/120-volt derived from a 3-phase, 4-wire, wye-connected system.*
2. *The grounded conductor is not smaller than 10 AWG copper or 8 AWG aluminum.*
3. *The grounded conductor is insulated, or the grounded conductor is uninsulated and part of a*

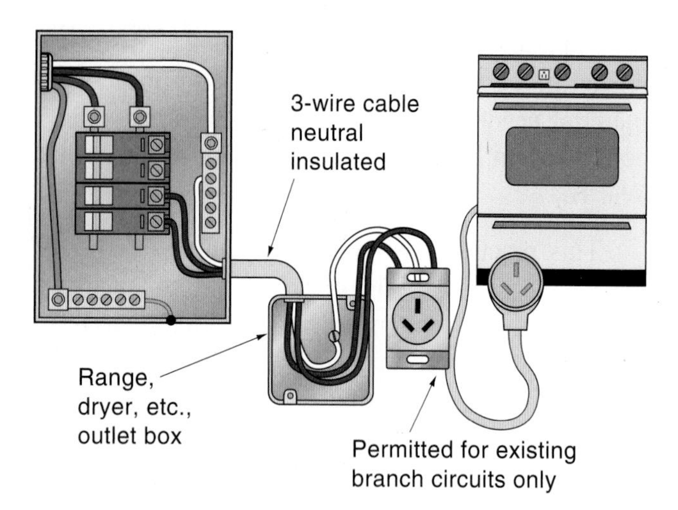

FIGURE 7-8 Grounding of ranges and dryers for existing branch circuits installed before the adoption of the 1996 *NEC, 250.140, Exception.* (*© Cengage Learning 2012*)

Type SE service-entrance cable and the branch circuit originates at the service equipment.

4. *Grounding contacts of receptacles furnished as part of the equipment are bonded to the equipment.*

See Figure 7-8.

Discussion: As can be seen, very stringent conditions must be met before the frame of the range and dryer and associated equipment are permitted to be connected to the neutral. It also is clear that this exception applies to existing branch circuits only. New branch-circuit installations are required to comply with the general rule. The authority having jurisdiction (AHJ) should be contacted to determine how a branch-circuit extension for a remodel project will be treated.

When compliance with all of the applicable conditions has been ensured, a 3-conductor branch circuit, 3-pole receptacle, and 3-wire pigtail are permitted. Note that the neutral conductor is required to be insulated or, if the branch circuit originates at service equipment, it is permitted to have two insulated conductors with one bare conductor serving as the neutral. All conductors of the branch circuit are required to be insulated if the existing branch circuit originates at a feeder panel, often referred to as a subpanel. Presence of the insulation avoids the likelihood that a bare neutral conductor will unintentionally bond the neutral to the panelboard enclosure.

CAUTION: A dangerous condition can be created when a range or a dryer is moved from one location to another if the wiring schemes are different. Where an appliance that has been wired with a 4-wire receptacle and pigtail and moved to a location where the receptacle is 3-pole, a 3-pole pigtail must be installed. Recall that the bonding jumper must be installed between the neutral terminal in the appliance junction box and the appliance frame. Otherwise, a line-to-ground fault will simply energize the frame and create a serious shock or electrocution hazard. If an appliance has been wired for 3 wires and the bonding jumper is in place and is moved to a location where the receptacle is 3-pole with ground, a 4-wire pigtail must be installed. The bonding jumper should be removed and the equipment grounding conductor connected to the appliance frame. If the bonding jumper is left in place, the appliance still will be grounded, but through the neutral.

250.142 USE OF GROUNDED CIRCUIT CONDUCTOR FOR GROUNDING EQUIPMENT

250.142(A) Supply-Side Equipment

The Requirements: A grounded circuit conductor (often a neutral) is permitted to be used to ground non-current-carrying metal parts of equipment, raceways, and other enclosures at any of the following locations:

1. On the supply side or within the enclosure of the ac service disconnecting means
2. On the supply side or within the enclosure of the main disconnecting means for separate buildings as provided in *250.32(B)*
3. On the supply side or within the enclosure of the main disconnecting means or overcurrent devices of a separately derived system where permitted by *250.30(A)(1)*

Discussion: Because this part of *Article 250* deals with grounding, these subsections recite other places in the article where it is specifically permitted to use the grounded system conductor to ground the equipment. As a reminder, "grounded" means connected to ground (earth), or to some conductive body that extends the ground (earth) connection. Refer to the specific section reference for the particulars related to the use of the grounded system (circuit) conductor to ground equipment.

250.142(B) Load-Side Equipment

The Requirements: Except as permitted in *250.30(A)(1)* and *250.32(B)* Exception, a grounded circuit conductor is not permitted to be used for grounding non-current-carrying metal parts of equipment on the load side of the service disconnecting means or on the load side of a separately derived system disconnecting means or the overcurrent devices for a separately derived system not having a main disconnecting means.

Exception No. 1: *The frames of ranges, wall-mounted ovens, counter-mounted cooking units, and clothes dryers are permitted to be connected to a grounded-circuit conductor under the conditions permitted for existing installations by 250.140.*

Discussion: The grounded circuit (system) conductor generally is not permitted to be used to ground equipment on the load side of the service. One reason for this rule is that if equipment is grounded to the neutral and the neutral conductor becomes loose or is broken, a single ground fault will energize the metal equipment on the load side of the break at the line-to-ground voltage.

Exception No. 2: *It is permissible to ground meter enclosures by connection to the grounded circuit conductor on the load side of the service disconnect if*

1. *Ground-fault protection of equipment is not installed for the service, and*
2. *All meter enclosures are located immediately adjacent to the service disconnecting means, and*
3. *The size of the grounded circuit conductor is not smaller than the size specified in Table 250.122 for equipment grounding conductors.*

See Figure 7-9.

FIGURE 7-9 Use of grounded-circuit conductor for grounding equipment on load side of service, *250.142(B), Exception 2.* (*© Cengage Learning 2012*)

Discussion: This exception has application for multimeter installations that are commonly installed for multiple-occupancy buildings. When the number of disconnecting means exceeds six, it is common to put a main disconnecting means on the line side of multimeter equipment. Usually, the serving utility will lock or seal the disconnecting means to prevent energy diversion (theft). The multimeter equipment is permitted to be grounded by the grounded circuit (system) conductor because it is immediately adjacent to the service disconnecting means.

This provision applies whether or not the neutral in the meter equipment is bonded to the meter enclosure. If the neutral is bonded to the meter enclosure that is located on the load side of the service disconnecting means, a parallel path for neutral current is established between the points of bonding of the neutral. This parallel path for neutral current is permitted by these *Code* rules.

250.146 CONNECTING RECEPTACLE GROUNDING TERMINAL TO BOX

The Requirement: An equipment bonding jumper is required to be used to connect the grounding terminal of a grounding-type receptacle to a grounded

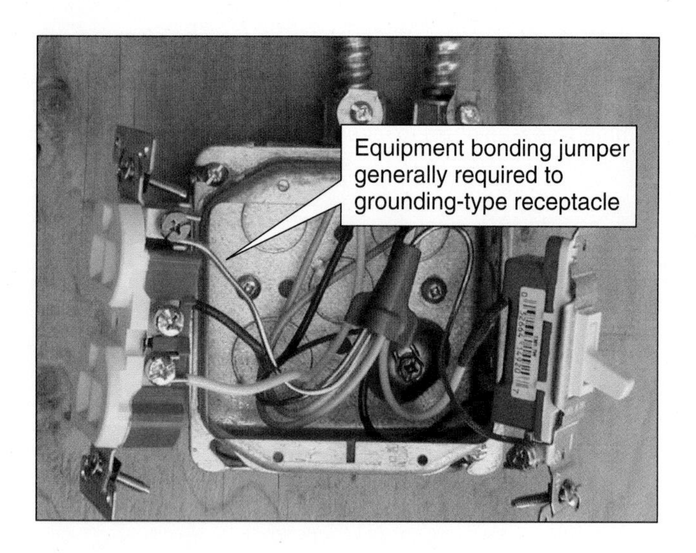

FIGURE 7-10 Connecting receptacle grounding terminal to box, *250.146.* (*© Cengage Learning 2012*)

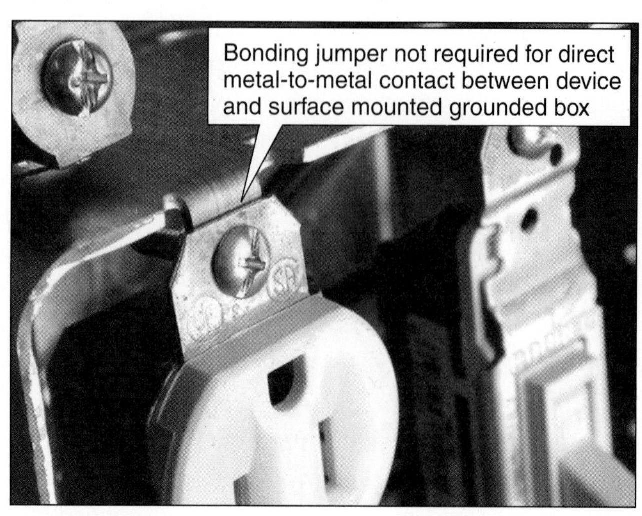

FIGURE 7-11 Receptacle in surface-mounted box, *250.146(A).* (*© Cengage Learning 2012*)

box unless grounded as in *250.146(A)* through *(D).* See Figure 7-10. The equipment bonding jumper is required to be sized in accordance with *Table 250.122,* based on the rating of the overcurrent device protecting the circuit conductors.

Discussion: The requirement to install a bonding jumper from a grounded box to the receptacle is a general one, with several exceptions in *(A)* through *(D).* The bonding jumper ensures a positive connection between the receptacle and the box as well as to the equipment grounding conductor. Some receptacles are installed with the plaster ears on the surface of the wall finish; thus, there is not a metal-to-metal connection between the device and the box. The bonding jumper is sized the same as for the equipment grounding conductor of the circuit.

250.146(A) Surface-Mounted Box

The Requirements: Where the box is mounted on the surface, direct metal-to-metal contact between the device yoke and the box or a contact yoke or device that complies with *250.146(B)* is permitted to ground the receptacle to the box. At least one of the insulating washers must be removed from receptacles that do not have a contact yoke or device that complies with *250.146(B)* (self-grounding type) to ensure direct metal-to-metal contact.

Discussion: Handy boxes and similar formed steel boxes often are mounted on the surface. As shown in Figure 7-11, if the receptacle has direct metal-to-metal contact with the box, a bonding jumper is not required. Also, a self-grounding receptacle is not required to comply with this section by having a bonding jumper to the box. The ground-fault return path is from the grounding terminal on the receptacle to the metal yoke of the device to the grounded box over the equipment grounding conductor back to the source.

This section requires at least one of the nonmetallic washers that are designed to retain the mounting screws on the wiring device be removed. This mirrors the way the receptacles are tested to verify compliance with the product safety standard. It also ensures direct metal-to-metal contact between the receptacle and the grounded box.

Note that it is not permitted to ground the wiring device and then use the wiring device to ground the box, because the box would no longer be grounded if the wiring device was removed for any reason. In other words, the grounded box is permitted to ground the receptacle, but a grounded receptacle is not permitted to ground the box.

More Requirements: This provision does not apply to cover-mounted receptacles unless the box and cover combination are listed as providing

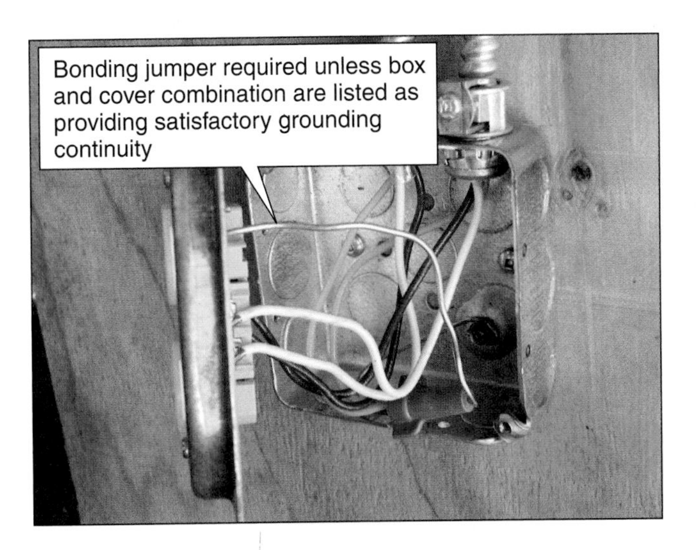

FIGURE 7-12 Boxes with surface-mounted covers, *250.146(A)*. (*© Cengage Learning 2012*)

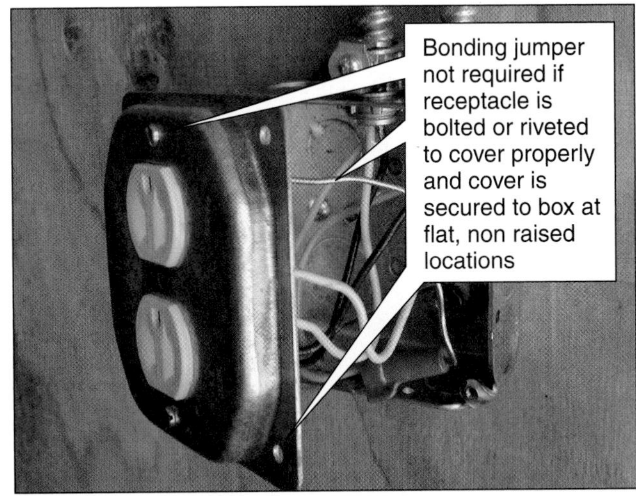

FIGURE 7-13 Exposed work cover listed for grounding without bonding jumper, *250.146(A)*. (*© Cengage Learning 2012*)

satisfactory ground continuity between the box and the receptacle.

Discussion: Cover-mounted receptacles commonly are installed for 4-in.-square or 4¹¹⁄₁₆-in.-square metal boxes that are grounded (see Figure 7-12). The covers are domed about ½ in. to provide additional space for the receptacle(s). If the box and cover are listed as an assembly, they are marked for that purpose and do not require a bonding jumper between the grounded box and the receptacle. If the box and cover are not so marked, a bonding jumper is required. So far as can be determined, no box-cover combinations are available that are listed for grounding. The size of the bonding jumper from the box to the receptacle is the same as for the equipment grounding conductor.

More Requirements: A listed exposed work cover is permitted to be the grounding and bonding means when (1) the device is attached to the cover with at least two fasteners that are permanent (such as a rivet) or have a thread locking or screw or nut locking means and (2) when the cover mounting holes are located on a flat nonraised portion of the cover.

Discussion: Covers have been designed and listed to provide a reliable connection between the receptacle and the grounded surface-mounted metal box. They can be recognized by the receptacle being connected to the raised cover by two or more rivets or bolts and locking-type nuts and by having a flat surface for connecting the cover to the box. (See Figure 7-13.) The grounded metal box thus serves to complete the effective ground-fault return path from the receptacle back to the source.

250.146(B) Contact Devices or Yokes

The Requirement: Contact devices or yokes designed and listed as self-grounding are permitted in conjunction with the supporting screws to establish the grounding circuit between the device yoke and flush-type boxes without a bonding jumper.

Discussion: These wiring devices commonly are referred to as self-grounding receptacles (see Figure 7-14). They are identified by a spring mechanism of some type that ensures positive connection between the metal yoke of the receptacle to the device mounting screw and to the grounded box. These self-grounding receptacles are not required to have a bonding jumper between the device and the metal box.

The self-grounding receptacles are permitted to be used where the box is mounted on the surface, at the surface, or recessed back from the surface no more than ¼ in. as permitted in *314.20* for walls or ceilings with a surface of concrete, tile, gypsum, plaster, or other noncombustible material.

FIGURE 7-14 Self-grounding receptacle, *250.146(B)*. (*© Cengage Learning 2012*)

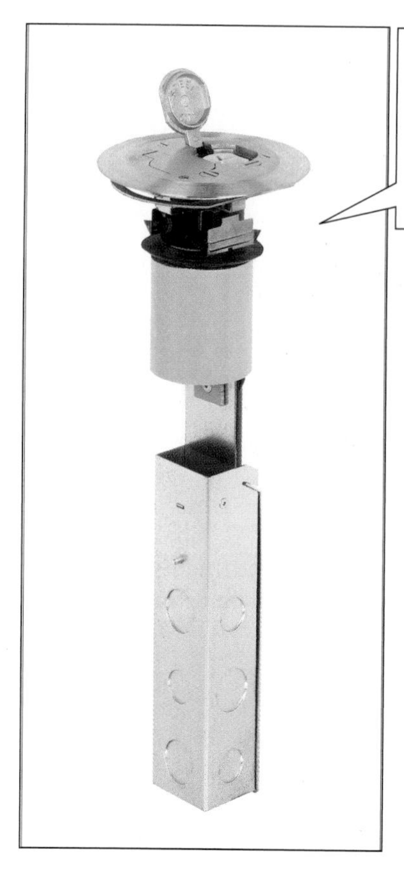

FIGURE 7-15 Receptacles in listed floor boxes permitted without bonding jumper, *250.146(C)*. (*Courtesy of Thomas and Betts*)

250.146(C) Floor Boxes

The Requirement: Floor boxes designed for and listed as providing satisfactory ground continuity between the box and the device are permitted without the device-to-box bonding jumper. See Figure 7-15.

Discussion: The construction of these floor boxes includes either direct metal-to-metal contact or a factory-provided bonding jumper to ensure a reliable ground-fault return path for cord-connected equipment.

250.146(D) Isolated Receptacles

The Requirements:

1. Where installed for the reduction of electrical noise (electromagnetic interference) on the grounding circuit, a receptacle in which the grounding terminal is purposely insulated from the receptacle mounting means is permitted without a bonding jumper to the box.
2. The receptacle grounding terminal is required to be grounded by an insulated equipment grounding conductor run with the circuit conductors.
3. This equipment grounding conductor is permitted to pass through one or more panelboards without connection to the panelboard grounding terminal as permitted in *408.20, Exception.* The isolated grounding conductor must terminate within the same building or structure directly at an equipment grounding conductor terminal of the applicable derived system or service.
4. If installed in accordance with the provisions of this section, this equipment grounding conductor is also permitted to pass through boxes, wireways, or other enclosures without being connected to such enclosures.

Informational Note: The use of an isolated equipment grounding conductor does not relieve the requirement for grounding the raceway system and outlet box. See Figure 7-16.

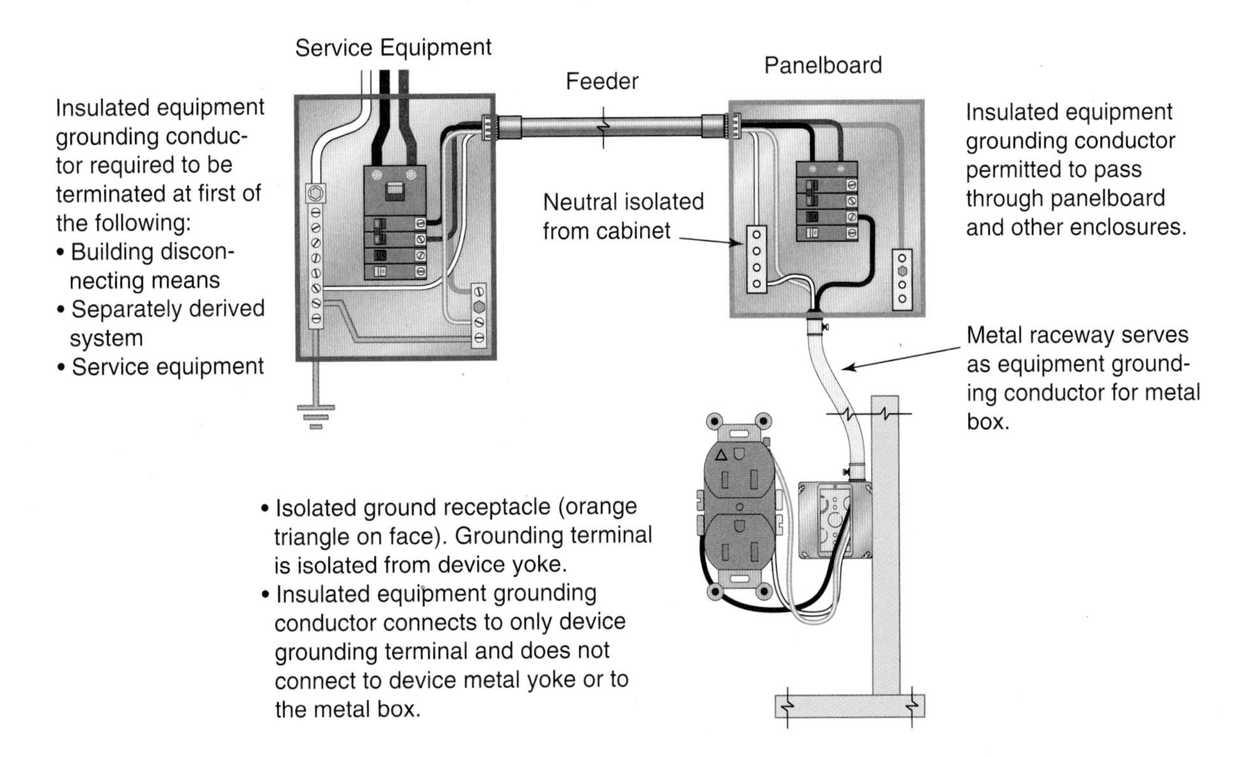

FIGURE 7-16 Installation of isolated-ground receptacle, *250.146(D)*. (© *Cengage Learning 2012*)

Discussion: It is common to install these receptacles for personal computers and peripherals such as printers and scanners, as well as for cash registers or terminals. Once the decision is made to install isolated grounding receptacles, the rules in this and other sections must be followed. A bonding jumper is not required from the receptacle to the metal box. This concept is sometimes referred to as providing a "clean ground," whereas the building ground is called a "dirty ground" because of the associated electromagnetic interference.

If this concept or grounding scheme is chosen, an insulated equipment grounding conductor is required to be installed (see Figure 7-16). It is sized from *Table 250.122* on the basis of the rating of the overcurrent device on the supply side of the branch circuit.

The equipment grounding conductor is permitted to bypass the equipment grounding conductor terminal bar in one or more panelboards, boxes, wireways, and other enclosures on its way back toward the service equipment. As indicated, it is required to terminate the isolated equipment grounding conductor within the same building or structure at the nearest of the following: the source of a separately derived system, the building disconnecting means, or the service. It is not permitted to bypass the separately derived system with the equipment grounding conductor if that is the source. This is because the secondary windings of the transformer or other applicable separately derived system is the "home" or "source" for that circuit. Likewise, it is not permitted to bypass the building or structure disconnecting means on the way back to the service that is located in another building. This prohibition is an effort to keep all conductive surfaces related to the electrical system at the same or close to the same potential. An isolated ground receptacle is shown in Figure 7-17.

FIGURE 7-17 Isolated-ground receptacles, *250.146(D)*. (© *Cengage Learning 2012*)

250.148 CONTINUITY AND ATTACHMENT OF EQUIPMENT GROUNDING CONDUCTORS TO BOXES

The Requirements: If circuit conductors are spliced within a box, or terminated on equipment within or supported by a box, any equipment grounding conductor(s) associated with those circuit conductors must be connected within the box or to the box with devices suitable for the use in accordance with *250.148(A)* through *(E)*.

Exception: *The isolated equipment grounding conductor permitted in 250.146(D) is not required to be connected to the other equipment grounding conductors or to the box.*

See Figures 7-18 and 7-19.

Discussion: The term *circuit conductors* is intended to include feeder as well as branch-circuit conductors. The language of the section is intended to allow feeder conductors to pass through a pull or junction box without breaking the equipment grounding conductor(s), splicing them, and connecting them to the box, provided that the pull or junction box is grounded and bonded by other means such as by a metal conduit or EMT. However, if the circuit conductors are spliced within the box, or terminated on equipment, such as a wiring device, the equipment grounding conductors associated with the circuit conductors are required to also be spliced within the box or connected to the metal box with suitable connectors.

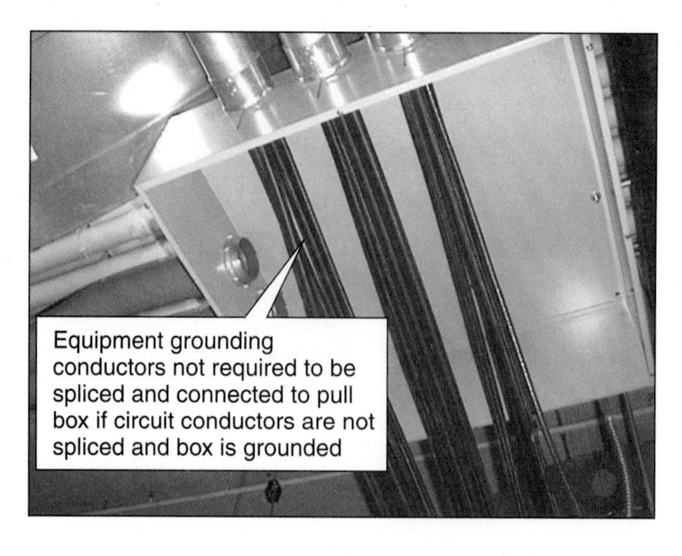

FIGURE 7-19 Continuity and attachment of equipment grounding conductors on pull boxes, *250.148.* (*© Cengage Learning 2012*)

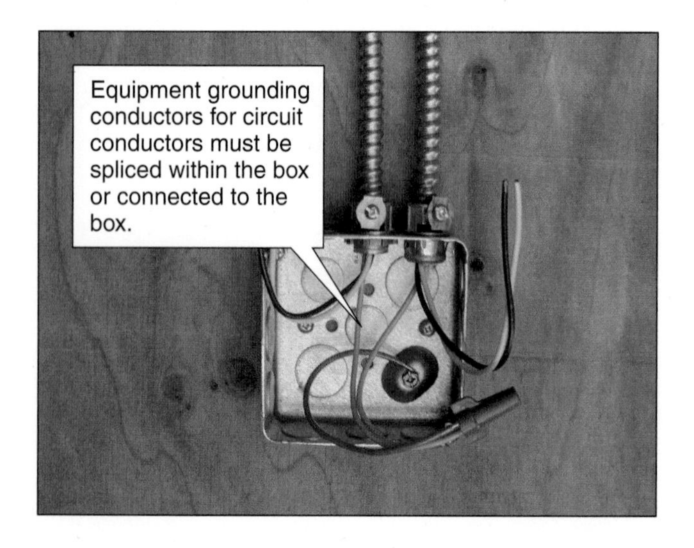

FIGURE 7-18 Continuity and attachment of equipment grounding conductors to boxes, *250.148.* (*© Cengage Learning 2012*)

250.148(A) Connections

The Requirement: Connections and splices are required to be made in accordance with *110.14(B)* except that insulation is not required.

Discussion: Compliance with all of the rules for connections in *110.14(B)* is required, including the number of conductors to be connected and not mixing copper and aluminum conductor materials in the same connector unless it is designed for that purpose. This section covers the acceptable methods for making splices of conductors. Included is the use of devices identified for the purpose or by methods such as brazing, welding, or soldering with a fusible metal or alloy. In addition to *110.14,* the rules of *250.8* must be complied with. *NEC 250.8* provides specific methods permitted to be used for splicing and connecting equipment grounding and bonding conductors. Included are listed pressure connectors, terminal bars, pressure connectors listed as grounding and bonding equipment, exothermic welding, machine screw-type fasteners that engage not less than two threads or are secured with a nut, thread-forming machine screws that engage not less than two threads in the enclosure, connections that are part of a listed assembly, and other listed means. Connection devices or fittings that depend solely on solder are not permitted to be used. Sheet metal screws are not permitted to be used for connecting grounding conductors or terminal bars to enclosures.

250.148(B) Grounding Continuity

The Requirement: The arrangement of grounding connections must be made so the disconnection or the removal of a receptacle, luminaire, or other device fed from the box will not interfere with or interrupt the grounding continuity. See Figure 7-20.

Discussion: Complying with this requirement results in splicing all of the equipment grounding conductors together and including a bonding jumper to connect to the wiring device. By doing so, a wiring device or other equipment such as a luminaire can be removed without interrupting the equipment grounding conductor path to other devices or equipment. An alternative method is to connect all of the equipment grounding conductors to the metal box with ground screws or listed ground clips and then to install a bonding jumper to the wiring device or other equipment.

250.148(C) Metal Boxes

The Requirement: A connection is required to be made between the one or more equipment grounding conductors and a metal box by means of a grounding screw that must be used for no other purpose, equipment listed for grounding, or a listed grounding device. See Figure 7-21.

Discussion: As indicated, the connection method from the equipment grounding conductor to the box is carefully controlled. Three methods are provided: (1) a ground screw, (2) equipment listed for grounding, and (3) a listed grounding device such as a ground clip.

The ground screw is not a listed connection device, so its use must be approved by the AHJ. The ground screw must be of the proper size to connect the ground wire under the head and have the proper thread pitch for connecting in the metal box. The rules in this section prevent the use of screws for cable clamps from being used for securing the equipment grounding conductor or bonding jumper.

Equipment listed for grounding and bonding has been evaluated to Underwriters Laboratories

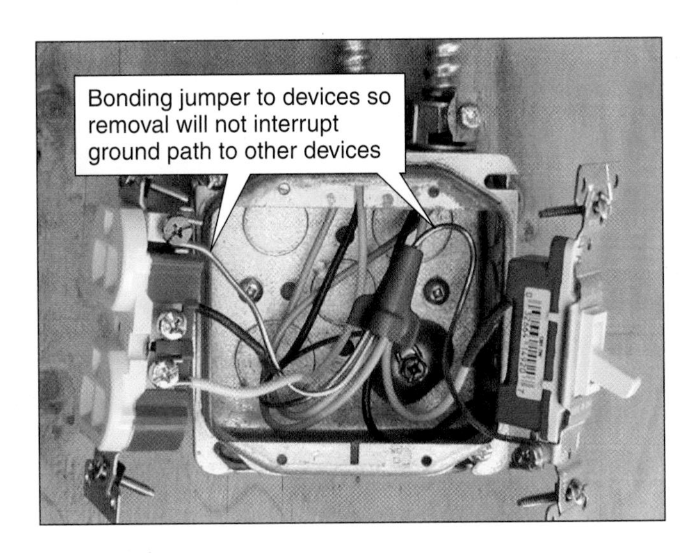

FIGURE 7-20 Grounding continuity at outlet boxes, *250.148(B)*. (*© Cengage Learning 2012*)

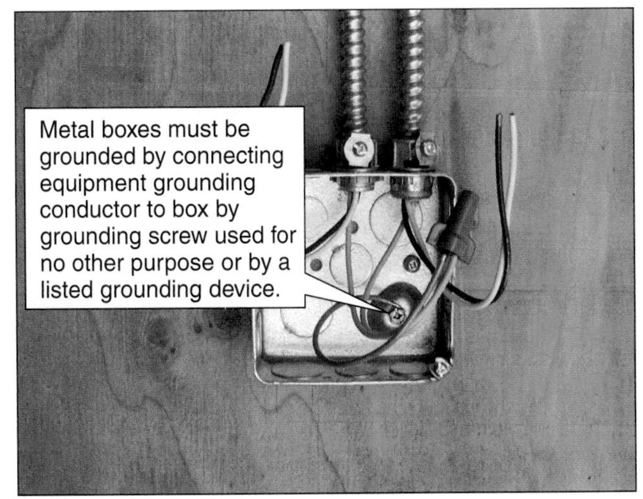

FIGURE 7-21 Grounding metal boxes, *250.148(C)*. (*© Cengage Learning 2012*)

Standard 467. To be listed to this standard, equipment must satisfactorily endure a high-current test. For additional information, see the guide card information in the *UL White Book* under the heading "Grounding and Bonding Equipment (KDER)."

Listed ground clips are available that have been evaluated to *UL 467, Grounding and Bonding Equipment,* to ensure their satisfactory performance.

250.148(D) Nonmetallic Boxes

The Requirement: One or more equipment grounding conductors brought into a nonmetallic outlet box are required to be arranged so a connection can be made to any fitting or device in that box requiring grounding.

Discussion: This rule results in splicing together all of the equipment grounding conductors in the box by means of proper connecting methods such as a twist-on wire connector or a compression-type splicing sleeve. If compression-type sleeves are used, be certain to use a matching compression tool to result in a reliable electrical connection. Preferably, the compression tool should be from the same manufacturer as the compression sleeve and identified in the manufacturer's installation instructions for the splice sleeve. The tool and sleeve should be of the proper size for the number and size of conductors being connected. The sleeve must also be identified for the wire being connected, either copper or aluminum. A bonding jumper is to be provided for grounding all wiring devices. (See Figure 7-22.)

250.148(E) Solder

The Requirement: Connections depending on solder alone are not permitted to be used.

Discussion: Soldering of connections used in lighting and power branch circuits is almost a thing of the past. Many listed connection devices are readily available and are faster as well as easier to use. It also is usually much easier to remove splicing devices such as twist-on wire connectors. Be certain to use connectors that are properly sized for the size, type, and number of conductors to be connected.

If solder connections are to be used, the splices first must be spliced or joined so they are mechanically and electrically secure without solder, and then they can be soldered. (See Figure 7-23.)

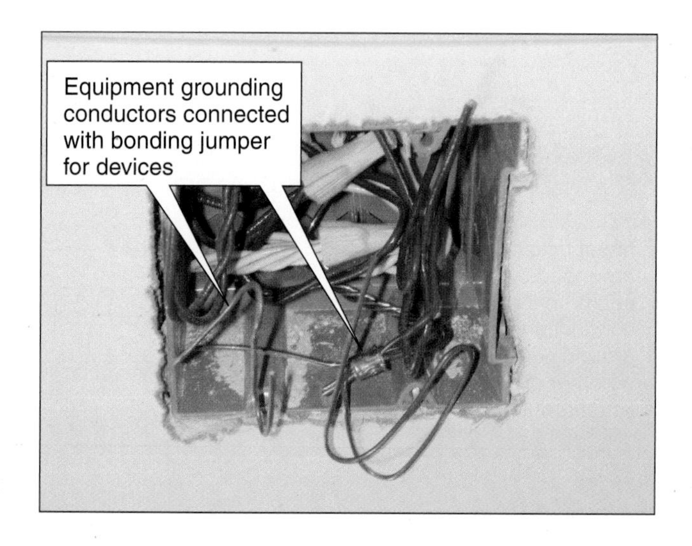

FIGURE 7-22 Grounding and nonmetallic boxes, *250.148(D)*. (*© Cengage Learning 2012*)

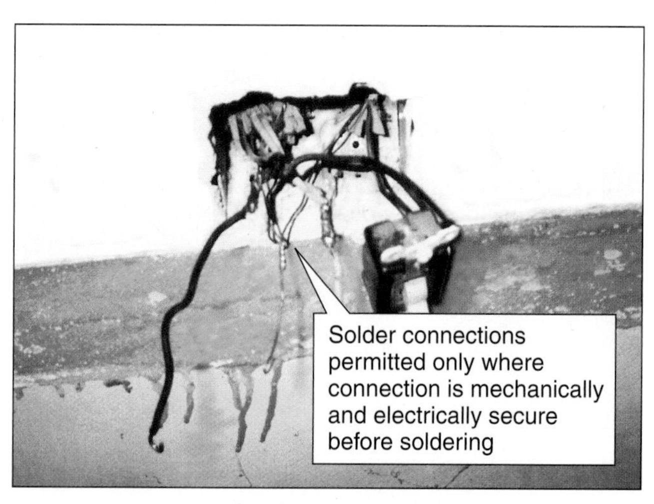

FIGURE 7-23 Solder connections, *250.148(E)*. (*© Cengage Learning 2012*)

REVIEW QUESTIONS

1. For grounded systems, equipment grounding conductor connections are required to be made in accordance with which of the following statements?
   a. The equipment grounding conductor is bonded to the grounded service conductor.
   b. The equipment grounding conductor is bonded to the grounding electrode conductor.
   c. The equipment grounding conductor is bonded to the grounded service conductor and the grounding electrode conductor.
   d. The equipment grounding conductor is routed back to the utility transformer, where it is connected to the neutral.
2. For ungrounded systems, equipment grounding conductor connections are required to be made in accordance with which of the following statements?
   a. The equipment grounding conductor is bonded to the ungrounded service conductor.
   b. The equipment grounding conductor is bonded to the grounding electrode conductor.
   c. The equipment grounding conductor is bonded to the grounded service conductor and the grounding electrode conductor.
   d. The equipment grounding conductor is routed back to the utility transformer, where it is connected to the neutral.
3. For replacement of nongrounding receptacles and branch-circuit extensions, the equipment grounding conductor of a grounding-type receptacle or a branch-circuit extension is permitted to be connected to which of the following?
   a. Any accessible point on the grounding electrode system
   b. Any accessible point on the grounding electrode conductor
   c. The equipment grounding terminal bar within the enclosure where the branch circuit for the receptacle or branch circuit originates
   d. Any of the above
4. For replacement of ungrounded receptacles, which of the following statements is *not* true?
   a. If a grounding means exists in the box, a grounding-type receptacle is required.
   b. If a grounding means does not exist in the box, a nongrounding-type receptacle is generally permitted.
   c. A GFCI-type receptacle is permitted to replace a nongrounding-type receptacle even though there is not an equipment grounding conductor in the box.
   d. An equipment grounding conductor is permitted to be installed from the GFCI receptacle to grounding-type receptacles on the load side.
5. Which of the following statements is true about supplying a grounding-type receptacle from a GFCI device?
   a. A grounding-type receptacle is permitted only when an equipment grounding conductor is present in the branch circuit.
   b. A grounding-type receptacle is permitted when supplied from a GFCI device even though an equipment grounding conductor is not present in the branch circuit.

c. An equipment grounding conductor is required from the GFCI device to the grounding-type receptacle.

d. The replacement grounding-type receptacle is required to be marked "Caution, No Equipment Ground."

6. Which of the following statements about grounding of equipment is true?

   a. Electrical equipment is considered grounded if secured to and in electrical contact with a metal rack or structure that is properly grounded.

   b. Rack-mounted electrical equipment is not considered grounded unless an equipment grounding conductor is connected to each piece of equipment.

   c. The structural metal frame of a building that is bonded is permitted to be used as the required equipment grounding conductor for ac equipment.

   d. The structural metal frame of a building that is not bonded is permitted to be used as the required equipment grounding conductor for ac equipment.

7. Which of the following statements about the grounding and bonding of electric ranges and dryers is *not* true?

   a. Generally, these appliances and related junction boxes that are in the supply circuit must be grounded by means of an equipment grounding conductor run with the branch circuit.

   b. For installation of new branch circuits only, the appliances and related equipment are permitted to be grounded by the grounded branch-circuit conductor.

   c. For existing branch circuits only, the appliances and related equipment are permitted to be grounded by the grounded branch-circuit conductor.

   d. When appropriate, a 3-wire pigtail is permitted for an existing branch circuit, but a 4-wire pigtail is required for a new branch circuit.

8. The grounded-circuit conductor is permitted to ground equipment under which of the following conditions?

   a. On the supply side or within the enclosure of the ac service disconnecting means

   b. On the supply side or within the enclosure of the main disconnecting means for separate buildings

   c. On the supply side or within the enclosure of the main disconnecting means or overcurrent devices of a separately derived system

   d. At all of the above locations

9. Which of the following statements about grounding equipment to the grounded circuit conductor (neutral) on the load side of the service equipment is *not* true?

   a. It is permitted to ground electric ranges and dryers for existing branch circuits only under specific conditions.

   b. It is permitted to ground the disconnecting means for separate buildings that are supplied by a feeder under specific conditions.

   c. It is permitted to ground meter sockets if they are immediately adjacent to the service disconnecting means.

   d. It is not permitted to ground equipment to the grounded-circuit conductor on the load side of the service.

10. A bonding jumper is required from a box to a grounding-type receptacle *except* under which of the following applications?

    a. Self-grounding receptacles

    b. Isolated ground installations

    c. Surface-mounted boxes

    d. None of the above equipment

11. A bonding jumper is not required between a grounded metal box and a grounding-type receptacle in a listed exposed work cover if

    (1) the device is attached to the cover with at least two specific fasteners.

    (2) the cover is attached to the box at a flat nonraised portion of the cover.

    a. Requirement (1) only

    b. Requirement (2) only

    c. Both requirement (1) and (2)

    d. Neither requirement (1) nor (2)

12. Listed floor boxes are required to have a device-to-box bonding jumper.

    a. True

    b. False

13. Equipment grounding conductors are required to be bonded to a grounded pull box under which of the following conditions?

    (1) When the circuit conductors are spliced within the box

    (2) Irrespective of whether the circuit conductors are spliced within the box

    a. Condition (1) only

    b. Condition (2) only

    c. Both conditions (1) and (2)

    d. Neither condition (1) nor (2)

UNIT 8

Grounding of Specific Equipment Covered in Chapter 5 of the *NEC*

OBJECTIVES

After studying this unit, the reader will understand

- requirements for grounding and bonding in hazardous (classified) locations.
- grounding and bonding requirements in patient care areas of health care facilities.
- wiring methods permitted in patient care areas and the result of mechanical protection requirements for patient care areas of health care facilities.
- grounding requirements for agricultural buildings and structures supplied overhead and underground from a site-isolation device.
- requirements for an equipotential plane and for GFCI protection in agricultural buildings or structures.
- grounding and bonding requirements for services and feeders for mobile homes and manufactured homes.

CLASS I, CLASS II, AND CLASS III HAZARDOUS (CLASSIFIED) LOCATIONS

500.5 Hazardous (Classified) Locations: Hazardous (classified) locations are rated or classified by the type of hazard presented by the chemical or substance and by the degree of risk of fire or explosion present. The locations are then classified according to the likelihood that a flammable or combustible mixture is present by division, such as Division 1 or Division 2. Division 1 areas are more likely to have flammable or combustible atmospheres, dusts or fibers present than are Division 2 areas.

Generally, it is not intended that the *NEC* be used to perform classification studies to establish the type of hazard involved or the extent of the boundaries. Many other standards are useful for these purposes. See the *Informational Notes* in *500.4(B)* for a helpful list of standards. The *NEC* is very valuable for providing the wiring methods required for wiring in the various classified areas.

Class I Locations: Class I locations are those in which flammable gases or vapors are or may be present in the air in quantities sufficient to produce explosive or ignitable mixtures. Class I, Division 1 locations are more likely to have these gases or vapors present and Class I, Division 2 locations are less likely to have the gases or vapors present in quantities sufficient to produce an explosion.

Class II Locations: Class II locations are those that are hazardous because of the presence of combustible dust. Class II locations are divided into Division 1 and Division 2 areas on the basis of the risk of explosion present.

Class III Locations: Class III locations are those that are hazardous because of the presence of easily ignitable fibers or flyings, but in which such fibers or flyings are not likely to be in suspension in the air in quantities sufficient to produce ignitable mixtures. Class III locations are also divided into Division 1 and Division 2 areas on the basis of the risk of fires being initiated by electrical equipment.

250.100 BONDING IN HAZARDOUS (CLASSIFIED) LOCATIONS

The Requirement: Regardless of the voltage of the electrical system, the electrical continuity of non-current-carrying metal parts of equipment, raceways, and other enclosures in any hazardous (classified) location as defined in *500.5* is required to be ensured by any of the bonding methods specified in *250.92(B)(2)* through *(4)* that are approved for the wiring method used. One or more of these bonding methods must be used whether or not equipment grounding conductors of the wire type are installed.

Discussion: Because of the risk of fire or explosion in hazardous locations, additional precautions for bonding are required (see Figure 8-1). This bonding attempts to ensure the reliability of the ground-fault

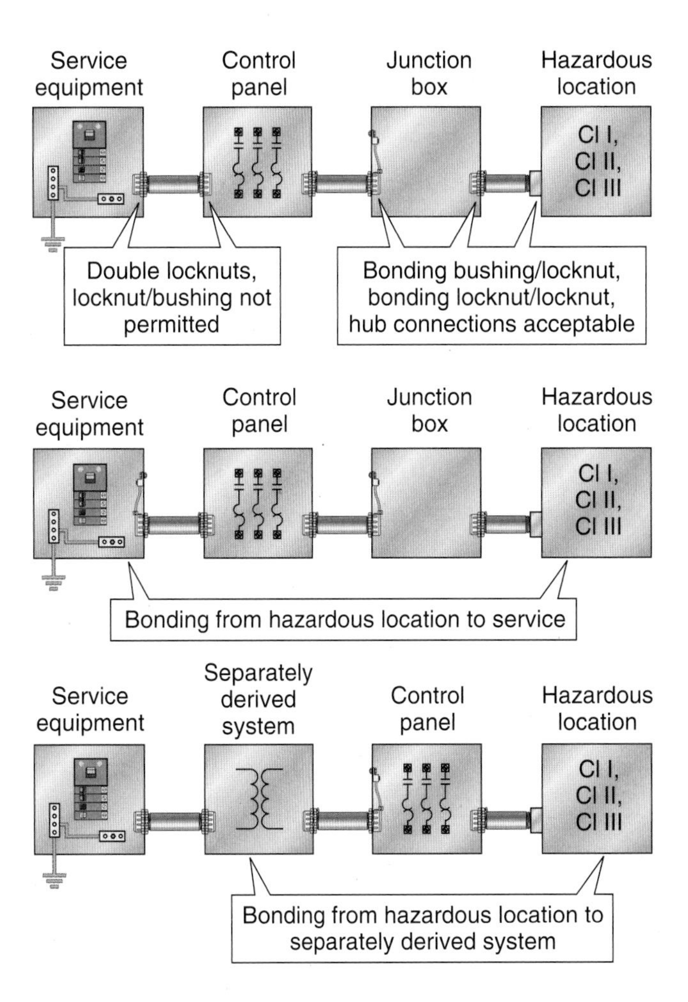

FIGURE 8-1 Bonding in hazardous (classified) locations, *250.100*. (*© Cengage Learning 2012*)

return path of the metal conduit or other metallic wiring method and to prevent arcing or sparking where connections are made at enclosures. These rules do not require a supplemental equipment grounding conductor in metal raceways in hazardous (classified) locations. The purpose of the bonding requirements is to ensure the reliability and performance of the single-fault current path provided by the wiring method. Arcing or sparking in a hazardous location could be the source of ignition. It is fairly common for consulting and design electrical engineers to require a supplemental equipment grounding conductor in all metal raceways. If these equipment grounding conductors are installed, they must be installed in compliance with all applicable rules, including the sizing requirements of *250.122.* The installation of an equipment grounding conductor in a metal raceway does not relieve the installer from the requirement to use the bonding methods otherwise required for the metal conduit or other metal raceway.

Specific requirements for bonding and providing the effective ground-fault return path are contained in the following *Code* locations:

501.30 Class I locations (classified due to flammable vapors)

502.30 Class II locations (classified due to combustible dust)

503.30 Class III locations (classified due to fibers and flyings)

To emphasize the importance of bonding in Class I locations, *501.30(A)* contains the following requirements (identical requirements are contained in *502.30(A)* for Class II locations and in *503.30(A)* for Class III locations):

1. The locknut-bushing and double-locknut types of contacts are not permitted for bonding purposes.
2. Bonding jumpers with proper fittings or other approved means of bonding are required to be used.
3. Such means of bonding must be provided at all intervening raceways, fittings, boxes, enclosures, and so forth between Class I locations and the point of grounding for service equipment or point of grounding of a separately derived system.

These requirements are very specific and very clear.

In addition to the general classification of hazardous (classified) locations in *NEC 500.5,* specific articles in the *NEC* that describe *hazardous locations* include the following:

Article 511 Commercial Garages, Repair, and Storage
Article 513 Aircraft Hangars
Article 514 Motor Fuel–Dispensing Facilities
Article 515 Bulk Storage Plants
Article 516 Spray Application, Dipping, and Coating Processes
Article 517 Health Care Facilities (rooms or areas where flammable inhalation anesthetics are administered)

These articles apply to occupancies that commonly have hazardous (classified) areas. Included in these articles are classification of specific areas that are hazardous because of flammable vapors, combustible dusts, or fibers and flyings. On the basis of the degree of hazard, the areas are classified as Division 1 or Division 2. Division 1 areas are more likely to be hazardous in normal operations, and Division 2 areas are less likely. Specific requirements are included in these articles, with the goal of allowing the electrical equipment to operate safely and not become the source of ignition. Specific wiring methods are provided where necessary to achieve minimum safety. The types of equipment that are permitted to be used in these areas are carefully controlled. Protection techniques and conduit seals are implemented for Class I areas to contain explosions inside enclosures and prevent igniting flammable vapors on the outside of the conduit systems. Other requirements for Class II areas include limiting the temperature of enclosures and equipment to prevent the equipment from igniting flammable vapors. Steps are taken in Class III areas to prevent fibers or flyings from entering enclosures and to control temperatures below the ignition points of the fibers or flyings.

Electrical installations in other locations classified as Class I, such as may be found in a petrochemical plant, must have grounding and bonding systems that comply with the applicable requirements of the *NEC.* A hazardous location area classification study is required for those occupancies not included in *Articles 511* through *517.* This study must be documented as required by *500.4(A)* and be available to those authorized to design, install, inspect, maintain, or operate electrical equipment at the location.

Examples of Class I locations can be found in *500.5(B)(1)*, Informational Note Nos. 1 and 2. Also see NFPA 497, *Recommended Practice for the Classification of Flammable Liquids, Gasses, or Vapors and of Hazardous (Classified) Locations for Electrical Installations in Chemical Process Areas.* Examples of Class II locations can be found in *500.6(B)* and in NFPA 499, *Recommended Practice for the Classification of Combustible Dusts and of Hazardous (Classified) Locations for Electrical Installations in Chemical Process Areas.* Examples of Class III hazardous (classified) locations can be found in *500.5(D)(1)*, Informational Notes Nos. 1 and 2.

An Exception to 501.30(A): *The specific bonding means is only required to the nearest point where the grounded circuit conductor and the grounding electrode are connected together on the line side of the building or structure disconnecting means as specified in 250.32(B), provided the branch circuit overcurrent protection is located on the load side of the disconnecting means. (A similar exception is found in 502.30(A) for Class II areas and in 503.30(A) for Class III areas.)*

Discussion: This exception applies to complexes that have one or more buildings or structures that are supplied by one or more feeders or branch circuits from another building or structure. (See Figure 8-2.) Note that *225.30* generally limits the supply from one building or structure to another to one feeder or branch circuit. Provisions are made in *225.30(A)* through *(E)* for additional supplies under the specific conditions provided. These rules for a single supply, except for the specific exceptions provided, are not modified in *Articles 500* through *516*, so they apply to buildings or structures having a hazardous (classified) location.

As provided in the exception to *250.32(B)*, the grounded circuit conductor is permitted to be used for grounding at the building or structure disconnecting means only for existing installations if the requirements of the exception are complied with. The grounded circuit conductor (often a neutral) is not permitted to be grounded again at the building or structure disconnecting means for installations made to comply with recent editions of the *NEC*. An equipment grounding conductor is required to be installed with the feeder or branch circuit. The requirement in *250.32(B)* has the effect of requiring the positive bonding required by *501.30(A)* and identical rules in other articles to extend to the service equipment or separately derived system even though the hazardous location is in another building or structure. As indicated in the exception to *501.30(A)*, the bonding for hazardous locations is not required to extend beyond the separately derived system that is the source of power for the branch circuit or feeder to the hazardous location.

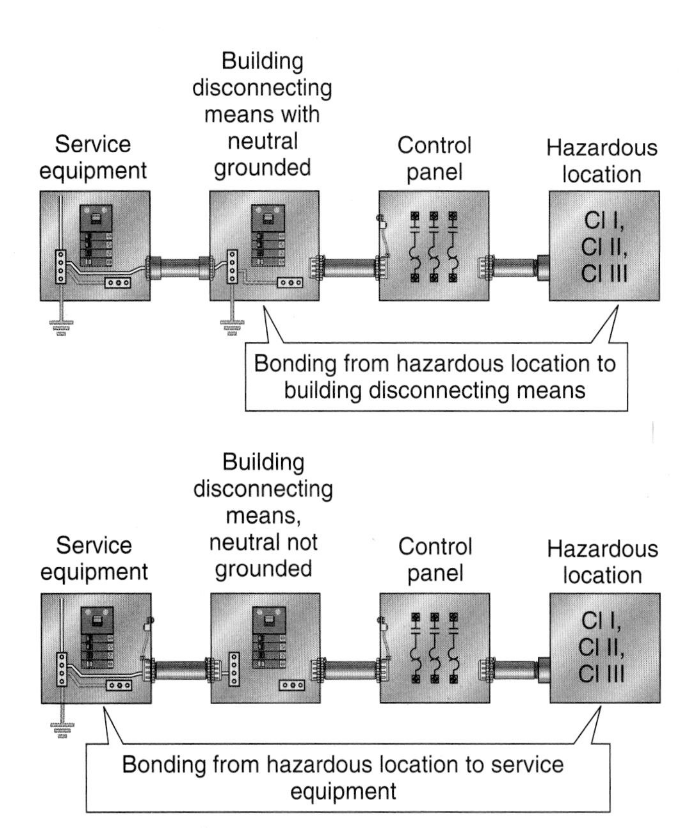

FIGURE 8-2 Bonding hazardous (classified) locations in building or structure separate from service, *501.30, 502.30, 503.30.* (*© Cengage Learning 2012*)

ARTICLE 517 HEALTH CARE FACILITIES

To determine the applicability of the following requirements, see the definitions in *517.2,* including but not limited to *health care facility, patient care vicinity, life safety branch, critical branch, emergency system, essential electrical system, exposed conductive surfaces,* and any others specifically related to your installation.

The following section includes information on the purposes behind the requirements for specific rules related to grounding and bonding in health care facilities.

517.11 GENERAL INSTALLATION—CONSTRUCTION CRITERIA

The purpose of this article is to specify the installation criteria and wiring methods that minimize electrical hazards by the maintenance of adequately low potential differences only between exposed conductive surfaces that are likely to become energized and could be contacted by a patient.*

Discussion: This section serves as a preamble to Part II of *Article 517,* which covers wiring and protection in health care facilities. It is similar to a scope statement in other articles in that it sets out the purpose to be achieved by proper grounding of equipment as well as specific circuits, receptacles, and performance requirements of the emergency system. The following *Informational Note* contains valuable insight into the philosophy of grounding and bonding in patient care areas.

> "**Informational Note:** In a health care facility, it is difficult to prevent the occurrence of a conductive or capacitive path from the patient's body to some grounded object, because that path may be established accidentally or through instrumentation directly connected to the patient. Other electrically conductive surfaces that may make an additional contact with the patient, or instruments that may be connected to the patient, then become possible sources of electric currents that can traverse the patient's body. The hazard is increased as more apparatus is associated with the patient, and, therefore, more intensive precautions are needed.
>
> "Control of electric shock hazard requires the limitation of electric current that might flow in an electric circuit involving the patient's body by raising the resistance of the conductive circuit that includes the patient, or by insulating exposed surfaces that might become energized, in addition to reducing the potential difference that can appear between exposed conductive surfaces in the patient care vicinity, or by combinations of these methods.
>
> "A special problem is presented by the patient with an externalized direct conductive path to the heart muscle. The patient may be electrocuted at current levels so low that additional protection in the design of appliances, insulation of the catheter, and control of medical practice is required."*

Previous editions of the *NEC* contained requirements for equipotential bonding in critical care patient care areas. Although no longer required in individual patient rooms, this bonding may continue to be required by specifications of consulting electrical engineers or for special procedure rooms.

517.13 GROUNDING OF RECEPTACLES AND FIXED ELECTRIC EQUIPMENT IN PATIENT CARE AREAS

Wiring in patient care areas is required to comply with *517.13(A)* and *517.13(B).*

517.13(A) Wiring Methods

The Requirement: All branch circuits serving patient care areas must be provided with an effective ground-fault current path by installation in a metal raceway system, or a cable having a metallic armor or sheath assembly. The metal raceway system, metallic cable armor, or sheath assembly is required to itself qualify as an equipment grounding conductor in accordance with *250.118.*

Discussion: Two grounding paths are required to be provided to ensure one effective ground-fault current path is available and functional in patient care areas. This can be thought of as the primary and secondary ground-fault return paths. Some have referred to this second path as a *redundant* path. Because secondary meanings of the word *redundant* include "superfluous, surplus, unneeded, unnecessary, or uncalled for," a "secondary path" is a better

*Reprinted with permission from NFPA 70-2011.

description of the intended additional safeguard. The primary purpose of requiring two equipment ground-fault return paths is to ensure that one path is always available in the event the other path is damaged or interrupted for any reason. The first of those paths is required to be the wiring method itself (see Figure 8-3).

Section 250.118 includes a list of wiring methods permitted to be used as an equipment grounding conductor. Acceptable wiring methods typically used in patient care areas include the following.

The Requirements for Wiring Methods: *250.118(4).* Electrical metallic tubing (EMT).

Discussion: Electrical metallic tubing (EMT) is considered to provide adequate mechanical protection for the wiring in patient care areas as well as providing an equipment ground-fault return path. It is permitted to be used in any length and in any size. Good workmanship in the installation of EMT is necessary to provide a reliable ground-fault return path, because a single loose fitting or coupling can impair the integrity of the path.

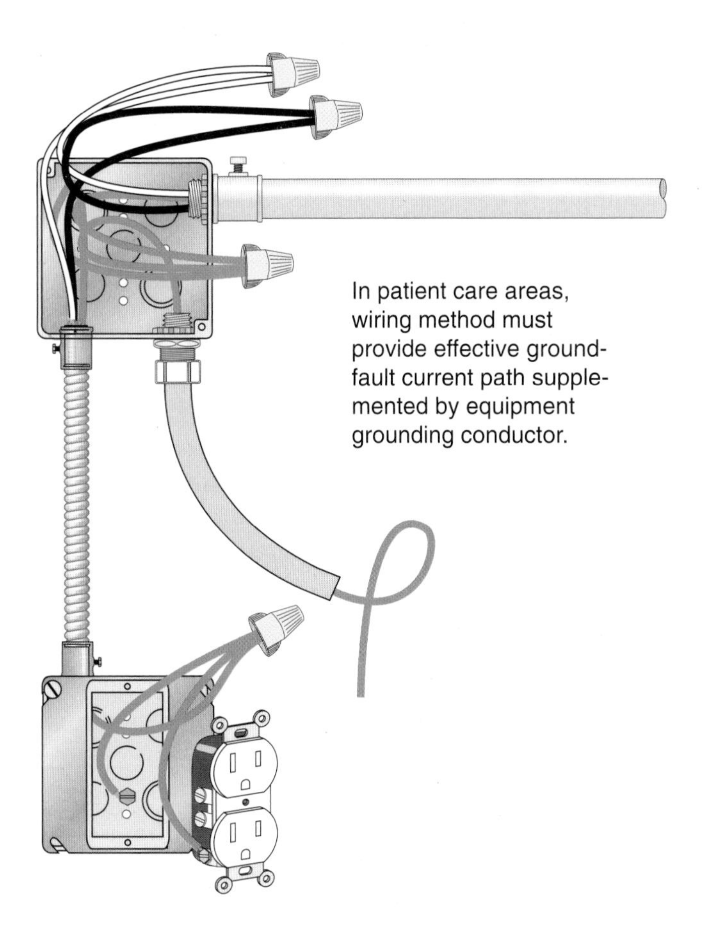

FIGURE 8-3 Grounding of receptacles and fixed electric equipment in patient care areas, *517.13.* (© *Cengage Learning 2012*)

Another Requirement: *250.118(5)*

Listed flexible metal conduit is permitted if it meets all the following conditions:

a. The conduit is terminated in listed fittings.
b. The circuit conductors contained in the conduit are protected by overcurrent devices rated at 20 amperes or less.
c. The combined length of flexible metal conduit and flexible metallic tubing and liquidtight flexible metal conduit in the same ground return path does not exceed 6 ft (1.8 m).
d. If used to connect equipment where flexibility is necessary to minimize the transmission of vibration from equipment or to provide flexibility for equipment that requires movement after installation, an equipment grounding conductor is required to be installed.

Discussion: As indicated in this section, the use of listed FMC as an equipment grounding conductor is limited to accommodate its ability to carry fault current (see Figure 8-4). If flexible metal conduit is installed in patient care areas of health care facilities, the rules in *NEC 517.13* as well as those in *250.118(5)* must be complied with. *NEC 250.118(5)* recognizes FMC as an equipment grounding conductor up to 6 ft (1.8 m) long with overcurrent protection of contained conductors not exceeding 20 amperes. As a result, 6 ft (1.8 m) becomes the maximum length of FMC in patient care areas even with an insulated equipment grounding conductor inside the FMC. The reason is because if the FMC is longer than 6 ft (1.8 m), it is not recognized as an equipment grounding conductor. If FMC longer than 6 ft (1.8 m) were to be used, the patient care area wiring would then be supplied by only one of the two equipment grounding conductors required (the insulated equipment grounding conductor required in *517.13[(B)]*).

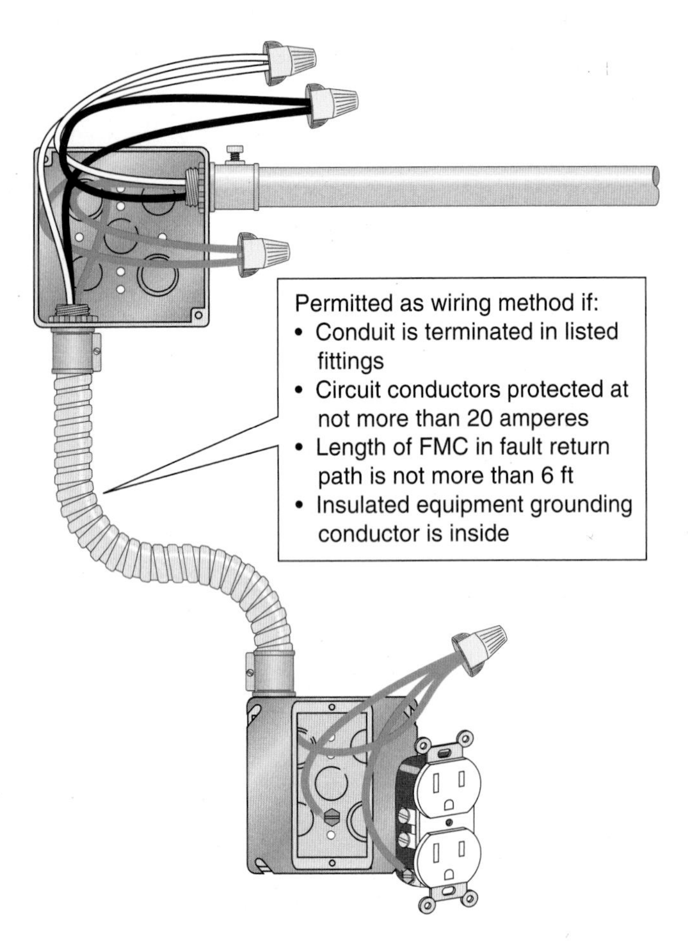

FIGURE 8-4 FMC as equipment grounding conductor, *250.118(5)*. (*© Cengage Learning 2012*)

More Requirements: *250.118(6)*

Listed liquidtight flexible metal conduit (LFMC) is permitted to be installed if it meets all the following conditions:

a. The conduit is terminated in listed fittings.

b. For trade sizes ⅜ through ½ (metric designators 12 through 16), the circuit conductors contained in the conduit are protected by overcurrent devices rated at 20 amperes or less.

c. For trade sizes ¾ through 1¼ (metric designators 21 through 35), the circuit conductors contained in the conduit are protected by overcurrent devices rated not more than 60 amperes and there is no flexible metal conduit, flexible metallic tubing, or liquidtight flexible metal conduit in trade sizes ⅜ through ½ (metric designators 12 through 16) in the grounding path.

d. The combined length of flexible metal conduit and flexible metallic tubing and liquidtight flexible metal conduit in the same ground return path does not exceed 6 ft (1.8 m).

e. If used to connect equipment where flexibility is necessary to minimize the transmission of vibration from equipment or to provide flexibility for equipment that requires movement after installation, an equipment grounding conductor shall be installed.

Discussion: Listed liquidtight flexible metal conduit (LFMC) is identified as an equipment grounding conductor in *250.118(6)*. However, requirements are included to recognize its limited capability of carrying fault current. The permitted use of this wiring method as an equipment grounding conductor applies only to the product that is listed in compliance with UL-360, which is the product safety standard. Additional installation requirements or limitations are contained in the Underwriters Laboratories *General Information Directory* in category DXHR.

Unlisted LFMC should not be used as an equipment grounding conductor because it is not constructed, nor has it been tested to perform as an equipment grounding conductor. Figure 8-5 summarizes the use of Type LFMC as an equipment grounding conductor. The rules on using listed LFMC as an equipment grounding conductor are as follows:

- The fittings must be listed. Listed fittings have been tested to ensure their ability to carry fault current as provided in the product standard.
- For trade sizes through ½ in. (metric designators 12 through 16), the circuit conductors contained in the conduit must be protected by overcurrent devices rated at 20 amperes or less.
- For trade sizes ¾ through 1¼ (metric designators 21 through 35), the circuit conductors contained in the conduit must be protected by overcurrent devices rated not more than 60 amperes. Flexible metal conduit, flexible metallic tubing, or liquidtight flexible metal conduit in trade sizes through ½ (metric designators 12 through 16) are not permitted in this grounding path.
- The combined length of flexible metal conduit and flexible metallic tubing and liquidtight flexible metal conduit in the same ground return path does not exceed 6 ft (1.8 m).

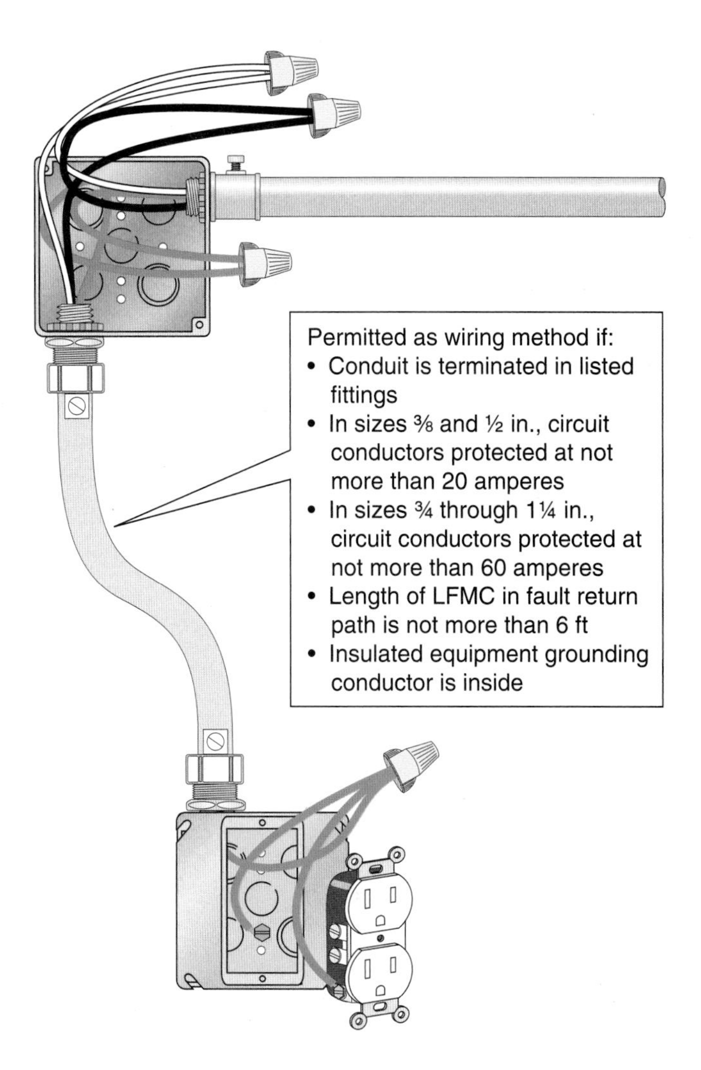

FIGURE 8-5 LFMC as equipment grounding conductor, *250.118(6)*. (*© Cengage Learning 2012*)

- If used to connect equipment where flexibility is necessary to minimize the transmission of vibration from equipment or to provide flexibility for equipment that requires movement after installation, an equipment grounding conductor is required to be installed.

If LFMC is installed in patient care areas of health care facilities, the rules in *NEC 517.13* as well as those in *250.118(6)* must be complied with. *NEC 250.118(6)* recognizes LFMC as an equipment grounding conductor up to 6 ft (1.8 m) in length. As a result, 6 ft (1.8 m) becomes the maximum length of LFMC in patient care areas, even with an insulated equipment grounding conductor inside the LFMC. If LFMC longer than 6 ft (1.8 m) were to be used, the patient care area wiring would then be supplied by only one of the two equipment grounding conductors required (the insulated equipment grounding conductor required in *517.13(B)*).

More Requirements: *250.118(8)*

The armor of Type AC cable is permitted in patient care areas as an equipment grounding conductor as provided in *320.108.*

Discussion: Type AC cable is a factory assembly of insulated conductors protected in an overall metallic sheath of the spiral, interlocking armor type. The metallic sheath may be of aluminum or steel material. Armored cable having an aluminum sheath is suitable for use in alternating current circuits only. Type AC cable was originally known as BX cable. Type AC cable is listed in accordance with UL-4, the applicable product safety standard.

Type AC cable is permitted to have from two to four conductors in sizes 14 through 1 AWG copper, with or without an equipment grounding conductor. All insulated conductors have an individual moisture-resistant fiber wrap and are cabled together in the manufacturing process.

Typical branch circuit Type AC cables have copper conductors with THHN insulation and are suitable for installation in dry locations. See *320.10* for uses permitted and *320.12* for uses not permitted. Figure 8-6 shows Type AC cable with and without an insulated equipment grounding conductor.

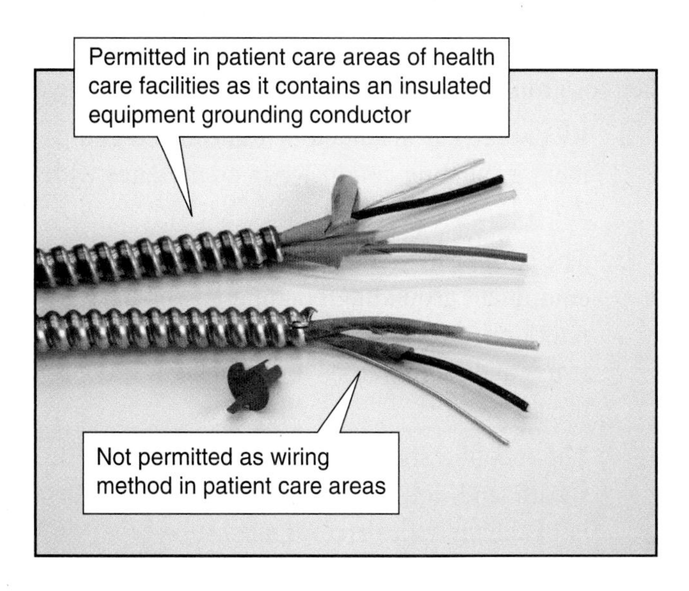

FIGURE 8-6 Type AC cable as equipment grounding conductor, *250.118(8)*. (*© Cengage Learning 2012*)

Type AC cable has a bare 16 AWG aluminum bonding strip installed between the conductors and the outer armor and is in intimate contact with the armor for the full length of the cable. The outer armor of the cable in conjunction with this bonding strip is acceptable as an equipment grounding conductor. This construction allows Type AC cable to qualify as an equipment grounding conductor and to satisfy the requirement to serve as a wiring method in patient care areas in *517.13(A)*. Because this 16 AWG aluminum bonding conductor is not itself recognized as an equipment grounding conductor, it is not necessary to terminate this conductor in boxes, to wiring devices, or to other enclosures such as panelboard cabinets. At times, this bonding conductor is bent back over the insulator required to be installed at the termination of Type AC cables. This action is sometimes referred to a "back-wrapping."

Some Type AC cables also contain, in addition to the circuit conductors, an insulated equipment grounding conductor. AC cables of this type are permitted for installation in patient care areas of health care facilities because they satisfy the requirement of both *517.13(A)* and *(B)*. The cable armor assembly provides one equipment grounding conductor path, and the insulated equipment grounding conductor provides the second path.

More Requirements: *250.118(10)*

Type MC cable is permitted to be used as an equipment grounding conductor if listed and identified for grounding in accordance with the following:

a. It contains an insulated or uninsulated equipment grounding conductor in compliance with *250.118(1)*.

b. The combined metallic sheath and uninsulated equipment grounding/bonding conductor is of interlocked metal tape–type MC cable that is listed and identified as an equipment grounding conductor.

c. The metallic sheath or the combined metallic sheath and equipment grounding conductors is of the smooth or corrugated tube-type MC cable that is listed and identified as an equipment grounding conductor.

Discussion: Type MC cable is a factory assembly of one or more current-carrying conductors and one or more equipment grounding conductors (if required) in an overall metallic sheath. Type MC cable is manufactured with three different types of armor:

1. Interlocking metal-tape sheath (steel or aluminum)
2. Smooth sheath (aluminum only)
3. Corrugated sheath (copper or aluminum)

Type MC cable consists of one or more current-carrying conductors, one or more equipment grounding conductors if required, and in some cases optical fibers. Type MC cables containing optical fibers are designated Type MC-OF and are considered *composite cables* in accordance with *Article 770*. Type MC cables are permitted to contain conductors from 18 AWG through 2000 kcmil for copper and 12 AWG through 2000 kcmil for aluminum. Typical branch circuit Type MC cables have copper conductors with THHN, THHN/THWN, or XHHW insulation and are suitable for circuits up to 600 volts.

The armor of the "traditional" Type MC cable with an interlocking metal tape sheath is not itself suitable as an equipment grounding conductor. A nonmetallic tape is wound around the circuit conductors and insulates the armor from the conductors. The cable includes an equipment grounding conductor that is either bare or insulated. For branch-circuit sizes, the most common type of equipment grounding conductor is insulated with an unstriped (solid) green color. This "traditional" Type MC cable with interlocked armor is not permitted to be installed in patient care areas of health care facilities because the armor itself is not suitable as an equipment grounding conductor as required by *517.13(A)*. Figure 8-7 illustrates varieties of Type MC cable that are suitable for wiring in patient care areas.

Type MC cable has recently become available that has an interlocking metal tape sheath and a 10 AWG aluminum equipment grounding/bonding conductor that is in continuous contact with the outer armor. The circuit conductors are spiraled together during the manufacturing process and wrapped with a nonmetallic tape or the individual circuit conductors have an additional plastic film extruded on the circuit conductors. The aluminum grounding/bonding conductor is installed on the outside of the nonmetallic tape (if present) and is spiraled to match the lay of the circuit conductors. It is thus in continuous

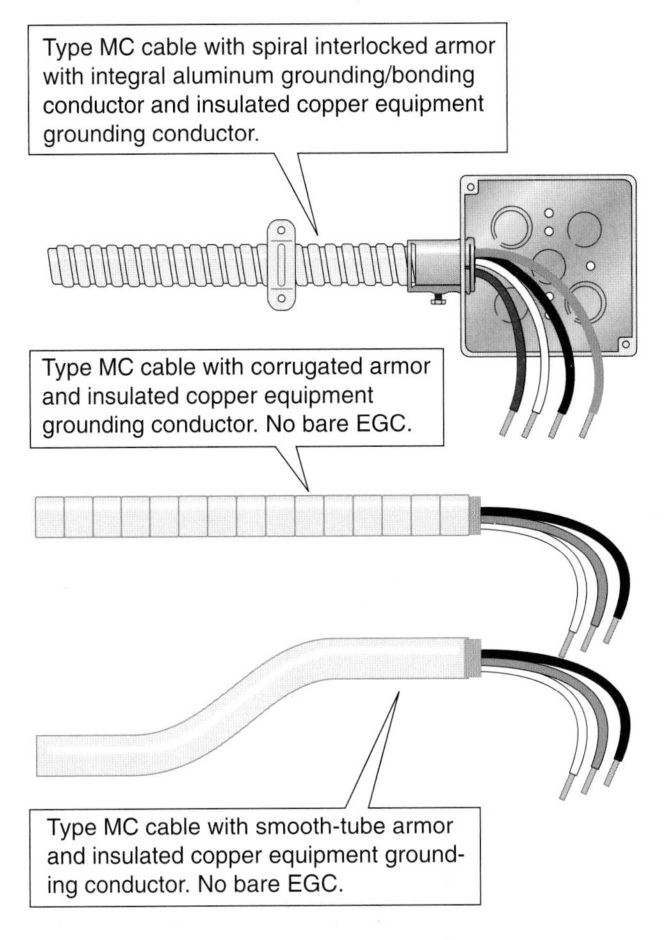

FIGURE 8-7 Type MC cable as equipment grounding conductor, *250.118(10).* (*© Cengage Learning 2012*)

contact with the outer armor. The combination of the interlocking metal tape sheath and the equipment grounding/bonding conductor constitutes an equipment grounding conductor. This Type MC cable is suitable for installation in patient care areas of healthcare facilities if it contains an insulated equipment grounding conductor inside the cable with the circuit conductors. The wiring method satisfies the requirement in *517.13(A)* and the insulated equipment grounding conductor satisfies the requirement in *517.13(B).* The equipment grounding conductor must be sized in accordance with *Table 250.122.*

More Requirements: *250.118(14)*

Surface metal raceways listed for grounding are permitted in patient care areas.

Discussion: Surface metal raceways are often used for remodeling in existing facilities to avoid the need to fish wiring inside finished walls. Surface metal raceways are typically listed for grounding. Care and good workmanship practices must be exercised during the installation of the product to ensure an effective ground-fault current path is maintained at every connection. The UL product certification directory (White Book) for surface metal raceways contains the following information under [RJBT], "Surface metal raceways are considered suitable for grounding for use in circuits over and under 250V and where installed in accordance with the *NEC*."

517.13(B) Insulated Equipment Grounding Conductor

The Requirement: (1) General. The following are required to be directly connected to an insulated copper equipment grounding conductor that is installed with the branch circuit conductors in the wiring methods as provided in *517.13(A).*

1. The grounding terminals of all receptacles
2. Metal boxes and enclosures containing receptacles
3. All non-current-carrying conductive surfaces of fixed electrical equipment likely to become energized that are subject to personal contact, operating at over 100 volts.

Exception: *An insulated equipment bonding jumper that directly connects to the equipment grounding conductor is permitted to connect the box and receptacle(s) to the equipment grounding conductor.*

Discussion: The requirement for the insulated equipment grounding conductor is intended to ensure one of the two equipment grounding conductor fault return paths is complete and functional. The metallic wiring method required in *517.13(A)* provides the first ground-fault return path, and the insulated equipment grounding conductor required in *517.13(B)* provides the second path. Figure 8-8 illustrates the required two equipment grounding conductor paths.

The grounding terminals of receptacles in the patient care area are required to be connected to the insulated equipment grounding conductor. *Section 250.148* requires that if circuit conductors

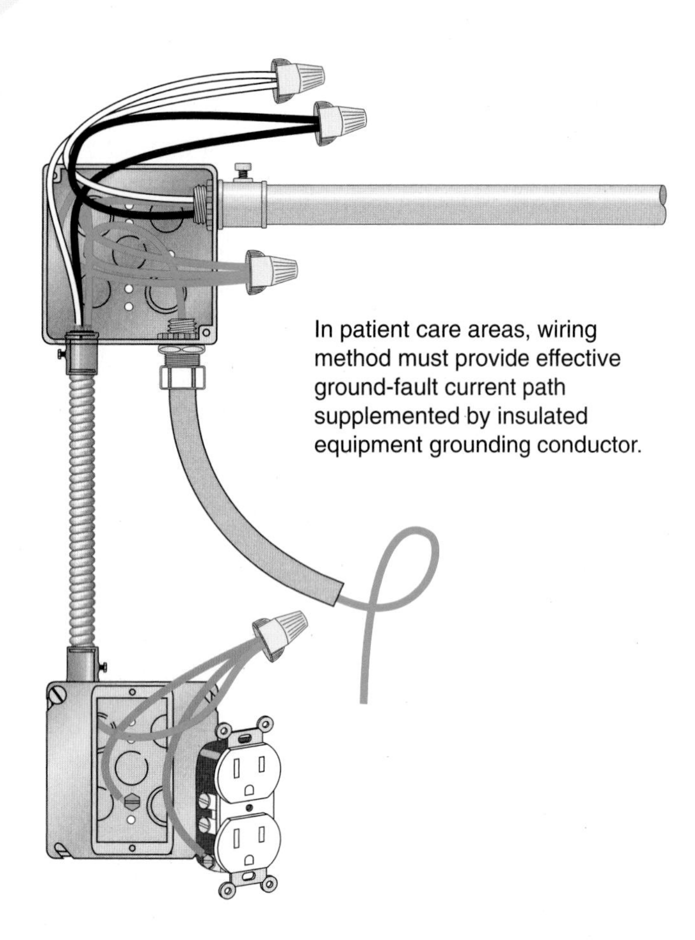

FIGURE 8-8 Insulated equipment grounding conductor, *517.13(B)*. (© *Cengage Learning 2012*)

are spliced within a box, or terminated on equipment within or supported by a box, any equipment grounding conductor(s) associated with those circuit conductors must be connected within the box or to the box with devices suitable for the use.

Section 250.148(B) requires the arrangement of grounding connections be such that the disconnection or the removal of a receptacle, luminaire, or other device fed from the box does not interfere with or interrupt the grounding continuity.

Finally, *250.148(C)* requires a connection be made between the one or more equipment grounding conductors and a metal box by means of a grounding screw that is used for no other purpose, by equipment listed for grounding, or by a listed grounding device.

When these rules are considered, it becomes obvious that the best way to comply with all these requirements is to splice two bonding jumpers to the equipment grounding conductor installed with the branch circuit conductors. One bonding jumper is connected to the metal box and the second bonding jumper is connected to the grounding terminal of the receptacle. Given the specific requirement in *517.13(B)* for the equipment grounding conductor to connect to the terminal of the receptacle, a self-grounding receptacle without an equipment grounding jumper would not be acceptable.

The equipment grounding conductor is required to be sized in accordance with *Table 250.122*. For 20-ampere branch circuits, the equipment grounding conductor is required to be not smaller than 12 AWG copper.

Exception No. 1 to (B)(1)(3): *Metal faceplates are permitted to be connected to the equipment grounding conductor by means of a metal mounting screw(s) securing the faceplate to a grounded outlet box or grounded wiring device.*

Discussion: This exception provides a commonsense solution to grounding requirements for metal faceplates that are often used in patient care areas. These faceplates are not provided with a terminal for connecting a bonding jumper. Were it not for this exception, it could be interpreted that a hole would have to be drilled through the plate for attaching a lug to which a bonding jumper could be connected. Figure 8-9 shows metal faceplates installed on receptacles in patient care areas.

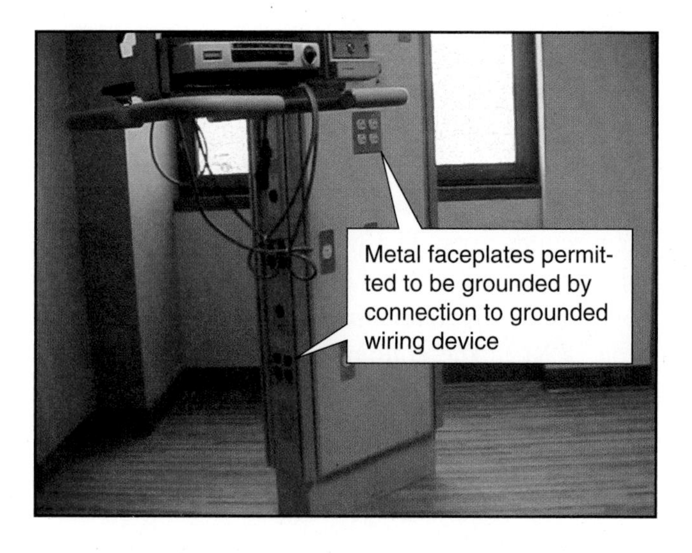

FIGURE 8-9 Grounding metal faceplates, *517.13(B)(1)(3), Exception*. (© *Cengage Learning 2012*)

Exception No. 2 to (B)(1)(3): *Luminaires more than 7½ ft (2.3 m) above the floor and switches located outside of the patient care vicinity are not required to be connected to an insulated equipment grounding conductor.*

Discussion: This exception recognizes the dimensions of the *patient care vicinity* as defined in *517.2.* It reads, "In an area in which patients are normally cared for, the *patient care vicinity* is the space with surfaces likely to be contacted by the patient or an attendant who can touch the patient. Typically in a patient room, this encloses a space within the room not less than 6 ft (1.8 m) beyond the perimeter of the bed in its nominal location, and extending vertically not less than 7½ ft (2.3 m) above the floor."*

The luminaires are required to be grounded by the normal wiring method or equipment grounding conductor but are not required to be grounded by an insulated equipment grounding conductor required by *517.13(B).* Likewise, snap switches and metal covers are required to be grounded in accordance with *404.9(B).* A snap switch with a metal yoke that is connected to a grounded box with a metal cover secured to the snap switch satisfies these requirements. These requirements are shown in Figure 8-10.

*Reprinted with permission from NFPA 70-2011.

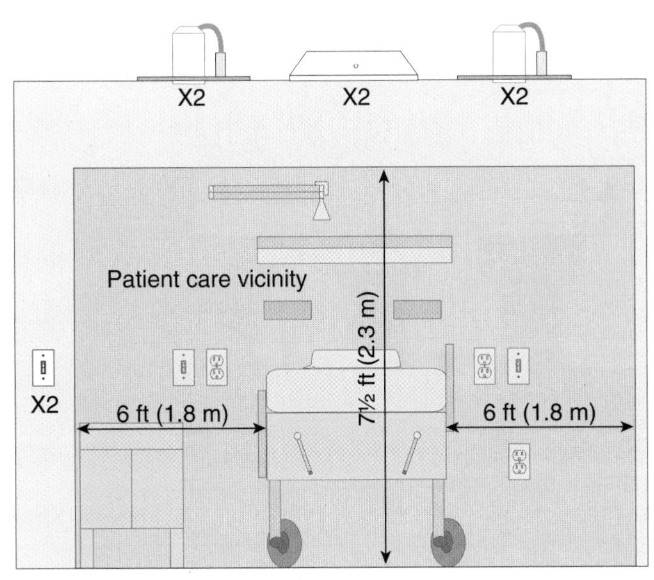

FIGURE 8-10 Insulated equipment grounding conductor not required, *517.13(B)(1)(3), Exception 2.* (*© Cengage Learning 2012*)

The Requirement: 517.13(B)(2) **Sizing.** Equipment grounding conductors and equipment bonding jumpers are required to be sized in accordance *with 250.122.*

Discussion: This is not a new requirement but now appears in a section by itself due to the reorganization of *517.13(B).* The reference to *250.122* is to the section and is not limited to the table. Thus all the requirements of the section apply. Review the application of the requirements by referring to the requirements, illustration, and discussions in *250.122.*

517.14 PANELBOARD BONDING

The Requirement: The equipment grounding terminal buses of the normal and essential branch-circuit panelboards serving the same individual patient care vicinity are required to be connected together with an insulated continuous copper conductor not smaller than 10 AWG. If two or more panelboards serving the same individual patient care vicinity are served from separate transfer switches on the emergency system, the equipment grounding terminal buses of those panelboards must be connected together with an insulated continuous copper conductor not smaller than 10 AWG. This conductor is permitted to be broken in order to terminate on the equipment grounding terminal bus in each panelboard.

Discussion: This rule incorporates the concept of equipotential bonding in *Article 517.* Although the two equipment grounding paths required in *517.13* also perform a bonding function (see the definition of *equipment grounding conductor* in *Article 100*), the 10 AWG insulated copper bonding conductor serves to reduce potential (voltage) differences between these equipment grounding conductors to a minimum level. Panelboard bonding is shown in Figure 8-11. Branch circuits serving the same patient care vicinity can come from panelboards supplied from different transfer switches and feeders. Metallic equipment in the patient vicinity may experience a voltage (potential) above ground for the brief period of time after a ground fault occurs until the overcurrent device opens. This is due to voltage drop on the ground-fault current path. The bonding

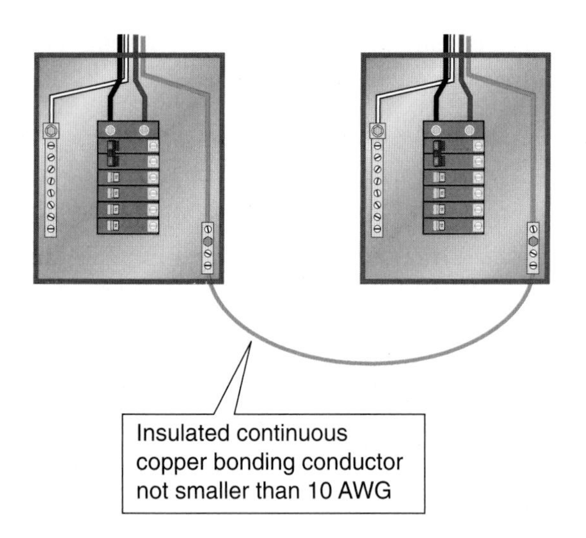

FIGURE 8-11 Panelboard bonding in health care facilities, *517.14*. (*© Cengage Learning 2012*)

jumper also tends to keep electrically connected equipment at the same potential if a ground fault occurs on one branch circuit.

517.17 GROUND-FAULT PROTECTION

The Requirement: (A) Applicability. The requirements of *517.17* shall apply to hospitals and other buildings (including multiple-occupancy buildings) with critical care areas or utilizing electrical life-support equipment, and buildings that provide the required essential utilities or services for the operation of critical care areas or electrical life-support equipment.

Discussion: This requirement for ground-fault protection of equipment has broad application. This requirement should not be confused with GFCI protection for personnel. Type A GFCI protection is required to trip or open on a ground-fault in the range of 4 to 6 mA of current imbalance. (1 mA = 1/1000 ampere.) See the discussion in the Introduction to Grounding and Bonding in this text. Figure I-29 illustrates the principle of operation of GFCI protection.

Ground-fault protection of equipment is required for services and the disconnecting means for buildings or structures supplied by a feeder in specific voltage and current ranges. Such protection is intended to reduce the chances of destructive burndown from an arcing fault. The protection is required for services and building or structure disconnecting means rated 1000 amperes or greater for systems rated more than 150 volts to ground and less than 600 volts phase to phase.

The requirements of this section clearly apply to hospitals and other buildings with critical care areas or that have electrical life-support equipment. This includes multiple occupancy buildings where a common service supplies more than one occupancy. See Figure 8-12. If the common service equipment requires ground-fault protection of equipment, all of the feeders are required to be provided with ground-fault protection of equipment to comply with the rules in this section. The reason for this rule is to preserve

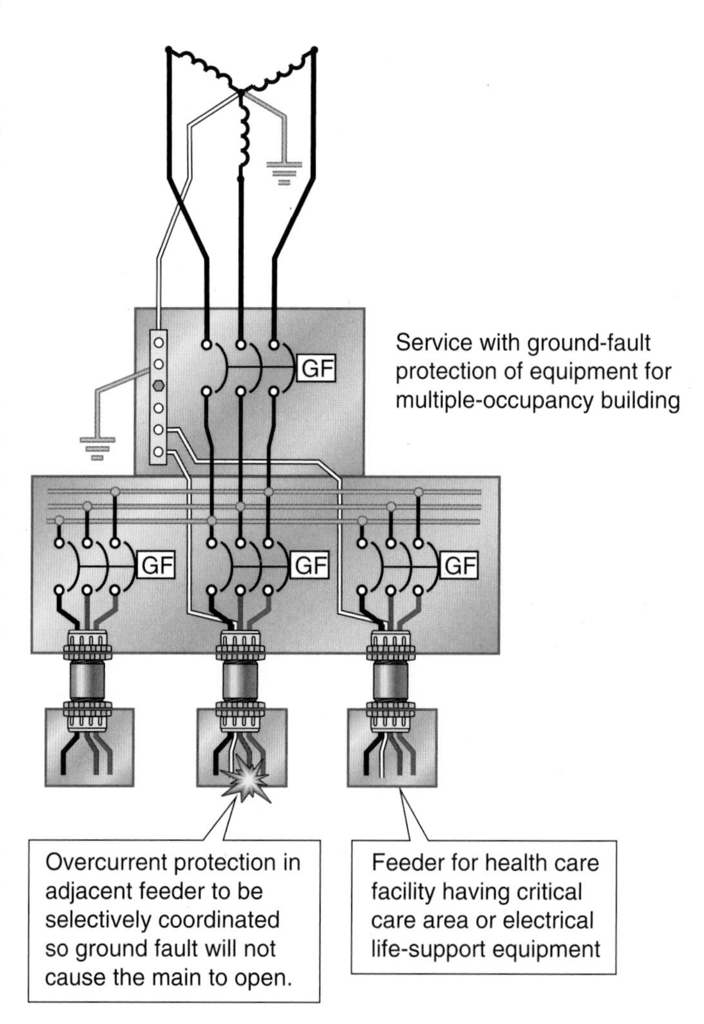

FIGURE 8-12 Application of equipment ground-fault protection for health care facilities having critical care area or electrical life-support equipment, *517.17(A)*. (*© Cengage Learning 2012*)

the operational integrity of the electrical system to the healthcare occupancy. If a ground fault should occur in one of the feeders to a non–health care occupancy, the feeder overcurrent device must open, and not open the upstream main. This preserves normal power to the health care occupancy that has a critical care area or that has life-support equipment.

The Requirement: **(B) Feeders.** Where ground-fault protection is provided for operation of the service disconnecting means or feeder disconnecting means as specified by *230.95* or *215.10*, an additional step of ground-fault protection is required in all next level feeder disconnecting means downstream toward the load. Such protection is required to consist of overcurrent devices and current transformers or other equivalent protective equipment that will cause the feeder disconnecting means to open.

The additional levels of ground-fault protection are not permitted to be installed on the load side of an essential electrical system transfer switch.

Discussion: The requirement for ground-fault protection of equipment has been expanded to include feeders for the alternate source of power, which are very often provided by one or more generators. If ground-fault protection is provided on the normal service, one additional level of ground-fault protection is required downstream or at the first feeder level toward the load. The same requirement applies to overcurrent devices installed from the alternate source of power.

It is clear that an additional level of ground-fault protection is not permitted on the load side of an essential electrical system transfer switch. The essential electrical system is made up of two branches: the life safety and critical branches, which constitute the emergency system, and the equipment system. See *517.30(B)(1)*.

The Requirement: **(C) Selectivity.** Ground-fault protection for operation of the service and feeder disconnecting means is required to be fully selective such that the feeder device, but not the service device, will open on ground faults on the load side of the feeder device. Separation of ground-fault protection time-current characteristics must conform to manufacturer's recommendations and shall consider

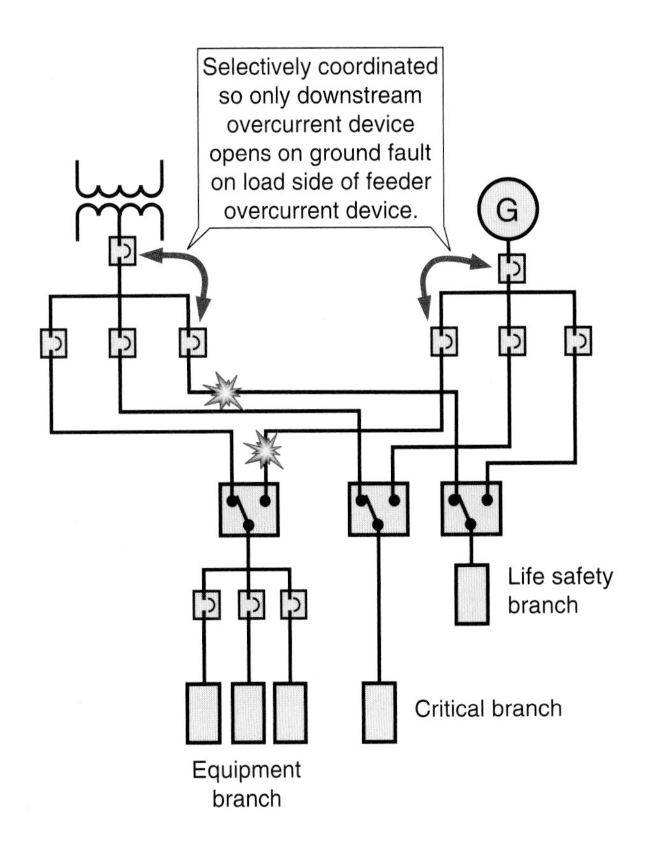

FIGURE 8-13 Selective coordination of equipment ground-fault protection system, *517.17(B)* and *(C)*. (*© Cengage Learning 2012*)

all required tolerances and disconnect operating time to achieve 100% selectivity.

Discussion: This section requires that the settings of the ground-fault protection system at the two levels provide a fully selected system to ensure that only the overcurrent device closest to a ground fault, and not the upstream main overcurrent device, will open. This section previously required a minimum of a six-cycle separation to achieve 100% selectivity. This rule now requires compliance with manufacturer's recommendations for time delay or ground-fault pickup settings to achieve 100% selectivity. See Figure 8-13.

(D) Testing. When equipment ground-fault protection is first installed, each level shall be performance tested to ensure compliance with *517.17(C)*.

Discussion: The requirement wire to the requirement for testing ground-fault protection systems as provided in *230.95*.

517.30(C)(3) Mechanical Protection of the Emergency System

The Requirement: *517.30(C)(3)(3).* Listed flexible metal raceways and listed metal-sheathed cable assemblies are permitted for protection of the emergency system in any of the following conditions:

a. If used in listed prefabricated medical headwalls
b. If used in listed office furnishings
c. If fished into existing walls or ceilings, not otherwise accessible and not subject to physical damage
d. If necessary for flexible connection to equipment

Discussion: The requirement for mechanical protection of the emergency system in patient care areas of hospitals has the effect of limiting the use of flexible metal conduit, liquidtight flexible metal conduit, and Types AC and MC cables in those areas. Nonflexible metal raceways such as EMT are permitted for wiring in patient care areas without restriction. Figure 8-14 shows wiring methods permitted for mechanical protection of the emergency system under the restrictions of *517.30(C)(3)(3).*

Section 517.30(B)(2) designates the critical branch of the hospital electrical system as a part of the emergency system. The critical branch is required to be supplied from a transfer switch that will transfer the system to the emergency power supply (usually a generator) within 10 seconds should the normal power supply fail for any reason. See Figure 8-15 in this book and *Figure 517.30, Informational Note No. 1* in the *NEC* for an overview of typical electrical systems for hospitals.

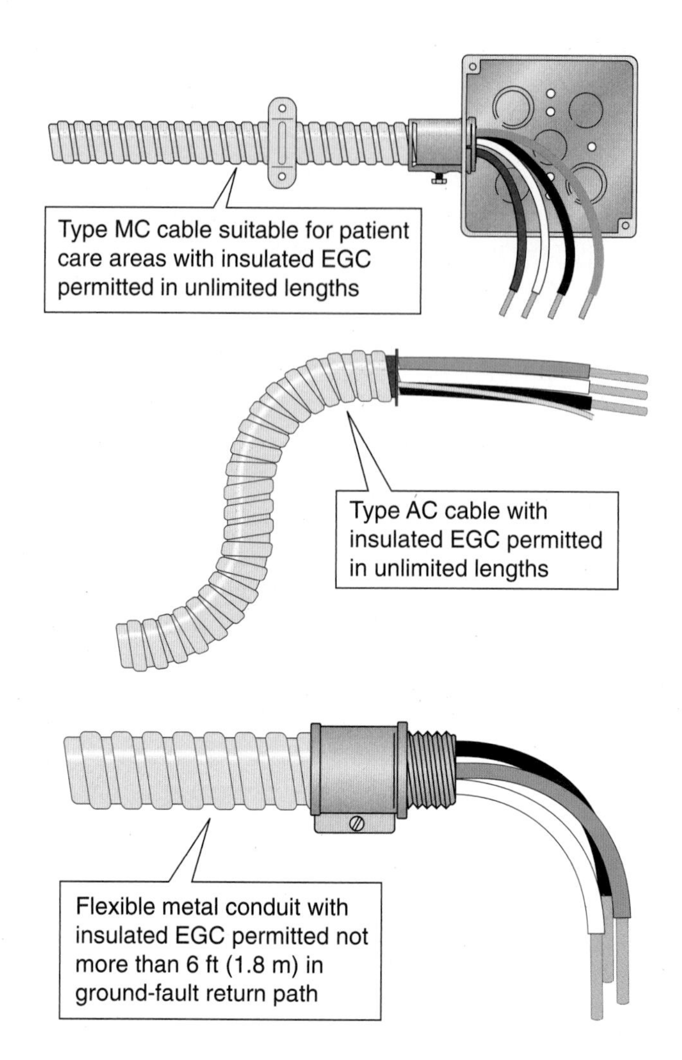

FIGURE 8-14 Mechanical protection of the emergency system, *517.30(C)(3).* (*© Cengage Learning 2012*)

Sections 517.18(A) for general care areas and *517.19(A)* for critical care areas require each patient bed location to be supplied by at least two branch circuits, one or more from the emergency system and one from the normal system. The branch circuits from the emergency system are from the critical branch. The typical supply to the patient care area is shown in Figure 8-15.

An exception to these rules allows all the branch circuits in the general and critical patient care areas to be from the emergency system if they are supplied from different transfer switches. These rules and the exceptions recognize the need for more than one supply should a single source fail. They also recognize that transfer switches, being mechanical devices, are subject to failure.

The result of these rules is flexible metal conduit (FMC), liquidtight flexible metal conduit (LFMC), and Types AC and MC cables that comply with the rules in *517.13(A)* and *(B)* are limited to serving as the wiring method for those branch circuits that serve patient care areas that are from the normal source. These flexible conduits and cables are permitted as the wiring method for branch circuits from the emergency system only as provided in the section.

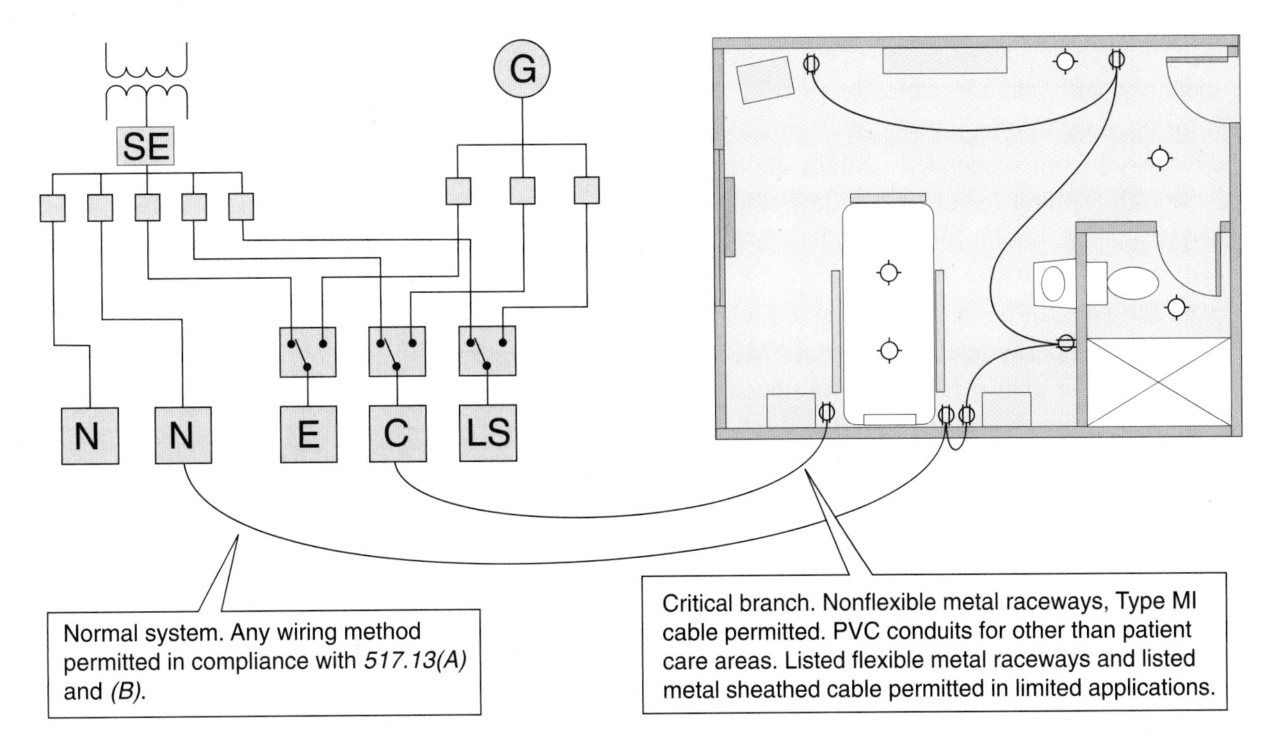

FIGURE 8-15 Wiring methods for normal system and critical branch of emergency system for health care facilities, *517.30(C)(3)(3)*. (*© Cengage Learning 2012*)

517.30(C)(3)(3)a Wiring in Prefabricated Medical Headwalls

The Requirement: Listed flexible metal raceways and listed metal-sheathed cable assemblies are permitted as the wiring method in listed prefabricated medical headwalls.

Discussion: Note that this rule will allow the flexible conduits and cables to be used as the wiring method in listed prefabricated medical headwalls but does not permit the flexible conduits or cables to be used to connect the headwalls to the branch circuit. The UL Safety Standard controls the construction and internal wiring of the headwalls. However, because these headwalls are secured in a fixed location, the wiring method to connect them to supply circuits is not permitted to be flexible metal conduit or Types AC or MC cables. See the discussion below on the use for the flexible wiring methods for fishing into existing finished surfaces such as walls or ceilings.

517.30(C)(3)(3)b Wiring in Listed Office Furnishings

The Requirement: Listed flexible metal raceways and listed metal-sheathed cable assemblies are permitted as the wiring method in listed office furnishings.

Discussion: Listed office furnishings are often used at nurses' stations in patient care areas. Office furnishings are covered in *Article 605*. FMC, LFMC, and Types AC and MC cables are permitted to be used as the wiring method at the factory in listed office furnishings. However, because of the restrictions covered earlier in *517.30(C)(3)(3)a*, these wiring methods are not permitted to be used to connect these office furnishings to the supply branch circuit in patient care areas. See the following discussion on the use for the flexible wiring methods for fishing into existing finished surfaces such as walls or ceilings.

517.30(C)(3)(3)c Fishing Wiring in Existing Walls or Ceilings

The Requirement: Listed flexible metal raceways and listed metal-sheathed cable assemblies are permitted to be fished into existing walls or ceilings, not otherwise accessible and if the wiring methods are not subject to physical damage.

Discussion: Fishing involves inserting a wiring method into a finished wall or ceiling. This operation is often performed by cutting access openings at strategic locations and snaking cables or flexible conduits from opening to opening. At other times, tiles can be removed from suspended ceilings and holes drilled through framing members to allow cables to be fished or snaked though finished walls.

Type AC and MC cables that comply with *517.13(A)* and *(B)* can be used in unlimited lengths for these fishing operations. FMC and LFMC are limited to not more than 6 ft for fishing operations because the wiring methods are not recognized as an equipment grounding conductor in longer lengths. See the discussion at *517.13(A)* for additional information.

517.30(C)(3)(3)d Wiring for Flexible Connections

The Requirement: The flexible wiring methods can be used if necessary for flexible connection to equipment.

Discussion: The *Code* is silent on the intent of this provision. Other sections such as *250.118(5), 250.118(6), 348.60,* and *350.60* include requirements that an equipment grounding conductor be installed in these flexible conduits if they are used to connect equipment "where flexibility is necessary after installation." No provision is made in these sections for convenience of connection. It seems the use of these wiring methods is not permitted for connection to fixed equipment when also intended to provide mechanical protection of the hospital emergency system.

ARTICLE 547 AGRICULTURAL BUILDINGS

Special wiring requirements are contained in *Article 547* for agricultural buildings. These requirements include specific wiring methods, type of equipment grounding conductor, equipotential bonding, and GFCI protection. Several of these requirements are aimed at ensuring the performance of equipment grounding conductors in a harsh, wet, and corrosive environment. Other requirements are intended to limit voltage differences between exposed conductive surfaces. Many studies have been done that conclude dairy cattle, in particular, are very sensitive to voltage differences on exposed conductive surfaces of a few volts. Animal behavior problems, such as dairy cows failing to release milk, occur at these extremely low potential differences. Serious health problems can result if this occurs for some period of time.

547.5(F) Separate Equipment Grounding Conductor

The Requirement: Where an equipment grounding conductor is installed within a location falling under the scope of *Article 547*, it is required to be a copper conductor. If an equipment grounding conductor is installed underground, it is required to be insulated or covered copper.

Discussion: Copper equipment grounding conductors are required in agricultural buildings because they are more resistant to corrosive influences common in these areas than aluminum conductors. Equipment grounding conductors installed underground are required to be insulated or covered copper. To qualify as insulation, the material will have a voltage rating such as 600 V. Covered conductors have an insulating material over the conductor, but it does not have a voltage rating as insulation does.

Underground installations are designated as wet locations in *Article 100.* As a result, conductors intended for installation in underground conduits are required to have a "W" in their type designation, such as THW or THWN. See *310.10(C).* Conductors or cables suitable for direct earth burial have a "U" in their type designation, such as Type UF or Type USE. See the Underwriters Laboratories *Cable Marking Guide* for additional information on marking for underground use.

547.9 ELECTRICAL SUPPLY TO BUILDING(S) OR STRUCTURE(S) FROM A DISTRIBUTION POINT

The Requirement: Overhead electrical supply to agricultural buildings or structures must comply with *547.9(A)* and *547.9(B),* or with *547.9(C).* Underground electrical supply must comply with *547.9(C)* and *547.9(D).*

547.9(A)(4) **Bonding Provisions.** The site-isolating device enclosure is required to be connected to the grounded circuit conductor and the grounding electrode system.

547.9(A)(5) **Grounding.** At the site-isolating device, the system grounded conductor is required to be connected to a grounding electrode system via a grounding electrode conductor.

547.9(B) Service Disconnecting Means and Overcurrent Protection at the Building(s) or Structures(s)

If the service disconnecting means and overcurrent protection are located at the building(s) or structure(s), the requirements of *547.9(B)(1)* through *(B)(3)* apply.

547.9(B)(3) **Grounding and Bonding.** For each building or structure, grounding and bonding of the supply conductors must comply with *250.32,* and the following conditions must be met:

1. The equipment grounding conductor must be not smaller than the largest supply conductor if of the same material, or adjusted in size in accordance with the equivalent size columns of *Table 250.122* if of different materials.
2. The equipment grounding conductor must be connected to the grounded circuit conductor and the site-isolating device at the distribution point.

Discussion: As can be seen, different rules apply depending on whether the supply to agricultural buildings is overhead or underground and on whether the service equipment is located at the site-isolating device or is located at the building or structure served. For overhead supply to buildings or structures, such as with open individual conductors or with triplex or quadruplex cables shown in Figure 8-16, the requirements can be summarized as follows:

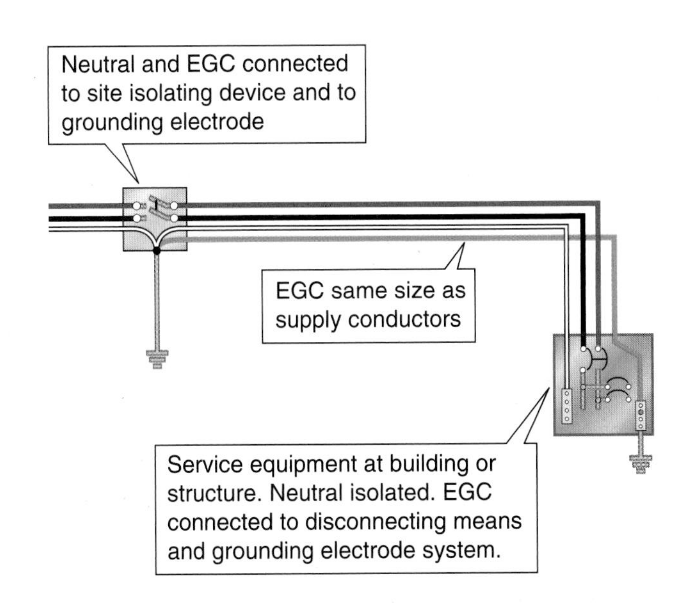

FIGURE 8-16 Grounding and bonding requirements with overhead electrical supply to buildings or structures, *547.9(A)* and *(B)*. (*© Cengage Learning 2012*)

1. The site-isolating device must be connected to a grounding electrode system at the pole where the device is located. One or more grounding electrodes must be installed at this location if none exist. Choose from the list of grounding electrodes acceptable in *250.52(A)(1)* through *(A)(8)* and install them in accordance with the rules in *250.53.*
2. The grounded neutral service conductor must be connected to the metal enclosure of the site-isolating device. This has the result of connecting the grounded service conductor to the grounding electrodes at this location. Note that overcurrent protection is not required at the site-isolating device for the service conductors that supply the individual agricultural buildings or structures.
3. An equipment grounding conductor must be run with the service conductors from the site-isolating device to the individual agricultural building or structure. The equipment grounding conductor is required to be not smaller than the supply conductors if of the same material, or adjusted in size according to *Table 250.122* if of different material.

Here's How: Assume a 200-ampere supply is run to an agricultural building or structure. The ungrounded and grounded conductors are copper, and it is desired to install an aluminum equipment grounding conductor. The copper conductors are 4/0 AWG copper to compensate for voltage drop. By reference to

Table 250.122, it can be seen that a 350-kcmil conductor is equivalent to the 4/0 copper conductor and would have to be installed to comply with this rule.

4. The equipment grounding conductor run with the supply conductors is required to be connected to the grounded service conductor at the site-isolating device and to a grounding electrode system at the individual agricultural building or structure. Bond all the grounding electrodes together at each building or structure served to form a grounding electrode system, as required by *250.50.* One or more grounding electrodes must be installed at each agricultural building or structure if none exist. Choose from the list of grounding electrodes acceptable in *250.52(A)(1)* through *(A)(8).* Install the grounding electrodes to comply with the rules in *250.53.*
5. The grounded supply conductor to the agricultural building or structure is not permitted to be connected to the building or structure disconnecting means and must be isolated electrically from the enclosures.
6. Overcurrent protection for the supply conductors to the buildings or structures must be provided at the disconnecting means for the individual building or structure.

547.9(C) Service Disconnecting Means and Overcurrent Protection at the Distribution Point

The Requirements: If the service disconnecting means and overcurrent protection for each set of feeder or branch circuit conductors are located at the distribution point, feeders, and branch circuits to building(s) or structure(s) must meet the requirements of *250.32* and *Article 225, Parts I and II.*

> ***Informational Note:*** Methods to reduce neutral-to-earth voltages in livestock facilities include supplying buildings or structures with 4-wire, single-phase services, sizing 3-wire, single-phase service and feeder conductors to limit voltage drop to 2%, and connecting loads line to line.

Discussion: *Section 547.9* permits overhead supply to either have the installation comply with *547.9(A)* and *(B),* with no overcurrent protection located at the site-isolating equipment and the building or structure disconnecting means considered the service disconnecting means, or to comply with *547.9(C),* in which the service disconnecting means are located at the site-isolating equipment. This "scheme" is permitted for either overhead or underground supply to the agricultural building or structure. The requirements are shown in Figure 8-17 and can be summarized as follows:

1. Service disconnecting means are located at the site-isolation equipment. All the rules in *Article 230* must be complied with for these and other factors: the number of service disconnecting means at one location, the rating of service disconnecting means, grouping disconnecting means, the equipment being suitable for use as service equipment, and overcurrent protection.
2. A connection of the neutral (grounded conductor) to the grounding electrode system must be made at the site-isolation equipment.
3. The equipment grounding conductor(s) from the disconnecting means at the buildings or structures served must be connected to the neutral at the service equipment, which is located at the site-isolation location.
4. The neutral conductor is not permitted to be connected to the building or structure disconnecting means, to the grounding electrode system, or to equipment grounding conductors at the agricultural building or structure for any electrical system installed after the adoption of the 2008 *NEC*. The grounded system conductor (probably a neutral) is permitted to remain grounded at the service and at the building or structure disconnecting means for previously installed electrical systems, as provided in *250.32(B), Exception.*
5. The feeders from the site-isolation equipment to the individual agricultural building or structure are permitted to be installed overhead or underground. The clearance requirements of *225.18* and *225.19* must be met for overhead supply and the cover requirements of

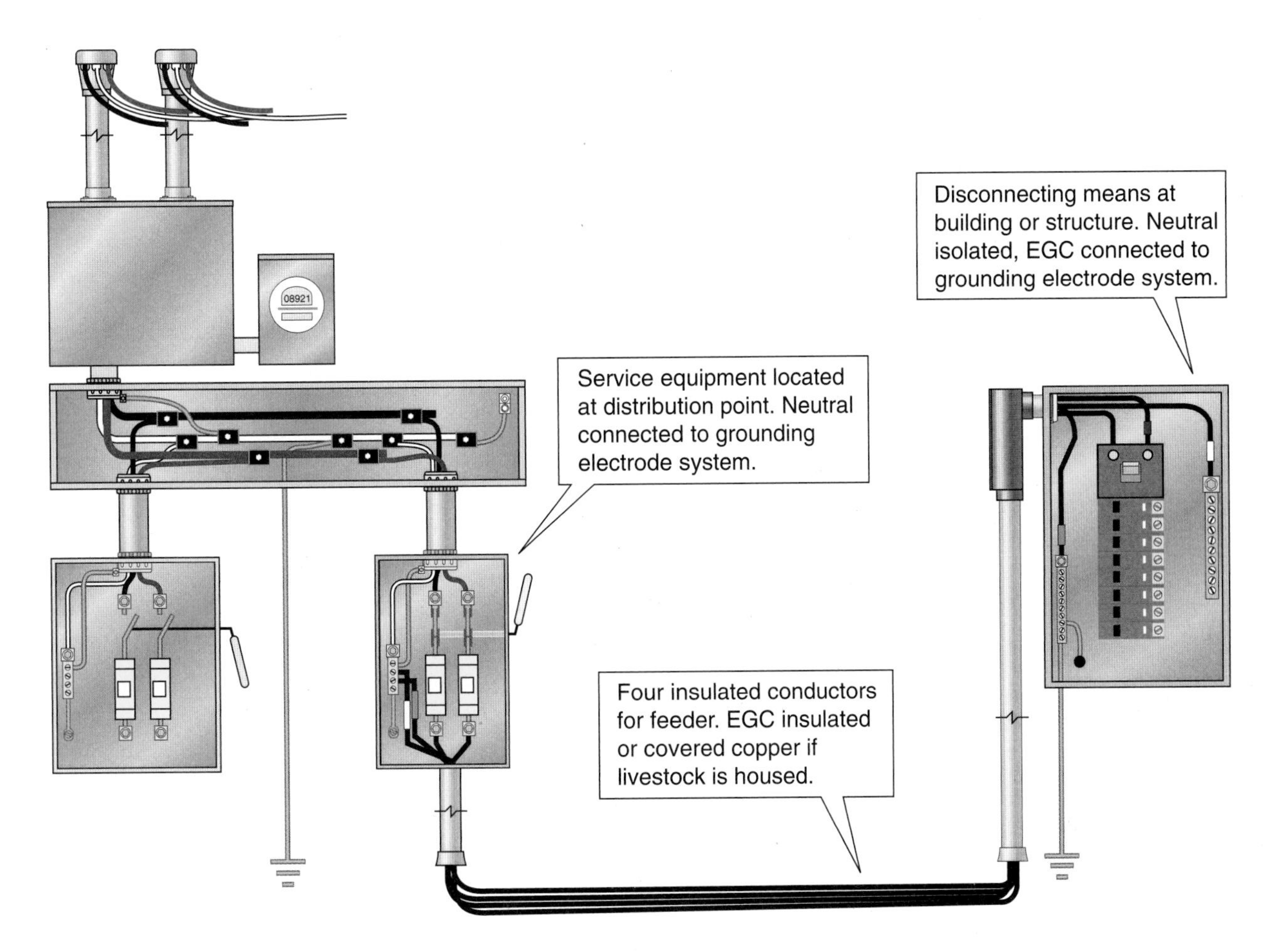

FIGURE 8-17 Underground supply with service disconnecting means and overcurrent protection at the distribution point, *547.9(C)*. (*© Cengage Learning 2012*)

300.5 must be complied with for underground conductors.

6. The requirements in *225.30* for the number of supplies as well as in *225.32* for the location of disconnecting means at the agricultural building or structure must be met.
7. Several other requirements are in *Part II* of *Article 225* for disconnecting means at the individual agricultural building or structure.

547.9(D) Direct-Buried Equipment Grounding Conductors

The Requirement: If livestock is housed, any portion of a direct-buried equipment grounding conductor run to the building or structure is required to be insulated or covered copper.

Discussion: This requirement is intended to help ensure the continuity of directly buried equipment grounding conductors that provided the fault-current return path. Underground areas of agricultural buildings or structures can be highly corrosive due to animal excrement. Standard sizing rules for equipment grounding conductors in *250.122* apply. Notice *547.10* explains that the term *livestock* does not include "poultry."

547.10 EQUIPOTENTIAL PLANES AND BONDING OF EQUIPOTENTIAL PLANES

The installation and bonding of equipotential planes is required to comply with *547.10(A)* and *547.10(B)*. For the purposes of this section, the term *livestock* does not include poultry.

(A) Where Required. Equipotential planes must be installed if required in (1) and (2).

(1) Indoors. Equipotential planes must be installed in confinement areas with concrete floors if metallic equipment is located that may become energized and is accessible to livestock.

(2) Outdoors. Equipotential planes must be installed in concrete slabs if metallic equipment is located that may become energized and is accessible to livestock. The equipotential plane is required to encompass the area where the livestock stands while accessing metallic equipment that may become energized.

(B) Bonding. Equipotential planes are required to be connected to the electrical grounding system. The bonding conductor must be solid copper, insulated, covered or bare, and not smaller than 8 AWG. The means of bonding to wire mesh or conductive elements must be accomplished by pressure connectors or clamps of brass, copper, copper alloy, or an equally substantial approved means. Slatted floors that are supported by structures that are a part of an equipotential plane are not required to be bonded.

Discussion: The concept of providing an equipotential plane for exposed conductive parts in agricultural buildings or structures is intended to limit the voltage between these parts to the lowest practicable level. This concept has not been without controversy during recent years. Some have maintained that with any current flowing, the voltage gradients of connected parts will be at a different potential due to the voltage drop as current flows. Others have stated installing an equipotential plane simply masks the tingle-voltage problem rather than correcting the source of these stray voltages that can cause low-level voltages to appear on conductive surfaces. No doubt, the stray voltage problem is a complex one for agricultural complexes, because the source of stray voltages may be from a neighboring facility. The practice of electrical utilities grounding their primary neutral at many locations is sometimes blamed as the source of the tingle voltage problem because the neutral conductor carries current with the resultant voltage drop. The installation of the equipotential plane may not entirely equalize voltage differences but should limit the voltage between these parts to the lowest practicable level. An equipotential plane is illustrated in Figure 8-18.

The term *equipotential plane* is defined in *547.2.* The definition reads, "An area where wire mesh or other conductive elements are embedded in or placed under concrete, bonded to all metal structures

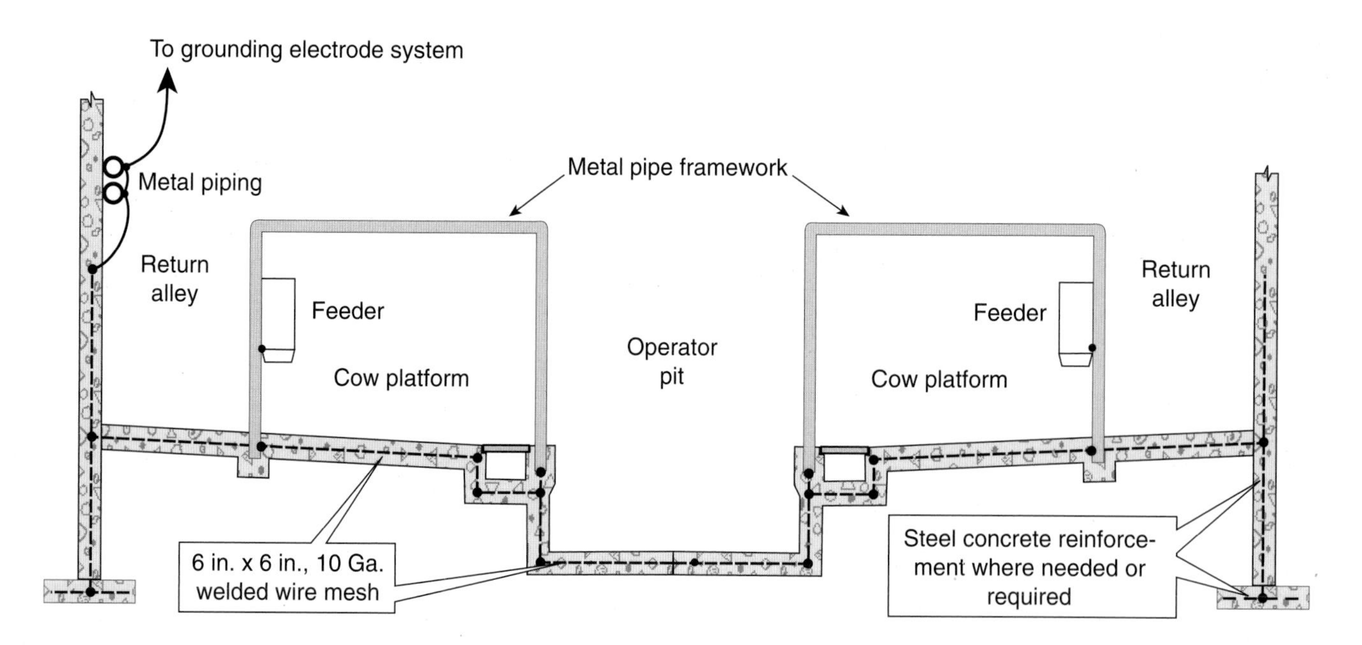

FIGURE 8-18 Equipotential planes and bonding of equipotential planes, *547.10.* (*© Cengage Learning 2012*)

and fixed nonelectrical equipment that may become energized, and connected to the electrical grounding system to prevent a difference in voltage from developing within the plane."*

An *Informational Note* to *547.10* points out, "Methods to establish equipotential planes are described in American Society of Agricultural and Biological Engineers (ASABE) EP473.2-2001, *Equipotential Planes in Animal Containment Areas.*" The second *Informational Note* states, "Methods for safe installation of livestock waterers are described in American Society of Agricultural and Biological Engineers (ASABE) EP342.2-1995, *Safety for Electrically Heated Livestock Waterers.*"

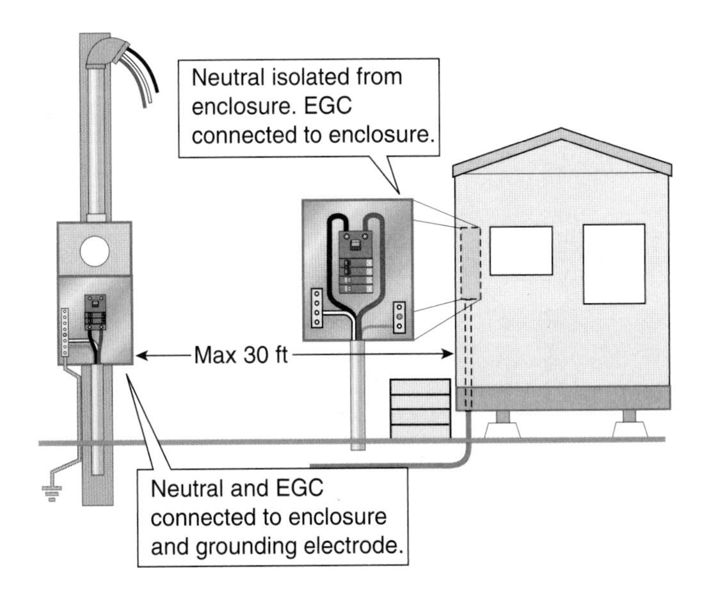

FIGURE 8-19 Grounding mobile home service equipment, *550.32(A)*. (© *Cengage Learning 2012*)

ARTICLE 550, MOBILE HOMES, MANUFACTURED HOMES, AND MOBILE HOME PARKS

Article 550 provides several requirements for the construction of manufactured homes as well as for the sizing of service equipment for mobile home parks and the connection of manufactured and mobile homes to the source of electricity. The manufacturing of mobile and manufactured homes has been under the authority of the federal government since the mid 1970s. The electrical requirements in *Part II* of *Article 550* are usually applied when wiring is done after the manufactured home leaves the factory where it was originally produced. Connection of the electrical system to the source of electricity is required to comply with *Article 550*.

550.32 SERVICE EQUIPMENT

550.32(A) Mobile Home Service Equipment

The Requirement: The mobile home service equipment is required to be located adjacent to the mobile home and not mounted in or on the mobile home. The service equipment must be located in sight from and not more than 30 ft (9.0 m) from the exterior wall of the mobile home it serves. The service equipment is permitted to be located elsewhere on the premises, provided that a disconnecting means, suitable for use as service equipment, is located within sight from and not more than 30 ft (9.0 m) from the exterior wall of the mobile home it serves and is rated not less than that required for service equipment per *550.32(C)*. Grounding at the disconnecting means is required to be in accordance with *250.32*.

Discussion: Note that the requirements of this section apply equally to mobile homes and to manufactured homes. Generally, mobile homes were manufactured before the federal HUD program was initiated many years ago. The terms *mobile home* and *manufactured home* are defined in *550.2*. As stated in the definition of *mobile home,* the term includes *manufactured homes* unless stated otherwise. So, the requirements for location of the service equipment apply to both mobile homes and manufactured homes.

As shown in Figure 8-19, service equipment is generally required to be located adjacent to the mobile home or manufactured home it serves and within 30 ft (9.0 m) of the structure. The neutral is required to be connected to the disconnecting means and is also connected to the grounding electrode(s). Four insulated conductors, including an insulated equipment grounding conductor, is run to the mobile home. The neutral conductor is electrically isolated from the distribution panelboard in the mobile or manufactured home. The equipment grounding conductor is connected to the equipment grounding conductor terminal bar in the distribution panelboard in the structure.

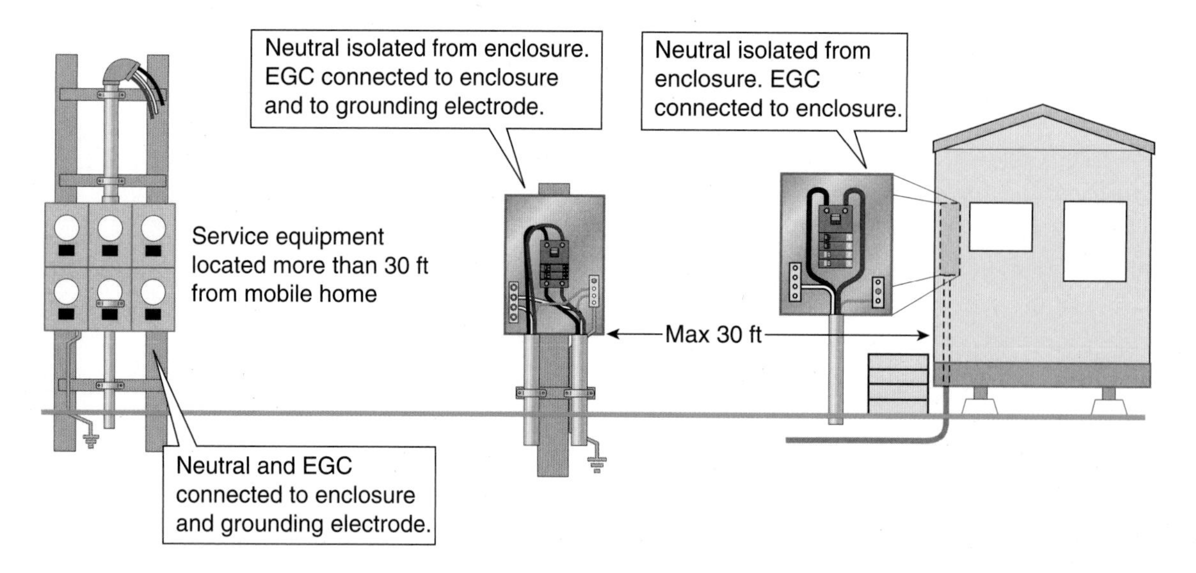

FIGURE 8-20 Grounding if service equipment is located more than 30 ft from mobile home, *550.32(A)*. (*© Cengage Learning 2012*)

To maintain the isolation of the neutral conductor in the structure, 4-pole wiring devices are installed for appliances such as ranges and clothes dryers and 4-wire pigtails are used for these appliances.

Discussion: As shown in Figure 8-20, service equipment is permitted to be located elsewhere if disconnecting means is installed within 30 ft (9.0 m) of the mobile or manufactured home being served. Service equipment is sometimes installed at a central location such as a laundry or recreational structure and feeders are run to each mobile or manufactured home disconnecting means.

This section requires the grounding to be in compliance with *250.32.* That section was changed for the 2008 *NEC* to no longer permit the neutral to be grounded at buildings or structures supplied by a feeder or branch circuit. This rule now requires the neutral to be electrically isolated from the disconnecting means enclosure. The equipment grounding conductor run with the feeder is required to be connected to the metallic enclosure for the disconnecting means.

550.32(B) Manufactured Home Service Equipment

The Requirement: The manufactured home service equipment is permitted to be installed in or on a manufactured home, provided all of the following conditions are met:

1. The manufacturer includes in its written installation instructions information indicating that the home is secured in place by an anchoring system or installed on and secured to a permanent foundation.
2. The installation of the service equipment complies with *Part I* through *Part VII* of *Article 230.*
3. Means is provided for the connection of a grounding electrode conductor to the service equipment and routing it outside the structure.
4. Bonding and grounding of the service complies with *Part I* through *Part V* of *Article 250.*
5. The manufacturer includes in written installation instructions one method of grounding the service equipment at the installation site. The instructions must clearly state that other methods of grounding are found in *Article 250.*
6. The minimum-size grounding electrode conductor must be specified in the instructions.
7. A red warning label must be mounted on or adjacent to the service equipment. The label is to state the following:

> **WARNING:** Do not provide electrical power until the grounding electrode(s) is installed and connected (see installation instructions).

If the service equipment is not installed in or on the unit, the installation is required to comply with the other provisions of this section.

Discussion: This section applies to only *manufactured homes* as defined in *550.2.* It does not apply to units identified on the nameplate as *mobile homes.* As can be seen, the manufacturer can design the electrical system so service equipment is mounted directly on the manufactured home. The neutral terminal bar is electrically isolated from the enclosure for the distribution panelboard and from the terminal bar for the equipment grounding conductors if the mobile or manufactured home is supplied by a feeder. However, *550.16* requires the neutral to be bonded to the metallic enclosure and the terminal bar for the equipment grounding conductors if the distribution panelboard is used as the service equipment, as permitted by *550.32(B).* The installation is shown in Figure 8-21.

In addition to complying with the requirements of this section of *Article 550,* the requirements of the serving electrical utility must be complied with as well. These utility requirements typically include the location and elevation of metering equipment as well as providing adequate clearance for service drop conductors.

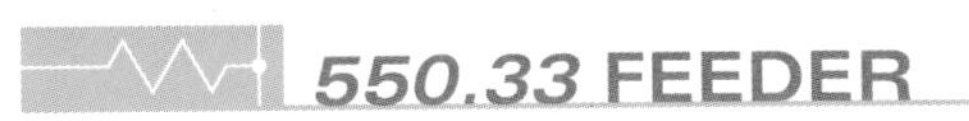

550.33 FEEDER

550.33(A)(1) Feeder Conductors

The Requirements: Feeder conductors are required to consist of either a listed cord, factory installed in accordance with *550.10(B),* or a permanently installed feeder consisting of four insulated, color-coded conductors that must be identified by the factory or field marking of the conductors in compliance with *310.12.* Since equipment grounding conductors are required to be insulated, they are not permitted to be identified by stripping or removing the insulation.

Discussion: This is a long-standing requirement that goes back to the day mobile homes were manufactured with aluminum outer skins. The insulated neutral helped ensure the conductor did not energize the frame of the mobile home and thus create a shock hazard. Conductors suitable for directly buried feeders are often available with only black insulation. These conductors are typically marked at each end with phase tape to comply with the rules in *310.12.* Typical marking is shown in Figure 8-22. Ungrounded conductors are permitted to be marked with any color other than white, gray, or green. Neutral conductors must be marked with white or gray tape or paint and equipment grounding conductors are required to be identified with green tape or paint.

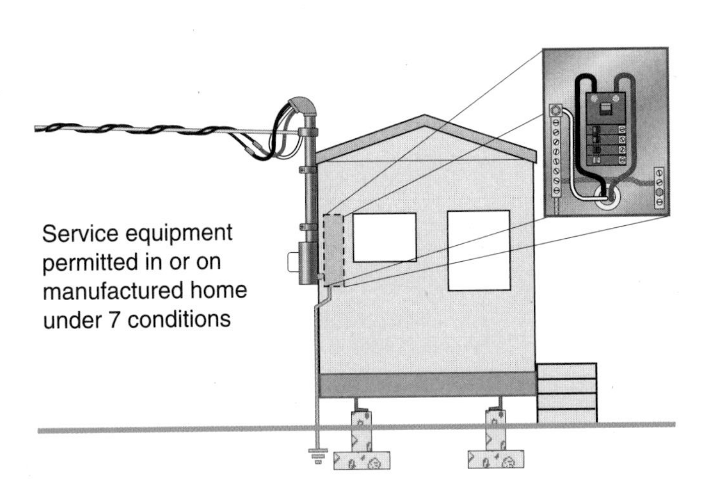

FIGURE 8-21 Grounding if service equipment is mounted on manufactured home, *550.32(B).* (*© Cengage Learning 2012*)

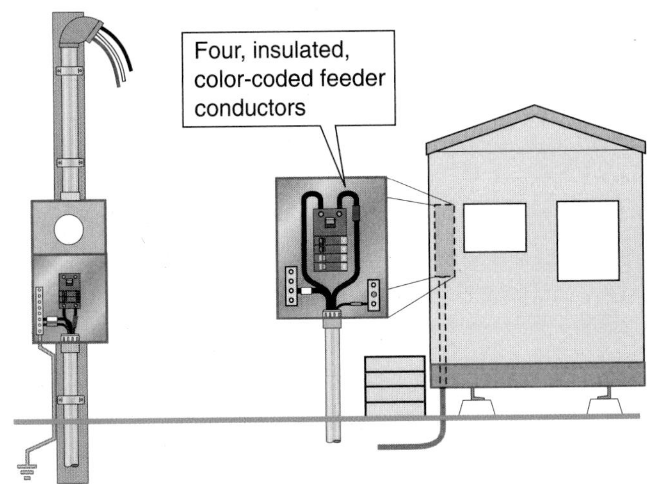

FIGURE 8-22 Feeder requirements to mobile homes, *550.32(B).* (*© Cengage Learning 2012*)

REVIEW QUESTIONS

1. Which of the following statements is true about the requirements for bonding in hazardous (classified) locations?
   a. No special requirements for bonding exist so long as an effective ground-fault return path is provided.
   b. Metal conduit is not required to be bonded so long as an equipment grounding conductor is installed inside for bonding equipment.
   c. Bonding is required from the hazardous location equipment to the source of a separately derived system.
   d. All the methods for bonding service equipment in *250.92(B)(1)* through *(4)* are permitted to be used for bonding.
2. Which of the following statements is most correct about the methods required for bonding metal conduits for wiring in hazardous (classified) locations?
   a. The locknut-bushing and double-locknut types of contacts are not permitted for bonding purposes.
   b. Bonding jumpers with proper fittings or other approved means of bonding are required to be used.
   c. Specific bonding must be provided at all intervening raceways, fittings, boxes, enclosures, and so forth between Class I locations and the point of grounding for service equipment or point of grounding of a separately derived system.
   d. All of the above.
3. Which of the following occupancies typically are required to have the hazardous (classified) locations documented by study?
   a. Commercial Garages, Repair and Storage
   b. Aircraft Hangars
   c. Motor Fuel–Dispensing Facilities
   d. Petrochemical plants
4. Which of the following methods is *not* permitted to be used for bonding in hazardous (classified) locations?
   a. Bonding equipment to the grounded service conductor
   b. Connections using threaded couplings or threaded bosses on enclosures
   c. Threadless couplings and connectors when made up tight for metal raceways and metal-clad cables
   d. Other approved devices, such as bonding-type locknuts and bushings
5. Which of the following statements most accurately describes the bonding requirements for hazardous (classified) locations?
   a. Bonding is required from the hazardous location to the control panel.
   b. Bonding is required from the hazardous location to the panelboard where the branch circuit originates.

c. Bonding is required from the hazardous location to the source of the separately derived system.

d. If locknuts and bushings are tightened securely, extra bonding is not required in hazardous locations.

6. Bonding is required for which of the following hazardous (classified) locations?

   a. Class I locations only

   b. Class I and Class II locations only

   c. Class I, Class II, and Class III locations

   d. If locknuts and bushings are tightened securely, extra bonding is not required in hazardous locations.

7. Which of the following wiring methods and equipment grounding conductor combinations are permitted in a patient care area of a health care facility?

   a. Electrical metallic tubing (EMT) with a 14 AWG copper equipment grounding conductor

   b. Rigid polyvinyl chloride conduit (PVC) conduit with a 12 AWG copper equipment grounding conductor

   c. "Traditional" Type MC cable with a 12 AWG equipment grounding conductor

   d. Type AC cable with a 12 AWG equipment grounding conductor

8. Which of the following wiring methods and equipment grounding conductor combinations are permitted in a patient care area of a health care facility?

   a. One-half-inch flexible metal conduit (FMC) 6 ft long with a 12 AWG equipment grounding conductor

   b. One-half-inch flexible metal conduit (FMC) 12 ft long with a 12 AWG equipment grounding conductor

   c. One-half-inch liquidtight flexible nonmetallic conduit (LFNC) 12 ft long with a 12 AWG equipment grounding conductor

   d. One-half-inch liquidtight flexible metallic conduit (LFMC) 12 ft long with a 12 AWG equipment grounding conductor

9. Which of the following wiring methods is recognized as providing adequate mechanical protection for wiring of the emergency system in the patient care area of a hospital under the conditions stated?

   a. Electrical metallic tubing installed exposed above a suspended ceiling

   b. Type AC cable with a 12 AWG equipment grounding conductor fished into an existing wall

   c. Type MC cable with a 10 AWG aluminum equipment grounding/bonding conductor and a 12 AWG equipment grounding conductor that is fished above a finished ceiling

   d. All of the above

10. Explain in your own words the difference between GFCI protection and ground-fault protection of equipment.

    Answer: ____

    ____

11. Ground-fault protection of equipment is required for which of the following systems?

    a. 208Y/120 volt, 3-phase, 4-wire with an 800-ampere main breaker

    b. 208Y/120-volt, 3-phase, 4-wire with a 1200-ampere main breaker

    c. 480Y/277-volt, 3-phase, 4-wire with an 800-ampere main breaker

    d. 480Y/277-volt, 3-phase, 4-wire with a 1200-ampere main breaker

12. A second level of ground-fault protection of equipment is required for which of the following health care facilities?

    a. A hospital supplied by a 480Y/277-volt service having a 600-ampere, and three 800-ampere main breakers.

    b. A nursing home that has a 1200-ampere, 208Y/120-volt, 3-phase, 4-wire service.

    c. An 800-ampere feeder to an outpatient surgery from a 1600-ampere, 480Y/120-volt service for a multi-occupancy building that has ground-fault protection of equipment.

    d. A retirement facility supplied by a 1200-ampere, 480Y/277-volt, 3-phase, 4-wire service.

13. Aluminum equipment grounding conductors are permitted for an underground feeder to an agricultural building so long as it is listed for direct burial applications.

    a. True

    b. False

14. For overhead supply to an agricultural building or structure from a site-isolating device, which of the following requirements applies?

    a. The system grounded (neutral) conductor must be isolated from the site-isolating device.

    b. The system grounded (neutral) conductor must be grounded at each of the disconnecting means at each agricultural building or structure.

    c. A grounding electrode system is required at the site-isolating device *and* at each agricultural building or structure served.

    d. None of the above.

15. A 200-ampere supply consisting of 3/0 copper conductors is run from a site-isolating device to the service disconnecting means at an agricultural building or structure. What is the minimum size of the equipment grounding conductor?

    a. 6 AWG copper

    b. 4 AWG aluminum

    c. 2/0 copper

    d. 3/0 copper

16. Which of the following statements about equipotential planes at or in agricultural buildings or structures is *not* true?
    a. They are a good idea but are not required.
    b. Components must be connected with a minimum 8 AWG copper conductor.
    c. Connections must be made with suitable pressure connectors.
    d. Slatted floors supported by bonded structures are *not* required to be bonded.
17. Which of the following statements about the general requirements for service equipment for mobile homes is *not* true?
    a. The service equipment must be located not less than 30 ft from the mobile home it serves.
    b. The service equipment must also be located within sight of the mobile home it serves.
    c. The service equipment is permitted elsewhere on the premises if another disconnecting means is located within 30 ft of the mobile homes.
    d. The neutral is permitted to be grounded again at the disconnecting means if the feeder complies with *250.32.*
18. Which of the following statements about the requirements for feeders to mobile homes is *not* true?
    a. The feeder conductors are required to be installed with copper conductors.
    b. Four insulated conductors are required to be installed.
    c. The feeder conductors are required to be color-coded in accordance with *310.12.*
    d. Service equipment is permitted to be installed in or on the manufactured home if it complies with the rules.

Appendix A: Some Fundamentals of Equipment-Grounding Circuit Design

R. H. Kaufmann
Fellow AIEE

Synopsis: An effective equipment-grounding system should, under conditions of maximum ground-fault current flow, accomplish the following objectives: 1. maintain a low potential difference, perhaps 50 volts maximum, between machine frames, equipment enclosures, conductor enclosures, building metallic structure, and metallic components contained therein to avoid electric shock hazard and unwanted circulating current, and 2. incorporate adequate conductance to carry this maximum ground short-circuit current without thermal distress and the attendant fire hazard. There is good reason to believe that current flow in the equipment grounding system in ac power systems will not stray far from the power cable over which the outgoing current flows. It follows that the installation of conductive material in an equipment ground system unless properly located can be ineffective and wasteful, and can create a false sense of security. This paper presents the results of a special series of full-scale tests dealing with this specific problem and a general analysis of the circuit behavior.

A much better understanding of protective grounding systems is necessary to ensure freedom from hazard to life and property. The objective of this paper is to present the significant factors which control the behavior of protective grounding circuits in ac industrial power distribution circuits during short-circuit conditions. It is hoped that this presentation may emphasize that improper application of material in equipment-grounding systems creates the wastefulness and unwarranted sense of security just mentioned.

Under a short-circuit to ground condition in an ac distribution circuit, grounded or ungrounded systems alike, inductive reactance will exert a powerful influence in directing the return current to flow in a path closely paralleling the outgoing power conductor. The enclosing conduit or metallic raceway would constitute such a path. To attempt to relieve the conductor enclosure by installation of an external conductor is quite ineffective. Connections to near-by building structural members are equally ineffective. Only through the installation of an internal

Paper 54-244, recommended by the AIEE Industrial Power Systems Committee and approved by the AIEE Committee on Technical Operations for presentation at the AIEE Summer and Pacific General Meeting, Los Angeles, Calif., June 21–25, 1954. Manuscript submitted March 23, 1954; made available for printing April 27, 1954.

R. H. Kaufmann is with the General Electric Company, Schenectady, N.Y.

The generous assistance of J. M. Schmidt and his staff in Facilities Engineering of the General Electric Company was invaluable to the test portion of this investigation. To him goes the credit for making the test installation, arranging for the loan of power supply transformers, together with switching and relaying equipment, and, in addition, authorizing the short-circuit tests from the 13,800-volt feeder lines serving the Schenectady plant. To L. J. Carpenter of Industrial Power Engineering goes credit for assistance in conducting the tests and the analysis of results.

grounding conductor can the current which must be carried by the enclosure be noticeably reduced. Joints in conduits and raceways require special consideration to avoid a shower of sparks and attendant fire hazard during the short circuit's duration. Any thought of keeping the conduit or raceway insulated from ground except at one point seems totally impractical. Furthermore, such practice or a contemplated omission of the metallic enclosure may well lead to large induced voltages in near-by metallic structures which may appear either as dangerous shock hazards or unwanted circulating currents. Ungrounded distribution systems require equally careful treatment. Very often a second ground fault occurs before the previous one has been located and corrected. The problems discussed in this paragraph then appear simultaneously on each of the two circuits.

TEST PROCEDURE

A special installation of 2½-inch conventional heavy-wall steel conduit and 4/0 copper cable conductors was made for this investigation. It was installed in a building previously used for short-circuit testing because this building contained a heavy steel column construction and all columns were tied to an extensive grounding mat composed of 250,000-circular-mil copper cables. The test installation is illustrated in Figure 1. The conduit was supported on insulators throughout the 100-foot length. The conduit run was about 5 feet from a line of building columns. The external 4/0 cable was spaced about 1 foot from the conduit on the side opposite the building columns.

The particular arrangement of components makes possible the measurement of all currents and voltages at one location. This is especially valuable in that it avoids the running of length voltage-measuring leads. The setup is intended to simulate an electric feeder circuit with power source at the left end and various simulate fault conditions at the right end. In all cases here reported the test current was caused to flow over the *A* cable to the far end. A variety of different possible return paths were examined, controlled by the connections made at the left end.

One series of tests was made at low current, 200 and 350 amperes, using an ac welding transformer as a source of 60-cycle power. At these low-current magnitudes, the current flow could be maintained for extended periods. Voltage measurements were made with high quality indicating voltmeters. Current measurement was made with a clip-on ammeter.

A second series of tests was made at high current, around 10,000 amperes, using a 450-kva 3-phase 60-cycle transformer with a 600-volt secondary as a

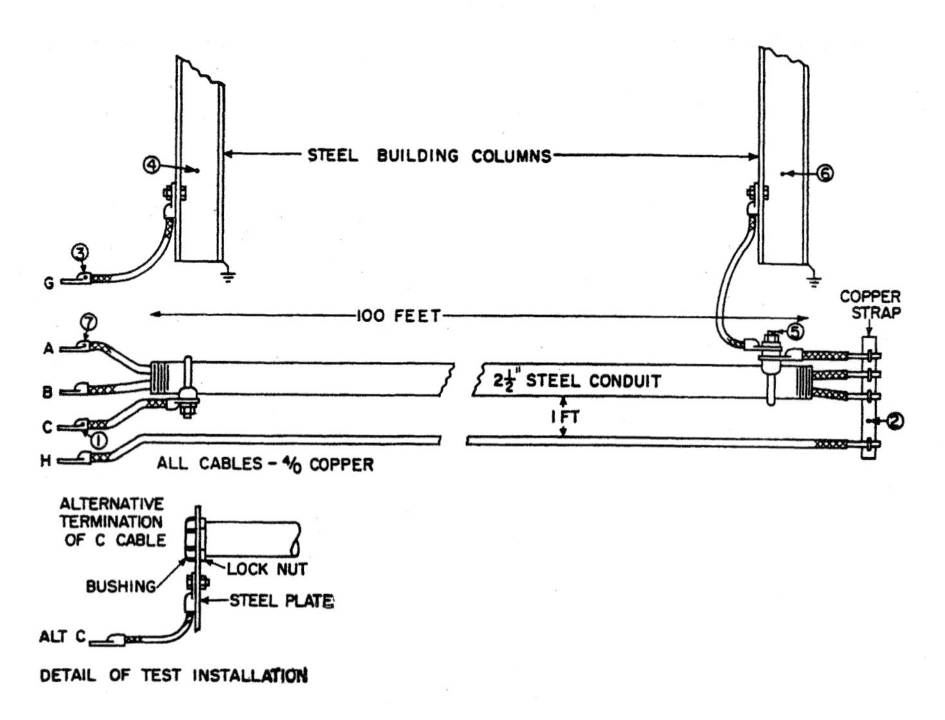

FIGURE 1 Identification of the test installation.

source of power. Switching was done at primary voltage, 13,800 volts, and an *IAC* induction relay was used to control the duration of current flow to an interval of about ¼ second. An oscillograph was used for all measurement of current and voltage.

RESULTS

The magnitude of voltage and current obtained in the entire series of 60-cycle ac tests are presented in original form in Table I. The test number is an arbitrary assignment for reference purposes. The next two columns identify the connections used and indicate the possible paths of current flow. Next the current values are tabulated: first, the total input current into the *A* conductor and, next, the return current in the conduit and its percentage of the total; and then the magnitudes of current in the other possible paths are given.

In all cases the tabulated values taken from oscillograph records are the crest magnitudes divided by $\sqrt{2}$ which is the true rms only if the wave form is sinusoidal. The wave shape of E_{CG} was commonly quite different from a sine wave, but in the extreme cases the magnitude also was small. See Figure 3 pertaining to test BS.

Measurements of dc resistance were made of the several return paths comprising the equipment-grounding system. See Table II. A controlled magnitude of direct current was caused to flow through the circuit and the dc potential drop across the desired section was measured with an indicating millivoltmeter. The points of measurement refer to those indicated in Figure 1.

No further analysis is needed to show conclusively that only by the use of an internal grounding conductor can any sizable fraction of the return current be diverted from the conduit or raceway. In spite of the extremely low resistance of the building structural frame, it was ineffective in reducing the magnitude of return current in the conduit. See tests A6, A7, B6, and B7. Further analysis is desirable to

TABLE II

Measured DC Resistances

Element	Millivolt Drop Between	Ohms
100-foot length of 4/0 conductor	2–7	0.00515
100-foot length of 2½-inch steel conduit	1–5	0.00475
Steel building frame	4–6	0.00011
	3–5	0.0006

TABLE I

Measured Electrical Quantities

Test No.	Current Flow: Out On	Current Flow: Return On	Current Magnitudes: I_A Total	I_C Amperes	I_C Per Cent of Total	I_B	I_H	I_G	Voltage Magnitudes: E_{AC}	E_{CG}	E_{AB}	E_{AH}	E_{AG}	E_{GB}
Low-Current Tests														
A1	A	B	350	0		350	0	0	2.47		4.85			
A2	A	C	350	350	100	0	0	0	15.9	0.45	2.5			
A3	A	C	200	200	100	0	0	0	9.05	0.15	1.51			
A4	A	CH	350	340	97	0	12	0	16.0	0.05	2.55			
A5	A	CH	200	190	95	0	8	0	9.13	nil	1.55			
A6	A	CG	350	340	97	0	0	12	14.6		2.54	14.4		
A7	A	CG	200	180	90	0	0	8	9.5		1.50	9.4		
A8	A	CB	350	62	18	290	0	0	4.55	nil	4.55			
A9	A	CB	200	40	20	150	0	0	2.68	nil	2.68			
A10	A	GH	350	0	0	0	160	160	14.0	12.5	2.5	26.4	26.4	24.6
A11	A	GH	200	0	0	0	98	98	9.2	8.1	1.5	17.1	17.1	15.1
High-Current Tests														
B2	A	C	11,200	11,200	100	0	0	0	168	36*				
B3	A	C	11,070	11,070	100	0	0	0	173	38*				
B4	A	CH	11,070	11,200	101	0	1,140	0	173	18*				
B5	A	CH	11,080	11,090	100	0	1,220	0	173	17*				
B6	A	CG	10,830	10,770	99	0	0	1,080	168		71			
B7	A	CG	10,910	10,780	99	0	0	1,145	173	9*				
B8	A	CB	11,620	5,810	50	5,660	0	0		27*			155	
B9	A	CB	11,380	6,070	53	5,620	0	0	146	25*				
B10	A	GH	8,710	0	0	0	4,300	4,500		146			268	

* Distorted wave shape. Tabulated values are crest/$\sqrt{2}$.

establish a better understanding of the nature of the circuits involved.

Some interesting secondary effects were observed in the course of the tests. The first high-current test produced a shower of sparks from about half the couplings in the conduit run. From one came a blowtorch stream of sparks which burned out many of the threads. Several small fires set in near-by combustible material would have been serious if not promptly extinguished. The conduit run had been installed by a crew regularly engaged in such work and they gave assurance that the joints had been pulled up to normal tightness and perhaps even a little more. A short 4/0 copper jumper was bridged around this join but, even so, some sparks continued to be expelled from this coupling on subsequent tests. Other couplings threw no more sparks during subsequent tests. Apparently small tack welds had occurred on the first test.

In one high-current test the conduit termination was altered to simulate a connection to a steel cabinet or junction box. See bottom of Figure 1. The bushing was applied finger-tight. On test with about 11000 amperes flowing for about ¼ second, a fan-shaped shower of sparks occurred parallel to the plate. In the process, a weld resulted and the parts were separated only with considerable difficulty, with the use of wrenches and a hammer. This suggested that a repeat shot would have produced no disturbance.

During high-current test B10 (conduit circuit open) a shower of sparks was observed at an intermediate building column. Careful inspection disclosed that the origin was at a spot at which a water pipe passed through an opening cut in the web of the steel beam involved. The pipe had been loosely in contact with the edges of the hole. Here is evidence of the objectionable effects of forcing the short-circuit current to seek return paths remote from the outgoing conductor. The large spacing between outgoing and returning current creates a powerful magnetic field which extends far out in space around the current-carrying conductors.

CIRCUIT ANALYSIS

The test installation as identified in Figure 1 allows a study of a wide variety of equipment-grounding arrangements. The *A* conductor in all cases is used to represent the power conductor which has faulted to ground. One side of the test power was connected directly to the *A* conductor in all tests. The *B* conductor can be connected to act as an internal grounding conductor. The *H* conductor can represent an external grounding cable run parallel to the power cables with about 1 foot separation.

In addition to a resistance value associated with every conductor element is an inductive reactance value of ac impedance. The reactance will increase as the spacing increases.Thus, for conditions in which the outgoing current will in all cases be carried by the *A* conductor, the reactances of the various elements will increase as their spacing from the *A* conductor increases.

The upper sketch in Figure 2 defines the spacing characteristics of the circuit elements being studied. The steel conduit represents an annular area of high magnetic permeability. Until saturation occurs,

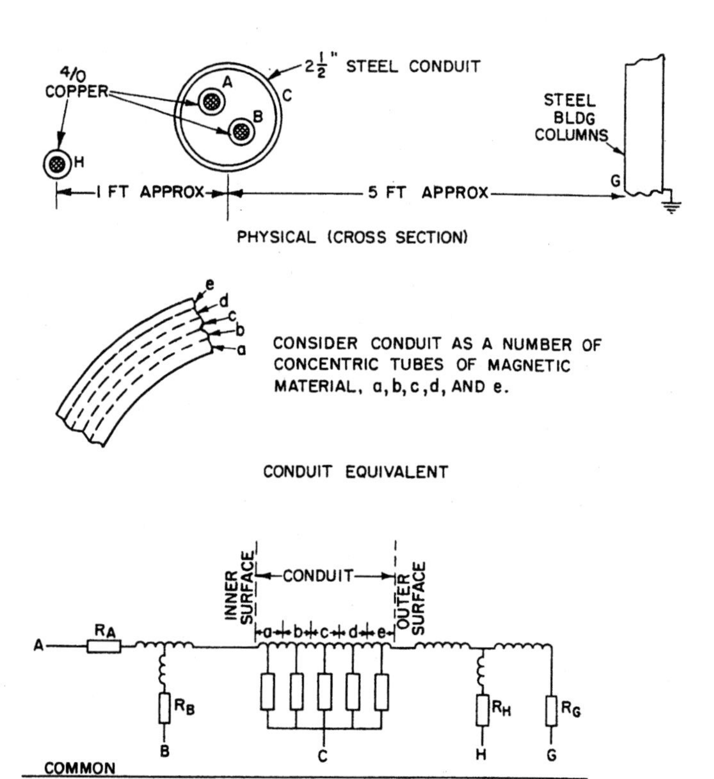

FIGURE 2 Test circuit conductor geometry and equivalent electric circuit.

the magnetic flux produced in the steel pipe may be 500 or more times what it would have been in nonmagnetic material. The conduit may be considered as a series of thin-wall magnetic tubes, one within the other, such as that indicated in the center sketch in Figure 2. The resulting equivalent circuit of the system under investigation takes the form indicated at the bottom of Figure 2. The reactance of the circuit including the *B* conductor will be the lowest. Next will be the innermost tube of the conduit, followed by the others in successive order until finally the outer tube is reached. The inductance of these tubular elements of the steel conduit assumes unusual importance because of the high magnetic permeability. Next in spacing is the external grounding conductor (*H* cable) and, last, the structural members of the building frame and their interconnecting grounding cables buried below floor level. Any given circuit element is brought into use by connecting its terminal to the common bus.

The effectiveness of the conduit in confining the flow of fault current within the conduit can be visualized with the aid of this equivalent circuit. Consider only the conduit (*C* terminal) connected to the common bus. This forces all the current to return on the conduit alone. The equivalent circuit shows that the current by returning on the inner surface of the conduit will avoid the reactance of the other annular sections. Furthermore, the conduit reactance will impede the diversion of current to the conduit outer surface. The voltage drop in a circuit path external to the conduit should be expected to be low until magnetic saturation occurs. Observe this effect in tests A2 and A3. The voltage E_{CG} is is a voltage drop measured around a circuit external to the conduit. The value of 0.45 volt with 200 amperes flowing is less than the product of the current magnitude and the conduit dc resistance. The same is true of the values obtained in tests B2 and B3 at high current. The oscillogram associated with test B8 is reproduced in Figure 3. Notice the lag in the appearance of voltage in the circuit external to the conduit (E_{CG}). Magnetic saturation has been occurring in the steel conduit starting at the inner surface and progressing toward the outer wall. As complete saturation occurs the voltage appearing in the external circuit rises abruptly. Note that the current flow in the conduit had nearly reached crest value when this occurred.

Another interesting effect of the conduit is displayed when the conduit is not a part of the return electric circuit. Suppose that only the *H* cable is available as a return current path (all other terminals open). The current flowing out over the *A* conductor and returning over the *H* cable links the magnetic conduit as a transformer winding links its core. There tends to be reflected into the electric circuit a high magnetizing reactance modified only by the circulating current in the conduit metal. The conduit, not being laminated, allows a circulating current to flow much as a short-circuited secondary winding. This circulating current flows in a direction opposite to that in the *A* conductor on the conduit's inside surface and returns on the outside surface. In the equivalent circuit this circulating current is accounted for by a highly resistive circuit in parallel with the magnetizing reactance. The net impedance reflected into the electric circuit is much higher than would prevail if the conduit were part of the return electric circuit. These effects in iron conduits are described elsewhere.[1]

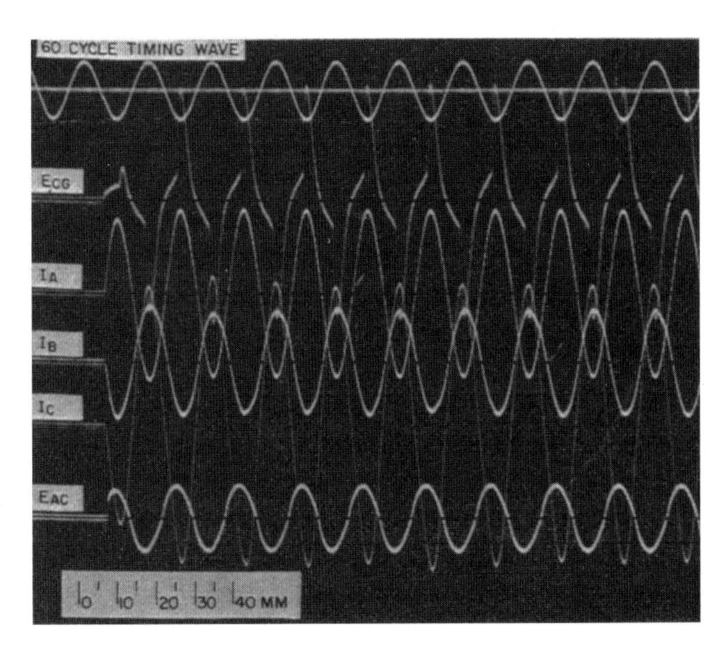

E_{CO} conduit to ground volts, 1.26 volts per millimeter
I_A line current, 785 Angstrom units per millimeter
I_B ground cable current, 623 Angstrom units per millimeter
I_C conduit current, 247 Angstrom units per millimeter
E_{AC} input volts, 25.8 volts per millimeter

FIGURE 3 Oscillogram of electrical quantities in test B8.

Figures 4 and 5 give the phasor relationship of voltages and currents, together with the waveform pattern of the voltage appearing along the exterior of the conduit as derived from several representative oscillograms.

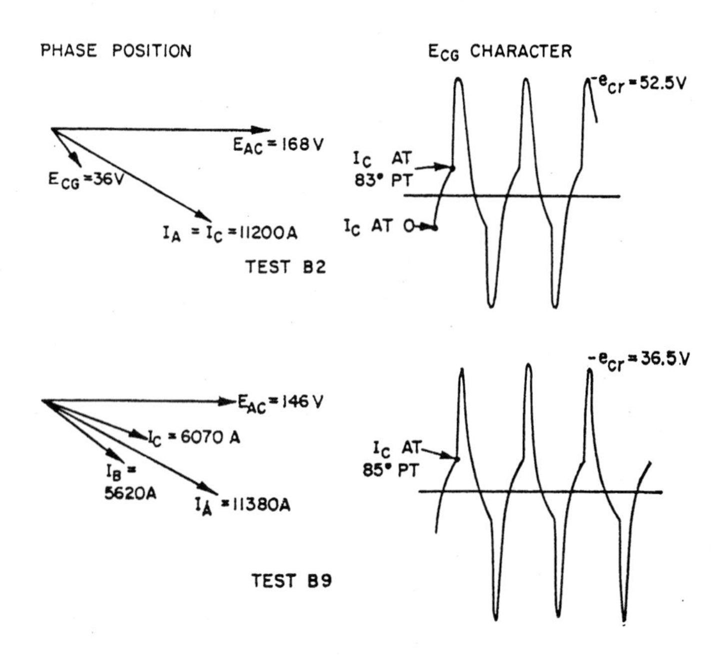

FIGURE 4 Oscillogram analysis, tests B2 and B9.

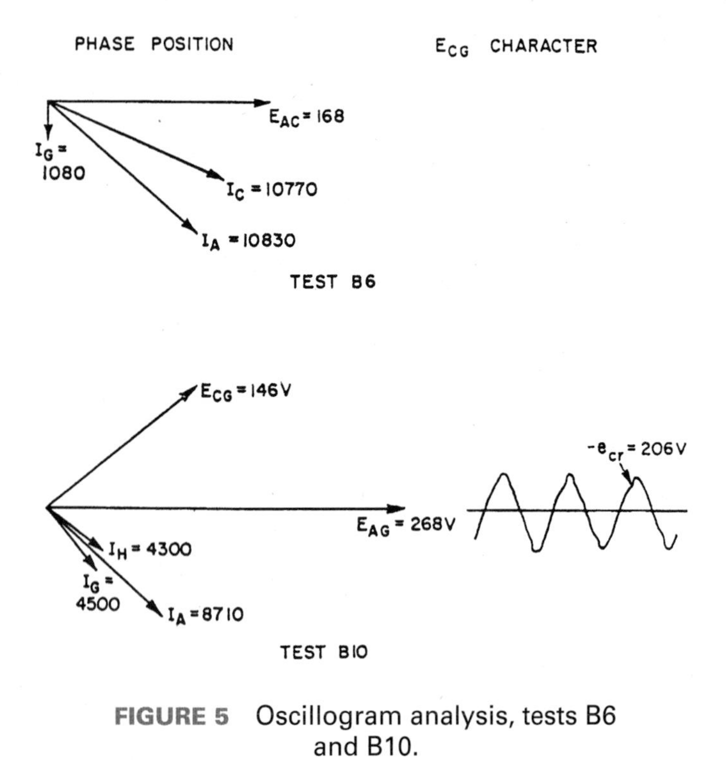

FIGURE 5 Oscillogram analysis, tests B6 and B10.

In test B2, Figure 4. the entire current was forced to return over the conduit alone. Note that the voltage E_{CG} appearing along the conduit exterior lags considerably behind the conduit current I_C, as was predicted. The wave shape of this voltage E_{CG} clearly shows the abrupt rise at the time of complete magnetic saturation.

In test B9, Figure 4, the internal *B* cable was in parallel with the conduit and the return current divided nearly equally. The fact that the conduit current I_C leads the *B* cable current by a substantial angle is indicative that the conduit current is taking the highly resistive path along the inner conduit surface. The externally measured voltage E_{CG} exhibits the pattern found in test B2. After conduit saturation the external voltage closely approaches the current resistance drop computed from conduit dc resistance.

In test B6, Figure 5, the building structure was paralleled with the conduit. The large angle between the conduit current and the building frame current identifies the building structure as a highly inductive circuit while the conduit path appears as highly resistive. Although the current in the building frame was almost 1100 amperes, it resulted in only a slight reduction in the conduit current because of the large phase displacement between the two current components. The voltage appearing along the conduit exterior in this test was impressed on the building structure and would appear as potential difference along beams and columns.

In test B10, Figure 5, both the exterior grounding conductor (*H* cable) and the building frame (*G* terminal) were connected to provide parallel paths for the return current, but the conduit circuit was left open (*C* terminal not connected). The test results clearly evidence the powerful forces tending to maintain current flow in the conduit circuit. Note that across the open connection at the *C* terminal appeared a voltage of 146 volts (or, more significantly, over 50 percent of the total impressed driving voltage), that required to force the current to return via the *H* cable and building frame in parallel. Such a voltage could be a serious shock hazard. Furthermore, unless the conduit were well insulated throughout the entire length, there would be a sufficient number of sparks at stray contact points to constitute a serious fire hazard. It was during this test that a shower of sparks observed between metallic members in the building system was caused simply by the strong magnetic field extending far out from the power conductors.

CONCLUSIONS

The significance of this investigation points unmistakably to the conclusions presented earlier in this paper. Effective use of the conduit or raceway in the equipment-grounding system is paramount. Additional work is needed to develop joints which will not throw fire during ground fault. To improve effectiveness requires greater conductivity in the conductor enclosure or the use of an internal grounding conductor. Grounding cables connecting building structure with ground electrodes (connection to earth) are needed to convey lightning currents or similar currents seeking a path to earth, but these conductors will play a negligible part in the performance of the equipment-grounding system. Of course, the importance of proper equipment-grounding becomes greater with the larger size feeder circuits and the availability of higher short-circuit currents.

REFERENCE

1. Iron Conduit Impedance Effects in Ground Circuit Systems, A. J. Bisson, E. A. Rochau. *AIEE Transactions*, vol. 73, pt. II, July 1954, pp. 104–07.

DISCUSSION

J. A. Geinger (Eastman Kodak Company, Rochester, N.Y.): The material covered in this paper is timely. A number of cases of faulty operation in this plant have been identified with the characteristics of grounding circuits presented. We believe that there is a real need for curves or tables of volts versus amperes per unit length of conduit in the common sizes and types, and that the greatest interest would be for data on the circuit in which the current flow is *A* to *C* as illustrated in Figure 1.

The bad showing of the conduit joints in throwing fire when conducting high current, we believe, is due entirely to improper assembly or unsuitable fittings. A number of cases of sparking at conduit couplings or terminals in this plant have come to our attention; at least one fire was started by such sparking. In every case involving couplings, however, we found the coupling loose or conclusive evidence of previous looseness.

We recently tested two 2½-inch couplings at 8500 amperes. The current was maintained for approximately two seconds. Uncleaned factory threads were employed in each case. The joints were assembled by one man utilizing 24-inch pipe wrenches. In the two couplings (four threads) tested, no sparking resulted in this tight condition. No sparking, resulted when the joints had been loosened one turn. However, in all cases, when the threads were backed off two turns, which is equivalent to "hand-tight," extensive sparking resulted. This corresponds to results of our tests made several years ago on smaller conduit couplings.

Many electricians assume that the conduit is used essentially for mechanical protection and do not realize the likelihood of sparking when the conduit is called upon to carry fault currents. We believe there are other electricians who may know better but are rather careless in their work and this results in loosely assembled conduit systems which will definitely throw sparks if they are called upon to carry fault currents.

We should be interested in a practical design of joint which an unskilled mechanic could not possibly assemble in an inadequate way. In the meantime, however we intend to emphasize the proper assembly of our conduit system. We believe there is no practical substitute for a well-designed and well-assembled conduit system.

The inadequate current-carrying capacity of ordinary lock nuts and bushings has been recognized for many years. The National Electric Code currently specifies grounding bushings and bonding under specific situations. In our recent tests, properly bonded conduit terminals and properly assembly grounding bushings in the case of 2½-inch conduit carried 8500 amperes without emitting sparks. Again, improperly assembled terminals emitted generous showers of sparks.

Our tests were limited by our test equipment to 8500 amperes. From the results of the tests on partly loosened couplings, a properly tightened joint would probably carry without sparking several times this value of current.

L. Brieger (Consolidated Edison Company of New York, Inc., New York, N.Y.): This excellent paper should be welcomed for its wealth of reliable information on an important subject on which very little factual material has been published.

The author demonstrates quite clearly that where an iron conduit is part of an electrical circuit, the routine notions of circuit behavior may lead to considerable error of judgment. The theoretical analysis offered in explanation of the tendency of an iron conduit to restrain the flow of return current within itself and prevent leakage to external paths is valuable indeed. Similar conditions exist in other applications of a somewhat different nature, such as, in the case of single-phase faults on pipe-type high tension cable.[1]

Although the paper is specifically addressed to the problem of equipment grounding, it may not be amiss to emphasize here that care must be exercised also in dealing with the grounding of the circuit neutral. The circuit grounding conductor is often installed in a conduit bonded to the conductor at both ends as required by the National Electrical Code. The conduit is thought of primarily as mechanical protection rather than as an element of the circuit, but a little theorizing on skin effect would indicate, and tests prove, that the current set up under accidental conditions while the grounding conductors are performing their intended protective function is carried almost entirely by the conduit. Table III shows test data on several combinations of circuit grounding conductors and conduits.

TABLE III

Test Data

Conductor	Conduit, Inches	Total Current, Amperes	Current in Conductor, Amperes	Current in Conduit
6	1/2	100	3	97
6	1/2	300	5	295
2	3/4	90	7	83
2	3/4	350	10	340
2/0	1	150	15	135
2/0	1	590	5	585
4/0	1 1/4	225	15	210
4/0	1 1/4	885	15	870

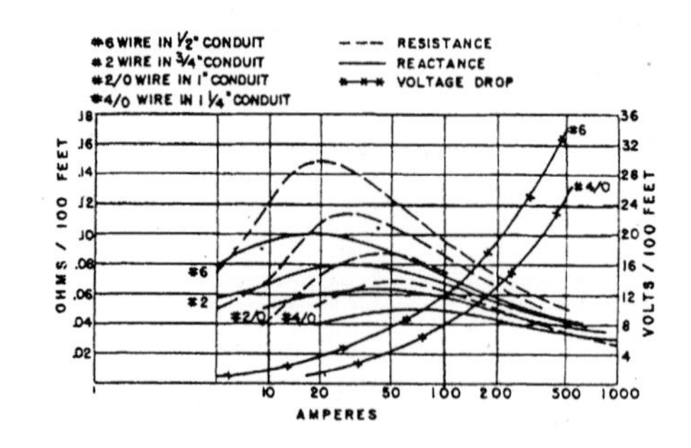

FIGURE 6 Characteristics of grounding conductors. Rubber-insulated single conductor in conduit, conductor bonded to conduit at both ends. Return circuit spaced at 30 feet from grounding conductor in conduit.

It is apparent that the conduit must be treated as a conductor, and therefore joints must be clean and tight. Mr. Kaufmann's reference to a fan-shaped shower of sparks brings out very vividly the inherent danger of poor contact. In our own tests, erratic current division between conductor and conduit was occasionally noted unless the conduit ends were cleaned before joining. During a 450-ampere test on no. 4/0 cable in 1¼-inch conduit, which was not cleaned before joining, one of the joints became blue because of high resistance at that point.

Figure 6 shows typical impedance and voltage drop characteristics of circuit-grounding conductors in conduits. Such curves are helpful for estimating the amount of voltage rise in the grounding circuit and for other problems in connection with the application of grounds.

As proper equipment and circuit grounding is a paramount measure of safety, it is to be hoped that Mr. Kaufmann and others will keep on contributing to the understanding of this subject.

REFERENCE

1. Single-Phase Impedance to Ground in Pipe-Type Cable, E. R. Thomas. *AIEE Transactions,* vol. 73, pt. III, Apr. 1954, pp. 336–44.

R. H. Kaufmann: The intense interest which has been exhibited in this paper is most gratifying and it constitutes ample reward for the work which was involved in its creation.

Mr. Gienger presents the results of tests and experience which are very interesting. His experience confirms the contention that the ground return current will, to a large extent, follow the metallic enclosure surrounding the power cables. The report on his test work indicating that adequate conductivity can be secured in threaded couplings of conduit runs is reassuring. This work indicates that screw joints, if properly cut and tightened, will probably represent no limitation in the conductivity of short-circuit currents.

Mr. Brieger brings an interesting reverse significance in the effect of a conduit surrounding a grounding conductor. In the instance reported, the outgoing current will be flowing on conductors external to the conduit rather than internal within the conduit. Hence, the grounding conductor with its enclosing conduit functions as an independent electrical conductor with the outgoing current flowing on a separate path external to the conduit. Under these circumstances, the current flow in the grounding conductor combination will encounter the lowest circuit reactance if it flows along the outside surface of the conduit. Thus, the conduit functions as a magnetic shield preventing the copper conductor within the conduit from carrying substantial current.

Mr. Brieger also observes the necessity for careful attention in joining and terminating sections of conductor enclosure to avoid thermal distress at these joints.

It is gratifying to note that the fruits of these investigations will lead directly to improved grounding techniques, better utilization of material, and to ways of avoiding application of inactive material. So far, this comprehensive subject has been treated only superficially. There remains great opportunity for fruitful investigation and it is urged that those with technical understanding and test facilities become active participants in expanding the useful knowledge relating to this field.

Appendix B: Investigation and Testing of Footing-Type Grounding Electrodes for Electrical Installations

H. G. Ufer
Associate Member IEEE

Summary: Footing-type grounding electrodes installed in the concrete foundations of residential and small commercial buildings designed to meet a maximum ground resistance value of 5 ohms are described. Bare solid copper electrode wires of various lengths, steel reinforcing rods, and 10-foot lengths of hot galvanized rigid steel conduit were used to determine the resistance values.

City Electrical Inspector members reported to the Southwestern Section of the International Association of Electrical Inspectors that a continuous metallic water piping system, as recommended by the National Electrical Code, is not always available. A committee was appointed to investigate this condition and report its findings at subsequent meetings.

The National Electrical Code, which is prepared by a committee of the National Fire Protection Association, is the recognized American standard for the safe installation of electrical equipment in the United States. It is based on the combined experience of all groups in the electrical industry and all factual information that is available at the time of each edition's preparation. One of the oldest basic requirements of the code is the protection of electric installations by grounding; it recommends that a metallic underground water piping system, either local or supplying a community, should always be used as the grounding electrodes when such a piping system is available.

There is increasing concern over the fact that a continuous metallic water piping system is not presently available in some areas for electrical grounding purposes. This is partially a result of the use of nonmetallic water pipe mains and laterals to bring water into a building, in addition, it has become standard practice in the installation of cast-iron water pipe mains and laterals to use neoprene gaskets to join the pipe sections and discontinue the poured-lead joint formerly used.

Inspection authorities have also reported that the ground is interrupted by the installation of insulating joints on the plumbing system. When copper pipe is employed for the hot water side and galvanized steel pipe for the cold water side, such joints are used to prevent decomposition of the copper pipe resulting from electrolysis. These insulating joints are also used for water-softening tanks. Bonding jumpers are not always used, and often they are removed.

Paper 63-1505, recommended by the IEEE Safety Committee and approved by the IEEE Technical Operations Committee for presentaton at the IEEE Western Appliance Technical conference, Los Angeles, Calif., November 4, 1963. Manuscript submitted August 1, 1963; made available for printing May 5, 1964.

H. G. Ufer is with Underwriters' Laboratories, Inc., Santa Clara, Calif.

The driven grounds required by the code in lieu of connection to a water system are not always dependable. In the paper "Grounding of Electric Services to Water Piping Systems" presented to the American Water Works Association on October 26, 1960, A. G. Clark of Los Angeles reported that the use of driven grounds had not proven satisfactory. The minimum resistance obtained on representative ground rods was 10 ohms, and most of the driven representative rods tested were several times the minimum value of 10 ohms.

In view of these conditions and failures, it seemed that an adequate means could be provided for grounding, one which is not likely to be disturbed, which requires very little maintenance, and which does not require connection to the water pipe systems to provide an adequate low resistance ground. The purpose of this paper is to report the development and testing of such a grounding method and to record the results of tests on installations in residential and small commercial buildings since 1961.

BACKGROUND

A number of military installations of ammunition and pyrotechnic storage facilities were built during World War II and provided with lightning protection systems which required a permanent ground connection. The author conducted the field inspection of the lightning protection for Underwriters' Laboratories, Inc., and had first-hand knowledge of the construction details. Installations in Arizona have been selected for this paper because the climate is normally hot and dry during most of the year.

Tucson Installation

The installation at the Davis-Monthan Air Force Base, Tucson, Ariz., consists of six bomb-storage vaults and four pyrotechnic storage sheds. The bomb storage vaults are steel-reinforced concrete, while the storage sheds are steel angle frame construction covered with corrugated galvanized sheet steel.

The soil at this location is sand and gravel, and the average rainfall is 10.91 inches. Low resistance grounding had to be provided for each of these buildings in order to discharge to ground any static charge of electricity caused by wind and sandstorms or by exposure to lightning during a rain storm or thunder shower.

In the areas where these bomb storage vaults and storage sheds were located, an underground water piping system is usually not available. Driven-ground rods or a bare copper counterpoise could be used, but at the time these installations were being rushed to completion, strategic materials were being conserved. It was decided to provide grounding by the use of the steel angle iron or the reinforcing rods of the building structure.

For the bomb storage vaults (igloos as designated by the Armed Forces) with dimensions 10 by 30 or 40 feet, the two parallel exterior walls were provided with footings. The front and rear walls of these vaults were not provided with footings.

The footings were dug to a depth of 2 feet. In these footings, 1 2-inch vertical reinforcing rods approximately 30 inches long were spaced about 12 inches apart. They were pushed a few inches into the earth to maintain their position and spacing while the concrete was poured. Across the upper ends of these vertical rods, ½-inch steel reinforcing bars were laid horizontally and welded to every alternate vertical rod embedded in the concrete of the footings.

From this base, an umbrella or igloo-shaped structure was formed of ½-inch steel reinforcing rods welded to the rod at the top of the footing. The lightning rod terminals were erected on the top of each igloo by welding them to the reinforcing rod assembly. The umbrella was then covered with concrete.

For the 10- by 4-foot storage sheds, grounding was obtained through the angle iron framework, which had the ends of the vertical angle iron at each corner of the building embedded in concrete to a depth of 1 foot.

Complete foundations were not provided for these small storage sheds. As a bond between the various metal parts of the frame and metal covering for these buildings, welding or bolts were used; the adjacent edges of the corrugated steel sheets were overlapped and secured by rivets or sheet metal screws.

Flagstaff Installation

The installation at the Navajo Ordnance Depot is located adjacent to U.S. Highway 66, approximately

15 miles west of Flagstaff, Ariz. This ordnance depot has an area of about 56 square miles. The soil is clay, shale, gumbo, and loam, with small-area stratas of soft limestone. The average rainfall is 20.27 inches.

The original contract in 1942 called for the construction of 800 bomb storage vaults. These were of the same construction and dimensions as those erected at the Davis-Monthan Air Force Base at Tucson.

A water piping system either local or serving a community was not available for grounding at this location; accordingly, the method used at Tucson was utilized.

GROUND RESISTANCE TESTS

Each bomb storage vault and storage shed erected in Tucson was inspected, and ground resistance tests were made on each of these structures. The 800 bomb storage vaults erected at the Navajo Ordnance Depot were inspected during the construction, and then individual ground resistance tests were performed.

Ground resistance readings were taken between each lightning protection terminal and ground on all igloos and storage sheds. Tests were also made between all exposed conductive material and ground on each installation described in this paper.

Readings were taken by using a heavy-duty megger ground tester with a scale from 0 to 50 ohms graduated in increments of 1 ohm for each scale division. Each vault and shed was required to have a ground resistance reading of not more than 5 ohms.

The method of measuring the ground resistance values was a 3-point method.[1] The sequence of ground-resistance tests for the two lengths of conduit recorded in Table I were as follows: Two lengths of rigid steel ¾-inch galvanized conduit were laid 5 inches apart in a trench 36 inches deep, 11 feet long, and 15 inches wide. One length of conduit was enclosed in 2 inches of concrete (electrode 1), while the second length was not enclosed in concrete (electrode 2); the second length was then enclosed in 2 inches of concrete and placed back in the trench beside the other length of conduit, and the dirt was replaced in the trench. The resistance measurements were repeated with the results recorded in Table I.

TABLE I

Beverly Hills Tests

Type of Ground Electrode	Installation Date	Testing Date	Resistance (Ohms)
Electrode 1	11/1/60	11/1/60	5.1
	11/1/60	11/2/60	9
	11/2/60	11/10/60	9
	11/2/60	11/30/60	9
Electrode 2		11/1/60	90
	11/1/60	11/2/60	
	11/1/60	11/10/60	100
		11/30/60	
Bare pipe and pipe in concrete	11/1/60	11/1/60	
		11/2/60	14
		11/10/60	
		11/30/60	
Electrode 2 in concrete	1/9/61	1/9/61	8.7
	1/9/61	1/11/61	9
Electrodes 1 and 2 in concrete (series)	1/9/61	1/9/61	5
		1/11/61	
Cast-iron sewer and gas pipes-bonded		1/11/61	30
Sewer, gas, and water pipes-bonded		1/11/61	1
Sewer, gas, and water pipes connected to electrode 1		1/11/61	2

Check Tests

In order to determine the adequacy of the ground resistance values which were made in 1942, it was decided to conduct check tests. Permission was obtained from the office of Chief of Operations in Washington, D.C., to check the grounds at Flagstaff Army Base and Tucson Air Force Base.

Check tests were made at the Navaho Ordnance Depot on July 12, 1960, and at the Davis-Monthan Air Force Base on August 23, 1960. Additional check tests were made at the Navajo Ordnance Depot on October 20, 1960.

At both bases, the readings taken measured from 2 to 5 ohms in all instances. All readings taken in 1942 when the original installations were made did not exceed 5 ohms to ground.

The Surveillance Officer at the Navajo Ordnance Depot must regularly check and record the ground resistance for each igloo, shell loading, and storage building. These tests are made at approximately 30–60-day intervals. During the 20 years since these ground systems were installed and accepted by the author, the maintenance department at the Navajo Ordnance Depot has not been requested to replace or repair any grounding components on the 800 bomb storage vaults or igloos. As far as could be determined, there was no report of repairs or maintenance on grounding systems which had been installed on igloos and sheds at the Davis-Monthan Air Force Base.

PRELIMINARY INVESTIGATION

Beverly Hills Tests

Test A. In view of the previous findings, an installation was made using conduit electrodes laid horizontally in the ground and enclosed in 2 inches of concrete at a depth of 3 feet.

Two 10-foot lengths of ¾-inch galvanized steel rigid conduit were laid side by side horizontally at a single trench 15 inches wide, 11 feet long, and 36 inches deep. The installation was made in Beverly Hills, Calif., on property which has not been cultivated during the past 30 years.

The resistivity of the soil in the area was 3,830 ohms per cm^3 (cubic centimeter). A ground rod driven to a depth of 20 feet was calculated to have a resistance of 6 ohms based on this resistivity test. (Table II shows the soil resistivity of the various sites mentioned in the paper.)

In the first series of the conduit electrode tests, one length of conduit (electrode 1) was enclosed in 2 inches of concrete; the other (electrode 2) was not enclosed in concrete. These two lengths of conduit were then covered with the soil removed to prepare the test trench.

In the second series of tests, the soil in the trench was removed and the bare conduit electrode 2 was enclosed in 2 inches of concrete.

The ground resistance of the water, gas, and sewer drainage piping systems bonded together was 1 ohm. The resistance was also 2 ohms when connected to electrode 1. This low resistance value is apparently due to the fact that the local water pipe system is galvanized steel pipe for the mains and the lateral. The lateral extends underground 40 feet from the water meter to the supply connection in the building.

The overhead electric service is a 3-wire ac 110–240-volt system. The neutral of this service is grounded to the water pipe in the building. The utility ground for this service is provided at the utility pole which is 25 feet from the water meter.

The results of the ground resistance tests on the two conduit electrodes are recorded in Table I.

TABLE II

Resistivity of Soils

Installation	Soil	Resistivity (Ohms/Meter^3)
Beverly Hills	Loam and sand	38.30
Bishop	Gravel and sand	120
Riverside	Red adobe	50
Livermore	Loam and sand	35
Hayward	Large rocks and gravel	150
Portland, Ore.	Loam and sand	40
Long Beach	Heavy loam and sand	25* 20†
Twenty-nine Palms	Decomposed granite and sand	70
Burbank	Sandy loam	40

* Metal sign on 12-inch steel I-beams.
† Concrete floor slab in garage.

Test B. On January 16, 1960, one 10-foot length of 3½-inch galvanized rigid steel conduit was installed vertically to a depth of 5 feet and enclosed in 2 inches of concrete.

The ground resistance was 10 ohms, this value has remained constant during the 2-year period of these tests, and the readings are apparently stable.

In 1961, during a discussion of these supplemental tests, it was suggested that a test should be made on the property where the supplemental tests were conducted to be sure that the ground did not contain any foreign items which would favor these tests. In September 1961, the same Biddle ground megger, electrodes, and wire leads were used for this test as were used to measure the ground resistance of electrodes 1 and 2 (Table I).

The megger was placed adjacent to the ground in which pipe electrodes 1 and 2 were placed for the supplemental tests, and it was connected to these electrodes for a ground resistance test. The two 25-foot *No. 14* American wire gage (AWG) stranded wire leads were connected to the proper terminals on the megger. The probes were 36-inch ⅜-inch copper weld rods, and two were now placed in the ground 50 feet apart. On a 30-foot arc, ground resistance readings were taken at 3-foot intervals until probe 1 was adjacent to probe 2.

The probes were returned to their original positions. The readings were repeated with the wire leads reversed on the terminals of the megger. Six readings measured 11 ohms, two measured 12 ohms, and two measures 13 ohms; the average reading was 11.1 ohms. The measurements were repeated with the wire leads reversed on the megger, with no change recorded in these measurements.

The megger was then connected to electrodes 1 and 2 with the probes separated 50 feet and attached to the proper terminals on the ground megger. The ground resistance of electrodes 1 and 2 measured 5 ohms, as has been reported in Table I.

BUILDING INSTALLATIONS

The military installation tests demonstrate that a low-resistance ground electrode can be obtained in an area where conditions are unfavorable and that such a ground can be continuously effective over many years without maintenance. Encasement of electrodes in concrete and compaction provided by building weight may be the controlling factors in these installations. The supplementary tests on a limited-size electrode indicates that a concrete encasement contributes to the reduction of ground resistance.

A number of tests on footing-type electrode installations in building constructions were made during 1961, 1962, and 1963. Arrangements were made to have this type of electrode installed in conventional residential and commercial buildings. Locations selected were widely separated and these buildings were provided with a footing-type concrete building foundation. *No. 4* AWG solid bare copper wires were embedded in the center of the concrete footing approximately 2 inches above the base of the footing.

The concrete building foundations are T-shaped. The base is about 12 inches wide, while the top of the footing where the foundation still is attached is 6 inches wide. The depth of these footings varies from 2 to 4 feet below the grade level of the building plot.

Two separate lengths of wire were installed in each foundation: one wire approximately 30 feet and the other 100 feet in length. The ends of each wire extended out of the foundation walls to provide ready access for testing and for connection to the neutral terminal in the service switch if and when authorized by the local electrical department.

The area of these buildings on which these installations were made varied from 500 to 8,000 square feet. They are 1-story residential- and commercial-type buildings.

An installation was made with five 20-feet lengths of *No. 4* AWG bare solid copper wire embedded in a concrete building foundation. The ends of each 20-foot length of wire extended from the walls of the concrete footing to be available for measurements. This special installation was made to determine the length of the grounding wires required to provide 5-ohm maximum resistance values.

Bishop Installation

An experienced electrical contractor was contacted in Bishop, Calif., and requested to assist in this program of installing footing-type grounding electrodes in a residential or small commercial building. He had previously received a contract to wire a new residence at Bishop, and the owner agreed to have the footing-type grounding electrode installed.

The California Electric Power Company is the serving public utility for this area. They were notified that this installation was to be made and their engineer conducted the ground resistance tests when the installation was completed.

The electrical contractor installed two *No. 4* AWG bare copper wires in the concrete forms for the building foundation, one 76 feet long, the other 82 feet. The ends of both wires were brought out of the side of the concrete foundation below the location for the installation of the main switch. The grounding wires were run in opposite directions, and the ends were brought out of the side of the foundation at a location diametrically opposite the wall on which the main switch was installed.

The contractor was requested to install one long wire in the foundation about 100 feet long and one short wire about 25 feet long. This was suggested so that test results would show how much wire was required to provide a ground resistance of not more than 5 ohms.

This installation was made during September 1961. The building is a one-story masonry residence with an area of approximately 1,500 square feet. The concrete footing is T-shaped. The base is 14 inches wide, while the top of the footing where the wood sills are installed is 8 inches wide. The footing is approximately 12 inches below grade level. The *No. 4* AWG bare copper grounding conductor was installed in the center of the concrete foundation about 2 inches above the base of the footing.

On November 1, 1961, 60 days after the concrete had been poured for the foundation, the test

TABLE III

Location	Installation Date	Testing Date	Ground-Wire Resistance (Ohms)	Ground-Wire Length (Feet)	Soil Condition	Average Rainfall (Inches)
Bishop	9/61	11/1/61 12/2/62	4.0, 4.6 1.2	76, 82 76 + 82	Gravel, sand	11
Riverside	10/61	11/30/61 11/16/62	2.8, 2.8 0.3, 0.3	100, 100	Red adobe	8
Livermore	3/62	4/62 9/62	1.3, 1.3 1.55	25, 75 25 + 75	Loam, sand	14
Hayward	9/62	9/62 10/62	4.2, 2.8 3.2, 1.8	22, 73 22, 22 + 73	Gravel, large rocks	14
Hayward	3/26/62	4/3/62	1.29, 1.25	25, 75	Adobe, topsoil	14
Portland	2/62	7/62 12/62	1.1, 0.8 1.1, 0.8	78, 90	Loam, sand	42.67
Long Beach*	12/61	12/62 3/25/63†	0.6 0.8		Heavy loam, sand	13.32
Long Beach‡	8/61	8/62 3/25/63†	4.5 5.4	30	Heavy loam, sand	13.32
Twenty-nine Palms	2/8/63	3/22/63	4.3, 2.2§ 6.5, 4.4‖	50, 100	Decomposed granite, sand	0.5
Palm Springs, Calif.¶	9/20/63	11/21/63**	1	100	Loam, sand	6.74

* Billboard of two 12-inch steel I-beams 8 feet in ground.
† Four days after rain.
‡ No. 6 wire embedded in garage floor.
§ Biddle megger.
‖ Siemens and Halske megger.
¶ More recent data.
**Heavy rain 10/20/63.

engineer for the California Electric Power Company measured the ground resistance with a Siemens & Halske ground megger and recorded a ground resistance of 4.6 and 4.0 ohms (See Table III).

The serving public utility is the authority responsible for the inspection and test of all electrical installations in the area. Their engineer authorized the electrical contractor to use this footing-type grounding electrode for grounding the electrical installation of this building and permitted the contractor to connect the free ends of both grounding conductors together to provide a grounding conductor with a total length of approximately 158 feet in the concrete footing. The ground resistance for the 158 feet measured 4.6 ohms.

During a regular trip to Bishop in December 1962, the engineer for the California Electric Power Company made the second test on the ground resistance of the footing-type grounding electrode and recorded a reading of 1.2 ohms.

Riverside Installation

The Chief Electrical Inspector of Riverside, Calif., arranged for the installation of footing-type grounding electrodes in the concrete footings of a residence erected in Riverside. Two *No. 4* AWG bare copper wire electrodes, each 100 feet in length, were installed in the concrete footings of this residence. In all locations, the electrode wire was placed in the concrete forms to be spaced about 2 inches above the base of the footing. The author was at this location during construction to see that the electrode wires were properly placed above the base of the foundation footing before the concrete was poured for the foundation.

The Chief Electrical Inspector of Riverside supervised these installations. The first ground resistance readings were taken November 30, 1961, about 60 days after the concrete was poured, and the second readings were taken 1 year later. All readings are recorded as 2.8 ohms, for the years 1961 and 1962.

Livermore Installation

The City Engineer for Alameda County, Calif., selected the locations for the installation of footing-type grounding electrodes. Two residential buildings were provided with these grounding electrodes. One building was in Livermore, and two electrode wires, 25 and 71 feet long, were installed in the concrete foundation.

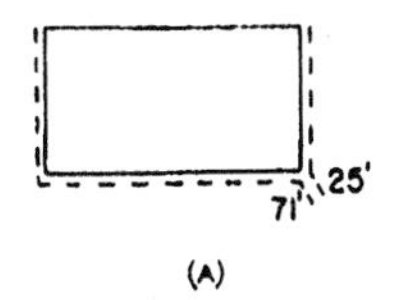

LOCATION	APRIL 3, 1962			SEPT. 20, 1962		
	25'	71'	BOTH	25'	71'	BOTH
#1	1.2Ω	1.15Ω	115Ω	1.55Ω	1.55Ω	1.55Ω
#2	1.27Ω	1.25Ω	1.22Ω	1.55Ω	155Ω	1.55Ω
#3	1.29Ω	1.26Ω	125Ω	1.55Ω	155Ω	155Ω

FIGURE 1 Livermore installation.

Ground resistance measurements were made by the Testing Engineer for the Pacific Gas and Electric Company. The first measurements were made April 3, 1962, 1 week after the concrete was poured and 12 days after rain. The second measurements were made September 20, 1962, 6 months after the concrete was poured. The engineer reported that the adobe and topsoil was very dry and cracked, since the last rain in the area consisted of 0.22 inch in April.

All readings were recorded as less than 2 ohms with very slight difference between the long and short wires in the first readings. For the second measurements, the resistance for the short and long wires were measures separately and when connected together were recorded as 1.5 ohms. Figure 1(A) shows the position of the wires, while Figure 1(B) gives the results of the measurements.

Hayward Installations

The City Engineer of Alameda County selected these locations in Hayward for a footing-type grounding electrode, since the soil was mostly rock and not dependable for a good ground. The last rain in this area was in April 1962.

Two *No. 4* AWG bare copper wires, 22 and 73 feet long, were placed in the concrete foundation for this residence, as shown in Figure 2(A). The foundation was poured on September 1, 1962.

Ground resistance readings were measured on September 20, 1962, by a test engineer of the Pacific Gas and Electric Company, the serving public utility in this area. As shown in Figure 2(B), the 22-foot wire measured 4.2 ohms; the 73-foot wire, 33 ohms; both wires (95 feet), 2.75 ohms.

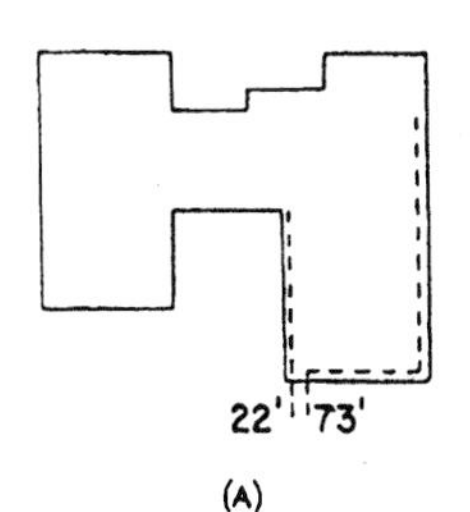

LOCATION	SEPT. 20, 1962		
	22'	73'	BOTH
#1	4.2Ω	3.3Ω	2.75Ω
#2	4.2Ω	3.25Ω	2.8Ω
#3	3.2Ω	2.3Ω	1.8Ω

(B)

FIGURE 2 Hayward installation.

Measurements taken October 1962 were 3.2 ohms for the 22-foot wire, 1.8 for the 73-foot wire, and 1.8 ohms for both wires.

At a second location, where the earth was a mixture of adobe and topsoil, two wires were placed in the foundation wall as shown in Figure 3. Readings were made by an engineer from the Pacific Gas and Electric Company on April 3, 1962, 1 week after the footing was poured; the weather was dry, since the last rain had been on March 22, 1962. See Figure 3 and Table III for the results. A 25-foot ground wire is apparently adequate in this location as the resistance measured less than 2 ohms for both the long and short wires.

Portland Installation

The Chief Electrical Inspector of Portland, Ore., selected the location for a footing-type grounding electrode. Two *No. 4* AWG bare copper wires, approximately 78 and 100 feet long, were installed in the concrete foundation of a 1-story commercial warehouse with dimensions of 80 by 100 feet. The building foundation, the floor slab and the 6-inch precast walls were reinforced concrete, and a wood truss roof on wood columns was provided.

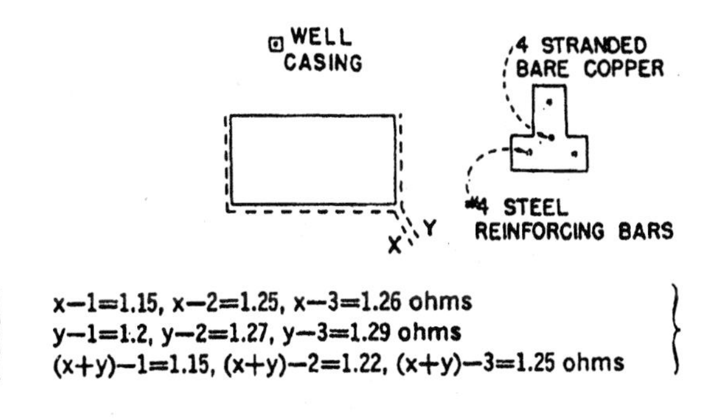

FIGURE 3 Second Hayward installation.

The ground resistance tests recorded 0.8 ohm. The power company then tested the water pipe ground and recorded 1.1 ohms. They decided that the probable cause of the 0.8-ohm reading was that the footing ground wire might possibly have been touching the water pipe. The author visited this installation to check the test results but found that the ground wires had been cut off where they extended beyond the faces of the concrete foundation.

The *No. 4* AWG wires installed in the concrete foundation contacted the steel reinforcing rods, but apparently this did not affect the values of the ground resistance readings.

One of the engineers of the serving pubic utility cut away the concrete where the wire had been cut and was able to attach the testing conductor to this wire; he recorded a ground resistance of about 1 ohm. Because of the location of this building and the availability of an adequate water pipe system for grounding the electrical installation, the water pipe is being used for the ground since the ends of the *No. 4* grounding wires are not accessible.

Long Beach Installations

The first installation consists of a steel frame sign mounted on two 12-inch steel I-beams spaced 25 feet apart. The ends of the I-beams stand 8 feet in the ground. The ground resistance measured 0.6 ohms.

The second Long Beach installation consists of 30 feet of *No. 6* AWG bare copper wire embedded in the floor slab of a 2-car garage. The wire is in the shape of a circular loop with one end extending beyond the face of the floor slab for connection to the neutral terminal of the service switch for this installation.

Ground resistance reading were made by the Chief Electrical Inspector of Long Beach with a *Model 263A* Vibroground. The resistance measured 4.5 ohms.

The second resistance tests for both Long Beach installations were made 4 days after rain.

Twenty-Nine Palms Installation

The Chief Electrical Inspector for San Bernardino County arranged to have the footing-type ground electrode installed in the concrete foundation of the Methodist Church of Twenty-nine Palms. This is desert location, and ground resistance tests recorded for this area by other engineers indicated that an adequate ground is not available.

The church is a 1-story building approximately 35 by 70 feet with a wood-frame stucco construction on a poured concrete foundation. The concrete footing is T-shaped 12 inches below grade; the base is 14 inches wide, while the upper end is 6 inches wide with ½-inch bolts embedded in the concrete for attaching the foundation sills.

Two *No. 4* AWG bare copper wire electrodes were used, one 50 and the other 100 feet long. One end of the 50-foot wire extended beyond the face of the foundation wall on the north elevation of the building (point C), while the other end of this 50-foot conductor extended beyond the face of the foundation on the west elevation (point A). At point A, one end of the 100-foot electrode wire extends from the face of the foundation wall, while the other end of this 100-foot wire extends itself from the face of the east elevation (point B) of the concrete foundation wall of this building (refer to Figure 4).

Ground resistance measurements were taken by the Test Engineer for the California Electric Power Company using a Siemens & Halske ground megger *No. 2585269* with a range of 0–25 ohms and 0–2500 ohms. These readings were checked with a Biddle ground megger *No. 167273* having three ranges, 0–3, 0–30, and 0–300 ohms.

As bare wire leads were used to record the readings taken with the Siemens & Halske megger, it was decided to use bare and insulated wire for connections from the megger to each of two reference grounds and to connect the third lead to the ground being tested. The results are recorded in Table IV.

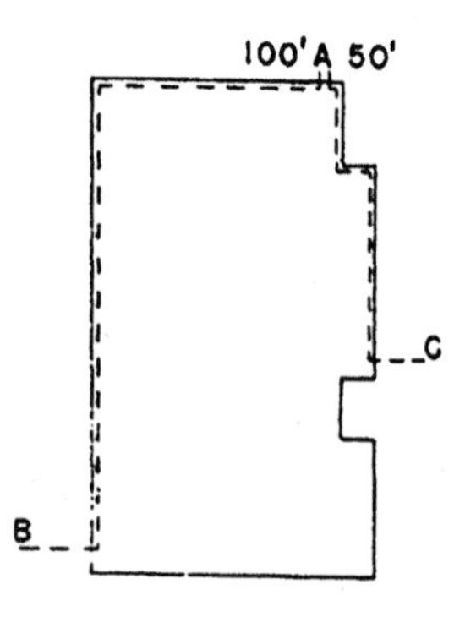

FIGURE 4 Twenty-nine Palms installation.

TABLE IV

Location of Ground Lead	Type of Lead	Length of Grounding Conductor (Feet)	Resistance (Ohms) Siemens and Halske Megger	Biddle Megger
Point C	Insulated wire	50	6.0	4.3
		150*	4.4	2.2
		50	6.5	4.3
Point A	Bare wire	50	3.1	5.5
		100	2.0	3.2
		150*	2.0	3.2

* Short and long wires.

An engineer for the California Electric Power Company reported a grounding installation in a residence in San Bernardino County where a 4-inch concrete floor slab was poured over a membrane. The hot and cold water pipes enclosed in concrete were the grounding conductors. An insulating joint was installed in the water supply pipe lateral. The engineer for the California Electric Power Company measured the ground resistance of the copper water pipes enclosed in the concrete floor slab as over 10 ohms. He then measured the ground resistance of the underground water pipe lateral a short distance from the insulating joint as 1.3 ohms.

LENGTH OF ELECTRODES

Installations were made at Alameda and San Bernardino counties where the soil conditions were not favorable for an adequate ground. At Hayward and Livermore, the soil was poor and filled with large rocks. In these locations, electrode wires 22 and 25 feet long were installed (Table III).

An installation was also made in the desert at Twenty-nine Palms where the soil conditions are not favorable for an adequate ground. The ground resistance readings for these installations, including the 50-foot electrode at Twenty-nine Palms, record measurements of 5 ohms or less using a Biddle megger and a recognized test method.[2]

Burbank Installation

During 1952, a steel floodlight pole was installed in the athletic field of John Burroughs High School, Burbank, Calif. The pole is 14 inches in diameter at the base and tapers to a diameter of 8 inches at the top; it is 100 feet long over-all and is set in a centering sleeve from the base to a height of 16 feet. This assembly is set in a reinforced concrete footing 8 by 8 by 18 feet. (See Figure 5)

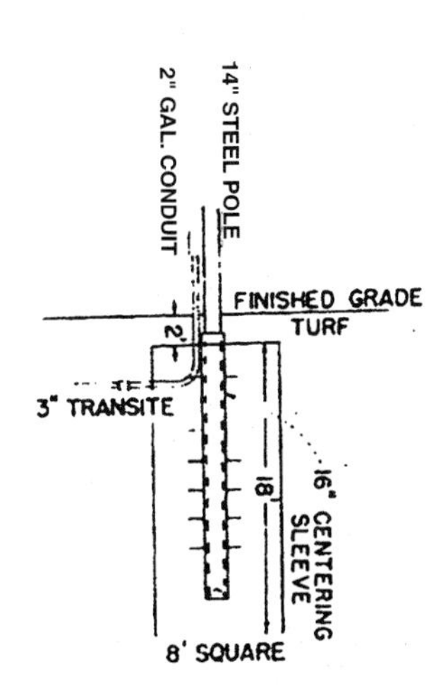

FIGURE 5 Burbank installation.

Since installation, the resistance to ground has been checked at least once each year by the Burbank city engineers. On July 2, 1963, the ground resistance was checked with their ground megger and measured 1.8 ohms. The soil is sandy loam, the soil resistivity in this area is 4,000 ohms per cm^3, (Table II), and the average rainfall is about 10 inches.

CONCLUSION

The 20-year history of the military installation in Arizona, together with the shorter records of installations in conventional residential and small commercial buildings, apparently establishes the adequacy and advisability of footing-type grounding electrodes where a continuous underground water system is not available or dependable; in addition, it indicates that the grounding conductor ought to be placed about 2 inches above the base of the concrete foundation footing.

REFERENCES

1. Guide for Measuring Ground Resistance and Potential Gradients in the Earth. *AIEE Report No. 81*, July 1960.
2. Master Test Code for Resistance Measurement. *AIEE Standard No. 550*, May 1949.

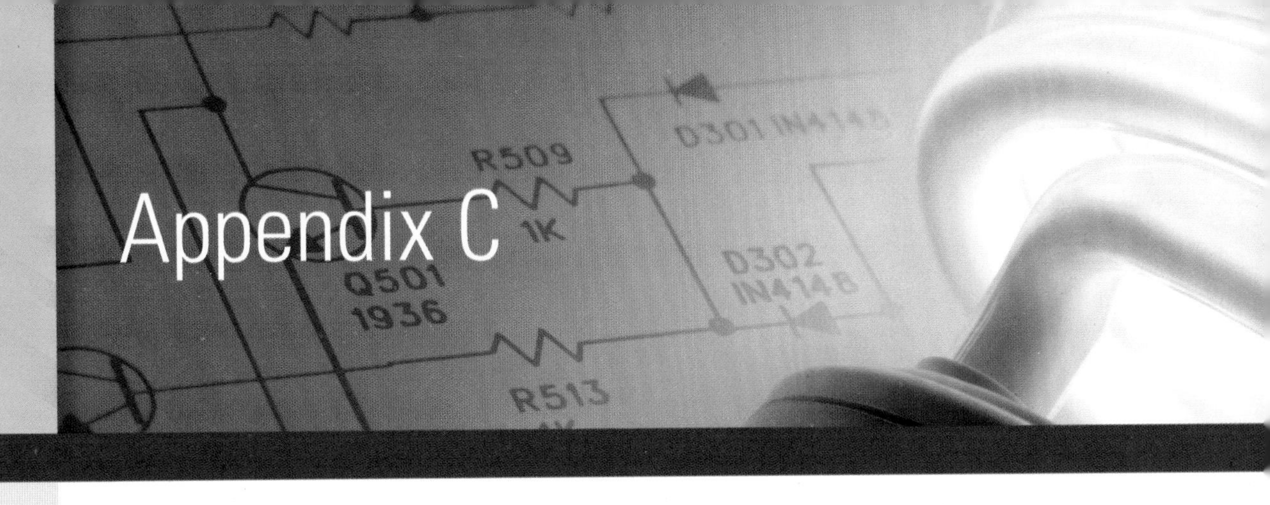

UNDERSTANDING
Ground Resistance Testing

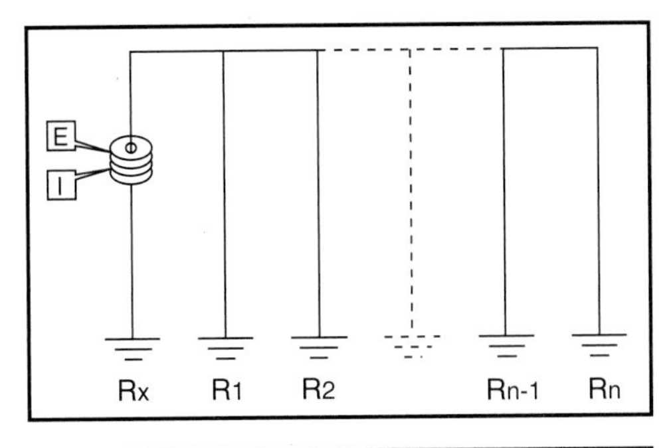

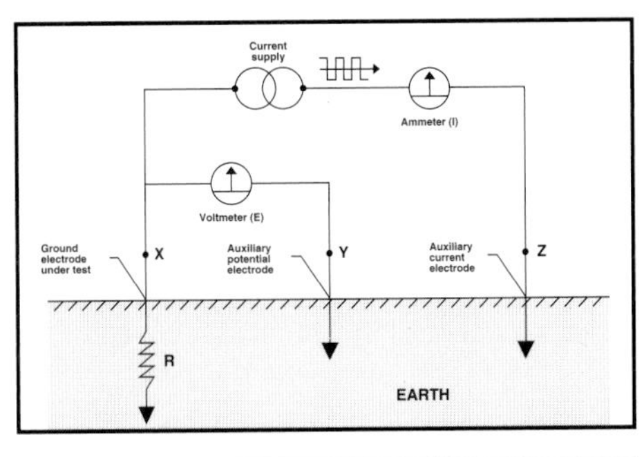

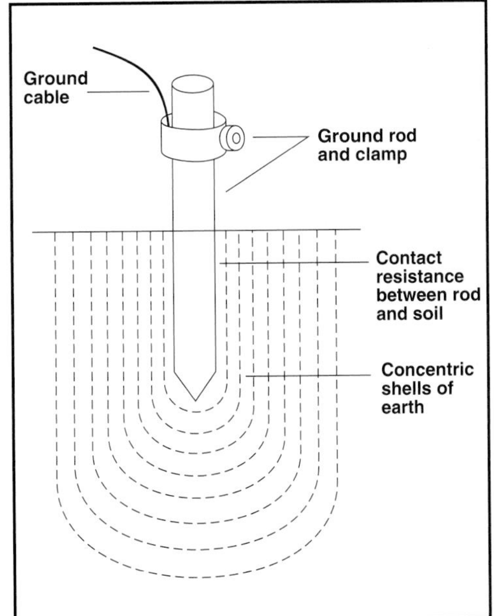

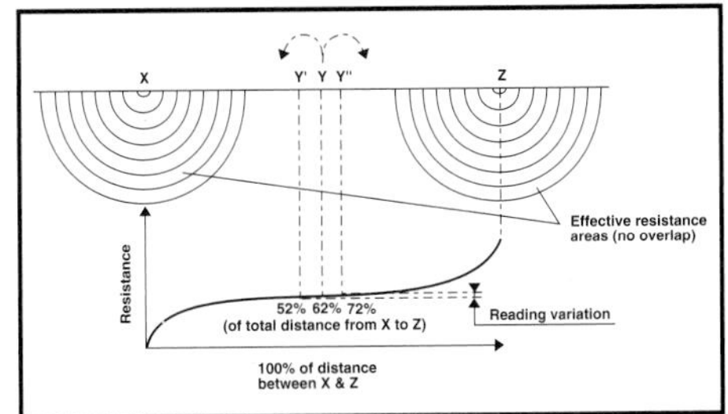

- *Soil Resistivity*
- *Ground Resistance*
- *3-Point Measurements*
- *4-Point Measurements*
- *Clamp-On Measurements*

Courtesy of Chauvin Arnoux® Inc./dba AEMC® Instruments, Foxborough, MA

Table of Contents

Models 3711 & 3731 have replaced Models 3710 & 3730

Workbook Edition 8.0
950.WKBK-GROUND 03/04

Notes

Soil Resistivity

Why Measure Soil Resistivity?

Soil resistivity measurements have a threefold purpose. First, such data are used to make sub-surface geophysical surveys as an aid in identifying ore locations, depth to bedrock and other geological phenomena. Second, resistivity has a direct impact on the degree of corrosion in underground pipelines. A decrease in resistivity relates to an increase in corrosion activity and therefore dictates the protective treatment to be used. Third, soil resistivity directly affects the design of a grounding system, and it is to that task that this discussion is directed. When designing an extensive grounding system, it is advisable to locate the area of lowest soil resistivity in order to achieve the most economical grounding installation.

Effects of Soil Resistivity on Ground Electrode Resistance

Soil resistivity is the key factor that determines what the resistance of a grounding electrode will be, and to what depth it must be driven to obtain low ground resistance. The resistivity of the soil varies widely throughout the world and changes seasonally. Soil resistivity is determined largely by its content of electrolytes, which consist of moisture, minerals and dissolved salts. A dry soil has high resistivity if it contains no soluble salts. (Figure 1)

	Resistivity (approx), Ω-cm		
Soil	**Min.**	**Average**	**Max.**
Ashes, cinders, brine, waste	590	2370	7000
Clay, shale, gumbo, loam	340	4060	16,300
Same, with varying proportions of sand and gravel	1020	15,800	135,000
Gravel, sand, stones with little clay or loam	59,000	94,000	458,000

Figure 1

Factors Affecting Soil Resistivity

Two samples of soil, when thoroughly dried, may in fact become very good insulators having a resistivity in excess of $10^9\Omega$-cm. The resistivity of the soil sample is seen to change quite rapidly until approximately 20% or greater moisture content is reached. (Figure 2)

Moisture content % by weight	Resistivity Ω-cm Top soil	Sandy loam
0	>10^9	>10^9
2.5	250,000	150,000
5	165,000	43,000
10	53,000	18,500
15	19,000	10,500
20	12,000	6300
30	6400	4200

Figure 2

The resistivity of the soil is also influenced by temperature. Figure 3 shows the variation of the resistivity of sandy loam, containing 15.2% moisture, with temperature changes from 20° to -15°C. In this temperature range the resistivity is seen to vary from 7200 to 330,000Ω-cm.

Temperature C	F	Resistivity Ω-cm
20	68	7200
10	50	9900
0	32 (water)	13,800
0	32 (ice)	30,000
-5	23	79,000
-15	14	330,000

Figure 3

Because soil resistivity directly relates to moisture content and temperature, it is reasonable to assume that the resistance of any grounding system will vary throughout the different seasons of the year. Such variations are shown in Figure 4. Since both temperature and moisture content become more stable at greater distances below the surface of the earth, it follows that a grounding system, to be most effective at all times, should be constructed with the ground rod driven down a considerable distance below the surface of the earth. Best results are obtained if the ground rod reaches the water table.

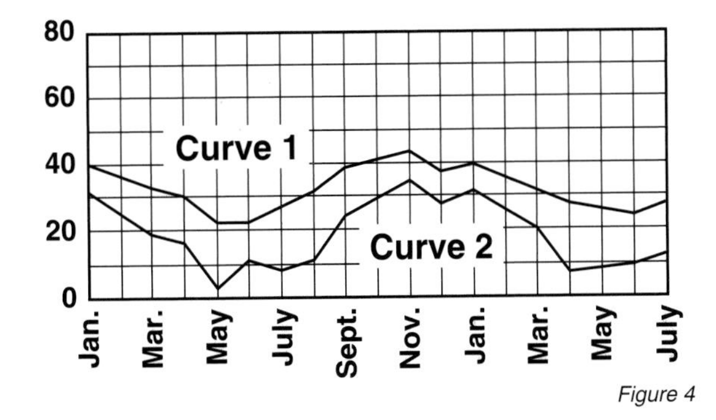

Figure 4

Seasonal variation of earth resistance with an electrode of ¾" pipe in rather stony clay soil. Depth of electrode in earth is 3 ft for Curve 1, and 10 ft for Curve 2

Notes

Notes

In some locations, the resistivity of the earth is so high that low-resistance grounding can be obtained only at considerable expense and with an elaborate grounding system. In such situations, it may be economical to use a ground rod system of limited size and to reduce the ground resistivity by periodically increasing the soluble chemical content of the soil. Figure 5 shows the substantial reduction in resistivity of sandy loam brought about by an increase in chemical salt content.

THE EFFECT OF SALT* CONTENT ON THE RESISTIVITY OF SOIL

(Sandy loam, Moisture content, 15% by weight, Temperature, 17°C)

Added Salt (% by weight of moisture)	Resistivity (Ω-cm)
0	10,700
0.1	1800
1.0	460
5	190
10	130
20	100

Figure 5

Chemically treated soil is also subject to considerable variation of resistivity with temperature changes, as shown in Figure 6. If salt treatment is employed, it is necessary to use ground rods which will resist chemical corrosion.

THE EFFECT OF TEMPERATURE ON THE RESISTIVITY OF SOIL CONTAINING SALT*

(Sandy loam, 20% moisture. Salt 5% of weight of moisture)

Temperature (Degrees C)	Resistivity (Ω-cms)
20	110
10	142
0	190
-5	312
-13	1440

Figure 6

**Such as copper sulfate, sodium carbonate, and others. Salts must be EPA or local ordinance approved prior to use.*

Soil Resistivity Measurements
(4-Point Measurement)

Resistivity measurements are of two types; the 2-Point and the 4-Point method. The 2-Point method is simply the resistance measured between two points. For most applications the most accurate is the 4-Point method which is used in the Model 4610 or Model 4500 Ground Tester. The 4-Point method (Figures 7 and 8), as the name implies, requires the insertion of four equally spaced and in-line electrodes into the test area. A known current from a constant current generator is passed between the outer electrodes. The potential drop (a function of the resistance) is then measured across the two inner electrodes. The Model 4610 and Model 4500 are calibrated to read directly in ohms.

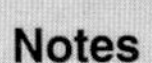

Notes

$$\rho = \frac{4\pi AR}{1 + \frac{2A}{\sqrt{(A^2 + 4B^2)}} - \frac{2A}{\sqrt{(4A^2 + 4B^2)}}}$$

Where: A = distance between the electrodes in centimeters
B = electrode depth in centimeters
If A > 20 B, the formula becomes:

$\rho = 2\pi$ AR (with A in cm)
$\rho = 191.5$ AR (with A in ft)
ρ = Soil resistivity (ohm-cm)

This value is the average resistivity of the ground at a depth equivalent to the distance "A" between two electrodes.

Soil Resistivity Measurements with 4-Point Ground Resistance Tester

Given a sizable tract of land in which to determine the optimum soil resistivity some intuition is in order. Assuming that the objective is low resistivity, preference should be given to an area containing moist loam as opposed to a dry sandy area. Consideration must also be given to the depth at which resistivity is required.

Example

After inspection, the area investigated has been narrowed down to a plot of ground approximately 75 square feet ($7m^2$). Assume that you need to determine the resistivity at a depth of 15 ft (450cm). The distance "A" between the electrodes must then be equivalent to the depth at which average resistivity is to be determined (15 ft, or 450cm). Using the more simplified Wenner formula ($\rho = 2\pi$ AR), the electrode depth must then be 1/20th of the electrode spacing or 8⁷⁄₈" (22.5cm).

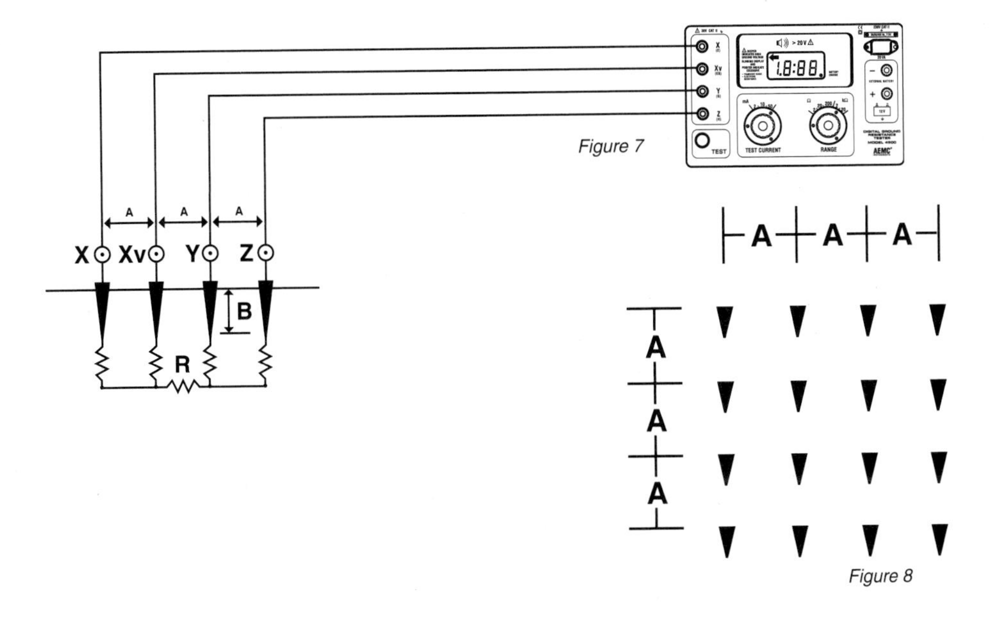

Figure 7

Figure 8

Notes

Lay out the electrodes in a grid pattern and connect to the Model 4500 as shown in Figure 8. Proceed as follows:

- Remove the shoring link between X and Xv (C1, P1)
- Connect all four auxiliary rods (Figure 7)

For example, if the reading is R = 15

ρ (resistivity) = 2π x A x R
A (distance between electrodes) = 450cm
ρ = 6.28 x 15 x 450 = 42,390Ω-cm

Ground Electrodes

The term "ground" is defined as a conducting connection by which a circuit or equipment is connected to the earth. The connection is used to establish and maintain as closely as possible the potential of the earth on the circuit or equipment connected to it. A "ground" consists of a grounding conductor, a bonding connector, its grounding electrode(s), and the soil in contact with the electrode.

Grounds have several protection applications. For natural phenomena such as lightning, grounds are used to discharge the system of current before personnel can be injured or system components damaged. For foreign potentials due to faults in electric power systems with ground returns, grounds help ensure rapid operation of the protection relays by providing low resistance fault current paths. This provides for the removal of the foreign potential as quickly as possible. The ground should drain the foreign potential before personnel are injured and the power or communications system is damaged.

Ideally, to maintain a reference potential for instrument safety, protect against static electricity, and limit the system to frame voltage for operator safety, a ground resistance should be zero ohms. In reality, as we describe further in the text, this value cannot be obtained.

Last but not least, low ground resistance is essential to meet NEC®, OSHA and other electrical safety standards.

Figure 9 illustrates a grounding rod. The resistance of the electrode has the following components:

(A) the resistance of the metal and that of the connection to it.
(B) the contact resistance of the surrounding earth to the electrode.
(C) the resistance in the surrounding earth to current flow or earth resistivity which is often the most significant factor.

More specifically:

(A) Grounding electrodes are usually made of a very conductive metal (copper or copper clad) with adequate cross sections so that the overall resistance is negligible.

Notes

(B) The National Institute of Standards and Technology has demonstrated that the resistance between the electrode and the surrounding earth is negligible if the electrode is free of paint, grease, or other coating, and if the earth is firmly packed.

(C) The only component remaining is the resistance of the surrounding earth. The electrode can be thought of as being surrounded by concentric shells of earth or soil, all of the same thickness. The closer the shell to the electrode, the smaller its surface; hence, the greater its resistance. The farther away the shells are from the electrode, the greater the surface of the shell; hence, the lower the resistance.

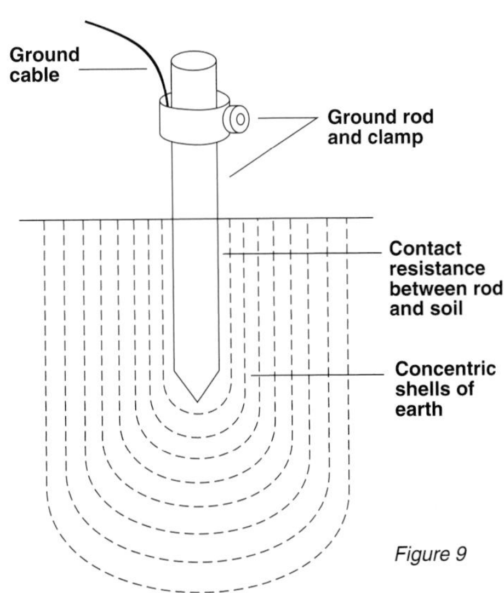

Figure 9

Eventually, adding shells at a distance from the grounding electrode will no longer noticeably affect the overall earth resistance surrounding the electrode. The distance at which this effect occurs is referred to as the effective resistance area and is directly dependent on the depth of the grounding electrode.

In theory, the ground resistance may be derived from the general formula:

$$R = \frac{\rho L}{A} \qquad \text{Resistance} = \text{Resistivity} \times \frac{\text{Length}}{\text{Area}}$$

This formula illustrates why the shells of concentric earth decrease in resistance the farther they are from the ground rod:

$$R = \text{Resistivity of Soil} \times \frac{\text{Thickness of Shell}}{\text{Area}}$$

In the case of ground resistance, uniform earth (or soil) resistivity throughout the volume is assumed, although this is seldom the case in nature. The equations for systems of electrodes are very complex and often expressed only as approximations. The most commonly used formula for single ground electrode systems, developed by Professor H. R. Dwight of the Massachusetts Institute of Technology, is the following:

$$R = \frac{\rho}{2\pi L}\left[\ln\left(\frac{4L}{r}\right) - 1\right]$$

R = resistance in ohms of the ground rod to the earth (or soil)
L = grounding electrode length
r = grounding electrode radius
ρ = average resistivity in Ω-cm

Notes

Effect of Ground Electrode Size and Depth on Resistance

Size: Increasing the diameter of the rod does not materially reduce its resistance. Doubling the diameter reduces resistance by less than 10%. (Figure 10)

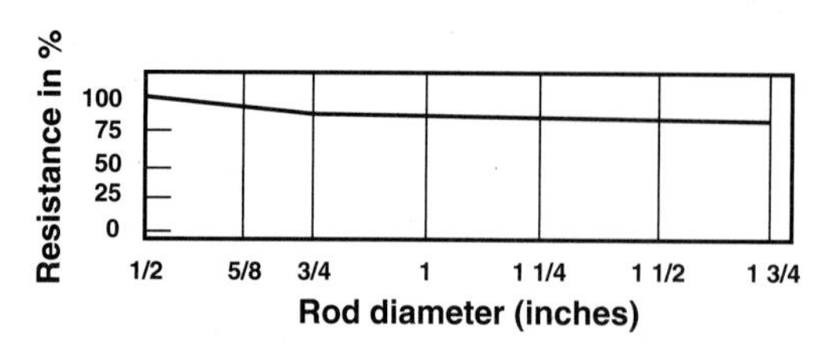

Figure10

Depth: As a ground rod is driven deeper into the earth, its resistance is substantially reduced. In general, doubling the rod length reduces the resistance by an additional 40% (Figure 11). The NEC® (1987, 250-83-3) requires a minimum of 8 ft (2.4m) to be in contact with the soil. The most common is a 10 ft (3m) cylindrical rod which meets the NEC® code. A minimum diameter of 5/8" (1.59cm) is required for steel rods and 1/2" (1.27cm) for copper or copper clad steel rods (NEC® 1987, 250-83-2). Minimum practical diameters for driving limitations for 10 ft (3m) rods are:

- 1/2" (1.27cm) in average soil
- 5/8" (1.59cm) in moist soil
- 3/4" (1.91cm) in hard soil or more than 10 ft driving depths

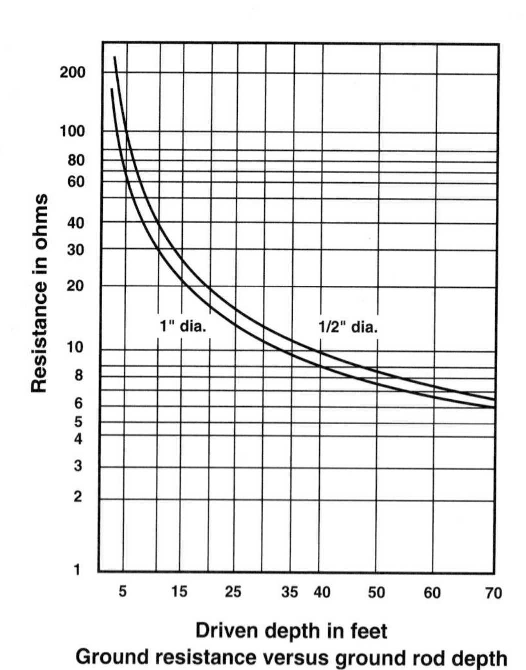

Ground resistance versus ground rod depth

Figure11

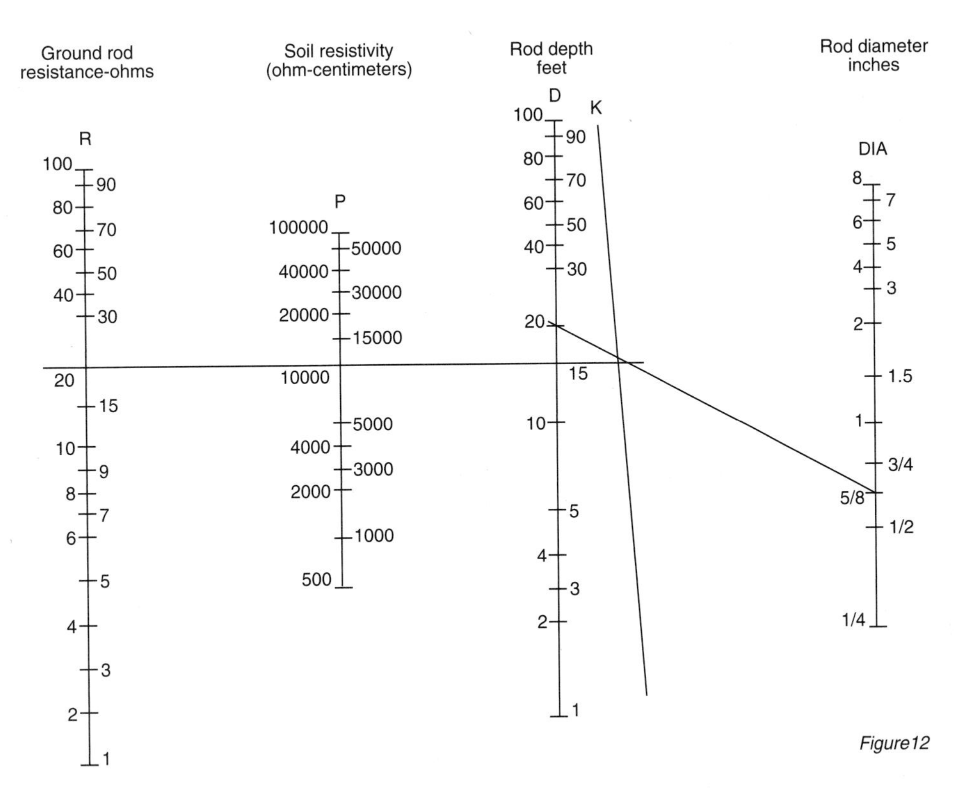

Figure12

Notes

Grounding Nomograph

1. Select required resistance on R scale
2. Select apparent resistivity on P scale
3. Lay straightedge on R and P scale, and allow to intersect with K scale
4. Mark K scale point
5. Lay straightedge on K scale point & DIA scale, and allow to intersect with D scale
6. Point on D scale will be rod depth required for resistance on R scale

Ground Resistance Values

NEC® 250-84 (1987): Resistance of man-made electrodes:

"A single electrode consisting of a rod, pipe, or plate which does not have a resistance to ground of 25Ω or less shall be augmented by one additional rod of any of the types specified in section 250-81 or 250-83. Where multiple rod, pipe or plate electrodes are installed to meet the requirements of this section, they shall be not less than 6 ft (1.83m) apart."

The National Electrical Code® (NEC) states that the resistance to ground shall not exceed 25Ω. This is an upper limit and guideline, since much lower resistance is required in many instances.

"How low in resistance should a ground be?" An arbitrary answer to this in ohms is difficult. The lower the ground resistance, the safer; and for positive protection of personnel and equipment, it is worth the effort to aim for less than one ohm. It is generally impractical to reach such a low resistance along a distribution system or a transmission line or in small substations. In some

Notes

regions, resistances of 5Ω or less may be obtained without much trouble. In other regions, it may be difficult to bring resistance of driven grounds below 100Ω.

Accepted industry standards stipulate that transmission substations should be designed not to exceed 1Ω. In distribution substations, the maximum recommended resistance is for 5Ω or even 1Ω. In most cases, the buried grid system of any substation will provide the desired resistance.

In light industrial or in telecommunication central offices, 5Ω is often the accepted value. For lightning protection, the arrestors should be coupled with a maximum ground resistance of 1Ω.

These parameters can usually be met with the proper application of basic grounding theory. There will always exist circumstances which will make it difficult to obtain the ground resistance required by the NEC® or other safety standards. When these situations develop, several methods of lowering the ground resistance can be employed. These include parallel rod systems, deep driven rod systems utilizing sectional rods, and chemical treatment of the soil. Additional methods discussed in other published data are buried plates, buried conductors (counterpoise), electrically connected building steel, and electrically connected concrete reinforced steel.

Electrically connecting to existing water and gas distribution systems was often considered to yield low ground resistance; however, recent design changes utilizing non-metallic pipes and insulating joints have made this method of obtaining a low resistance ground questionable and in many instances unreliable.

The measurement of ground resistances may only be accomplished with specially designed test equipment. Most instruments use the Fall-of-Potential principle of alternating current (AC) circulating between an auxiliary electrode and the ground electrode under test. The reading will be given in ohms, and represents the resistance of the ground electrode to the surrounding earth. AEMC has also recently introduced clamp-on ground resistance testers.

Note: The National Electrical Code® and NEC® are registered trademarks of the National Fire Protection Association.

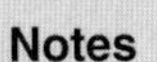

Ground Resistance Testing Principle
(Fall-of-Potential — 3-Point Measurement)

The potential difference between rods X and Y is measured by a voltmeter, and the current flow between rods X and Z is measured by an ammeter. (Note: X, Y and Z may be referred to as X, P and C in a 3-Point tester or C1, P2 and C2 in a 4-Point tester.) (Figure 13)

By Ohm's Law E = RI or R = E/I, we may obtain the ground electrode resistance R. If E = 20V and I = 1A, then

$$R = \frac{E}{I} = \frac{20}{1} = 20$$

It is not necessary to carry out all the measurements when using a ground tester. The ground tester will measure directly by generating its own current and displaying the resistance of the ground electrode.

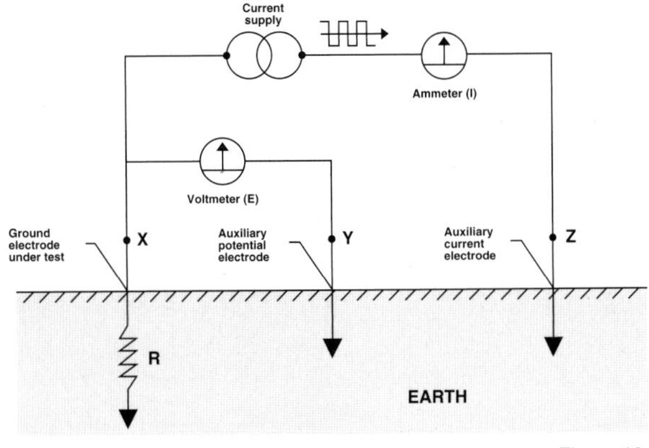

Figure 13

Position of the Auxiliary Electrodes on Measurements

The goal in precisely measuring the resistance to ground is to place the auxiliary current electrode Z far enough from the ground electrode under test so that the auxiliary potential electrode Y will be outside of the effective resistance areas of both the ground electrode and the auxiliary current electrode. The best way to find out if the auxiliary potential rod Y is outside the effective resistance areas is to move it between X and Z and to take a reading at each location. (See Figure 16) If the auxiliary potential rod Y is in an effective resistance area (or in both if they overlap, as in Figure 14), by displacing it the readings taken will vary noticeably in value. Under these conditions, no exact value for the resistance to ground may be determined.

Notes

Notes

On the other hand, if the auxiliary potential rod Y is located outside of the effective resistance areas (Figure 15), as Y is moved back and forth the reading variation is minimal. The readings taken should be relatively close to each other, and are the best values for the resistance to ground of the ground X. The readings should be plotted to ensure that they lie in a "plateau" region as shown in Figure 15. The region is often referred to as the "62% area." (See page 13 for explanation.)

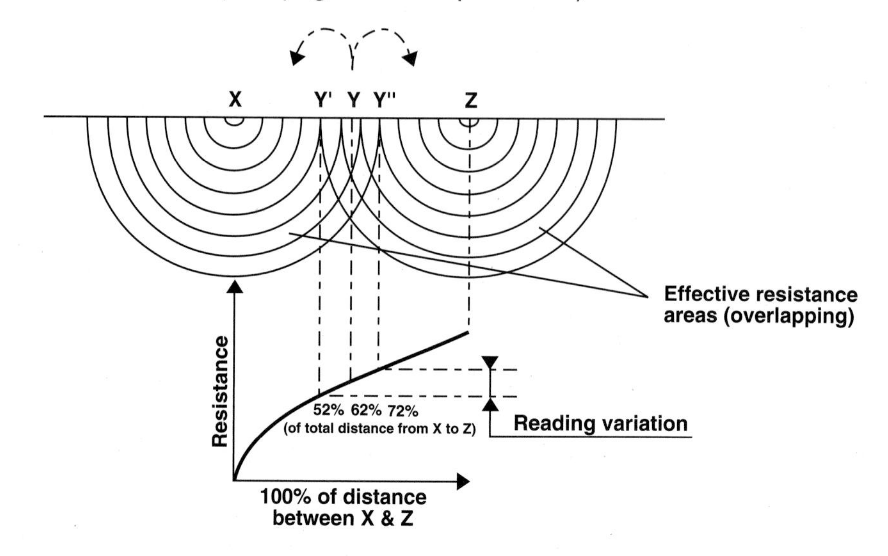

Figure 14

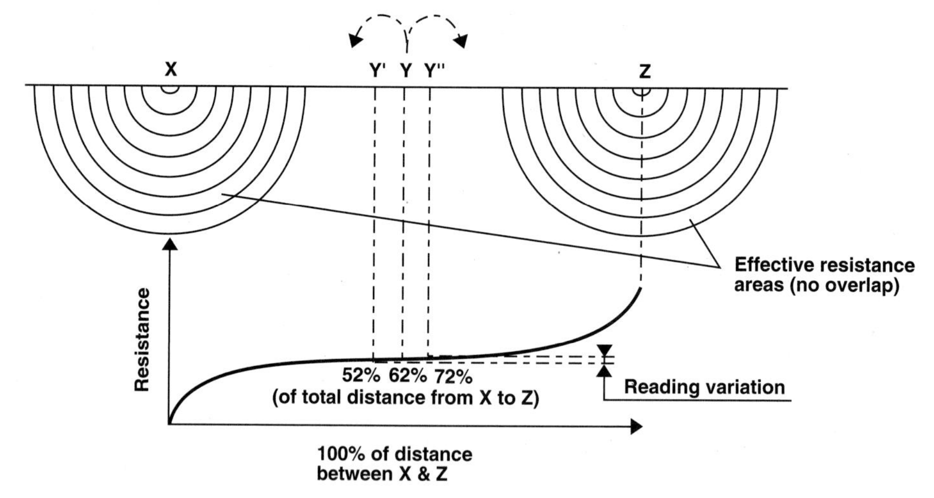

Figure 15

Measuring Resistance of Ground Electrodes (62% Method)

The 62% method has been adopted after graphical consideration and after actual test. It is the most accurate method but is limited by the fact that *the ground tested is a single unit.*

This method applies only when all three electrodes are in a straight line and the ground is a *single* electrode, pipe, or plate, etc., as in Figure 16.

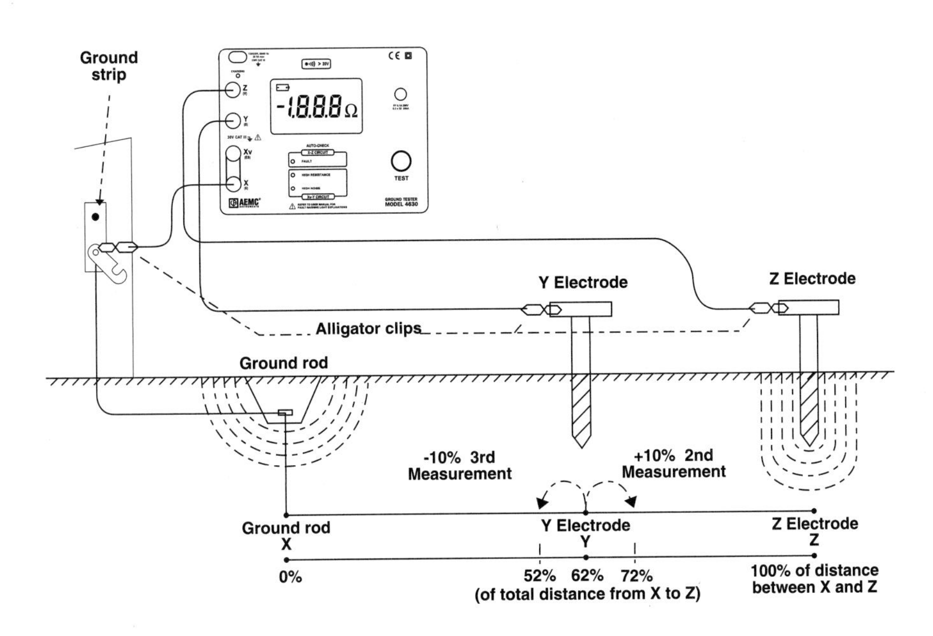

Figure 16

Consider Figure 17, which shows the effective resistance areas (concentric shells) of the ground electrode X and of the auxiliary current electrode Z. The resistance areas overlap. If readings were taken by moving the auxiliary potential electrode Y towards either X or Z, the reading differentials would be great and one could not obtain a reading within a reasonable band of tolerance. The sensitive areas overlap and act constantly to increase resistance as Y is moved away from X.

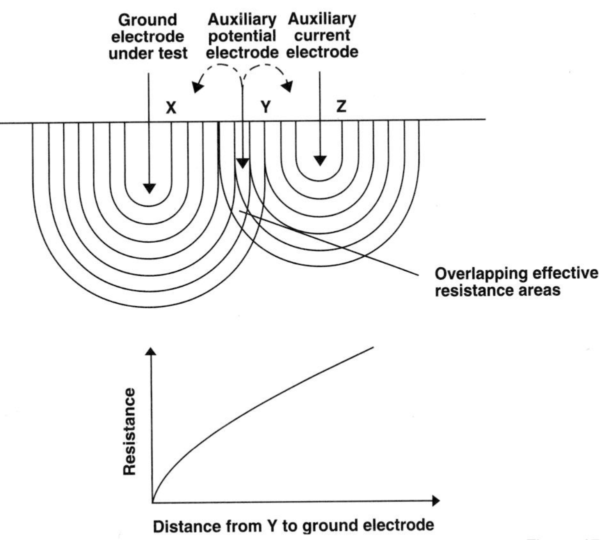

Figure 17

Notes

Notes

Now consider Figure 18, where the X and Z electrodes are sufficiently spaced so that the areas of effective resistance do not overlap. If we plot the resistance measured we find that the measurements level off when Y is placed at 62% of the distance from X to Z, and that the readings on either side of the initial Y setting are most likely to be within the established tolerance band. This tolerance band is defined by the user and expressed as a percent of the initial reading: ±2%, ±5%, ±10%, etc.

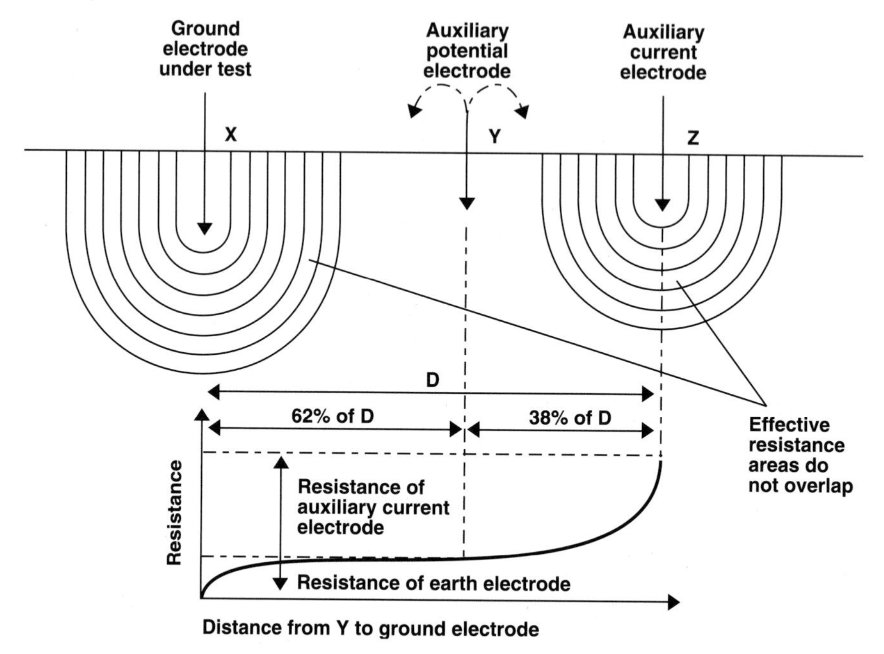

Figure 18

Auxiliary Electrode Spacing

No definite distance between X and Z can be given, since this distance is relative to the diameter of the electrode tested, its length, the homogeneity of the soil tested, and particularly, the effective resistance areas. However, an approximate distance may be determined from the following chart which is given for a homogeneous soil and an electrode of 1" in diameter. (For a diameter of ½", reduce the distance by 10%; for a diameter of 2" increase the distance by 10%.)

Approximate distance to auxiliary electrodes using the 62% method		
Depth Driven	**Distance to Y**	**Distance to Z**
6 ft	45 ft	72 ft
8 ft	50 ft	80 ft
10 ft	55 ft	88 ft
12 ft	60 ft	96 ft
18 ft	71 ft	115 ft
20 ft	74 ft	120 ft
30 ft	86 ft	140 ft

Multiple Rod Spacing

Parallel multiple electrodes yield lower resistance to ground than a single electrode. High-capacity installations require low grounding resistance. Multiple rods are used to provide this resistance.

A second rod does not provide a total resistance of half that of a single rod unless the two are several rod lengths apart. To achieve the grounding resistance place multiple rods one rod length apart in a line, circle, hollow triangle, or square. The equivalent resistance can be calculated by dividing by the number of rods and multiplying by the factor X (shown below). Additional considerations regarding step and touch potentials should be addressed by the geometry.

Multiplying Factors for Multiple Rods	
Number of Rods	**X**
2	1.16
3	1.29
4	1.36
8	1.68
12	1.80
16	1.92
20	2.00
24	2.16

Placing additional rods within the periphery of a shape will not reduce the grounding resistance below that of the peripheral rods alone.

Notes

Notes

Multiple Electrode System

A single driven ground electrode is an economical and simple means of making a good ground system. But sometimes a single rod will not provide sufficient low resistance, and several ground electrodes will be driven and connected in parallel by a cable. Very often when two, three or four ground electrodes are being used, they are driven in a straight line; when four or more are being used, a hollow square configuration is used and the ground electrodes are still connected in parallel and are equally spaced. (Figure 19)

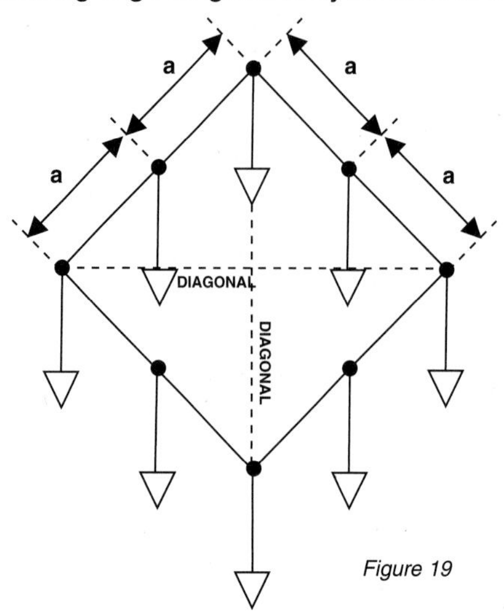

Figure 19

In multiple electrode systems, the 62% method electrode spacing may no longer be applied directly. The distance of the auxiliary electrodes is now based on the maximum grid distance (i.e. in a square, the diagonal; in a line, the total length. For example, a square having a side of 20 ft will have a diagonal of approximately 28 ft).

Multiple Electrode System		
Max Grid Distance	**Distance to Y**	**Distance to Z**
6 ft	78 ft	125 ft
8 ft	87 ft	140 ft
10 ft	100 ft	160 ft
12 ft	105 ft	170 ft
14 ft	118 ft	190 ft
16 ft	124 ft	200 ft
18 ft	130 ft	210 ft
20 ft	136 ft	220 ft
30 ft	161 ft	260 ft
40 ft	186 ft	300 ft
50 ft	211 ft	340 ft
60 ft	230 ft	370 ft
80 ft	273 ft	440 ft
100 ft	310 ft	500 ft
120 ft	341 ft	550 ft
140 ft	372 ft	600 ft
160 ft	390 ft	630 ft
180 ft	434 ft	700 ft
200 ft	453 ft	730 ft

Two-Point Measurement
(Simplified Method)

This is an alternative method when an excellent ground is already available.

In congested areas where finding room to drive the two auxiliary rods may be a problem, the 2-Point measurement method may be applied. The reading obtained will be that of the two grounds in series. Therefore, the water pipe or other ground must be very low in resistance so that it will be negligible in the final measurement. The lead resistances will also be measured and should be deducted from the final measurement.

This method is not as accurate as 3-Point methods (62% method), as it is particularly affected by the distance between the tested electrode and the dead ground or water pipe. This method should not be used as a standard procedure, but rather as a back-up in tight areas. (Figure 20)

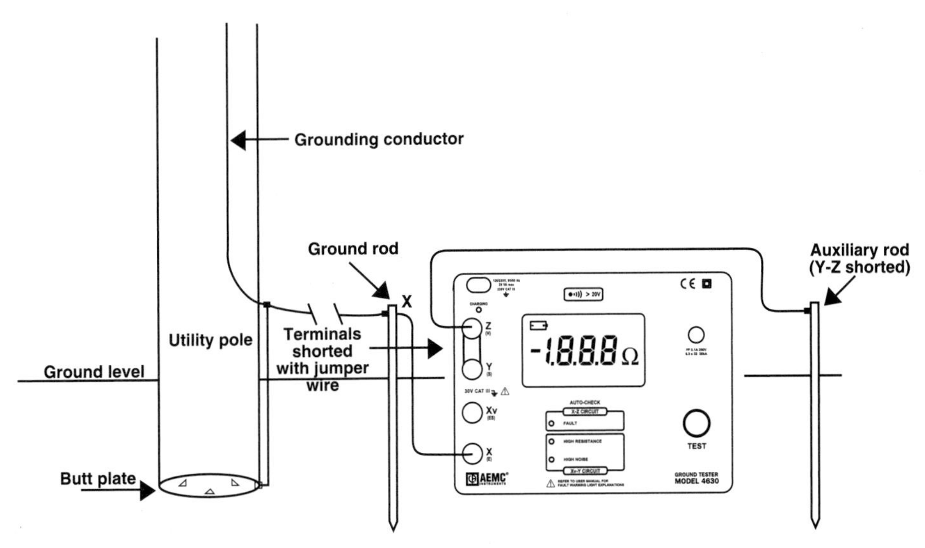

Figure 20

Notes

Continuity Measurement

Continuity measurements of a ground conductor are possible by using two terminals. (Figure 21)

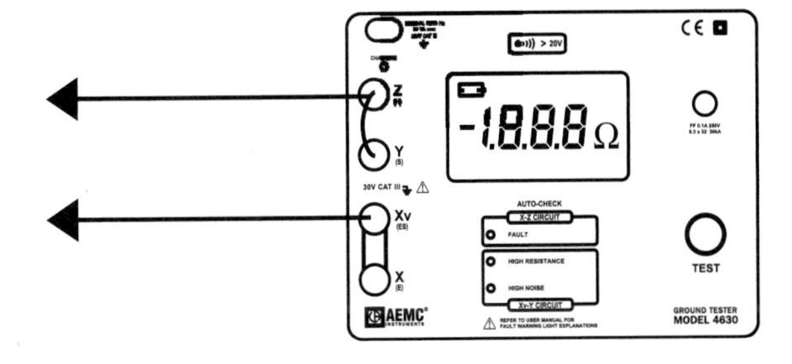

Figure 21

Notes

Tech Tips

Excessive Noise

Excessive noise may interfere with testing because of the long leads used to perform a Fall-of-Potential test. A voltmeter can be utilized to identify this problem. Connect the "X", "Y", and "Z" cables to the auxiliary electrodes as for a standard ground resistance test. Use the voltmeter to test the voltage across terminals "X" and "Z." (See Figure 22)

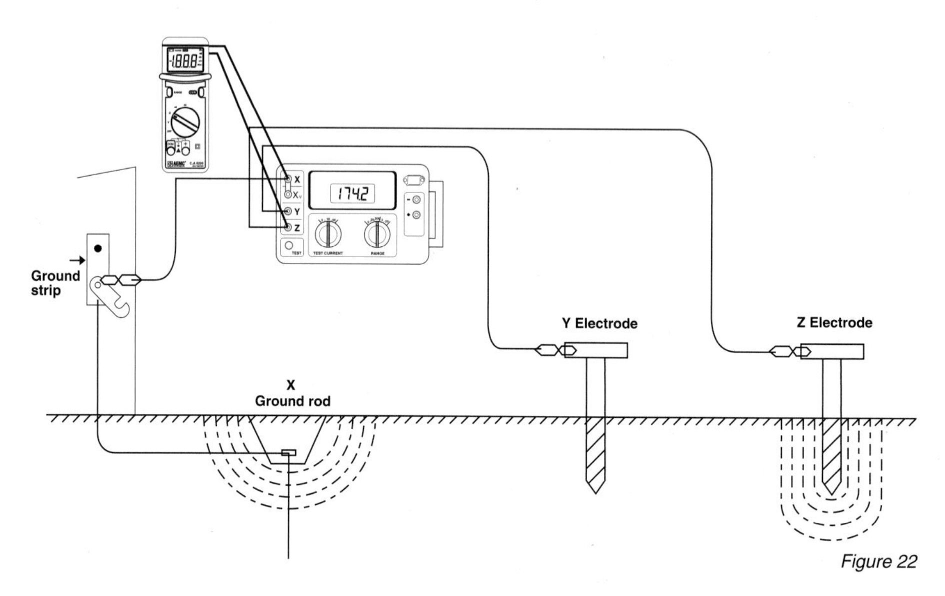

Figure 22

The voltage reading should be within stray voltage tolerances acceptable to your ground tester. If the voltage exceeds this value, try the following techniques:

A) Braid the auxiliary cables together. This often has the effect of canceling out the common mode voltages between these two conductors. (Figure 23)

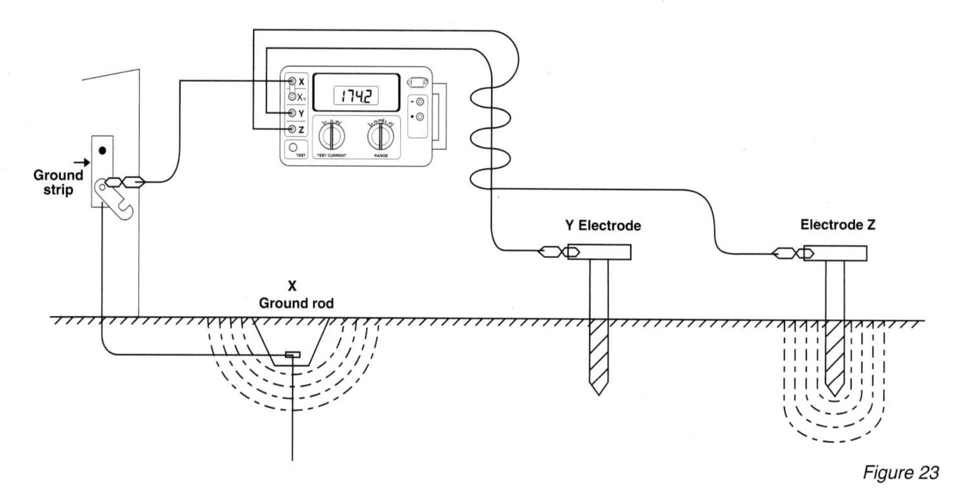

Figure 23

B) If the previous method fails, try changing the alignment of the auxiliary cables so that they are not parallel to power lines above or below the ground. (Figure 24)

C) If a satisfactory low voltage value is still not obtained, the use of shielded cables may be required. The shield acts to protect the inner conductor by capturing the voltage and draining it to ground. (Figure 25)

1. Float the shields at the auxiliary electrodes
2. Connect all three shields together at (but not to) the instrument
3. Solidly ground the remaining shield to the ground under test

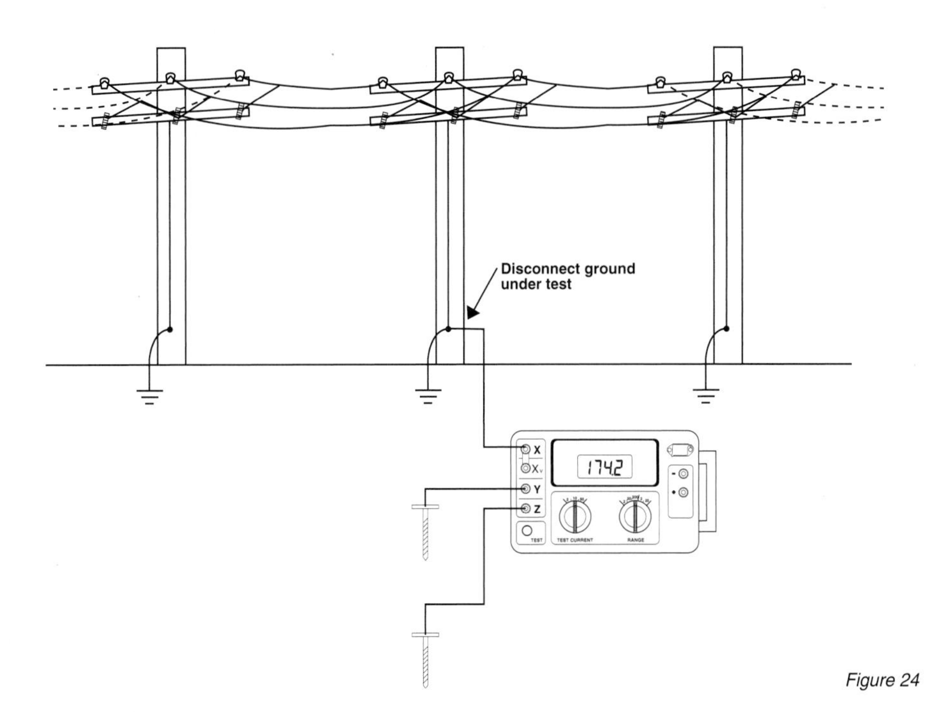

Figure 24

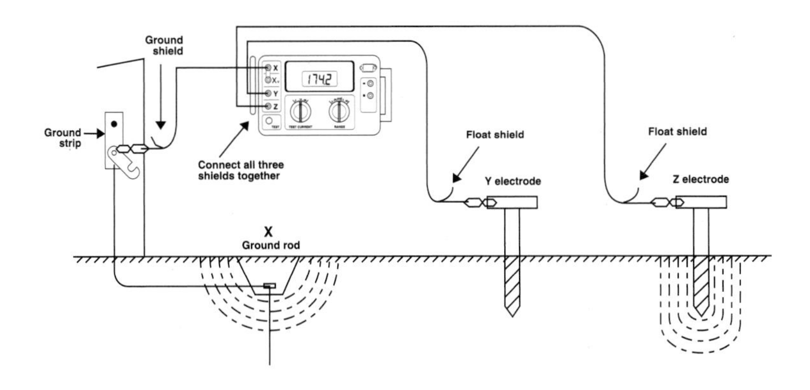

Figure 25

Notes

Notes

Excessive Auxiliary Rod Resistance

The inherent function of a Fall-of-Potential ground tester is to input a constant current into the earth and measure the voltage drop by means of auxiliary electrodes. Excessive resistance of one or both auxiliary electrodes can inhibit this function. This is caused by high soil resistivity or poor contact between the auxiliary electrode and the surrounding dirt. (Figure 26)

To ensure good contact with the earth, stamp down the soil directly around the auxiliary electrode to remove air gaps formed when inserting the rod. If soil resistivity is the problem, pour water around the auxiliary electrodes. This reduces the auxiliary electrode's contact resistance without affecting the measurement.

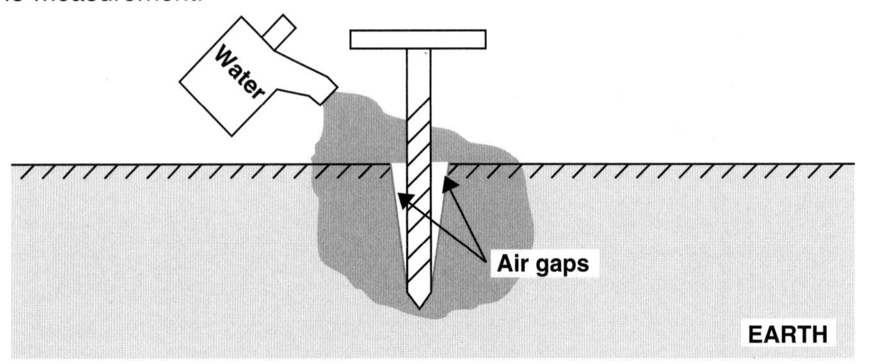

Figure 26

Tar or Concrete Mat

Sometimes a test must be performed on a ground rod that is surrounded by a tar or concrete mat, where auxiliary electrodes cannot be driven easily. In such cases, metal screens and water can be used to replace auxiliary electrodes, as shown in Figure 27.

Place the screens on the floor the same distance from the ground rod under test as you would auxiliary electrodes in a standard fall-of-potential test. Pour water on the screens and allow it to soak in. These screens will now perform the same function as would driven auxiliary electrodes.

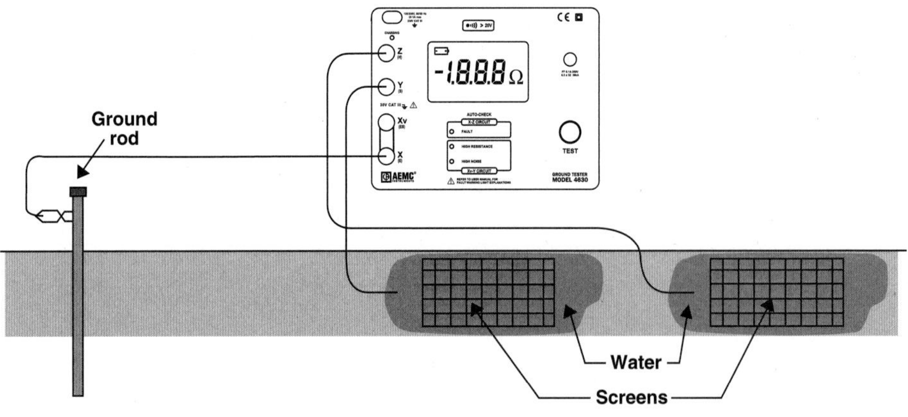

Figure 27

Touch Potential Measurements

Notes

The primary reason for performing Fall-of-Potential measurements is to observe electrical safety of personnel and equipment. However, in certain circumstances the degree of electrical safety can be evaluated from a different perspective.

Periodic ground electrode or grid resistance measurements are recommended when:

1) The electrode/grid is relatively small and is able to be conveniently disconnected.
2) Corrosion induced by low soil resistivity or galvanic action is suspected.
3) Ground faults are very unlikely to occur near the ground under test.

Touch potential measurements are an alternative method for determining electrical safety. Touch potential measurements are recommended when:

1) It is physically or economically impossible to disconnect the ground to be tested.
2) Ground faults could reasonably be expected to occur near the ground to be tested, or near equipment grounded by the ground to be tested.
3) The "footprint" of grounded equipment is comparable to the size of the ground to be tested. (The "footprint" is the outline of the part of equipment in contact with the earth.)

Neither Fall-of-Potential resistance measurements nor touch potential measurements tests the ability of grounding conductors to carry high phase-to-ground fault currents. Additional high current tests should be performed to verify that the grounding system can carry these currents.

When performing touch potential measurements, a four-pole ground resistance tester is used. During the test, the instrument induces a low level fault into the earth at some proximity to the subject ground. The instrument displays touch-potential in volts per ampere of fault current. The displayed value is then multiplied by the largest anticipated ground fault current to obtain the worst case touch potential for a given installation.

For example, if the instrument displayed a value of 0.100Ω when connected to a system where the maximum fault current was expected to be 5000A, the maximum touch potential would be: 5000 x 0.1 = 500V.

Touch potential measurements are similar to Fall-of-Potential measurements in that both measurements require placement of auxiliary electrodes into or on top of the earth. Spacing the auxiliary electrodes during touch potential measurements differs from Fall-of-Potential electrode spacing, as shown in Figure 28 on the following page.

Notes

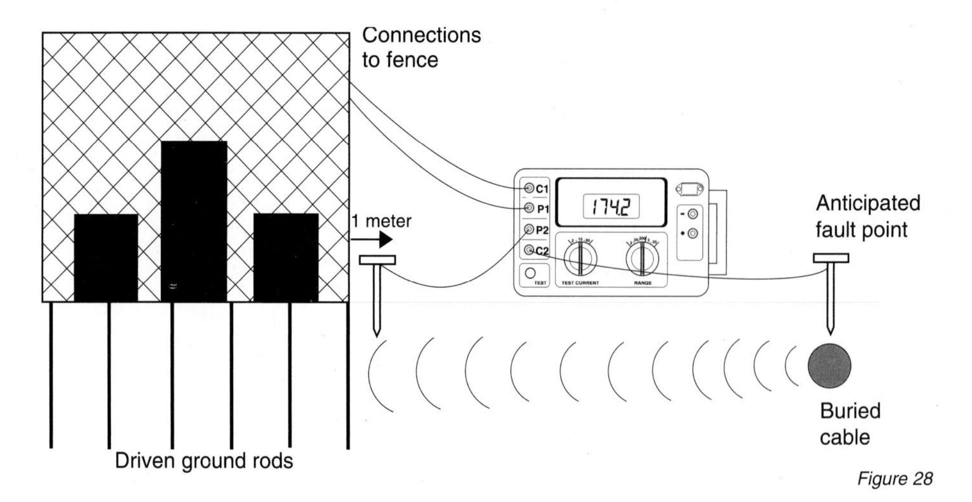

Figure 28

Consider the following scenario: If the buried cable depicted in Figure 28 experienced an insulation breakdown near the substation shown, fault currents would travel through the earth towards the substation ground, creating a voltage gradient. This voltage gradient may be hazardous or potentially lethal to personnel who come in contact with the affected ground.

To test for approximate touch potential values in this situation, proceed as follows: Connect cables between the fence of the substation and C1 and P1 of the four-pole earth resistance tester. Position an electrode in the earth at the point at which the ground fault is anticipated to occur, and connect it to C2. In a straight line between the substation fence and the anticipated fault point, position an auxiliary electrode into the earth one meter (or one arm's length) away from the substation fence, and connect it to P2. Turn the instrument on, select the 10mA current range, and observe the measurement. Multiply the displayed reading by the maximum fault current of the anticipated fault.

By positioning the P2 electrode at various positions around the fence adjacent to the anticipated fault line, a voltage gradient map may be obtained.

Clamp-on Ground Resistance Measurement
(Models 3711 & 3731)

This measurement method is innovative and quite unique. It offers the ability to measure the resistance without disconnecting the ground. This type of measurement also offers the advantage of including the bonding to ground and the overall grounding connection resistances.

Principle of Operation

Usually, a common distribution line grounded system can be simulated as a simple basic circuit as shown in Figure 29 or an equivalent circuit, shown in Figure 30. If voltage E is applied to any measured grounding point Rx through a special transformer, current I flows through the circuit, thereby establishing the following equation.

$$E/I = Rx + \frac{1}{\sum_{k=1}^{n} \frac{1}{Rk}} \quad \textit{where, usually} \quad Rx >> \frac{1}{\sum_{k=1}^{n} \frac{1}{Rk}}$$

Therefore, E/I = Rx is established. If I is detected with E kept constant, measured grounding point resistance can be obtained. Refer again to Figures 29 and 30. Current is fed to a special transformer via a power amplifier from a 2.4kHz constant voltage oscillator. This current is detected by a detection CT. Only the 2.4kHz signal frequency is amplified by a filter amplifier. This occurs before the A/D conversion and after synchronous rectification. It is then displayed on the LCD.

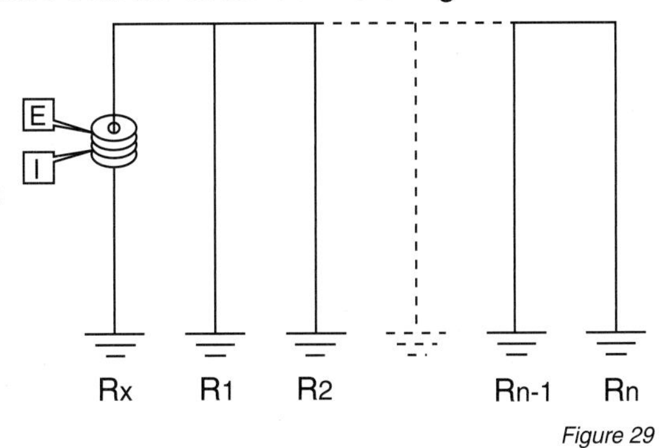

Figure 29

The filter amplifier is used to cut off both earth current at commercial frequency and high-frequency noise. Voltage is detected by coils wound around the injection CT which is then amplified, rectified, and compared by a level comparator. If the clamp is not closed properly, an "open jaw" annunciator appears on the LCD.

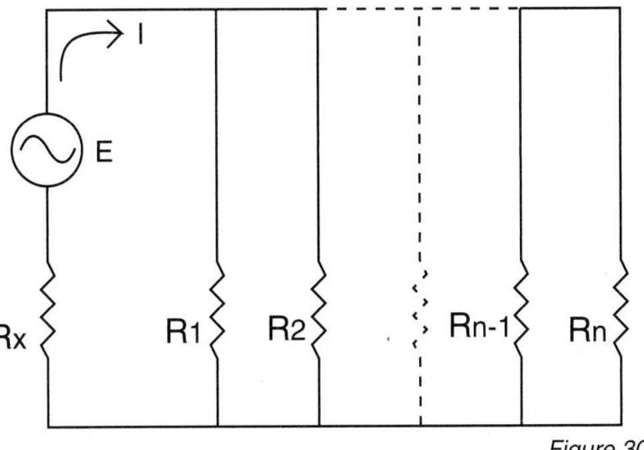

Figure 30

Notes

Notes

Examples: Typical In-Field Measurements

Pole Mounted Transformer

Remove any molding covering the ground conductor, and provide sufficient room for the Model 3711 & 3731 jaws, which must be able to close easily around the conductor. The jaws can be placed around the ground rod itself. **Note:** The clamp must be placed so that the jaws are in an electrical path from the system neutral or ground wire to the ground rod or rods as the circuit provides.

Select the current range "A." Clamp onto the ground conductor and measure the ground current. The maximum current range is 30A. If the ground current exceeds 5A, ground resistance measurements are not possible. *Do not proceed further with the measurement.* Instead, remove the clamp-on tester from the circuit, noting the location for maintenance, and continue to the next test location.

After noting the ground current, select the ground resistance range "Ω" and measure the resistance directly. The reading you measure with the Model 3711 & 3731 indicates the resistance not just of the rod, but also of the connection to the system neutral and all bonding connections between the neutral and the rod.

Note that in Figure 31 there is both a butt plate and a ground rod. In this type of circuit, the instrument must be placed above the bond so that both grounds are included in the test. For future reference note the date, ohms reading, current reading and point number. Replace any molding you may have removed from the conductor. **Note:** A high reading indicates one or more of the following:

A) poor ground rod

B) open ground conductor

C) high resistance bonds on the rod or splices on the conductor; watch for buried split bolts, clamps, and hammer-on connections.

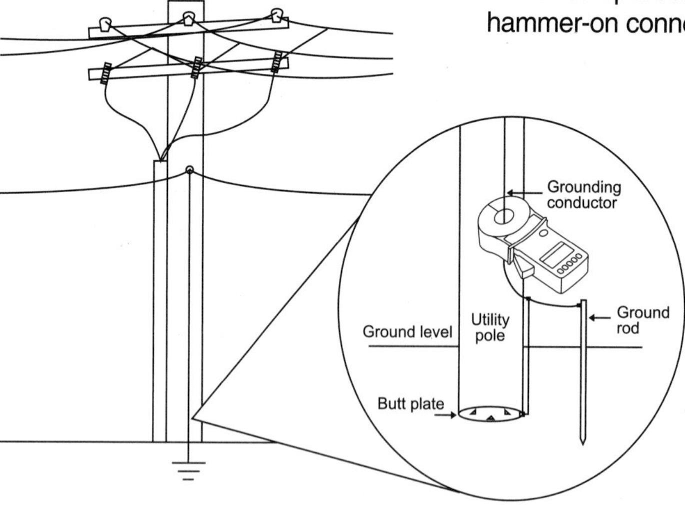

Figure 31

Service Entrance or Meter

Follow basically the same procedure as in the first example. Notice that Figure 32 shows the possibility of multiple ground rods, and in Figure 33 the ground rods have been replaced with a water pipe ground. You may also have both types acting as a ground. In these cases, it is necessary to make the measurements between the service neutral and both grounded points.

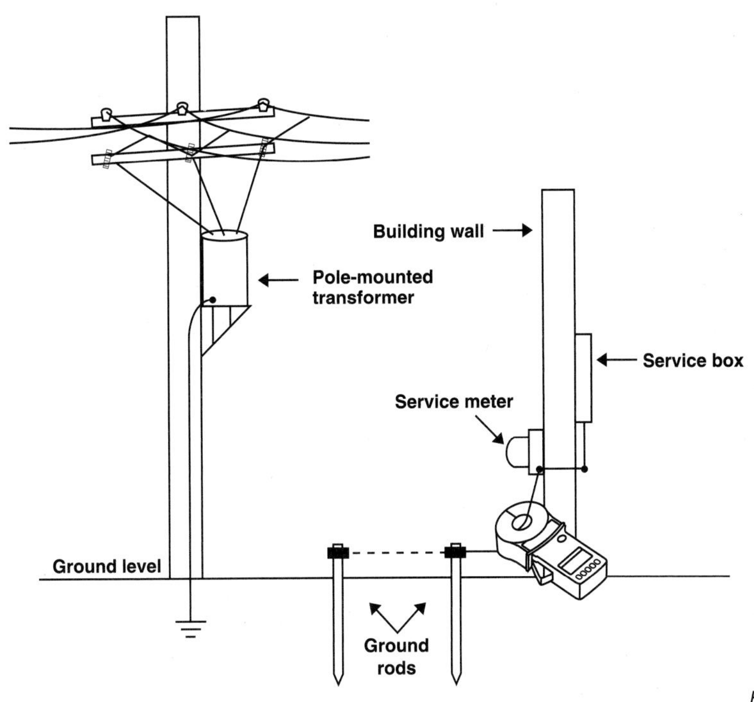

Figure 32

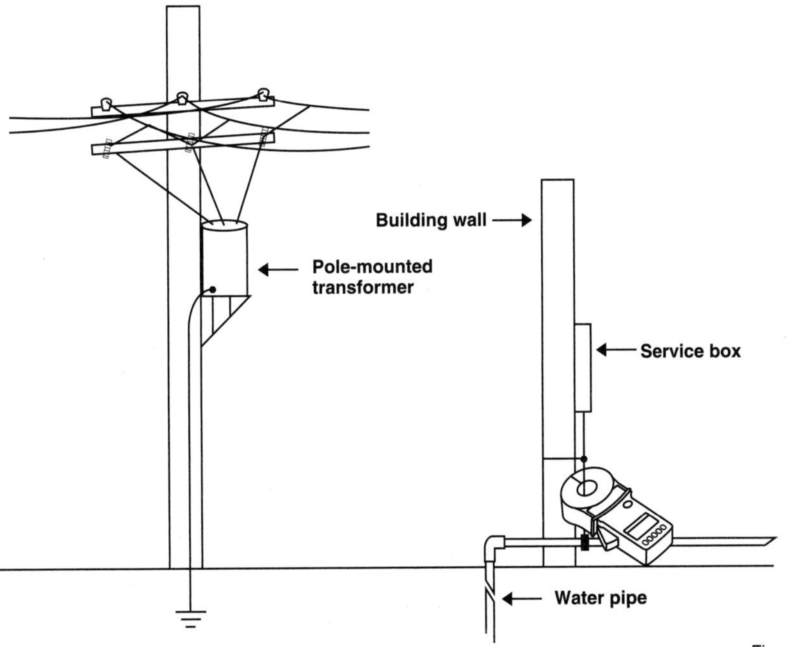

Figure 33

Notes

Pad Mounted Transformer

Note: Never open transformer enclosures. They are the property of the electrical utility. This test is for high voltage experts only.

Observe all safety requirements, since dangerously high voltage is present. Locate and number all rods (usually only a single rod is present). If the ground rods are inside the enclosure, refer to Figure 34 and if they are outside the enclosure, refer to Figure 35. If a single rod is found within the enclosure, the measurement should be taken on the conductor just before the bond on the ground rod. Often, more than one ground conductor is tied to this clamp, looping back to the enclosure or neutral.

In many cases, the best reading can be obtained by clamping the Models 3711 & 3731 onto the ground rod itself, below the point when the ground conductors are attached to the rod, so that you are measuring the ground circuit. Care must be taken to find a conductor with only one return path to the neutral.

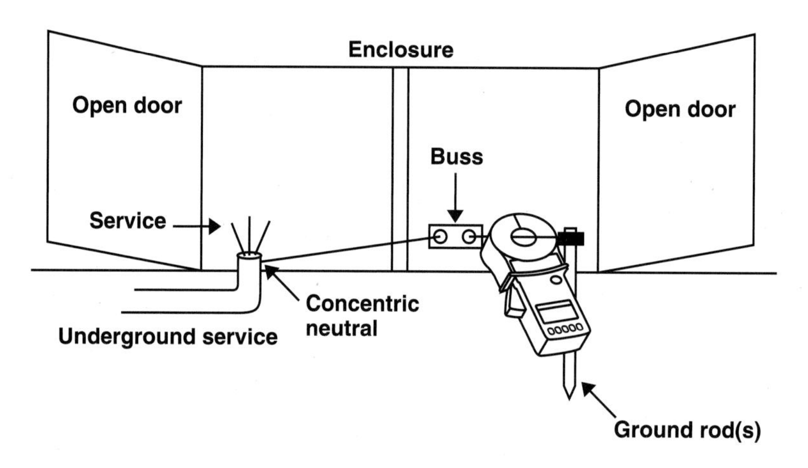

Figure 34

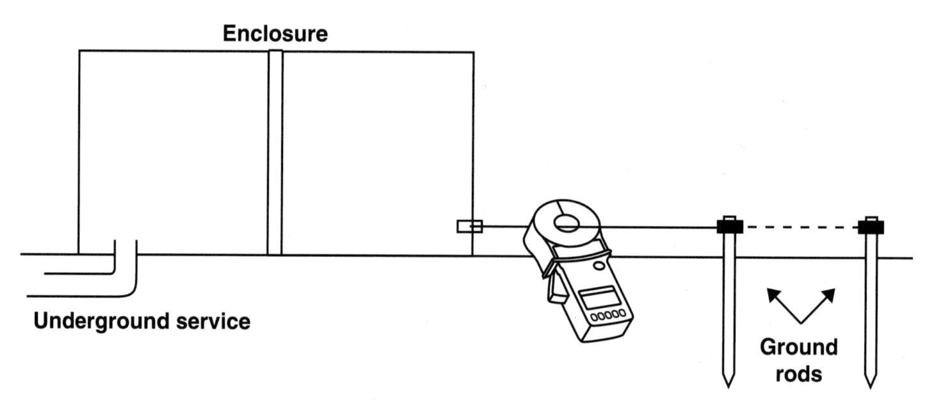

Figure 35

Transmission Towers

Observe all safety requirements, since dangerously high voltage is present. Locate the ground conductor at the base of the tower. **Note:** Many different configurations exist. Care should be taken when searching for the ground conductor. Figure 36 shows a single leg mounted on a concrete pad with an external ground conductor. The point at which you clamp the ground tester should be above all splices and connections which allow for multiple rods, butt wraps, or butt plates.

Central Office Locations

The main ground conductor from ground window or ground plane is often too large to clamp around. Due to the wiring practices within the central office, there are many locations at which you can look at the water pipe or counterpoise from within the building. An effective location is usually at the ground buss in the power room, or near the backup generator.

By measuring at several points and comparing the readings, both of current flow and resistance, you will be able to identify neutral loops, utility grounds, and central office grounds. The test is effective and accurate because the ground window is connected to the utility ground at only one point, according to standard practices.

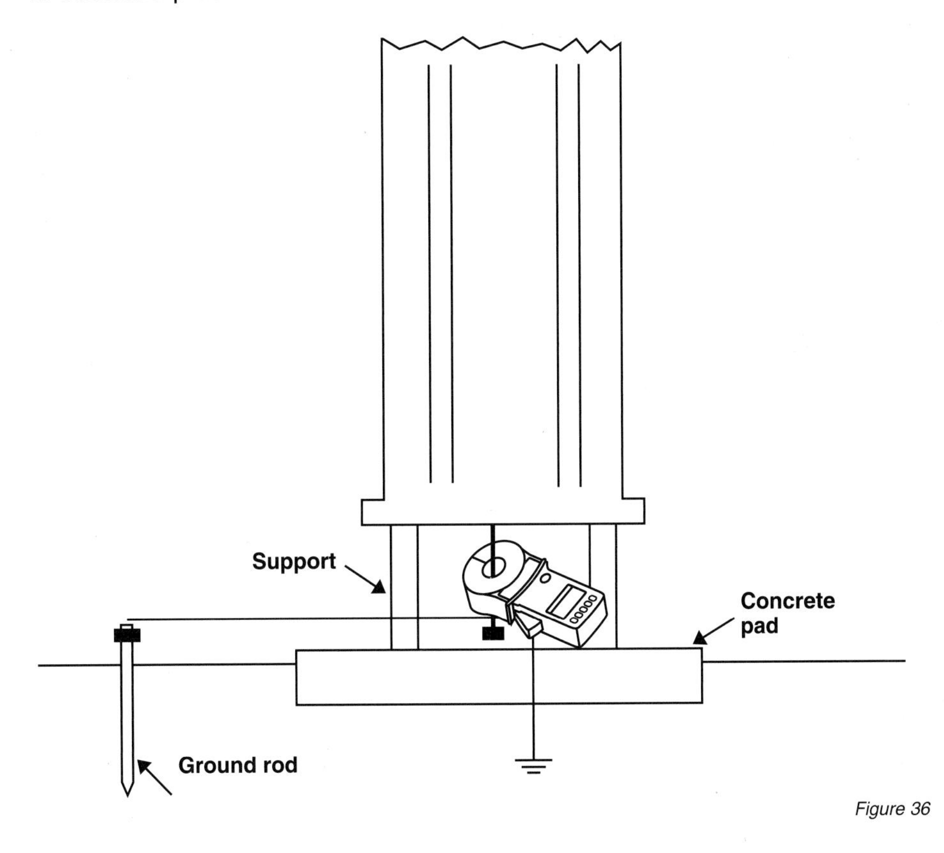

Figure 36

Notes

Notes

Telecommunications

The clamp-on ground tester developed by AEMC and discussed in the previous chapter has revolutionized the ability of power companies to measure their ground resistance values. This same proven instrument and technology can be applied to telephone industries to aid in detecting grounding and bonding problems. As equipment operates at lower voltages, the system's ability to remove any manmade or natural overpotentials becomes even more critical. The traditional Fall-of-Potential tester proved to be labor intensive and left much to interpretation. Even more important, the clamp-on ground test method allows the user to make this necessary reading without the risky business of removing the ground under test from service.

In many applications, the ground consists of bonding the two Utilities together to avoid any difference of potentials that could be dangerous to equipment and personnel alike. The clamp-on "Ohm meter" can be used to test these important bonds.

Here are some of the solutions and clamp-on procedures that have applications to the telephone industry.

Telephone Cabinets and Enclosures

Grounding plays a very important role in the maintenance of sensitive equipment in telephone cabinets and enclosures. In order to protect this equipment, a low resistance path must be maintained in order for any overvoltage potentials to conduct safely to earth. This resistance test is performed by clamping a ground tester, Model 3711 & 3731, around the driven ground rod, below any common telephone and power company bond connections.

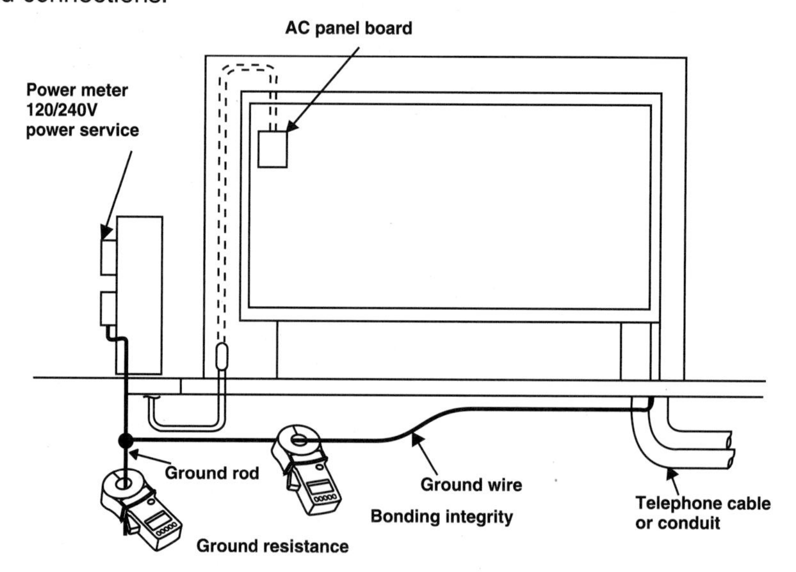

Figure 37

Notes

To avoid any high voltage potentials between the telephone and power companies, a low resistance bond is established. Bonding integrity is performed by clamping around the No. 6 copper wire between the master ground bar (MGB) and the power company's multigrounded neutral (MGN). The resistance value displayed on the tester will also include loose or poorly landed terminations that may have degraded over time.

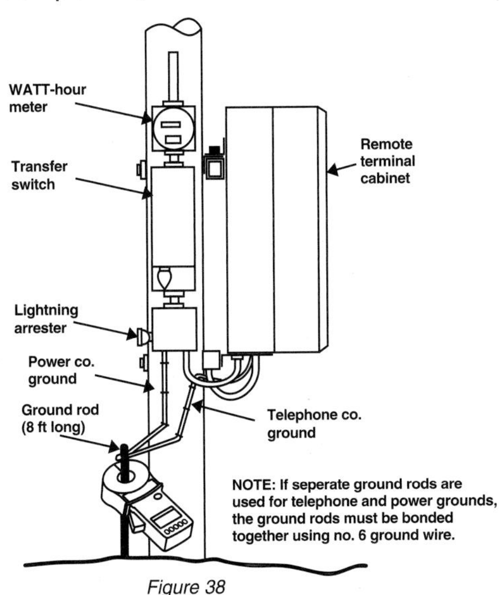

Figure 38

Additionally, the clamp-on ground tester can be used as a True RMS ammeter.

Pedestal Grounds

All cable sheaths are bonded to a ground bar inside each pedestal. This ground bar is connected to earth by means of a driven ground rod. The ground rod resistance can be found by using the instrument clamped around the ground rod or the No. 6 cable connecting these two points. (Figure 39)

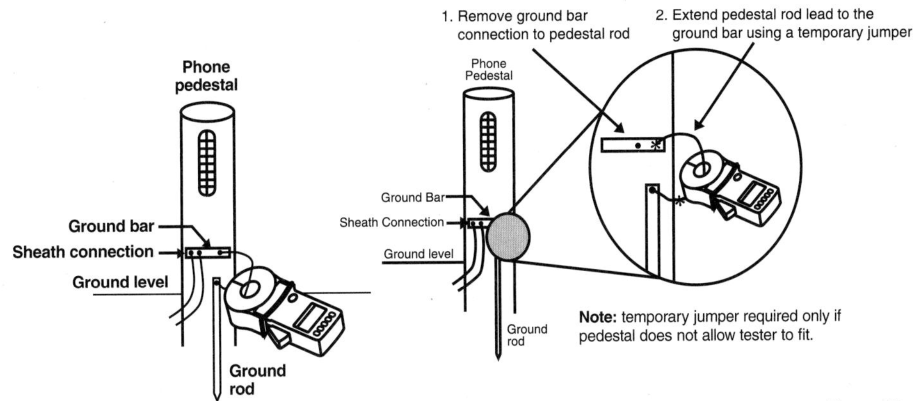

Figure 39

Cable Shield Bonds to MGN

The cable shields in a buried or above ground telephone enclosure may be grounded by means of the power company's multigrounded neutral. The clamp-on ground tester can be utilized to ensure that this connection has been successfully terminated. The low resistance return path for the instrument to make this measurement will be from this bond wire under test to the MGN back through all other bonds up and/or down stream (theory of parallel resistance).

The clamp-on ground tester also is a True RMS ammeter.

Notes

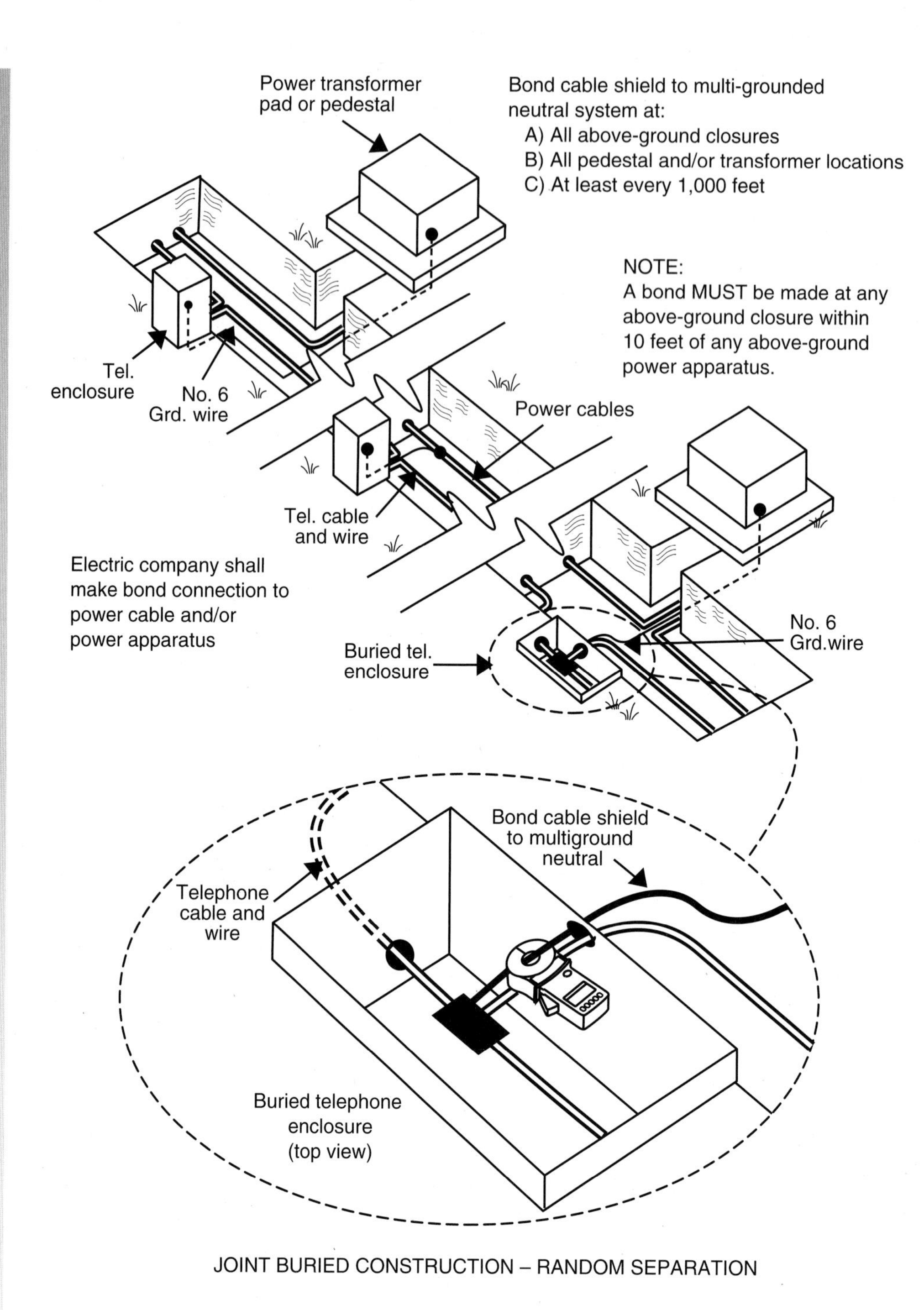

JOINT BURIED CONSTRUCTION – RANDOM SEPARATION

Figure 40

Notes

Network Interface Device (NID) with a Protector Block

The typical customer connection is achieved with the tip and ring drop cable pair. In order to protect against an overvoltage situation on the telephone wires, a protector block is installed inside the NID. This protector has two internal devices that conduct only when unwanted overvoltages are present. In order for the protector to function properly, it must have a low resistance path for any fault to conduct to earth. This bonding and ground resistance potential can be verified by using the clamp-on ground resistance tester. Simply take a short piece of wire and temporarily jumper the tip side (CO ground) to the ground connector on the protector block. By clamping around this jumper wire, you will now test the ground resistance potential including all terminations at this location. The return signal path required for the clamp-on ground tester to make this measurement will be the CO ground.

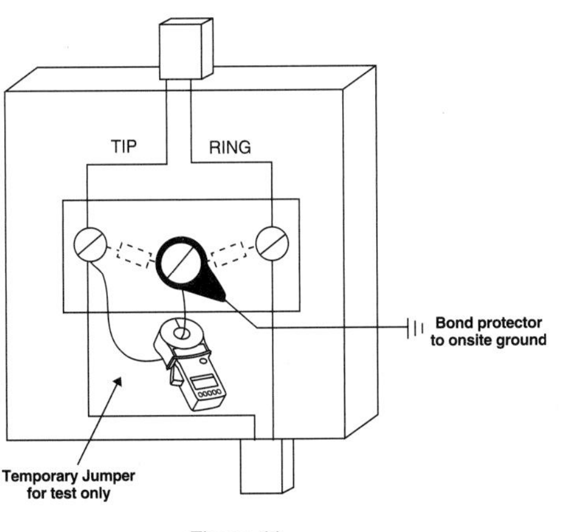

Figure 41

Overhead Telephone Distribution

Telephone systems delivered on overhead points must also be bonded to the MGN. This is typically performed by supplying a No. 6 copper wire connected to the grounding strand above telephone space. If power is not supplied on these points, driven ground rods must be installed at required point intervals and subsequently tested.

Note: Coil wire for attachment to power company MGN

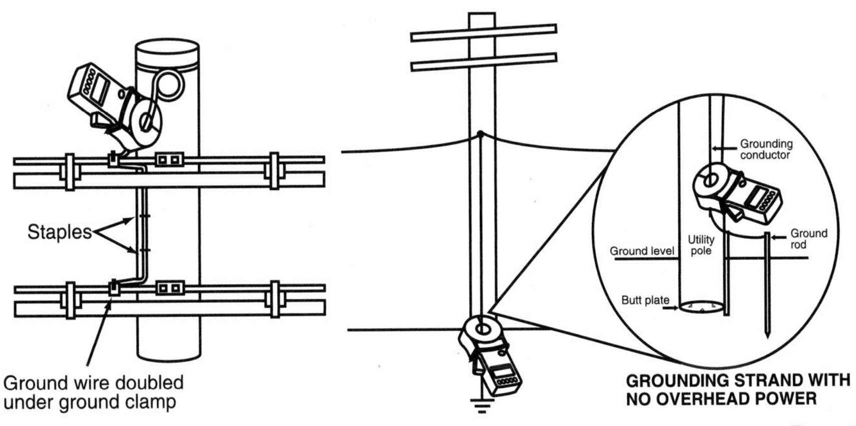

Figure 42

Summary Quiz

1. When using the simplified Wenner formula ($\rho = 2\pi AR$) for determining soil resistivity, four-pole electrode depth should be:
   a. 1/2 of the electrode spacing
   b. 1/20 of the electrode spacing
   c. 2 times the electrode spacing
   d. Equal to the electrode spacing

2. What factors determine soil resistivity?
   a. Soil type
   b. Amount of moisture in soil
   c. Amount of electrolytes in soil
   d. Temperature
   e. All of the above

3. When doing a soil resistivity test and placing auxiliary rods at a spacing of 15 ft, what depth of earth is being measured?
   a. 7.5 ft
   b. 15 ft
   c. 30 ft
   d. 60 ft

4. What results can be obtained by doing a soil resistivity measurement?
   a. Geophysical surveys
   b. Corrosion analysis
   c. Electrical grounding design
   d. All of the above

5. As the temperature of the soil decreases, what happens to the soil resistance?
   a. Decreases
   b. Increases
   c. No change

6. Doubling the diameter of the rod has what effect on the potential resistance of a ground rod to be installed?
   a. 100% reduction
   b. 50% reduction
   c. 25% reduction
   d. Less than a 10% reduction

7. As a general rule, doubling the depth of the rod length reduces the resistance by:
   a. 100%
   b. 40%
   c. Less than 10%

8. What is the most important reason for good grounding practices?
   a. Proper operation of electrical equipment
   b. Safety
   c. Meet National Electrical Code® requirements

9. If a 5/8-inch ground rod is supposed to measure 25Ω and the local soil resistivity measures 20kΩ-cm, approximately how deep must the rod be driven?
   a. 10 ft
   b. 25 ft
   c. 40 ft
   d. 50 ft

10. Fall-of-Potential ground resistance measurements are recommended when:
    a. The ground under test can be conveniently disconnected
    b. Ground faults are likely to occur near the ground under test
    c. The power system cannot be shut down

11. When performing Fall-of-Potential tests, the ground electrode should be:
    a. In service and energized
    b. Disconnected and de-energized
    c. It makes no difference

12. What is the minimum number of measurements needed to accurately perform a Fall-of-Potential test?
    a. 1
    b. 2
    c. 3
    d. 5

13. If when making a Fall-of-Potential test each test result is significantly different in value from previous measurements on the same rod, what corrective action should be attempted?
    a. Position the Z electrode farther from the rod under test
    b. Position the Z electrode closer to the rod under test

14. What is the maximum ground resistance required by the National Electrical Code®?
    a. 5Ω
    b. 15Ω
    c. 25Ω
    d. 1Ω

15. When testing a multiple electrode grid, auxiliary electrode spacing is determined by:
    a. Depth of the deepest rod
    b. Maximum internal grid dimension
    c. The VA rating of equipment being grounded

16. Touch potential measurements are recommended when:
    a. It is physically impossible to disconnect the subject ground from service
    b. Determining the degree of electrical safety under fault conditions is considered to be more important than measuring actual ground resistance
    c. The grounding system is extensive and undocumented
    d. All of the above

17. The clamp-on test method cannot be used on high tension towers due to their spacing.
    a. True
    b. False

18. The clamp-on tester must be clamped around the ground rod only.
    a. True
    b. False

19. The clamp-on tester can be used only if the system under test is energized.
    a. True
    b. False

20. The clamp-on method of testing should not be performed:
    a. When testing large substation grounds
    b. On ground electrodes disconnected from service
    c. On single-point, lightning protection grounds
    d. All of the above

References

IEEE Std 81-1983
— *EEE Guide for Measuring Earth Resistivity, Ground Impedance, and Earth Surface Potentials of Ground Systems*

IEEE Std 142-1991
— *IEEE Recommended Practice for Grounding of Industrial and Commercial Power Systems*

Blackburn/American Electric Co.
Memphis, TN 38119
— *A Modern Approach to Grounding Systems*

Grounding Nomograph

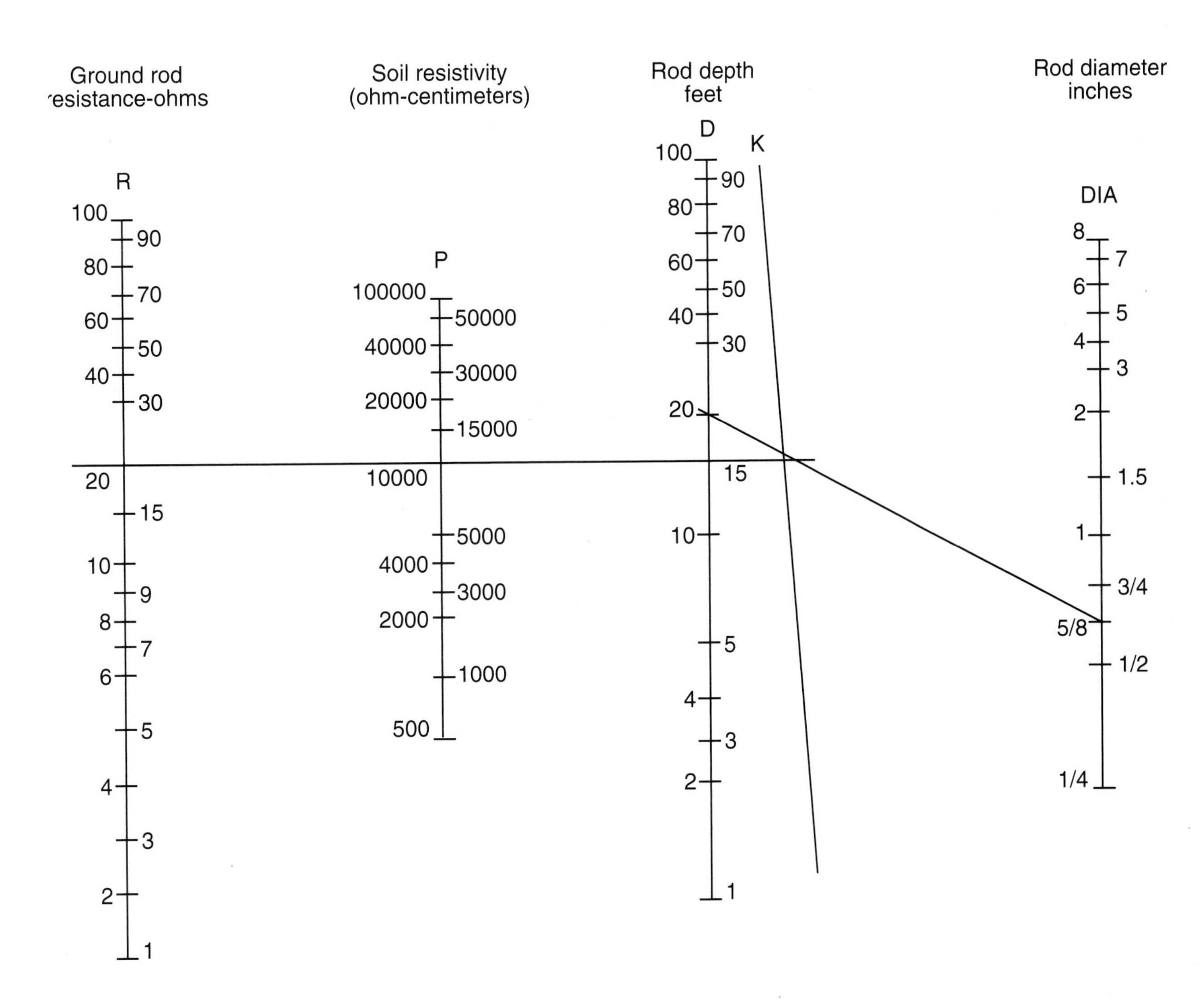

Represents example of a 20Ω, 20 ft ground rod

1. Select required resistance on R scale
2. Select apparent resistivity on P scale
3. Lay straightedge on R and P scale, and allow to intersect with K scale
4. Mark K scale point
5. Lay straightedge on K scale point and DIA scale, and allow to intersect with D scale
6. Point on D scale will be the rod depth required for resistance on R scale

Fall-of-Potential Plot

Instrument Mfg. ____________ Name of Operator ______________________

Model ____________ Location ______________________ Date ____________

Serial # ____________

Ground System Type:

Single Rod ☐ Rod Depth ☐

Multiple Rods ☐ Longest diagonal grid dimension ____________ ft

Voltage Electrode distance from Ground under Test	Resistance
FEET	OHMS
____	____
____	____
____	____
____	____
____	____
____	____
____	____
____	____
____	____
____	____
____	____
____	____
____	____
____	____
____	____
____	____
____	____
____	____
____	____
____	____

Test Conditions

Temp: ________ Soil: Moist ☐ Dry ☐

Soil Type

Loam ☐ Sand & Gravel ☐ Shale ☐ Clay ☐ Limestone ☐

Sandstone ☐ Granite ☐ Slate ☐ Other ____________

Resistance

50	100
45	90
40	80
35	70
30	60
25	50
20	40
15	30
10	20
5	10
0	0

0 10 20 30 40 50 60 70 80 90 100

0 5 10 15 20 25 30 35 40 45 50

Distance in Feet from Ground under Test to Voltage Electrode

Scale: ☐ 50 ☐ 100

Multiplier: ☐ x1 ☐ x10

Answers: 1 b; 2 e; 3 b; 4 d; 5 b; 6 d; 7 b; 8 b; 9 c; 10 a; 11 b; 12 c; 13 a; 14 c; 15 b; 16 b; 17 b; 18 b; 19 b; 20 d

Appendix D: Report of Research on Conduit Fitting Ground-Fault Current Withstand Capability

INTRODUCTION

It has been a long accepted practice to permit metal conduit and fittings as the required equipment grounding conductor according to the National Electrical Code. Recently, new constructions of conduit fittings including new alloys and securement means have been developed. These new fittings which depart in their configuration from traditional designs have prompted UL to propose the addition of a Current Test to the Second Edition of the *Standard for Fittings For Conduit And Outlet Boxes, UL 514B*, to evaluate the suitability of these and other fittings for grounding.

A UL Subjects 4 and 514B Bulletin dated December 3, 1990 (see Appendix A) proposed the addition of a Current Test to UL 514B. The proposal is intended to address concerns expressed regarding the ability of conduit fittings to carry ground-fault current. The proposal would require a conduit fitting to carry a specified current for a period of time based on conduit size. As a result of application of the test current, the conduit fitting shall not crack, break, or melt. After the conduit fitting has carried the test current, adequate continuity shall remain between the conduit and the enclosure or between lengths of conduit as established by the conduit fitting.

This Report of Research describes the work that was performed to study the ability of conduit fittings to carry ground-fault current, and maintain integrity of a metal equipment grounding system during and after a ground fault. The test results could be used to provide a rationale for accepting certain types of traditional conduit fittings for compliance with the proposed Current Test without extensive testing of a product that has an extensive use history.

The results of the work led to recommendations for improving the proposed Current Test method, and also provided some technical insight into how conduit fittings perform during a ground fault.

SCOPE OF WORK

The work included investigation of various types of conduit fittings for use with electrical metallic tubing (EMT), rigid metal conduit (RMC), and reduced-wall flexible metal conduit (RW FMC). Samples of conduit fittings were chosen from ten manufacturers to generally represent presently-marketed UL-Listed compression, set-screw, squeeze, and indenter constructions. The compression and squeeze fittings were formed of malleable iron, die-cast zinc, and steel. The set-screw fittings were formed of die-cast zinc and steel. The indenter fittings were formed of steel. Non-insulated straight connectors were chosen to represent couplings as well as offset, angled, elbowed, and insulated throat fittings from the standpoint of tolerance to ground-fault current.

The conduit fittings were investigated with steel EMT and RMC in the ½, ¾, 1¼, 2½, and 3 inch trade sizes; with aluminum EMT in the 2, 2½, and 3 inch trade sizes; and with both steel and aluminum RW FMC in the ⅜ inch trade size.

A sample test assembly consisted of a length of conduit with a fitting attached to a nominal 0.057 inch thick (16 MSG) painted steel or galvanized steel enclosure. The painted enclosures employed a grey baked-on acrylic enamel paint that was a nominal 0.001 inches thick. All of the fittings, conduit, and enclosures investigated were UL-Listed. Each sample assembly was subjected to the fault current for the test duration shown in proposed Table 36A.1 of the UL Subjects 4 and 514B Bulletin dated December 3, 1990.

The main objectives of this research work were to:

1. Accumulate data which could be used to determine an appropriate sampling of conduit fittings that would need to be tested during a UL Industry Review to demonstrate compliance with the proposed Current Test in the December 3, 1990 Bulletin – Proposed Requirements for the Second Edition of the *Standard For Fittings For Conduit And Outlet-Boxes, UL 514B.*
2. Assess the ability of certain types of conduit fittings to maintain the integrity of a metal grounding system during and after a ground fault.

SAMPLE SELECTION AND PREPARATION

A sample assembly consisted of a conduit fitting secured to one end of a two-foot length of conduit and attached to a metal enclosure. Each fitting was attached to the enclosure with the locknut provided with the fitting by the manufacturer. Compression type fittings were assembled to the conduit with a torque wrench in accordance with Table 29.1 of the *Standard For Fittings For Conduit And Outlet Boxes, UL 514B*, Second Edition. The set-screw and squeeze type fittings were secured to the conduit with the torque values specified in paragraph 29.4 of the UL-514B Standard. The tightening torque values are shown in Table 1.

After being attached to the conduit as described, the conduit fitting was secured to an enclosure with the locknut provided by the manufacturer. The locknut was first hand-tightened and then further tightened ¼ turn with a hammer and standard screwdriver. The fittings were installed through holes in the enclosures that were punched rather than being installed in preexisting knockouts. For each test, a newly punched hole was used in the enclosure. For test purposes, more than one fitting was installed in each enclosure. Care was taken to space each fitting far enough apart on the enclosure to avoid any undue influence on adjacent samples.

Three type-J thermocouples were used to monitor temperatures of the sample assembly during the ground-fault current testing.[1] A thermocouple was welded on a flat portion of the top uppermost

[1] The proposed Current Test does not require the measurement of temperatures. Temperatures were measured only as part of the methodology for this work.

TABLE 1

Tightening Torque for Various Sizes and Fitting Types

TORQUE FOR COMPRESSION TYPE FITTINGS		TORQUE FOR SET-SCREW AND SQUEEZE TYPE FITTINGS	
Trade Size of Fitting (in)	**Torque (lb-in)**	**Screw Type**	**Torque (lb-in)**
½	300	Slotted or Phillips No. 8	20
¾	500	Slotted or Phillips Greater Than No. 8	35
1¼	1000		
2, 2½, 3	1600	Bolthead	160

surface of the fitting and on a flat portion of the top uppermost surface of the locknut. In addition, a thermocouple was welded to the top uppermost outside surface of the conduit approximately three inches from the enclosure. Figures 1 and 2 show the thermocouple placement. Copper conductors of at least the minimum size required by the proposed Current Test were used to carry the ground-fault test current to and from the assembly.[2] One conductor was secured to a UL-Listed ground clamp that was attached to the end of the conduit opposite the fitting. The ground clamp was located approximately one inch from the end of the conduit. For the tests conducted with the ⅜ inch FMC, the ground clamp was located approximately four inches from the fitting so as to reduce the voltage drop between the fitting and the clamp.

[2] Since a large number of sample assemblies were tested in succession, a conductor larger than the minimum size required was often employed to reduce the effects of heating on the conductor.

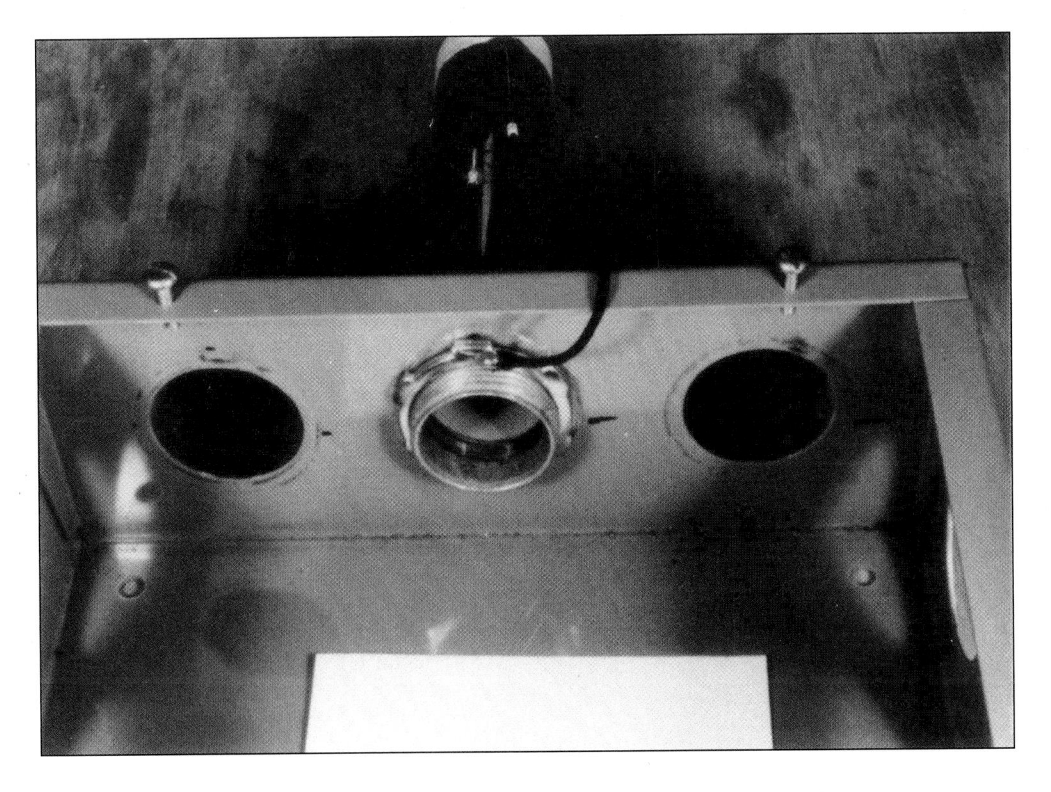

FIGURE 1 Thermocouple placement.

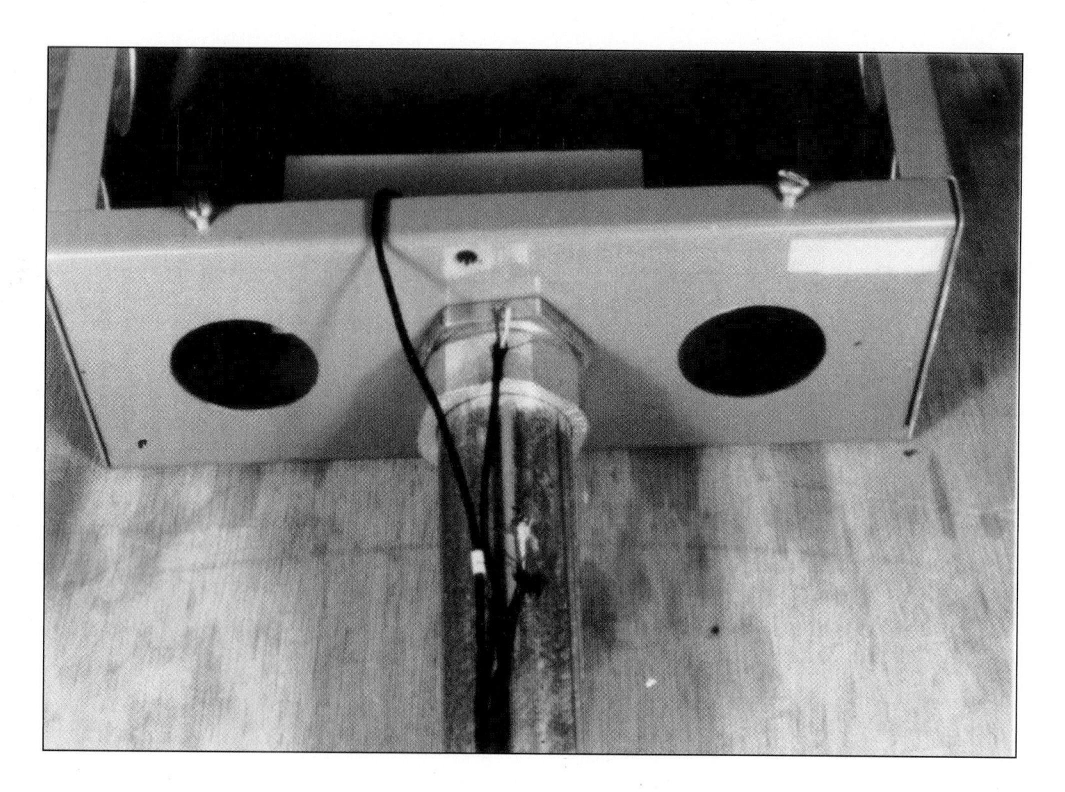

FIGURE 2 Thermocouple placement.

The second conductor was attached to a UL-Listed pressure wire connector that was secured to the center of the back of the enclosure with a nut and bolt.[3] On painted enclosures, a sufficient amount of paint was removed to assure there was a good electrical connection between the pressure wire connector and the enclosure. Figure 3 is a drawing of the sample assembly.

[3]To assure that there was a good electrical connection from the current source to the sample assembly, the wire connector and the ground clamp were tightened with set-screws and soldered in place on either end of the supply conductor.

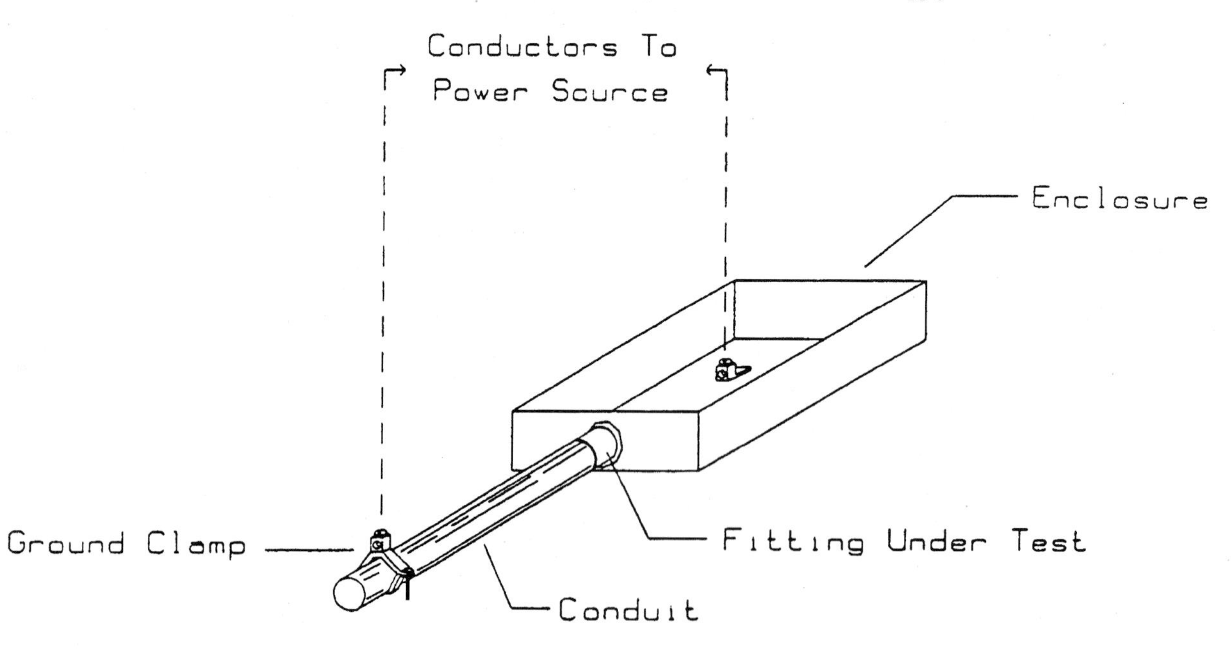

FIGURE 3 Sample assembly.

LABORATORY TESTS

Method

A Current Test was conducted in accordance with the parameters in proposed Table 36A.1 of the Subjects 4 and 514B December 3, 1990 Bulletin. Three of each identical sample assemblies were subjected to the fault current shown in Table 4.[4] The current was made to flow through the assembly for the specified test duration.

An adjustable voltage source was used to supply constant current to the sample assembly under test. During the test, the voltage was manually adjusted to compensate for the increase in resistance of the sample assembly due to heating. This manual adjustment allowed the fault current to be as close to the desired level as possible throughout the entire duration of the test. The mean value of actual fault current was typically within 1.5 percent of the specified value. Each sample assembly was subjected to the fault current only once.

A digital data acquisition system was used to control the duration of the test, and to measure the temperatures and fault current. The fault current was continuously monitored during the test, and temperatures were measured every two seconds during and after the test until peak temperatures were reached.

Following the Current Test, each fitting was visually examined for signs of cracking, breaking, or melting and then disassembled for further visual inspection.

[4]The fault current and test duration shown in Table 4 are the same as those in Table 36A.1 of the Subjects 4 and 514B December 3, 1990 Bulletin, and in the Standard For Grounding And Bonding Equipment, UL 467.

TABLE 4

Fault Current and duration

CONDUIT TRADE SIZE (in)	TEST DURATION (sec)	FAULT CURRENT[5] (A)
3/8	4	470
1/2	4	1180
3/4	6	1530
1 1/4	6	2450
2	6	3900
2 1/2	6	4900
3	9	5050

[5]Tests requiring fault currents of 3900, 4900, and 5050 A were conducted at the facilities of Littelfuse, Inc. in Des Plaines, IL. Tests requiring lower currents were conducted at the UI Northbrook, IL test facility.

RESULTS AND DISCUSSION

Test numbers 1 and 3 were conducted with painted enclosures and test numbers 2 and 4 were conducted with unpainted enclosures. The results suggest that the Current Test can be more difficult to complete with acceptable results when the fittings are installed in painted rather than in unpainted enclosures. All subsequent tests were conducted on sample assemblies having painted enclosures. The results of tests 2 and 4 were also used to compare the temperature rises of identical fittings with steel and aluminum EMT assembled to unpainted enclosures. The results showed that most of the fittings tested with steel EMT had higher temperature rises than fittings tested with aluminum EMT.

In some cases, it was observed that variability of temperature rise existed between identical sample assemblies. For example, the temperature rise of the fitting on sample assembly number 6a2a was 144°C while the temperature rises of the fitting on sample assembly numbers 6a2b and 6a2c were 56 and 41°C respectively. This variability in results with identical sample assemblies can probably be attributed to one or more of the following:

1. The method of assembling the locknut hand-tight with an additional 1/4 turn may not produce repeatable results from sample to sample.
2. Varying thicknesses or voids in the enclosure paint may allow for some assembly locknuts to penetrate the paint deeper than others.
3. Certain locknut designs may be such that the ability to fully penetrate the paint is marginal from sample to sample.

The results showed that for some sample assemblies such as 1d2b and 6c1c, elevated temperatures were recorded on the locknut and/or the

TABLE 6

Fittings Tested with Painted Enclosures That Sustained Damage

SAMPLE ASSEMBLY NO.	CONDUIT AND TRADE SIZE (in)	FITTING TYPE AND MATERIAL	LOCKNUT MATERIAL
1d1a	ST–EMT, 2	Compression, DCZ	DCZ
1d1b	ST–EMT, 2	Compression, DCZ	DCZ
1f1a	ST–EMT, 3	Compression, DCZ	ST
1f1c	ST–EMT, 3	Compression, DCZ	ST
1f2b	ST–EMT, 3	Compression, DCZ	DCZ
3d1c	AL–EMT, 2	Compression, DCZ	DCZ
3e1a	AL–EMT, 2½	Compression, DCZ	DCZ

fitting. These temperatures appeared to be the result of arcing between the locknut and the enclosure or between the fitting and conduit since evidence of arcing was noticed during the detailed visual examination of these sample assemblies. As a result of the arcing, however, the fitting or the locknut did not crack, break, or melt. The visual examination following the test showed that the locknut of sample assembly 6d2c welded to the enclosure.

As a result of the fault current, damage did not occur to any fitting that was assembled to an unpainted enclosure. Seven assemblies representing four different fittings that were assembled to painted enclosure did not successfully pass the fault current for the specified time duration. These fittings as indicated in Table 6 broke and/or melted as a result of the test. All of the fittings shown in Table 6 are the compression type with bodies formed of die-cast zinc. Except for fitting 1f1, all of these fittings had die-cast zinc locknuts. Fitting 1f1 used a steel locknut. The assemblies indicated in Table 6 represented fittings from three different manufacturers.

A visual examination of sample assemblies 1d1a, 1d1b, 1f2b, 3d1c, and 3e1a showed that melting of the die-cast zinc locknuts occurred as a result of the fault current as shown in Figures 4 and 5. Melting of the die-cast zinc body occurred on sample assemblies 1d1b, 1f1a, 1f1c, 1f2b, and 3e1a. The painted enclosures on which the fittings were tested were also examined. The examination indicated that melting of the die-cast zinc was probably due to the inability of the locknut to penetrate through the enclosure paint and provide good electrical contact between the fitting and metal of the enclosure.

A visual examination of all the conduit fittings with die-cast zinc locknuts showed that there were three different constructions of the locknuts. The three constructions differed in that the surface of the locknut contacting the enclosure was either flat, nibbed, or serrated. The sample assemblies with die-cast zinc locknuts that did not complete the Current Test with acceptable results had locknuts with flat or nibbed surfaces. All fittings having die-cast zinc locknuts with serrations completed the test with acceptable results. It appeared as though locknuts with serrations consistently penetrated through the enclosure paint and provided better electrical contact between the fitting and the metal of the enclosure than did the locknuts with flat or nibbed surfaces.

The fittings investigated in this work were formed of die-cast zinc, steel, and malleable iron. The melting point of zinc is 420°C while the melting point of steel and malleable iron is much higher, typically greater than 1400°C. Heat generated from the fault current in those sample assemblies in Table 6 was obviously greater than the melting point of the die-cast zinc fittings and locknuts, but not greater than the melting point of steel or malleable iron since no melting of the steel enclosure occurred. This was further evidenced by tests of sample assemblies 1f1a and 1f1c where the die-cast zinc body of the fittings melted but the steel locknuts did not.

FIGURE 4 Sample Assembly 1d1a After Current Test.

FIGURE 5 Sample Assembly 1f2b After Current Test.

All of the conduit fittings that were constructed of steel bodies and steel locknuts completed the test with acceptable results. A visual examination of the steel locknuts indicated that the nibs on these locknuts, which in most cases were sharp and well defined from the metal forming process, provided for better penetration through the enclosure paint than the nibs on the die-cast zinc locknuts.

CONCLUSIONS

1. Over 300 conduit fitting assemblies from ten different manufacturers were subjected to the Current Test to simulate performance under ground fault conditions. As a result of the tests, only seven assemblies representing four different conduit fittings and three different manufacturers did not withstand the fault current without breaking or melting of the conduit fitting assembly. All seven of these sample assemblies were compression type connectors with die-cast zinc bodies, and all but one of these assemblies utilized a die-cast zinc locknut.
2. An examination of the seven sample assemblies that did not complete the Current Test with acceptable results showed that the failures were probably due to high resistance from the inability of the fitting locknut to penetrate through the enclosure paint and provide good electrical continuity between the fitting and the metal enclosure. Heat generated by the high-resistance arcing was sufficient to melt the zinc, but not steel or iron.
3. Some of the sample assemblies that did not exhibit breaking or melting did show signs of arcing and welding between the locknut and the enclosure and/or the fitting and the conduit. These sample assemblies usually had higher temperatures during the Current Test, however, the temperatures were not sufficient to cause melting of the zinc or steel parts nor loss of continuity between the conduit, fitting, and enclosure.
4. Most of the sample assemblies that were subjected to the Current Test attained maximum temperatures on the fitting bodies and locknuts that were about the same as or less than the temperature of the conduit. For the tests with flexible metal conduit, the temperatures of the fittings were much less than the temperatures of the flexible conduit.
5. As a result of the tests, it was observed that if the fitting provides good electrical contact to both the enclosure and the conduit, the fitting will provide a suitable equipment ground path for fault current.

For most of the sample assemblies which completed the Current Test with acceptable results, the maximum temperatures on the fitting bodies and locknuts were about the same as or less than the temperature of the conduit. In the case of the flexible metal conduit, the temperatures on the fittings were much less than on the conduit. This would seem to indicate that if the fitting can provide good electrical contact to the enclosure metal, the fitting will provide for adequate equipment grounding.

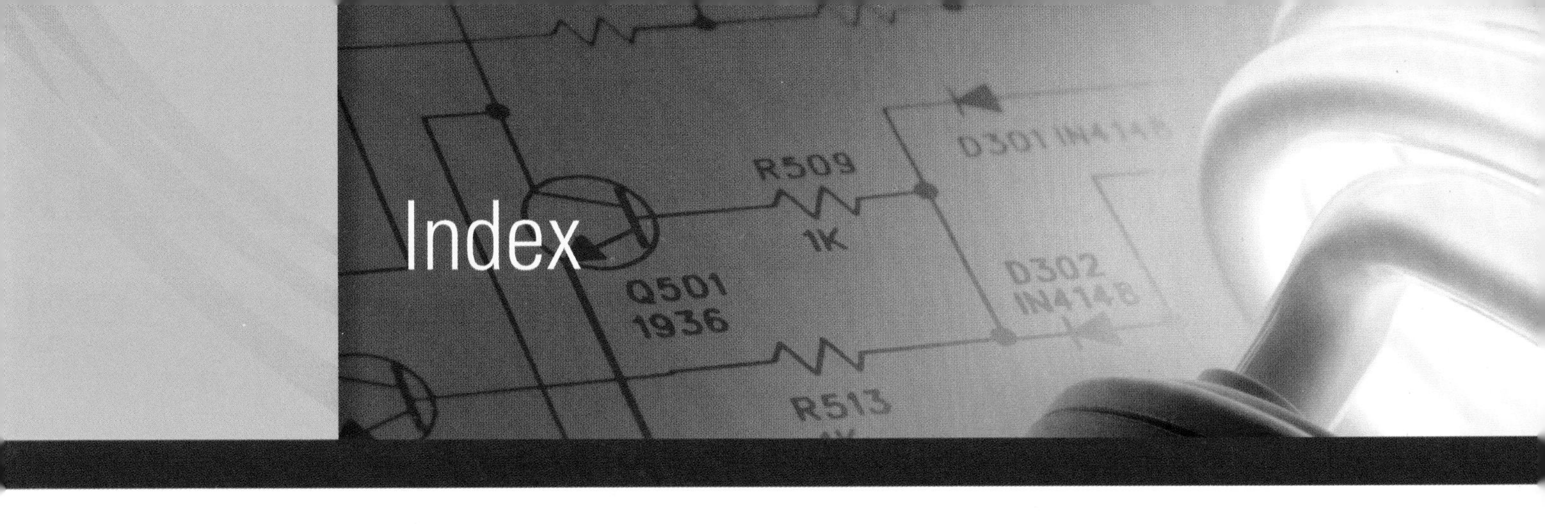
Index
R509
1K
Q501
1936
D301 IN4148
D302
IN4148
R513

K

L

M

N

O

P

R

S